Computational Physics

Computational Physics

Problem Solving with Python

Fourth Edition

Rubin H. Landau
Manuel J. Páez
Cristian C. Bordeianu (D)

With contributions by Guangliang He

WILEY-VCH

Authors

Prof. Rubin H. Landau
Oregon State University
Corvallis
OR
97331
United States of America

Prof. Manuel J. Páez
Departamento Fisica
Universad de Antioquia
Medellin
Colombia

Prof. Cristian C. Bordeianu[†]
University of Bucharest
Str. Mica nr. 7
jud. Suceav
725100
Romania

Cover Image: © PASIEKA/Getty Images

Library of Congress Card No.: applied for

British Library Cataloguing-in-Publication Data
A catalogue record for this book is available from the British Library.

Bibliographic information published by the Deutsche Nationalbibliothek The Deutsche Nationalbibliothek lists this publication in the Deutsche Nationalbibliografie; detailed bibliographic data are available on the Internet at <http://dnb.d-nb.de>.

© 2024 WILEY-VCH GmbH, Boschstraße 12, 69469 Weinheim, Germany

Print ISBN: 978 3 527-41425-3
ePDF ISBN: 978-3-527-84332-9
ePub ISBN: 978-3-527-84331-2

Typesetting Straive, Chennai, India
Printing and Binding CPI Group (UK) Ltd, Croydon, CR0 4YY

C9783527414253_031024

Contents

Preface

When the first edition of *Computational Physics* was published in 1997, who would have thought that we would be doing it again in 2024? Back then, we hoped that our writing might encourage the inclusion of more computation into the physics curriculum. Now, computational physics (CP) courses are taught widely (if only more with our book!). Yet, when Martin Preuss of Wiley asked if we might be interested in a fourth edition, the thought of getting to work with some recent developments in computation was just too appealing for us to resist. And so, here we are!

And who would have thought that our youngest coauthor, Christian Bordeianu, who joined us for the second and third editions, would not be around for this one! It is so sad that we must write without his enthusiasm, knowledge, good nature, and friendship.

This edition continues with the three main themes of previous ones:

1) That there is value when first learning CP to survey a broad range of topics in various specialties.
2) That the best way to learn CP is by doing it on a computer with examples, exercises, and problems.
3) That CP is part of computational science, which means that we are presenting CP as a mixture of physics, applied mathematics, and computer science, trying not to short-change the latter two.

This edition has entirely new chapters on neural networks and machine learning, quantum computing (with Guangliang He), and general relativity, as well as an expanded coverage of principal component analyses and Python programming. Finally, there has been editing throughout the text to improve clarity and organization. This edition also continues our advocacy of a compiled-type programming language, and especially Python, as the best vehicle for learning CP. This is in contrast to computing environments such as Matlab and Mathematica, where the mathematics, algorithms, and details are kept "under the hood." If computational results are to be scientifically sound, we believe it requires a physicist to understand the algorithms, their connections to mathematics, physics, the logic of a program, and, especially, the limits and uncertainties of the computation. Nevertheless, we appreciate how time-consuming and frustrating debugging programs may be, especially for beginners, and so we provide the reader with a large number of codes in the text and online at: sites.science.oregonstate.edu/~landaur/Books/Problems/Codes/. Our hope is that this leaves time for exploration, extensions, and analysis. It also provides experience in the modern work environment, in which one must incorporate new developments into the preexisting developments of others.

To make room for the new, we have (sadly) removed some of the old ones that were in previous editions. However, those materials can still be found in a Jupyter Notebook version of the previous edition online at sites.science.oregonstate.edu/~landaur/Books/CPbook/ eBook/.

There are also video lectures at sites.science.oregonstate.edu/~landaur/Books/CPbook/ eBook/Lectures/ and on YouTube's *Landau Computational Physics Course* at www.youtube. com/playlist?list=PLnWQ_pnPVzmJnp794rQXIcwJIjwy7Nb2U.

These videos and the concordant slides cover most of the topics in the text and may be helpful in blended or hybrid courses.

We hope you enjoy our work and we look forward to your comments.

Tucson, May 2024	*Rubin H. Landau, Tucson*
Medellin, May 2024	*Manuel J. Páez, Medellin*

Planning the first edition.

Acknowledgments

Immature poets imitate;
mature poets steal.

— T. S. Elliot

Previous editions of this book and our computational physics courses would not have been possible without financial support from the National Science Foundation's CCLI, EPIC, and NPACI programs, and the Physics Department of Oregon State University. Thank you, we hope we have made you proud.

Our CP developments have followed the pioneering paths paved by Thompson, Gould and Tobochnik, Christian, and Press et al. Indubitably, we have borrowed material from them and made it our own with no further thought.

We wish to acknowledge valuable contributions by Guangliang He, Hans Kowallik, Sally Haerer (video lecture modules), Paul Fink (deceased) Oscar A. Restrepo, Jaime Zuluaga, and Henri Jansen. It is our pleasure to acknowledge the invaluable friendship, encouragement, helpful discussions, and experiences we have had with many colleagues and students over the years. We are particularly indebted to Guillermo Avendaño-Franco, Saturo S. Kano, Melanie Johnson, Jon Maestri (deceased), David McIntyre, Shashikant Phatak, Viktor Podolskiy, C. E. Yaguna, and Zlatco Dimcovic. And, finally, it's been a pleasure to work with Martin Preuss, Aswini Murugadass, and Judy Howarth at Wiley again.

In spite of everyone's best efforts, there are still errors and confusing statements in the book and codes for which we are to blame.

Part I

Basics

1

Introduction

Beginnings are hard.   *Nothing is more expensive than a start.*
                — Chaim Potok   *— Friedreich Nietzsche*

We start this book with a description of how computational physics (CP) fits into the broader field of computational science, and how CP fits into physics. We describe the subjects we cover, the coordinated video lectures, and how the book may be used in a CP course. Finally, we get down to business by discussing the Python language and its many packages, some of which we'll use. In Chapter 2 we give an introduction to Python programming, and in Chapter 7 we examine Python's treatment of matrices.

1.1 Computational Physics and Science

As illustrated in Figure 1.1, we view CP as a bridge that connects physics, computer science (CS), and applied mathematics. Whereas CS studies computing for its own intrinsic interest and develops the hardware and software tools that computational scientists use, and while applied mathematics develops and studies the algorithms that computational scientists use, CP focuses on using all of that to do better and new physics. Furthermore, just as an experimentalist must understand many aspects of an experiment to ensure that her measurements are accurate and believable, so should every physicist undertaking a computation understand the CS and math well enough to ensure that her computations are accurate and precise.

As CP has matured, we see it not only as a bridge among disciplines, but also as a specialty containing core elements of its own, such as data-mining tools, computational methods, and a problem-solving mindset. To us, CP's commonality of tools and viewpoint with other computational sciences makes it a good training ground for students, and a welcome change from the overspecialization found in so much of physics.

As part of this book's emphasis on problem solving, we strive to present the subjects within a problem-solving paradigm, as illustrated on the right of Figure 1.1. Ours is a hands-on, inquiry-based approach in which there are problems to solve, a theory or an appropriate model to apply, an appropriate algorithm to use, and an assessment of the results.

Computational Physics: Problem Solving with Python, Fourth Edition.
Rubin H. Landau, Manuel J. Páez, and Cristian C. Bordeianu.
© 2024 WILEY-VCH GmbH. Published 2024 by WILEY-VCH GmbH.

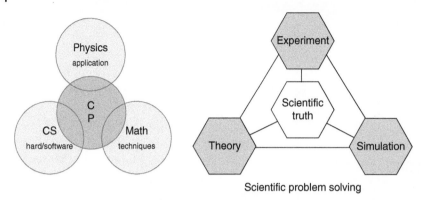

Figure 1.1 On the left a view of computational physics as a discipline encompassing physics, applied mathematics, and computer science. On the right is a broader view of computational physics fitting into various components of scientific problem solving.

This approach can be traced back to the post-World War II research techniques developed at US national laboratories. They deserve the credit for extending the traditional experimental and theoretical approaches of physics to also include simulation. Recent developments have also introduced powerful data mining tools, such as neural networks, artificial intelligence, and quantum computing.

1.2 This Book's Subjects

We do not intend this book to be a scholarly exposition of the foundations of CP. Instead, we employ a learn-by-doing approach with many exercises, problems, and ready-to-run codes. We survey many of the subjects that constitute CP at a level appropriate for undergraduate education, except maybe for the latter parts of some chapters. Our experience is that many graduate students and professionals may also benefit from this survey approach in which a basic understanding of a broad range of topics facilitates further in-depth study.

Chapters 1–8 cover basic numerics, ordinary differential equations with (many) applications, matrix computing using well-developed linear algebra libraries, and Monte-Carlo methods. Some powerful data mining tools such as discrete Fourier transforms, wavelet analysis, principal component analysis, and neural networks are covered in the middle of the book.

A traditional way to view the materials in this text is in terms of their use in courses. For a one-quarter class, we used approximately the first-third of the text, with its emphasis on computing tool familiarity with a compiled language [CPUG, 2009]. The latter two-thirds of the text, with its greater emphasis on physics, has typically been used in a two-quarter (20-week) course. What with many of the topics taken from research, these materials can easily be used for a full year's course, and for supplementary research projects.

1.3 Video Lecture Supplements

As an extension of the concept of a "text," we provide some 60 video lecture modules (as in Figure 1.2) that cover almost every topic in the book. The modules were originally

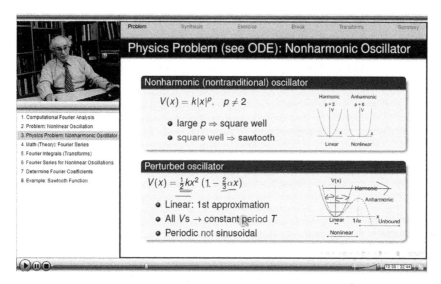

Figure 1.2 A screenshot from a lecture module showing a dynamic table of contents, a talking head, video controls, a slide with live scribbling, and some old man. (Originally in Flash, now as mpegs.)

a mix of Flash, Java, HTML, and mpeg, but with Flash no longer supported, we provide them as mp4 videos and PDF slides. They are available on our website: https://sites.science .oregonstate.edu/~landaur/Books/CPbook/eBook/Lectures, as well as on our YouTube channel under *Landau Computational Physics Course*: https://www.youtube.com/playlist? list=PLnWQ_pnPVzmJnp794rQXIcwJIjwy7Nb2U.

The video lectures can be used to preview or review materials, as part of an online course, or in a blended course in which they replace some lectures, thereby freeing up time for lab work with the instructor.

1.4 This Book's Codes and Problems

Separate from the problems and exercises throughout the text, almost every chapter starts off with a keynote "**Problem**" that leads into the various steps in computational problem solving (Figure 1.1). The additional problems and exercises distributed throughout the chapters are essential ingredients for learning, and are meant to be worked through. This entails studying the text, writing, debugging, and running programs, visualizing the results, and expressing in words what has been performed, and what can be concluded. We asked our students to write up mini lab reports containing

Equations solved	Numerical method	Code listing
Visualization	Discussion	Critique

Although we recognize that programming is a valuable skill for scientists, we also know that it is incredibly exacting and time-consuming. In order to lighten the workload, *we provide programs for most of the problems in the text*, both at the end of each chapter and online at: sites.science.oregonstate.edu/~landaur/Books/CPbook/Codes.

A complete list is given in the Appendix. We recommend that these codes be used as guides for the reader when writing their own programs, or, at the least, tested and extended to solve the problem at hand. We have been told that learning how to use someone else's code is a valuable workplace skill to develop; as with programs encountered in a workplace, they should be understood before use!

1.5 Our Language: The Python Ecosystem

The codes in this edition of *Computational Physics* employ the computer language *Python*. Previous editions have employed Java, Fortran, and C, and used post-computation tools for visualization.[1] Python's combination of language plus packages now makes it the standard for the explorative and interactive computing that typifies present-day scientific research.

Although valuable for research, we have also found Python to be the best language yet for teaching and learning CP. It is free, robust (programs don't crash), portable (programs run without modifications on various devices), universal (available for most every computer system), has a clean syntax that permits rapid learning, has dynamic typing (changes data types automatically as needed), has high-level, built-in data types (such as complex numbers), and built-in visualization. Furthermore, because Python is interpreted, students can learn the language by executing and analyzing individual statements within an interactive shell, or within a notebook environment, or by running an entire program in one fell swoop. Finally, it is easy to use the myriad of free Python packages supporting numerical algorithms, state-of-the-art visualizations, as well as specialized toolkits that rival those in Matlab and Mathematica/Maple. And did we mention, all of this is free?

Although we do not expect the readers to be programming experts, it is essential to be able to run and modify the sample codes in this book. For learning Python, we recommend the online tutorials [PyTut, 2023; Pguide, 2023; Plearn, 2023], the book [Langtangen, 2016], and the many books in the "Python for Scientists and Engineers" genre. For general numerical methods, [Press *et al.*, 2007] is the standard, and fun to read. The NITS Digital Library of Mathematical Functions [NIST, 2022] is a convenient reference for mathematical functions and numerical methods.

Python has developed rapidly since its first implementation in December 1989 [History, 2022]. The rapid developments of Python have led to a succession of new versions and the inevitable incompatibilities. The codes presented in the book are in the present standard, Python 3. The major difference from Python 2 is the print statement:

```
>>> print 'Hello, World!'      # Python 2
>>> print('Hello, World!')     # Python 3
```

1.6 The Easy Way: Python Distributions

The Python language plus its family of packages comprise a veritable ecosystem for computing. A *package*, or library, or module, is a collection of related methods, or classes of

---

1 All of our codes, even the old ones, are available online.

methods, that are assembled and designed to work together. Inclusion of the appropriate packages extends the language to meet the specialized needs of various science and engineering disciplines [CiSE, 2015]. The Python Package Index [PyPi, 2023], a repository of free Python packages, currently contains 425,320 projects and 7,313,641 files. In this book, we use:

Jupyter Notebooks: A web-based, interactive Python computing environment combining live code, type-set equations, narrative text, visualizations, and whatever. Some of our programs (`.ipynb` suffix) were developed in Jupyter, and our programs using Vpython work only within Jupyter. There is a previous edition of this text in notebook form at sites.science.oregonstate.edu/~landaur/Books/CPbook/eBook.
The interactive Python shell, *IPython* can also be used within Jupyter.

Numpy (Numerical Python): A comprehensive library of mathematical functions, random number generators, linear algebra routines, Fourier transforms, and most everything else. Permits the use of fast, high-level multidimensional arrays (explained in Chapter 7). The successor to both *Numeric* and *NumArray*, NumPy is used by Visual and Matplotlib.

Matplotlib (Mathematics Plotting Library): A 2D and 3D graphics library that uses NumPy, produces publication-quality figures in a variety of hard copy formats, and that permits interactive graphics. Similar to Matlab's plotting (except Matplotlib is free and doesn't need its license renewed yearly).

Pandas (Python Data Analysis Library): A collection of high-performance, user-friendly data structures, and data analysis tools (used in Chapter 11).

SymPy (Symbolic Python): A system for symbolic mathematics using pure Python (no external libraries) that provides a simple computer algebra system including calculus, differential equations, etc. Similar to Maple or Mathematica, with the *Sage* package being even more complete. Examples in Section 2.3.6.

Visual (Vpython): The Python language plus the no-longer-supported *Visual* graphics module (superseded by GlowScript). Particularly easy for creating educational 3D demonstrations and animations. Still useful as `Web Vpython` and within Jupyter Notebooks.

Although most Python packages are free, there is true value for both users and vendors to distribute a collection of packages that have been engineered and tuned to work well together, and that can be installed in one fell swoop. (This is similar to what Red Hat and Debian distributions do for Linux.) These distributions can be thought of as complete, Python ecosystems and are highly recommended. In particular, all you really need to do to get started with Python computing for this book is to load:

AnaConda: A free Python distribution including more than 8000 packages for science, mathematics, engineering, machine learning, and data analysis. Anaconda installs in its own directory and so runs independently from other Python installations on your computer. Go to https://www.anaconda.com/products/distribution to download Anaconda. Once you install *Anaconda*, the *Navigator* should open, and it will let you choose all that you will need.

Spyder IDE: The Scientific PYthon Development EnviRonment. An Integrated Development Environment (IDE) with advanced editing, interactive testing of code, debugging, and more.

Jupyter Notebook: The Web-based interactive computing notebook environment used for editing and running type-set-like documents, while also running Python code within the documents. As we have already said, a notebook (`.ipyn`) version of an earlier edition of this text is at sites.science.oregonstate.edu/~landaur/Books/CPbook/eBook.

Powershell Prompt: A powerful terminal that runs *conda* commands under the Windows shell environments `cmd.exe` (Command Prompt) and `powershell.exe`. Apple has a *Terminal* app where you will find a command prompt.

Conda: A package management and environment system included in Anaconda that finds, installs, and updates packages and their dependencies for you.

In Chapter 11 we describe how to load and run Google's *TensorFlow* package for machine learning, and in Chapter 12 we describe how to load and run the *Quantum Computing* packages, *Cirq, IBM Quantum*, and *Qiskit*.

2

Software Basics

This chapter discusses the computing basics of communications, number representations, Python programming, and visualizations. Since we want to do science, there is a particular emphasis on the limits of floating point arithmetic.

2.1 Making Computers Obey

The best programs are written so that computing machines can perform them quickly and so that human beings can understand them clearly. A programmer is ideally an essayist who works with traditional aesthetic and literary forms as well as mathematical concepts, to communicate the way that an algorithm works and to convince a reader that the results will be correct.

— Donald E. Knuth

As anthropomorphic as your view of your computer may be, keep in mind that computers always do exactly as they are told. This means that you must tell them exactly everything you want them to do. Of course, the programs you run may have to be at such a high level with such convoluted logic that you may not have the endurance to figure out the details of just what they are telling the computer to do, but it is always possible in principle (except maybe not with AI). So your first **problem** is to obtain enough understanding so that you feel sufficiently in control, no matter how illusionary, to figure out what the computer is doing.

Before you tell the computer to obey your orders, you need to understand that life is not simple for computers. The instructions they understand are in a *basic machine language*[1] that tells the hardware to do things like move a number stored in one memory location to another location, or to do some simple binary arithmetic. Very few computational scientists talk to computers in a language computers can understand. When writing and running programs, we usually communicate through *shells*, in *high-level languages* (Python, Java, Fortran, and C), or through *problem-solving environments* (Maple, Mathematica, and Matlab). Eventually, our commands or programs are translated into the basic machine language that the hardware understands.

---

1 The Beginner's All-Purpose Symbolic Instruction Code (BASIC) programming language of the original PCs should not be confused with basic machine language.

Computational Physics: Problem Solving with Python, Fourth Edition.
Rubin H. Landau, Manuel J. Páez, and Cristian C. Bordeianu.
© 2024 WILEY-VCH GmbH. Published 2024 by WILEY-VCH GmbH.

A *shell* is a *command-line interpreter*, that is, a set of small programs run by a computer that responds to the commands (the names of the programs) that you key in. Usually you open a special window to access the shell, and this window is called a shell as well. It is helpful to think of these shells as the outer layers of the computer's operating system (OS) in Figure 2.1, within which lies a *kernel* of elementary operations. (The user seldom interacts directly with the kernel, except possibly when installing programs or when building an OS from scratch.) It is the job of the shell to run programs, compilers, and utilities that do things like moving and copying files. There can be different types of shells on a single computer, or multiple copies of the same shell running at the same time.

Operating systems have names such as *Unix, Linux, DOS, MacOS*, and *MS Windows*. An *operating system* is no more than a group of programs used by the computer to communicate with users and devices, to store and read data, and to execute programs. The OS tells the computer what to do in an elementary way. The OS views you, other devices, and programs as input data for it to process; in many ways, it is the indispensable office manager. While all this may seem complicated, the purpose of the OS is to let the computer do the nitty-gritty work so that you can think higher-level thoughts and communicate with the computer in something closer to your normal everyday language.

When you submit a program to your computer in a *high-level language*, the computer may use a compiler to process it. A *compiler* is another program that treats your program as a foreign language and uses a built-in dictionary and set of rules to translate it into basic machine language. As you can probably imagine, the final set of instructions is quite detailed and long, and the compiler may make several passes through your program to decipher your logic and translate it into a fast code. The translated statements form an *object* or compiled code, and when *linked* together with other needed subprograms, form a *load module* that can be *loaded* into the computer's memory and read, understood, and followed by the computer.

Languages such as *Fortran* and *C* use compilers to read your entire program and then translate it into basic machine instructions. Interpreted languages such as Python, *BASIC*, and *Maple* translate or *interpret* each line of your program as it is entered, and thus permit

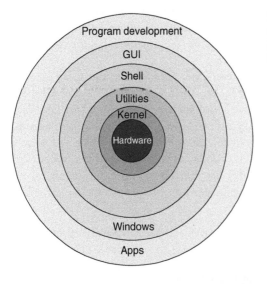

Figure 2.1   A schematic view of a computer's kernel and shells. The hardware is in the center surrounded by increasingly higher-level software.

line-by-line interactions. Compiled languages usually lead to more efficient programs and permit the use of vast subprogram libraries. Interpreted languages give a more immediate response to the user and thereby appear "friendlier." The Python and Java languages are actually a mix of the two. When you first compile your program, Python interprets it into an intermediate, universal *byte code*, which gets stored as a `.pyc` or `.pyo` file. This file can be transported to and used on other computers, although not with different versions of Python. Then, when you run your program, Python recompiles the byte code into a machine-specific and fast-running compiled code.

2.2 Computer Number Representations

Computers may be powerful, but they are finite. A problem in computer design is how to represent an arbitrary number using a finite amount of memory space, and then how to deal with the limitations arising from this representation. As a consequence of computer memories being based on the magnetic or electronic realizations of a spin pointing up or down, the most elementary units of computer memory are the two binary integers (*bits*) 0 and 1. This means that all numbers are stored in memory in *binary* form, that is, as long strings of zeros and ones. Accordingly, N bits can store integers in the range $[0, 2^N]$, yet because the sign of the integer is represented by the first bit (a zero bit for positive numbers), the actual range for N-bit integers decreases to $[0, 2^{N-1}]$.

Long strings of zeros and ones are fine for computers, but are awkward for humans. For this reason, binary strings are converted to *octal*, *decimal*, or *hexadecimal* numbers before the results are communicated to people. Octal and hexadecimal numbers are nice because the conversion maintains precision, but not all that nice because our decimal rules of arithmetic do not work for them. Converting to decimal numbers makes the numbers easier for us to work with, but unless the original number is a power of 2, some precision is lost. A description of a particular computer's system or language normally states the *word length*, that is, the number of bits used to store a number. The length is often expressed in *bytes*, (a mouthful of bits) where

$$1 \text{ byte} \equiv 1 \text{ B} \stackrel{\text{def}}{=} 8 \text{ bits.}$$

Memory and storage sizes are measured in bytes, kilobytes, megabytes, gigabytes, terabytes, and megabytes (10^{15}). Some care should be taken here by those who chose to compute sizes in detail because K does not always mean 1000:

$$1 \text{ K} \stackrel{\text{def}}{=} 1 \text{ KB} = 2^{10} \text{ bytes} = 1024 \text{ bytes.} \tag{2.1}$$

This is often (and confusingly) compensated for when memory size is stated in K, for example,

$$512 \text{ K} = 2^9 \text{ bytes} = 524,288 \text{ bytes} \times \frac{1 \text{ K}}{1024 \text{ bytes}}.$$

Conveniently, 1 byte is also the amount of memory needed to store a single letter like "a," which adds up to a typical printed page requiring $\sim 3 \text{ kB}$.

The memory chips in some older personal computers used 8-bit words, with modern PCs using 64 bits. This meant that the maximum integer was a rather small $2^7 = 128$ (7 because

1 bit is used for the sign). Using 64 bits permits integers in the range $1-2^{63} \simeq 10^{19}$. While at first this may seem like a large range, it really is not when compared to the range of sizes encountered in the physical world. As a case in point, the size of the universe compared to the size of a proton covers a scale of 10^{41}. Trying to store a number larger than the hardware or software was designed for (*overflow*) was common on older machines, but is less so now. An overflow is sometimes accompanied by an informative error message, and sometimes not.

2.2.1 IEEE Floating-Point Numbers

Real numbers are represented on computers in either *fixed-point* or *floating-point* notation. *Fixed-point notation* can be used for numbers with a fixed number of places beyond the decimal point (radix) or for integers. It has the advantages of being able to use *two's complement* arithmetic and being able to store integers exactly.[2] In the fixed-point representation with N bits and with a two's complement format, a number is represented as

$$N_{\text{fix}} = \text{sign} \times (\alpha_n 2^n + \alpha_{n-1} 2^{n-1} + \cdots + \alpha_0 2^0 + \cdots + \alpha_{-m} 2^{-m}), \tag{2.2}$$

where $n + m = N - 2$. That is, 1 bit is used to store the sign, with the remaining $(N-1)$ bits used to store the α_i values (the powers of 2 are understood). The particular values for $N, m,$ and n are machine-dependent. Integers are typically 4 bytes (32 bits) in length and in the range

$$-2147483648 \leq \text{4-B integer} \leq 2147483647. \tag{2.3}$$

An advantage of the representation (2.2) is that you can count on all fixed-point numbers having the same absolute error of 2^{-m-1} (the term left off the right-hand end of (2.2)). The corresponding disadvantage is that *small* numbers (those for which the first string of α values are zeros) have large *relative* errors. For that reason, relative errors in the real world tend to be more important than absolute ones; integers are usually used for counting purposes and in special applications (like banking).

Most scientific computations use double-precision floating-point numbers with $64\,\text{b} = 8\,\text{B}$. The *floating-point representation* of numbers on computers is a binary version of what is commonly known as *scientific* or *engineering notation*. For example, the speed of light $c = +2.99792458 \times 10^{+8}$ m/s in scientific notation and $+0.299792458 \times 10^{+9}$ or 0.299795498 E09 m/s in engineering notation. In each of these cases, the number in front is called the *mantissa* and contains nine *significant figures*. The power to which 10 is raised is called the *exponent*, with the plus sign in front as a reminder that these numbers may be negative.

Floating-point numbers are stored on the computer as a concatenation (juxtaposition) of a sign bit, an exponent, and a mantissa. Because only a finite number of bits are stored, the set of floating-point numbers that the computer can store exactly, *machine numbers* (the hash marks in Figure 2.2), is much smaller than the set of real numbers. In particular, machine numbers have a maximum and a minimum (the shading in Figure 2.2). If you

---

2 The *two's complement* of a binary number is the value obtained by subtracting the number from 2^N for an N-bit representation. Because this system represents negative numbers by the two's complement of the absolute value of the number, additions and subtractions can be made without the need to work with the sign of the number.

Figure 2.2 The limits of single-precision floating-point numbers and the consequences of exceeding these limits (not to scale). The hash marks represent the values of numbers that can be stored; storing a number in between these values leads to truncation errors. The shaded areas correspond to over- and underflow.

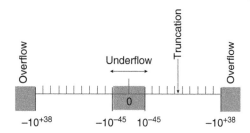

exceed the maximum, an error condition known as *overflow* occurs; if you fall below the minimum, an error condition known as *underflow* occurs. In the latter case, the software and hardware may be set up so that underflows are set to zero without your even being told. In contrast, overflows usually halt a program's execution.

The actual relation between what is stored in memory and the value of a floating-point number is somewhat indirect, with there being a number of special cases and relations used over the years. In fact, in the past, each computer OS and each computer language contained their own standards for floating-point numbers. Different standards meant that the same program running correctly on different computers could give different results. Although the results usually were only slightly different, the user could never be sure if the lack of reproducibility of a test case was as a result of the particular computer being used or to an error in the program's implementation.

In 1987, the Institute of Electrical and Electronics Engineers (IEEE) and the American National Standards Institute (ANSI) adopted the IEEE 754 standard for floating-point arithmetic. When the standard is followed, you can expect the primitive data types to have the precision and ranges given in Table 2.1. In addition, when computers and software adhere to this standard, and most do now, you are guaranteed that your program will produce identical results on different computers. Nevertheless, because the IEEE standard may not produce the most efficient code or the highest accuracy for a particular computer, sometimes you may have to invoke compiler options to demand that the IEEE standard be strictly

Table 2.1 The IEEE 754 standard for primitive data types.

Name	Type	Bits	Bytes	Range
`boolean`	Logical	1	$\frac{1}{8}$	`true` or `false`
`char`	String	16	2	$´\backslash$`u0000`$´ \leftrightarrow ´\backslash$`uFFFF`$´$ `(ISO Unicode)`
`byte`	Integer	8	1	$-128 \leftrightarrow +127$
`short`	Integer	16	2	$-32,768 \leftrightarrow +32,767$
`int`	Integer	32	4	$-2,147,483,648 \leftrightarrow +2,147,483,647$
`long`	Integer	64	8	$-9,223,372,036,854,775,808 \leftrightarrow 9,223,372,036,854,775,807$
`float`	Floating	32	4	$\pm 1.401298 \times 10^{-45} \leftrightarrow \pm 3.402923 \times 10^{+38}$
`double`	Floating	64	8	$\pm 4.94065645841246544 \times 10^{-324} \leftrightarrow \pm 1.7976931348623157 \times 10^{+308}$

followed for your test cases. After you know that the code is okay, you may want to run with whatever gives the greatest speed and precision.

There are actually a number of components in the IEEE standard, and different computer or chip manufacturers may adhere to only some of them. Furthermore, as Python develops, it may not follow all standards, but it probably will in time. Normally, a floating-point number x is stored as

$$x_{\text{float}} = (-1)^s \times 1.f \times 2^{e-\text{bias}}, \tag{2.4}$$

that is, with separate entities for the sign s, the fractional part of the mantissa f, and the exponential field e. All parts are stored in binary form and occupy adjacent segments of a single 32-bit word for singles, or two adjacent 32-bit words for doubles. The sign s is stored as a single bit, with $s = 0$ or 1 for a positive or a negative sign. Eight bits are used to store the exponent e, which means that e can be in the range $0 \le e \le 255$. The endpoints, $e = 0$ and $e = 255$, are special cases (Table 2.2). *Normal numbers* have $0 < e < 255$, and with them, the convention is to assume that the mantissa's first bit is a 1, so only the fractional part f after the *binary point* is stored. The representations for *subnormal numbers* and for the special cases are given in Table 2.2.

Note that the values $\pm$INF and NaN are not numbers in the mathematical sense, that is, objects that can be manipulated or used in calculations to take limits and such. Rather, they are signals to the computer and to you that something has gone awry and that the calculation should probably stop until you straighten things out. In contrast, the value -0 can be used in a calculation with no harm. Some languages may set unassigned variables to -0 as a hint that they have yet to be assigned, although it is best not to count on that!

As the uncertainty (error) is only in the mantissa and not the exponent, the IEEE representations ensure that all normal floating-point numbers have the same relative precision. Because the first bit of a floating point number is assumed to be 1, it does not have to be stored, and computer designers need only recall that there is a *phantom bit* there to obtain an extra bit of precision. During the processing of numbers in a calculation, the first bit of an intermediate result may become zero, but this is changed before the final number is stored. To repeat, for normal cases, the actual mantissa ($1.f$ in binary notation) contains an implied 1 preceding the binary point.

Finally, in order to guarantee that the stored biased exponent e is always positive, a fixed number called the *bias* is added to the actual exponent p before it is stored as the biased

Table 2.2 Representation scheme for normal and abnormal IEEE singles.

Number name	Values of s, e, and f	Value of single
Normal	$0 < e < 255$	$(-1)^s \times 2^{e-127} \times 1.f$
Subnormal	$e = 0, f \neq 0$	$(-1)^s \times 2^{-126} \times 0.f$
Signed zero (± 0)	$e = 0, f = 0$	$(-1)^s \times 0.0$
$+\infty$	$s = 0, e = 255, f = 0$	`+INF`
$-\infty$	$s = 1, e = 255, f = 0$	`-INF`
Not a number	$s = u, e = 255, f \neq 0$	`NaN`

exponent e. The actual exponent, which may be negative, is

$$p = e - \text{bias}. \tag{2.5}$$

2.2.1.1 Examples of IEEE Representations

There are two basic IEEE floating-point formats, singles and doubles. *Singles* or *floats* is shorthand for *single-precision floating-point numbers*, and *doubles* is shorthand for *double-precision floating-point numbers*. (In Python, however, floats are double precision.) Singles occupy 32 bits overall, with 1 bit for the sign, 8 bits for the exponent, and 23 bits for the fractional mantissa (which gives 24-bit precision when the phantom bit is included). Doubles occupy 64 bits overall, with 1 bit for the sign, 10 bits for the exponent, and 53 bits for the fractional mantissa (for 54-bit precision). This means that the exponents and mantissas for doubles are not simply double those of floats, as we see in Table 2.1. (In addition, the IEEE standard also permits *extended precision* that goes beyond doubles, but this is all complicated enough without going into that right now.)

To see the scheme in practice, consider the 32-bit representation (2.4):

	s	e	f	
Bit position	31	30 23	22 0	

The sign bit s is in bit position 31, the biased exponent e is in bits 30–23, and the fractional part of the mantissa f is in bits 22–0. Because 8 bits are used to store the exponent e and because $2^8 = 256$, e has the range

$$0 \le e \le 255. \tag{2.6}$$

The values $e = 0$ and 255 are special cases. With bias $= 127_{10}$, the full exponent

$$p = e_{10} - 127, \tag{2.7}$$

and, as indicated in Table 2.1, singles have the range

$$-126 \le p \le 127. \tag{2.8}$$

The mantissa f for singles is stored as the 23 bits in positions 22–0. For *normal numbers*, that is, numbers with $0 < e < 255$, f is the fractional part of the mantissa, and therefore the actual number represented by the 32 bits is

$$\text{Normal floating-point number} = (-1)^s \times 1.f \times 2^{e-127}. \tag{2.9}$$

Subnormal numbers have $e = 0$, $f \ne 0$. For these, f is the entire mantissa, so the actual number represented by these 32 bit is

$$\text{Subnormal numbers} = (-1)^s \times 0.f \times 2^{e-126}. \tag{2.10}$$

The 23 bits $m_{22} - m_0$, which are used to store the mantissa of normal singles, correspond to the representation

$$\text{Mantissa} = 1.f = 1 + m_{22} \times 2^{-1} + m_{21} \times 2^{-2} + \cdots + m_0 \times 2^{-23}, \tag{2.11}$$

with $0.f$ used for subnormal numbers. The special $e = 0$ representations used to store ± 0 and $\pm \infty$ are given in Table 2.2.

To see how this works in practice (Figure 2.2), the largest positive normal floating-point number possible for a 32-bit machine has the maximum value $e = 254$ (the value 255 being reserved) and the maximum value for f:

$$X_{max} = 01111\,1110\,1111\,1111\,1111\,1111\,1111\,111$$

$$= (0)(1111\,1110)(1111\,1111\,1111\,1111\,1111\,111), \tag{2.12}$$

where we have grouped the bits for clarity. After putting all the pieces together, we obtain the value shown in Table 2.1:

$$s = 0, \; e = 1111\,1110 = 254, \; p = e - 127 = 127,$$

$$f = 1.1111\,1111\,1111\,1111\,1111\,111 = 1 + 0.5 + 0.25 + \cdots \simeq 2,$$

$$\Rightarrow (-1)^s \times 1.f \times 2^{p=e-127} \simeq 2 \times 2^{127} \simeq 3.4 \times 10^{38}. \tag{2.13}$$

Likewise, the smallest positive floating-point number possible is subnormal ($e = 0$) with a single significant bit in the mantissa:

$$0\;0000\,0000\,0000\,0000\,0000\,0000\,0000\,001. \tag{2.14}$$

This corresponds to

$$s = 0, \; e = 0, \; p = e - 126 = -126$$

$$f = 0.0000\,0000\,0000\,0000\,0000\,001 = 2^{-23}$$

$$\Rightarrow (-1)^s \times 0.f \times 2^{p=e-126} = 2^{-149} \simeq 1.4 \times 10^{-45}. \tag{2.15}$$

In summary, single-precision (32-bit or 4-byte) numbers have six or seven decimal places of significance and magnitudes in the range

$$1.4 \times 10^{-45} \leq \text{single precision} \leq 3.4 \times 10^{38}. \tag{2.16}$$

Doubles are stored as two 32-bit words, for a total of 64 bits (8 B). The sign occupies 1 bit, the exponent e, 11 bits, and the fractional mantissa, 52 bits:

	s	e		f		f (cont.)	
Bit position	63	62	52	51	32	31	0

As we see here, the fields are stored contiguously, with part of the mantissa f stored in separate 32-bit words. The order of these words, and whether the second word with f is the most or least significant part of the mantissa, is machine-dependent. For doubles, the bias is quite a bit larger than for singles,

$$\text{Bias} = 1111111111_2 = 1023_{10}, \tag{2.17}$$

so the actual exponent $p = e - 1023$.

The bit patterns for doubles are given in Table 2.3, with the range and precision given in Table 2.1. To repeat, if you write a program with doubles, then 64 bits (8 bytes) will be used to store your floating-point numbers. Doubles have approximately 16 decimal places of precision (1 part in 2^{52}) and magnitudes in the range

$$4.9 \times 10^{-324} \leq \text{double precision} \leq 1.8 \times 10^{308}. \tag{2.18}$$

Table 2.3   Representation scheme for IEEE doubles.

Number name	Values of s, e, and f	Value of double
Normal	$0 < e < 2047$	$(-1)^s \times 2^{e-1023} \times 1.f$
Subnormal	$e = 0, f \neq 0$	$(-1)^s \times 2^{-1022} \times 0.f$
Signed zero	$e = 0, f = 0$	$(-1)^s \times 0.0$
$+\infty$	$s = 0, \ e = 2047, \ f = 0$	`+INF`
$-\infty$	$s = 1, \ e = 2047, \ f = 0$	`-INF`
Not a number	$s = u, \ e = 2047, \ f \neq 0$	`NaN`

If a single-precision number x is larger than 2^{128}, a fault condition known as an *overflow* occurs (Figure 2.2). If x is smaller than 2^{-128}, an underflow occurs. For overflows, the resulting number x_c may end up being a machine-dependent pattern, not a number (NAN), or unpredictable. For underflows, the resulting number x_c is usually set to zero, although this can usually be changed via a compiler option. (Having the computer automatically convert underflows to zero is usually a good path to follow; converting overflows to zero may be the path to disaster.) Since the only difference between the representations of positive and negative numbers on the computer is the sign bit of one for negative numbers, the same considerations hold for negative numbers.

In our experience, *serious scientific calculations almost always require at least 64-bit (double-precision) floats*. And if you need double precision in one part of your calculation, you probably need it all over, which means double-precision library routines for methods and functions.

2.2.2   Python and the IEEE 754 Standard

Python has been changing in recent years, and while in the past it did not adhere to all aspects of the IEEE 754 standard, it does now almost completely. Probably the most relevant difference from the standard is that *Python does not support single (32 bit) precision floating-point numbers*. So when we deal with a data type called a *float* in Python, it is the equivalent of a *double* in the IEEE standard. Because singles are inadequate for most scientific computing, this is not a loss for us. However be wary, if you switch over to Java or C you should declare your variables as `doubles` and not as `floats`. While Python eliminates single-precision floats, it adds a new data type *complex* for dealing with complex numbers. Complex numbers are stored as pairs of doubles and are quite useful in physics.

The details of how closely Python adheres to the IEEE 754 standard depend upon the details of Python's use of the C or Java language to power the Python interpreter. In particular, with the recent 64-bit architectures for CPUs, the range may even be greater than the IEEE standard, and the abnormal numbers (`±INF`, `NaN`) may differ. Likewise, the exact conditions for overflows and underflows may also differ. That being the case, the exploratory exercises to follow become all that more interesting because we cannot say that we know what results you should obtain!

2.3 Python Mini Tutorial

There is an official Python tutorial at docs.python.org/3/tutorial/ and that is a good place to go if you are starting with Python. In this section, we just highlight some basics that should help you better understand our programs. In fact, studying the programs is a good way to learn, and we recommend it! In addition, Chapter 7 discusses computing with matrices and how best to do it with Python.

2.3.1 Structure and Functions

We find Python to be the easiest programming language to learn and work with. This follows, in part, from its use of whitespace and indentation to construct code structures, in contrast to Java and C, which use braces and semicolons. For example, here we define and call a function:

```
def Defunct(x,j):          # Defines the function
    i = 1
    max = 10
    while (i < max):
        print(i)
        i = i + 1
    return i*x**j

Defunct(x,3)               # Calls the function
```

Here def is a reserved keyword and is followed by the function name, and two arguments are being passed to the function. Note how the spacings and indentations are used to define the structures within the program, that the colon: is needed to define the function and the control structure, and that the hash # is used for comments. The blank line is ignored by Python and is there for clarity; you should use many in order to make the program's structure evident. So, while there are no special characters here to separate statements (other than the newline control character), Python does use the backslash \ as a continuation character for long statements:

```
T[ix, 1] = T[ix, 0] +  cons*(T[ix+1, 0] \
 + T[ix-1, 0] - 2.*T[ix,0])
```

Python contains many built-in functions. Here are some of the ones from the C math library (in some cases needing a math. prefix):

factorial(x)	expo(x)	floor(x)	mod(x, y)	log(x[, base])	log10(x)
pow(x, y)	sqrt(x)	MacOS(x)	basin(x)	atan(x)	atan2(y, x)
cos(x)	sin(x)	tan(x)	degrees(x)	radians(x)	cosh(x)
sinh(x)	tanh(x)	cosh(x)	sinh(x)	tanh(x)	ref(x)
gamma(x)					

Also useful are the mathematical and computer constants:

math.pi   math.e   math.inf   math.nan

2.3.2 Variable Types and Operators

Variables in Python are symbols to which values are assigned. You can use almost any name for your variable, but not any of the built-in function names, reserved words, or these keywords:

False	def	if	raise	None	del
import	return	True	elf	in	try
and	else	is	while	as	except
lambda	with	assert	finally	nonlocal	yield
break	for	not	class	form	or
continue	global	pass			

Variable names can contain letters, numbers, and the underscore _, but they cannot start with a number or contain spaces. The variables' values are assigned with a single equal sign:

```
  Label = "Voltage"                      # A string
2 x = 25                                  # An integer
```

Python supports integers (int), floating point numbers (floats), complex numbers (complex), booleans (bool), and strings (str). Integers are created when they are assigned without a decimal point, while floats are created when assigned with decimal points. The division of two integers returns a float, as well as mixed arithmetic with floats and integers:

```
  >>> 6/3
2 2.0
  >>>3/6                                  # Note round off
  0.
  >>> 3./6                                # Mixed types
6 0.50000
```

Note that, in contrast to languages such as Java and C, you do not declare the variable type before using it. However, Python will change the variable type depending upon how it's used:

```
  >>> i = 3
2 >>> print (i)
  3
  >>> i = 12.*i
  >>> print (i)
6 36.0000
```

A complex number z uses two floats to store the real and imaginary parts:

```
  >>> import math                    # import math" for complex
2 >>> x = 2
  >>> y = 3
  >>> z = complex(x,y)                    # Assign a complex number
  >>> print(z.real , z.imag)
6 2., 3.
```

One can also represent a complex number in polar coordinates

```
cmath.phase(z)                          # The phase phi
>>> phase(complex(-1.0, 0.0))
3.141592653589793
abs(z)                                  # Modulus uses usual abs function
>>> 1.0
```

The function cmath.polar(z) converts a complex number into the polar representation (r, ϕ), while the function cmath.rect(r, phi) converts the polar representation into a Cartesian one.

A **string** is a series of characters that is to be understood literally. For example:

```
S = "A string using double quotes"
S = "A string using single quotes"
S = 'It\'s possible to escape a quote'
""" From "Computational Physics"
    Problem Solving with Python """     # Double within triple quotes
```

So either single, double, or triple quotes are used to set off the string. In the last two lines, the internal double quotes are kept as part of the string, and triple quotes are used to continue the string to an additional line. You can access individual elements of a string via an index beginning at 0:

```
S = ""Problem Solving With Python""
print(S[0])   # P
print(S[1])   # r
print(S[-1]) # n
```

Slicing is a technique to extract a substring from within a string:

```
S = ""Problem Solving With Python""
print(S[0:3])   # Pro
```

Math Operators:
+ addition, − subtraction, * multiplication, / division,
% modulus/remainder, ** exponentiation

Comparison Operators:
−− Equals, != Not equals, > Greater than, < Less than, >= Greater than or equal to, <= Less than or equal to.

Common Assignment Operators:
= += add and assign, −= subtract and assign, *= multiply and assign,
/= divide and assign

2.3.3 Boolean and Control Structures

Boolean variables can have the values True or False:

```
   >>> 1 < 2
 2 False
   >>> 2 > 1
   True
   >>> bool(2 > 1)
 6 True
```

The if statement executes a block if the condition is met (note colon: and indent):

```
   if condition:
 2     if-block
```

The if...else statement chooses which block to execute (note colon:, semicolon;, and indent):

```
   if condition:
 2     if-block;
   else:
       else-block;
```

Finally, the if...elif...else statement permits the choice from among several blocks:

```
   if-condition:
       if-block
   elif elif-condition1:
       elif-block1
 4 elif elif-condition2:
       elif-block2
   ...
 8 else:
       else-block
```

Python also supports for loops:

```
   >>> for index in range(1, 3):
           print(index)
 3 1
   2
   3
```

and while loops:

```
   while counter < max:
       print(counter)
 3     counter += 1
```

2.3.4 Python Lists as Arrays

A *list* is Python's built-in sequence of numbers or arbitrary objects. Although called a "list" it is similar to what other computer languages call an "array". (In Section 7.3.2, we will describe a higher-level `array` data type available with the *NumPy* package.) Python interprets a sequence of ordered items, $L = l_0, l_1, \ldots, _{N-1}$, as a *list* and represents it with a single symbol ʟ:

```
>>> L = [1, 2, 3]              # Create list                        1
>>> L[0]                       # Print element 0 (first)
1                              # Python output                      3
>>> L                          # Print entire list
 [1, 2, 3]                     # Output                             5
>>> L[0]= 5                    # Change element 0
>>> L                                                               7
[5, 2, 3]
>>> len(L)                     # Length of list                     9
3
>>> for items in L: print items    # For loop over items            11
5
2                                                                   13
3
```

Observe that square brackets with comma separators, [1, 2, 3], are used for lists, and that a square bracket is also used to indicate the index for a list item, as in line 2, L[0]. The items in lists are *mutable* or changeable. Immutable objects include integers, floats, strings, and tuples; after a string of them has been defined, its contents cannot be changed. As we see in line 7 in the L command, an entire list can be referenced as a single object, in this case, to print it.

Python also has another built-in list data type known as a *tuple* whose elements are not mutable (they also process faster than lists). Tuples are indicated by round parenthesis (.., .., .), with individual elements still referenced by square brackets:

```
>>> T = (1, 2, 3, 4)           # Create a tuple list
>>> T[3]                       # Print element 3                    2
4
>>> T                          # Print entire tuple                4
(1, 2, 3, 4)
>>> T[0] = 5                   # Attempt to change element 0        6
Traceable (most recent call last):
    T[0] = 5                                                        8
Error: 'tuple' object does not support item assignment
```

Note Python's error message when we tried to change an element of a tuple.

Many languages require you to specify the size of an array before you can store objects in it. In contrast, Python lists are *dynamic*, which means that their sizes adjust as needed. In addition, while a list is essentially one-dimensional because it is a sequence, a compound list can be created in Python with the individual elements themselves as lists:

```
>>> L = [[1,2], [3,4], [5,6]]  # A list of lists                   1
>>> L
[[1, 2], [3, 4], [5, 6]]                                           3
>>> L[0]                       # The first element
[1, 2]                                                             5
```

Python can perform a large number of operations on lists, for example:

Operation	Effect	Operation	Effect
L = [1, 2, 3, 4]	Form list	L1 + L2	Concatenate lists
L[i]	ith element	len(L)	Length of list L
i in L	True if i in L	L[i:j]	Slice from i to j

Operation	Effect	Operation	Effect
for i in L	Iteration index	L.append(x)	Append x to end of L
L.count(x)	Number of x's in L	L.index(x)	Location of 1st x in L
L.remove(x)	Remove 1st x in L	L.reverse()	Reverse elements in L
L.sort()	Order elements in L		

2.3.5 Python I/O

Outputting variables to the screen is easy, but as mentioned in Chapter 1, it differs in Python 2 and 3:

```
>>> print 'Hello, World!'      # Python 2
Hello , World
3 >>> print('Hello, World!')    # Python 3
'Hello, World!'
```

Here the ≫ indicates that we were working in an interactive shell. Inputting from the keyboard is accomplished with the `input` command:

```
name = input("Hello, What's your name?")
print("That's nice " + name + "thank you")
age = input("How old are you?")
4 print("So, you are already " + astr(age) + " years old, "
+ name + "!")
```

This is what we have done in the program `AreaFormatted.py` in Listing 2.11. This shows that we can print the value of a variable just by giving its name. We also see in `AreaFormatted.py` that we can input strings (literal numbers and letters) by either enclosing the string in quotes (single or double), or by using the `raw_input` (Python2) or `input` (Python 3) command without quotes. `AreaFormatted.py` also shows how to input both a string and numbers from a file.

Python uses its default format when you print a float by giving just its name, with the format varying depending on the precision of the number. You can control the format if you like. To print floats you need to specify how many digits (places) after the decimal point are desired, and how many spaces overall should be used for the number:

```
print("x=%6.3f,  Pi=%9.6f,  Age=%d \n") % (x, math.pi, age)
print ("x=%6.3f, %(x), "Pi=%9.6f," %(math.pi), "Age=%d "%(age)," \n)
x = 12.345,  Pi = 3.141593, Age=39      # Output from either
```

Here the `%6.3f` formats a float (which is a double in Python) to be printed in fixed-point notation (the `f`) with three places after the decimal point and with six places overall (one place for the decimal point, one for the sign, one for the digit before the decimal point, and three for the decimal). The directive `%9.6f` has six digits after the decimal place and nine overall. To print an integer, you need to specify only the total number of digits (there is no decimal part), and we do that with the `%d` (d for digits) format.

The % symbol in these output formats indicates a conversion from the computer's internal format to that used for output. Note above that we have also use a \n directive to indicate a new line. Other directives, some of which are demonstrated in `Directives.py` in Listing 2.12, are:

\"	double quote	\0NNN	octal NNN	\\	backslash
\a	alert (bell)	\b	backspace	\c	no more output
\f	form feed	\n	new line	\r	carriage ret
\t	horizontal tab	\v	vertical tab	%%	a single %

Notice in Listing 2.11 how we read from the keyboard, as well as from a file, and then output to both screen and file. Beware, if you do not create the file `Name.dat`, the program will issue ("throw") an error message of the sort:

```
Error: [Error 2] No such file or directory: 'Name.dat'.
```

2.3.6   Python's Algebraic Tools

While this book's focus is mainly on the use of Python for numerical simulations, that is not to discount the importance of computational symbolic manipulations. Python actually has (at least) two packages that can be used for symbolic manipulations, and they are quite different. As indicated in Section 1.6, the *Sage* package is very much in the same class as Maple and Mathematica. Sages's notebook interface lets users create publication-quality text, run programs, or manipulate equations symbolically. Yet Sage is a big and powerful package that goes beyond pure Python by including multiple computer algebra systems, as well as visualization tools, and more. Using the multiple features of Sage can get to be quite complicated, and, in fact, books have been written and workshops taught on the use of Sage. We refer the interested reader to the online Sage Documentation page www.sagemath.org/help.html.

The *SymPy* package for symbolic manipulations runs within a regular Python shell, very much like any other Python package. It can be downloaded from github.com/sympy/sympy/releases, or you can use the Canopy distribution that includes SymPy. Now we give some simple examples of SymPy's use, but really you should start with the *SymPy Tutorial*, docs.sympy.org/latest/tutorial/. To start, we'll take some derivatives to show that SymPy knows calculus:

```
1  >>> from SymPy import *
   >>> x, y = symbols('x y')
   >>> y = diff(tan(x),x); y            # y = derivative tan(x)
       $\tan^2(x) + 1$
5  >>> y = diff(5*x**4 + 7*x**2, x, 1); y    # Deriv, 1 optional
       $20 x^3 + 14 x$
   >>> y = diff(5*x**4+7*x**2, x, 2); y      # $d^2y/dx^2$
       $2\, (30 x^2 + 7)$
```

We see that we first import methods from SymPy, and then use the `symbols` command to declare the variables x and y as algebraic. The rest is rather obvious, with `diff` being the derivative operator, and the `x` argument indicating the derivative with respect to x. Now let's try expansions:

```
>>> from SymPy import *
>>> x, y = symbols('x y')
>>> z = (x + y)**8; z
    $(x + y)^8$
>>> expand(z)
    $x^8 + 8 x^7 y + 28 x^6 y^2 + 56 x^5 y^3 + 70 x^4 y^4 + 56 x^3 y^5 + 28 x^2 y^6
     + 8 x y^7 + y^8$
```

SymPy also knows about infinite series, and different expansion points:

```
>>> sin(x).series(x, 0)                      # $\sin x$ series about 0
    $x - x^3/6 + x^5/120 + \mathcal{O}(x^6)$
>>> sin(x).series(x,10)                      # $\sin x$ about x= 10
$\sin(10) + x\cos(10) - x^2 \sin(10)/2 - x^3 \cos(10)/6 + x^4 \sin(10)/24 + x^5
    \cos(10)/120 +\mathcal{O}(x^6)$
>>> z = 1/cos(x); z                          # Division, not inverse
    $1/\cos(x)$
>>> z.series(x, 0)                           # Expand $1/\cos x$ about $x=0$
    $1 + x^2/2 + 5 x^4/24 + \mathcal{O}(x^6)$
```

One of the classic difficulties with computer algebra systems is that even if the answer is correct, it may not look simple, and thus is not too useful. SymPy has the functions `simplify`, `factor`, `collect`, `cancel`, and `apart` to help make its output easier to understand:

```
>>> factor(x**2 -1)
    $(x - 1) (x + 1)$                        # A nice answer
>>> factor(x**3 - x**2 + x - 1)
    $(x - 1) (x^2 + 1)$                      # A nice answer
>>> simplify((x**3 + x**2 - x - 1)/(x**2 + 2*x + 1))
    $x - 1$                                  # Much better!
>>> simplify(x**3+3*x**2*y+3*x*y**2+y**3)
    $x^3 + 3 x^2 y + 3 x y^2 + y^3$          # No help!
>>> factor(x**3+3*x**2*y+3*x*y**2+y**3)
    $(x + y)^3$                              # Much better!
>>> simplify(1 + tan(x)**2)
    $\cos(x)^{(-2)}$
>>> simplify(2*tan(x)/(1+tan(x)**2))
    $\sin(2 x)$
```

2.4 Programming Warmup

Before we go on to serious CP work, we want to establish that your local computer is working right for you. Assume that calculators have not yet been invented, and that you need a program to calculate the area of a circle. You might try

```
read radius                    # Input
calculate area of circle       # Numerics
print area                     # Output
```

The instruction `calculate area of circle` has no meaning to most computers, so we need to specify an *algorithm*, that is, a set of rules for the computer to follow:

```
read radius                    # Input
PI = 3.141593                  # Set constant
area = PI * r * r              # Algorithm
print area                     # Output
```

This is better. Here is our Python program `Area.py`, and you should ensure that it runs for you. This is a simple program that outputs to the screen, with its input built into the program.

```
# Area.py: Area of a circle, simple program
from math import pi
N = 1
r = 1.
C = 2.* pi* r
A = pi * r**2
print ('Program number =', N, '\n r, C, A = ', r, C, A)
```

2.4.1 Program Design

Programming is a written art that blends elements of science, mathematics, and computer science into a set of instructions that permit a computer to accomplish a desired task. And now, with published scientific results increasingly relying on computation, it is increasingly important that the source version of your program itself be available to others so that they can reproduce your results. Reproducibility may not be as exciting as a new discovery, but it is an essential ingredient in science [Hinsen, 2013]. In addition to the grammar of a computer language, a scientific program should include a number of essential elements to ensure the program's validity and usability. As with other arts, we suggest that until you know better, you follow some simple rules. A good program should:

- Give the correct answers.
- Be clear and easy to read, with the action of each part easy to analyze.
- Document itself for the sake of readers and the programmer.
- Be easy to use.
- Be built up out of small programs that can be independently verified.
- Be easy to modify and robust enough to keep giving correct answers after modification and debugging.
- Document the data formats used.
- Use trusted libraries.
- Be published or passed on to others to use and to develop further.

One attraction of *object-oriented programming* is that it enforces these rules automatically. An elementary way to make any program clearer is to *structure* it with indentation, skipped lines, and strategically placed braces. This is done to provide visual clues as to the function of the different program parts (the "structures" in structured programming). Python actually uses indentations as structure elements. Although the space limitations of a printed page keep us from inserting as many blank lines as we would prefer, we recommend that you do as we say and not as we do!

We find *flowcharts*, such as the basic and the detailed ones in Figure 2.3 for projectile motion, useful in planning the chronological order for the essential steps in a program, and also providing a graphical overview of the computation. A flowchart is not meant to be a detailed description of a program, but instead is a visualization of a program's logical flow. We recommend that you draw a flowchart or (second best) write a pseudocode before you

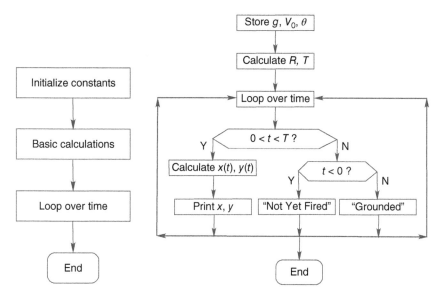

Figure 2.3 A flowchart illustrating a program to compute projectile motion. On the left are the basic components of the program, and on the right are some of its details. When writing a program, first map out the basic components, then decide upon the structures, and finally fill in the details. This is called *top-down programming*.

write a program. *Pseudocode* is like a text version of a flowchart that leaves out details and instead focuses on the logic and structures:

```
1 # A flowchart for projectile motion
  Store g, Vol, and theta
  Calculate R and T
  Begin time loop
5   Print out "not yet fired" if t < 0
    Print out "grounded" if t > T
    Calculate, print x(t) and y(t)
    Print out error message if x > R, y > H
9 End time loop    End program
```

2.4.2 First Programming Steps

1) To gain some experience with your computer system, use an editor to enter the program `Area.py` that computes the area of a circle (yes, we know you can copy and paste it, but don't). Save your program to a file in your home (personal) directory. *Note:* For those who are familiar with Python, you may want to enter the program `AreaFormatted.py` in Listing 2.11 that produces formatted output.

2) Compile and execute the appropriate version of `Area.py`.

3) Experiment with your program. For example, see what happens if you leave out decimal points in the assignment statement for *r*, if you assign *r* equal to a blank, or if you assign a letter to *r*. Remember, it is unlikely that you will "break" or hurt the computer by making a mistake, and it is good to see how the computer responds when something is wrong.

4) Change the program so that it computes the volume $\frac{4}{3}\pi r^3$ of a sphere and prints it out with the proper name. Save the modified program to a file in your personal directory and give it the name vol.py.

5) Open and execute vol.py and check that your changes are correct by running a number of trial cases. Good input data are $r = 1$ and $r = 10$.

6) Revise Area.py so that it takes input from a file name that you have made up, then outputs in a different format to another file you have created, and then reads from the latter file.

7) See what happens when the data type used for output does not match the type of data in the file (e.g., floating point numbers are read in as integers).

8) Revise Area.py so that it uses a main method (which does the input and output) and a separate function or method for the calculation. Check that you obtain the same answers as before.

2.4.3 Over and Underflow Exercises

1) Consider the 32-bit single-precision floating-point number A:

	s	e		f	
Bit position	31	30	23	22	0
Value	0	0000 1110		1010 0000 0000 0000 0000 000	

a) What are the binary values for the sign s, the exponent e, and the fractional mantissa f. (*Hint:* $e_{10} = 14$.)

b) Determine decimal values for the biased exponent e and the true exponent p.

c) Show that A's mantissa equals 1.625000.

d) Determine the full value of A.

2) Write a program that determines the **underflow** and **overflow** limits (within a factor of 2) for Python on your computer. Here's a sample pseudocode

```
    under = 1.
    over = 1.
3   begin do N times
        under = under/2.
        over = over * 2.
        write out: loop number, under, over
7   end do
```

You may need to increase N if your initial choice does not lead to underflow and overflow. If you want to be more precise regarding the limits of your computer, try multiplying and dividing by a number smaller than 2.

1) Check where under- and overflow occur for double-precision floating-point numbers. Give your answer in decimals.

2) Check where under- and overflow occur for floats.

3) Check where under- and overflow occur for integers. *Note:* There is no exponent stored for integers, so the smallest integer corresponds to the most negative one. To determine the largest and smallest integers, you must observe your program's output as you

explicitly pass through the limits. You accomplish this by continually adding and subtracting 1. (Inasmuch as integer arithmetic uses *two's complement* arithmetic, you should expect some surprises.)

2.4.4 Machine Precision

A recurring concern of computational scientists is that the floating-point representation used to store numbers is of limited precision. In general for a 32-bit-word machine, *single-precision numbers are good to 6–7 decimal places, while doubles are good to 15–16 places.* To see how limited precision affects calculations, consider the simple computer addition of two single-precision numbers:

$$7 + 1.0 \times 10^{-7} = ? \tag{2.19}$$

The computer fetches these numbers from memory and stores the bit patterns

$$7 = 0\ 10000010\ 1110\ 0000\ 0000\ 0000\ 0000\ 000, \tag{2.20}$$

$$10^{-7} = 0\ 01100000\ 1101\ 0110\ 1011\ 1111\ 1001\ 010, \tag{2.21}$$

in *working registers* (pieces of fast-responding memory). Because the exponents are different, it would be incorrect to add the mantissas, and so the exponent of the smaller number is made larger while progressively decreasing the mantissa by *shifting bits* to the right (inserting zeros) until both numbers have the same exponent:

$$10^{-7} = 0\ 01100001\ 0110\ 1011\ 0101\ 1111\ 1100101\ (0)$$
$$= 0\ 01100010\ 0011\ 0101\ 1010\ 1111\ 1110010\ (10) \tag{2.22}$$

$$\cdots$$

$$= 0\ 10000010\ 0000\ 0000\ 0000\ 0000\ 0000\ 000\ (0001101\cdots0$$
$$\Rightarrow \qquad\qquad 7 + 1.0 \times 10^{-7} = 7. \tag{2.23}$$

Because there is no room left to store the last digits, they are lost, and after all this hard work the addition just gives 7 as the answer; an example of the truncation error indicated in Figure 2.2. In other words, because a 32-bit computer stores only 6 or 7 decimal places, it effectively ignores any changes beyond the sixth decimal place.

The preceding loss of precision is categorized by defining the *machine precision* ϵ_m as the maximum positive number that, on the computer, can be added to the number stored as 1 without changing that stored 1:

$$1_c + \epsilon_m \stackrel{\text{def}}{=} 1_c, \tag{2.24}$$

where the subscript c is a reminder that this is a computer representation of 1. Consequently, an arbitrary number x can be thought of as related to its floating-point representation x_c by

$$x_c = x(1 \pm \epsilon), \quad |\epsilon| \leq \epsilon_m, \tag{2.25}$$

where the actual value of $\pm\epsilon$ is not known (but can be determined). In other words, except for powers of 2 that are represented exactly, we should assume that all single-precision numbers contain an error in the sixth decimal place, and that all doubles have an error in the 15th place. And, as is always the case with errors, we must assume that we really do not know what the error is, for if we knew, then we would eliminate it! Consequently, the arguments we are about to put forth regarding errors should be considered approximate, but that's typical for known unknowns.

2.4.5 Experiment: Your Machine's Precision

Write a program to determine the machine precision ϵ_m of your computer system within a factor of 2. A sample pseudocode is

```
1  eps = 1.
   begin do N times
      eps = eps/2.                        # Make smaller
      one = 1. + eps          # Write loop number, one, eps
5  end do
```

A Python implementation is given in Listing 2.13, while a more precise one would work at the byte level.

1) Determine experimentally the precision of double-precision floats.
2) Determine experimentally the precision of complex numbers.

It's good to remember that to print out a number in decimal format, the computer must make a conversion from its internal binary representation to decimal. This not only takes time, but unless the number is an exact power of 2, leads to a loss of precision. So if you want a truly precise indication of the stored numbers, you should avoid conversion to decimals and instead print them out in octal (\0NNN) or hexadecimal (0x) format.

2.5 Python's Visualization Tools

> *If I can't picture it, I can't understand it.*
>
> — Albert Einstein

In the sections to follow we discuss tools to visualize data produced by simulations and measurements. Whereas other books may choose to relegate this discussion to an appendix, or not to include it at all, we believe that visualization is such an integral part of CP, and so useful for your work in the rest of this book, that we have placed it here, right up front. We describe the use of Matplotlib [Matplotlib, 2023] and Vpython/Visual.

Generalities One of the most rewarding aspects of computing is visualizing the results. While in the past this was performed with 2D plots, in modern times it is regular practice to use 3D (surface) plots, volume rendering (dicing and slicing), animations, and virtual reality (gaming) tools. These types of visualizations are often breathtakingly beautiful and may provide deep insights into problems by letting us see and "handle" the functions with which we are working. Visualization also assists in the debugging process, the development of physical and mathematical intuition, and the all-around enjoyment of work.

In thinking about ways to view your results, keep in mind that the point of visualization is to make the physics clearer and to communicate your work to others. It follows then that you should make all figures as clear, informative, and self-explanatory as possible, especially if you will be using them in presentations without captions. This means labels for curves and data points, a title, and labels on the axes.[3] After this, you should study your visualization and ask whether there are better choices for units, ranges of axes, colors, style, and so on, that might get the message across better and provide more insight. And try to remember that those colors which look great on your monitor may turn into uninformative grays when printed. Considering the complexity of human perception and cognition, there may not be a single best way to visualize a particular data set, and so some trial and error may be necessary to "see" what works best.

2.5.1 Visual (VPython)'s 2D Plots

Vpython (Python plus the Visual package) is a simple way to get to create Python visualizations and it was used to create many of the visualizations in this book. Its development ended in 2006 and has been superseded by WebVpython. However, you can still run Vpython programs as WebVpython or within a Jupyter Notebook.

In Figure 2.4, we present two plots produced by the program `EasyVisual.py` in Listing 2.1. Notice that the plotting technique is to create first the plot objects `Plot1` and `Plot2`, and then to add the points to the objects, one-by-one, and then use the `plot` method to plot the objects. (In contrast, Matplotlib creates a vector of points and then plots the entire vector in one fell swoop.)

It is often a good idea to place several plots in the same figure. The program `3GraphVisual.py` in Listing 2.2 does that and produces the graph on the left of Figure 2.5. On your computer screen you will see white vertical bars created with `gears`, red dots created with `grots`, and a yellow curve created with `curve`.

Animations Creating animations with Visual is essentially just making the same 2D plot over and over again, with each one at a slightly differing time, and then placing the plots on top of each other. When performed properly, this gives the impression of motion. Several of our sample codes produce animations, for example, `HarmosAnimate.py` and `3Danimate.py`. Three frames produced by `HarmosAnimate.py` are shown on the right of

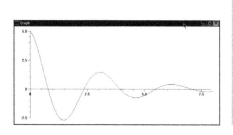
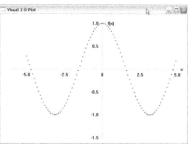

Figure 2.4 Screen dumps of two *x*-*y* plots produced by EasyVisual.py using the Visual package. The *left* plot uses default parameters while the *right* plot uses user-supplied options.

---

3 Although this may not need saying, place the independent variable *x* along the abscissa (horizontal), and the dependent variable $y = f(x)$ along the ordinate.

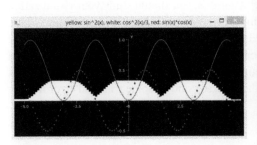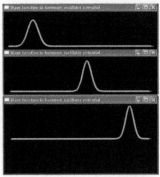

Figure 2.5 *Left*: Output from the program 3GraphVisual.py that places three different types of 2D plots on one graph using Visual. *Right*: Three frames from a Visual animation of a quantum mechanical wave packet produced with HarmosAnimate.py.

Figure 2.5. The major portions of these codes deal with the solution of partial differential equations, which need not concern us (yet). The part which makes the animation is simple:

```
PlotObj= curve(x=xs, color=color.yellow, radius=0.1)
...
3 while True:                                        # Runs forever
    rate(500)
    ps[1:-1] = ...
    psi[1:-1] = ..
7   PlotObj.y = 4*(ps**2 + psi**2)
```

Here `PlotObj` is a curve that gets continually built from within a while loop and thus appears to be moving. Note that being able to plot points individually without having to store them all in an array for all times keeps the memory demand of the program quite small and is fast.

2.5.2 Matplotlib's 2D Plots

Matplotlib is a powerful plotting package that lets you create 2D and 3D graphs, histograms, power spectra, bar charts, error charts, scatter plots, and what not, all directly from within your Python program. Matplotlib is free, uses the sophisticated numerics of NumPy and LAPACK, and, believe it or not, is easy to use. Since Matplotlib is not part of standard Python, you must import the entire Matplotlib package, or individual methods, into your program. We usually do that in our codes by importing `pylab`, which is a module that provides both Matplotlib and NumPy packages. Here, from `EasyMatPlot.py`, is how we do it:

```
1 from pylab import *                                # Load Matplotlib
  Min = -5.;    Max = +5.;   Npoints= 500
  Del = (Max - Min) / Points
  x = arrange(Min, Max, Del)
5 y =  sin(x) * sin(x*x)                             # f(x array)
  label('x');   label('f(x)');  title(' f(x) vs x')
  text(-1.75, 0.75, 'Matplotlib \n Example')         # Text on plot
```

```
   plot(x, y, '-', lw=2)
9  grid(True)                                           # Form grid
   show()
```

   Matplotlib commands are by design similar to the plotting commands of MATLAB, a commercial problem-solving environment that is particularly popular in engineering. As is true for MATLAB, Matplotlib assumes that you have placed the x and y values that you wish to plot into 1D arrays (vectors), and then plots the entire vectors in one fell swoop. Matplotlib uses the powerful NumPy `array` object to store the data, which we discuss further in Chapter 7. As you can see, NumPy's `arrange` method constructs an array covering "a range" between `Max` and `Min` in steps of `Del`. Because the limits are floating-point numbers, so too will be the x_i's. And because `x` is an array, `y = -sin(x)*cos(x)` is automatically one too! The actual plotting is performed with the `plot` command, with a dash '-' indicating a line, and `lw=2` setting the line width. The result is shown on the left of Figure 2.6, with the desired labels and title. The `show()` command produces the graph on your desktop. More commands are given in Table 2.4. We suggest you try out some of the options and types of plots possible. In Listing 2.5, we give the code `GradesMatplot.py`, and on the right of Figure 2.6 we show its output. This is not a simple plot. Here we repeat the `plot` command several times in order to plot several data sets on the same graph, and to plot both the data points and the lines connecting them. On Line 3 we import Matplotlib (pylab), and on Line 4 we import NumPy, which we need for the `array` command. Seeing that we have imported two packages, we add the `pylab` prefix to the `plot` commands so that Python knows which package to use.

   In order to place a horizontal line along $y = 0$, on lines 10 and 11 we create a data set as an array of x values, $-1 \leq x \leq 5$, and a corresponding array of y values, $y_i \equiv 0$. We then plot the horizontal on line 12. Next, we place four more curves on the figure. First on lines 14–15 we create data set 0, then plot the points as blue circles (gray on the page) `'bo'`, and connect the points with green (`'g'`) lines. On lines 19–21 we create and plot another data set as a red (`'r'`) line (gray on the page). Finally, on lines 23–25 we define unequal lower and upper error bars and place them on the plot. We finish by adding grid lines (Line 27) and *show*ing the plot on the screen.

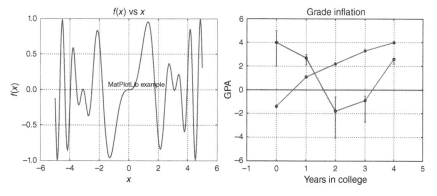

Figure 2.6 Matplotlib plots. *Left*: Output of EasyMatPlot.py (Listing 2.3) showing a simple, *x-y* plot. *Right*: Output from GradesMatPlot.py that places two sets of data points, two curves, and unequal upper and lower error bars, all on one plot.

Table 2.4 Some common Matplotlib commands.

Command	Effect	Command	Effect
plot(x, y, '-', lw=2)	x-y line width 2	myPlot.setYRange(−8., 8.)	Set y range
show()	Show graph	myPlot.setSize(500, 400)	Size in pixels
label('x')	x-axis label	pyplot.semilogx	Epilog x plot
label('f(x)')	y-axis label	pyplot.semilogy	Epilog y plot
title('f vs. x')	Add title	grid(True)	Draw grid
text(x, y, 's')	Add text s at (x, y)	myPlot.setColor(false)	Black & White
myPlot.addPoint (0,x,y,true)	Add (x, y) to 0 connect	myPlot.setButtons(true)	For zoom button
myPlot.addPoint (1,x,y, false)	Add (x, y) to 1, no connect	myPlot.fillPlot()	Fit ranges to data
pyplot.errorbar	Point + error bar	myPlot.setImpulses(true,0)	Vert lines, set 0
pyplot.clf()	Clear figure	pyplot.contour	Contour lines
pyplot.scatter	Scatter plot	pyplot.bar	Bar charts
pyplot.polar	Polar plot	pyplot.gca	For current axis
myPlot.setXRange (−1., 1.)	Set x range	pyplot.acorr	Autocorrelation

Often the science is clearer if there are several curves in one plot, and, several plots in one figure. Matplotlib lets you do this with the `plot` and the `subplot` commands. For example, in `MatPlot2figs.py` in Listing 2.6 and Figure 2.7, we have placed two curves in one plot, and then output two different figures, each containing two plots. The key here is a repetition of the `subplot` command:

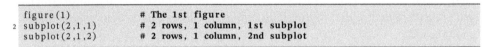

```
  figure(1)            # The 1st figure
2 subplot(2,1,1)       # 2 rows, 1 column, 1st subplot
  subplot(2,1,2)       # 2 rows, 1 column, 2nd subplot
```

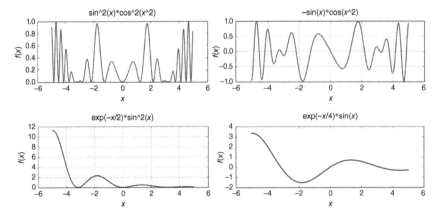

Figure 2.7 *Left* and *Right* Columns show two separate outputs, each of two figures, produced by MatPlot2figs.py. (We used the slider button to add some space between the upper and lower plots).

The listing is self-explanatory, with sections that set the plotting limits, that create each figure, and then create the grid.

Scatter Plots Sometimes we need a scatter plot of data, and maybe even a curve thrown in as well. In Figure 5.4, we show a scatter plot created with the code `PondMapPlot.py` in Listing 2.7. The key statements here are of the form `ax.plot(ox, yo, 'bo', markersize=3)`, which in this case adds a blue point (gray on the page) of size 3.

2.5.3 Matplotlib's 3D Surface Plots

A 2D plot of the potential $V(r) = 1/r$ *versus* r is fine for visualizing the radial dependence of the potential field surrounding a single charge, but if you want to visualize a dipole potential such as $V(x, y) = [B + C(x^2 + y^2)^{-3/2}]x$, you need a 3D, or surface, visualization. You get that by creating a world in which the z dimension (mountain height) is the value of the potential, and the x and y axes define the plane below the mountain. As the surface you are creating is a 3D object, it is not truly possible to draw it on a flat screen, and so different techniques are used to give the impression of three dimensions to our brains. That is accomplished by rotating the object (grabbing it with your mouse), shading it, employing parallax, and other tricks.

In Figure 2.8, we show a wire-frame plot (left) and a colored surface-plus-wire-frame plot (right). These are obtained from the program `simple3Dplot.py` in Listing 2.8. Note that there is an extra import of `Axes3D` from the Matplotlib tool kit needed for 3D plotting. Lines 8 and 9 are the usual creation of x and y arrays of floats using `arrange`. Line 11 uses the `meshed` method to set up the entire coordinate matrix grid from the x and y coordinate vectors with a vector operation, and line 12 constructs the entire Z surface with another vector operation. The remainder of the program is self-explanatory, with `fig` being the plot object, `ax` the 3D axes object, and `plot_airframe` and `plot_surface` creating wire frame wireframe and surface plots, respectively. Another type of 3D plot that is particularly useful when examining data of the form (x_i, y_j, z_k), is a scatter plot into a 3D volume. In Listing 2.9, we give the program `Scatter3dPlot.py` that created the plot in Figure 2.9. This program, which is taken from the Matplotlib documentation, uses the NumPy random number generator, with the 111 notation being a hand-me-down from MATLAB indicating a $1 \times 1 \times 1$ grid.

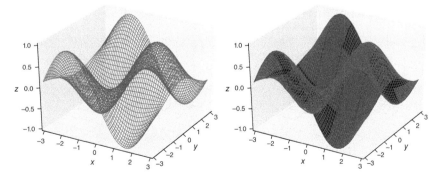

Figure 2.8 *Left*: A 3D wire frame. *Right*: A colored surface plot with wire frame. Both are produced by the program Simple3dplot.py using Matplotlib.

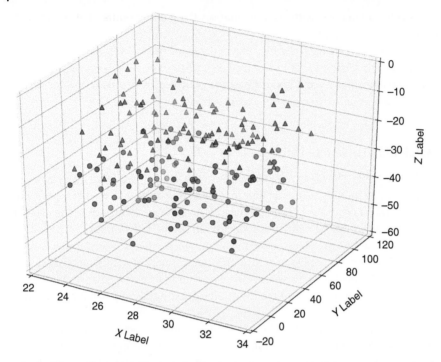

Figure 2.9 A 3D scatter plot produced by the program Scatter3dPlot.py using Matplotlib.

Finally, the program FourierMatplot.py, written by Oscar Estrepe, performs a Fourier reconstruction of a saw tooth wave, with the number of waves included controlled by the viewer via a slider bar, as shown in Figure 2.10. (We discuss Fourier transforms in Chapter 9.) The slider method is included via the extra lines:

```
1 from    matplotlib.widgets import Slider
  ...
  shortwaves = Slider(airwaves, '# Waves', 1, 20, valinit=T)
  ...
5 snumwaves.on_changed(update)
```

2.5.4 Matplotlib's Animations

Matplotlib can also create animations, although not as simply as Vpython. The Matplotlib examples page gives a number of them. We have included some Matplotlib animation codes in the Codes directory, and show a sample code for the heat equation in Listing 2.10. Here too, most of the code deals with solving a partial differential equation, which need not interest us yet. The animation is carried out at the bottom of the code.

2.6 Plotting Exercises

1) We encourage you to make your own plots and personalize them by trying out other commands and by including further options in the commands. The Matplotlib documentation is extensive and available on the Web. As an exercise, explore:

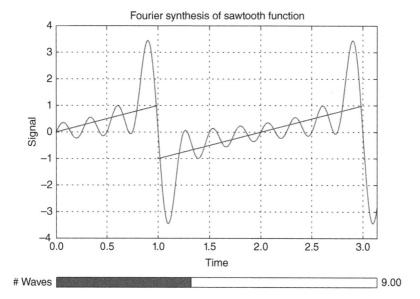

Figure 2.10 A comparison of a saw tooth function to the sum of its Fourier components, with the number of included waves varied interactively by a Matplotlib slider. FourierMatplot.py produced this output and was written by Oscar Estrepe.

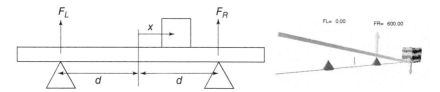

Figure 2.11 *Left*: A beam and a box supported at two points. *Right*: A screenshot from the animation showing the forces on the beam as the weight moves.

    a) how to zoom in and zoom out on sections of a plot,
    b) how to save your plots to files in various formats,
    c) how to print up your graphs,
    d) the options available from the pull-down menus,
    e) how to increase the space between subplots,
    f) and how to rotate and scale the surfaces.
2) As shown in Figure 2.11, a beam of length $L = 10$ m and weight $W = 400$ N rests on two supports at a distance $d = 2$ m apart. A box of weight $W_b = 800$ N, initially above the left support, slides frictionlessly to the right with a velocity $v = 7$ m/s.
    a) Write a program that calculates the forces exerted on the beam by the right and left supports as the box slides along the beam.
    b) Extend your program so that it creates an animation, or just a series of stills, showing the forces and the position of the block as the box slides along the beam. In Figure 2.11, left we present a screenshot captured from one of our animations.
    c) Extend the two-support problem to a box sliding to the right on a beam with a third support under the right edge of the beam.

2.7 Code Listings

Listing 2.1 **EasyVisual.py** Produces two different 2D plots using the Visual package.

```
# EasyVisual.py:        Simple graph object using Visual
                                                                    2
from visual.graph import *                      # Import Visual
Plot1 = gcurve(color = color.white)             # gcurve method     4
for x in arange(0., 8.1, 0.1):                  # x range
    Plot1.plot( pos = (x, 5.*cos(2.*x)*exp(-0.4*x)) ) # Plot pts     6
graph1 =  gdisplay(width=600, height=450,\
    title='Visual 2-D Plot', xtitle='x', ytitle='f(x)',\            8
    foreground = color.black, background = color.white)
Plot2 = gdots(color = color.black)              # Dots              10
for x in arange( -5., +5, 0.1 ):
    Plot2.plot(pos = (x, cos(x)))                                   12
```

Listing 2.2 **3GraphVisual.py** Produces a 2D x-y plot with the Matplotlib and NumPy packages.

```
# 3GraphVisual.py: 3 plots in the same figure, with bars, dots and curve  2

from visual import *                                                4
from visual.graph import*

                                                                    6
string = "blue: sin^2(x), white: cos^2(x), red: sin(x)*cos(x)"
graph1 = gdisplay(title=string, xtitle='x', ytitle='y')            8
y1 = gcurve(color=color.yellow, delta=3)        # Curve
y2 = gvbars(color=color.white)                  # Vertical bars     10
y3 = gdots(color=color.red, delta=3)            # Dots
for x in arange(-5, 5, 0.1):                    # arange for floats  12
    y1.plot( pos=(x, sin(x)*sin(x)) )
    y2.plot( pos=(x, cos(x)*cos(x)/3.) )                            14
    y3.plot( pos=(x, sin(x)*cos(x)) )
```

Listing 2.3 **3Dshapes.py** Produces a sample of VPython's 3D shapes.

```
# 3Dshapes.py: Some 3-D Shapes of VPython                           2

from visual import *                                                4

graph1 = display(width=500, height=500, title='VPython 3-D Shapes', range=10)  6
sphere(pos=(0,0,0), radius=1, color=color.green)
sphere(pos= (0,1,-3), radius=1.5, color=color.red)                  8
arrow(pos=(3,2,2), axis=(3,1,1), color=color.cyan)
cylinder(pos=(-3,-2,3), axis=(6,-1,5), color=color.yellow)          10
cone(pos=(-6,-6,0), axis=(-2,1,-0.5), radius=2, color=color.magenta)
helix(pos=(-5,5,-2), axis=(5,0,0), radius=2, thickness=0.4, color=color.orange)  12
ring(pos=(-6,1,0), axis=(1,1,1), radius=2, thickness=0.3, color=(0.3,0.4,0.6))
box(pos=(5,-2,2), length=5, width=5, height=0.4, color=(0.4,0.8,0.2))  14
pyramid(pos=(2,5,2), size=(4,3,2), color=(0.7,0.7,0.2))
ellipsoid(pos=(-1,-7,1), axis=(2,1,3), length=4, height=2, width=5,  16
    color=(0.1,0.9,0.8))
```

Listing 2.4 **EasyMatPlot.py** Produces a 2D *x-y* plot using the Matplotlib package (which includes the NumPy package).

```
# EasyMatPlot.py:        Simple use of matplotlib's plot command
                                                                    2
from pylab import *                              # Load Matplotlib
```

```
Xmin = -5.;   Xmax = +5.;   Npoints= 500                                    4
DelX = (Xmax - Xmin) / Npoints                                              6
x = arange(Xmin, Xmax, DelX)
y =  sin(x) * sin(x*x)                                 # F(x array)         8
print ('arange => x[0], x[1],x[499]=%8.2f %8.2f %8.2f' %(x[0],x[1],x[499]))
print ('arange => y[0], y[1],y[499]=%8.2f %8.2f %8.2f' %(y[0],y[1],y[499])) 10
print ("\n Now doing the plotting thing, look for Figure 1 on desktop" )
xlabel('x');        ylabel('f(x)');        title(' f(x) vs x')             12
text(-1.75,  0.75, 'MatPlotLib \n Example')            # Text on plot
plot(x, y, '-', lw=2)                                                       14
grid(True)                                             # Form grid
show()                                                                      16
```

Listing 2.5   GradesMatPlot.py Produces a 2D *x-y* plot using the Matplotlib package.

```
# Grade.py: Using Matplotlib's plot command with multi data sets & curves
                                                                            2
import pylab as p                                      # Matplotlib
from numpy import*                                                          4

p.title('Grade Inflation')                             # Title and labels   6
p.xlabel('Years in College')
p.ylabel('GPA')                                                             8

xa = array([-1, 5])                                    # For horizontal line 10
ya = array([0, 0])                                     # "          "
p.plot(xa, ya)                                         # Draw horizontal line 12

x0 = array([0, 1, 2, 3, 4])                            # Data set 0 points   14
y0 = array([-1.4, +1.1, 2.2, 3.3, 4.0])
p.plot(x0, y0, 'bo')                                   # Data set 0 = blue circles 16
p.plot(x0, y0, 'g')                                    # Data set 0 = line
                                                                            18
x1 = arange(0, 5, 1)                                   # Data set 1 points
y1 = array([4.0, 2.7, -1.8, -0.9, 2.6])                                     20
p.plot(x1, y1, 'r')
                                                                            22
errTop = array([1.0, 0.3, 1.2, 0.4, 0.1])              # Asymmetric error bars
errBot = array([2.0, 0.6, 2.3, 1.8, 0.4])                                   24
p.errorbar(x1, y1, [errBot, errTop], fmt = 'o')        # Plot error bars
                                                                            26
p.grid(True)                                           # Grid line
p.show()                                               # Create plot on screen 28
```

Listing 2.6   MatPlot2figs.py Produces the two figures shown in Figure 2.7. Each figure contains two plots with one Matplotlib figure.

```
# MatPlot2figs.py: plot of 2 subplots on 1 fig & 2 separate figs
                                                                            2
from pylab import *                                    # Load Matplotlib
                                                                            4
Xmin = -5.0;        Xmax =  5.0;        Npoints= 500
DelX= (Xmax-Xmin)/Npoints                              # Delta x             6
x1 = arange(Xmin, Xmax, DelX)                          # x1 range
x2 = arange(Xmin, Xmax, DelX/20)                       # Different x2 range   8
y1 =  -sin(x1)*cos(x1*x1)                              # Function 1
y2 =   exp(-x2/4.)*sin(x2)                             # Function 2          10
print("\n Now plotting, look for Figures 1 & 2 on desktop")
figure(1)                                              # Figure 1            12
subplot(2,1,1)                                         # 1st subplot in first figure
plot(x1, y1, 'r', lw=2)                                                     14
xlabel('x');        ylabel( 'f(x)' );        title( '-sin(x)*cos(x^2)' )
grid(True)                                             # Form grid           16
subplot(2,1,2)                                         # 2nd subplot in first figure
plot(x2, y2, '-', lw=2)                                                     18
xlabel('x')                                            # Axes labels
```

```
ylabel( 'f(x)' )                                                                     20
title( 'exp(-x/4)*sin(x)' )
figure(2)                                          #        Figure 2               22
subplot(2,1,1)                               # 1st subplot in 2nd figure
plot(x1, y1*y1, 'r', lw=2)                                                          24
xlabel('x');          ylabel( 'f(x)' );     title( 'sin^2(x)*cos^2(x^2)' )

     # form grid
subplot(2,1,2)                               # 2nd subplot in 2nd figure          26
plot(x2, y2*y2, '-', lw=2)
xlabel('x');          ylabel( 'f(x)' );     title( 'exp(-x/2)*sin^2(x)' )         28
grid(True)
show()                                              # Show graphs                 30
```

Listing 2.7 PondMatPlot.py Produces the scatter plot and the curve shown in Figure 5.4 in Chapter 5.

```
#  PondMatPlot.py: Monte–Carlo integration via vonNeumann rejection
                                                                                    2
import numpy as np, matplotlib.pyplot as plt
                                                                                    4
N = 100;   Npts = 3000;  analyt = np.pi**2
x1 = np.arange(0, 2*np.pi+2*np.pi/N,2*np.pi/N)                                       6
xi = [];  yi = [];  xo = [];  yo = []
fig,ax = plt.subplots()                                                             8
y1 = x1 * np.sin(x1)**2                         # Integrand
ax.plot(x1, y1, 'c', linewidth=4)                                                  10
ax.set_xlim ((0, 2*np.pi))
ax.set_ylim((0, 5))                                                                12
ax.set_xticks([0, np.pi, 2*np.pi])
ax.set_xticklabels(['0', '$\pi$','2$\pi$'])                                        14
ax.set_ylabel('$f(x) = x\,\sin^2 x$', fontsize=20)
ax.set_xlabel('x',fontsize=20)                                                     16
fig.patch.set_visible(False)
                                                                                   18
def fx(x):    return x*np.sin(x)**2                # Integrand
j = 0                                          # Inside curve counter              20
xx = 2.* np.pi * np.random.rand(Npts)             # 0 =< x <= 2pi
yy = 5*np.random.rand(Npts)                        # 0 =< y <= 5                    22
for i in range(1,Npts):
    if (yy[i] <= fx(xx[i])):                       # Below curve                   24
        if (i <=100): xi.append(xx[i])
        if (i <=100): yi.append(yy[i])                                             26
        j +=1
    else:                                                                          28
        if (i <=100): yo.append(yy[i])
        if (i <=100): xo.append(xx[i])                                             30
    boxarea = 2. * np.pi *5.                        # Box area
    area = boxarea*j/(Npts-1)                  # Area under curve                   32
    ax.plot(xo,yo,'bo',markersize=3)
    ax.plot(xi,yi,'ro',markersize=3)                                               34
    ax.set_title('Answers:  Analytic = %5.3f, MC = %5.3f'%(analyt,area))
plt.show()                                                                         36
```

Listing 2.8 Simple3Dplot.py Produces the Matplotlib 3D surface plots in Figure 2.8.

```
# Simple3Dplot.py: matplotlib 3D plot you can rotate and scale via mouse
                                                                                    2
import matplotlib.pylab as p
from mpl_toolkits.mplot3d import Axes3D                                             4

print ("Please be patient, I have packages to import & points to plot")            6
delta = 0.1
x = p.arange( -3., 3., delta )                                                     8
y = p.arange( -3., 3., delta )
X, Y = p.meshgrid(x, y)                                                            10
```

```
Z = p.sin(X) * p.cos(Y)                          # Surface height
fig = p.figure()                                 # Create figure            12
ax = Axes3D(fig)                                 # Plots axes
ax.plot_surface(X, Y, Z)                         # Surface                  14
ax.plot_wireframe(X, Y, Z, color = 'r')          # Add wireframe
ax.set_xlabel('X')                                                         16
ax.set_ylabel('Y')
ax.set_zlabel('Z')                                                         18
p.show()                                         # Output figure
```

Listing 2.9   Scatter3dPlot.py Produces a 3D scatter plot using Matplotlib 3D tools.

```
" Scatter3dPlot.py    from matplotlib examples"
                                                                            2
import numpy as np
from mpl_toolkits.mplot3d import Axes3D                                     4
import matplotlib.pyplot as plt
                                                                            6
def randrange(n, vmin, vmax):
    return (vmax-vmin)*np.random.rand(n) + vmin                            8
fig = plt.figure()
ax = fig.add_subplot(111, projection='3d')                                10
n = 100
for c, m, zl, zh in [('r', 'o', -50, -25), ('b', '^', -30, -5)]:          12
    xs = randrange(n, 23, 32)
    ys = randrange(n, 0, 100)                                             14
    zs = randrange(n, zl, zh)
    ax.scatter(xs, ys, zs, c=c, marker=m)                                 16
ax.set_xlabel('X Label')
ax.set_ylabel('Y Label')                                                  18
ax.set_zlabel('Z Label')
plt.show()                                                                20
```

Listing 2.10   EqHeatAnimateMat.py Produces an animation of a cooling bar using Matplotlib.

```
# EqHeat.py Animated heat equation soltn via fine differences
                                                                            2
from numpy import *                                                         4
import numpy as np
import matplotlib.pyplot as plt                                             6
import matplotlib.animation as animation
                                                                            8
Nx = 101
Dx = 0.01414                                                               10
Dt = 0.6
KAPPA = 210.                          # Thermal conductivity               12
SPH = 900.                            # Specific heat
RHO = 2700.                           # Density                            14
cons = KAPPA/(SPH*RHO)*Dt/(Dx*Dx);
T = np.zeros( (Nx, 2), float)         # Temp @ first 2 times               16

def init():                                                                18
    for ix in range (1, Nx - 1):      # Initial temperature
        T[ix, 0] = 100.0;                                                  20
    T[0, 0] = 0.0                     # Bar ends T = 0
    T[0, 1] = 0.                                                           22
    T[Nx - 1, 0] = 0.
    T[Nx - 1, 1] = 0.0                                                     24
init()
k = range(0,Nx)                                                            26
fig = plt.figure()                    # Figure to plot
# select axis; 111: only one plot, x,y, scales given                      28
```

```
ax = fig.add_subplot(111, autoscale_on=False, xlim=(-5, 105), ylim=(-5, 110.0))
ax.grid()                                          # Plot grid              30
plt.ylabel("Temperature")
plt.title("Cooling of a bar")                                              32
line, = ax.plot(k, T[k,0],"r", lw=2)
plt.plot([1,99],[0,0],"r",lw=10)                                           34
plt.text(45,5,'bar',fontsize=20)

                                                                           36
def animate(dum):
    for ix in range (1, Nx - 1):                                          38
        T[ix, 1] = T[ix, 0] + cons*(T[ix + 1, 0] + T[ix - 1, 0] - 2.0*T[ix, 0])
    line.set_data(k,T[k,1] )                                              40
    for ix in range (1, Nx - 1):
        T[ix, 0] = T[ix, 1]                        # 100 position row @ t = m   42
    return line,
ani = animation.FuncAnimation(fig, animate,1)  # Animation               44
plt.show()
```

Listing 2.11   AreaFormatted.py Does I/O to and from keyboard, as well as from a file. It works with either Python 2 or 3 by switching between *raw_input* and *input*. Note to read from a file using Canopy, you must right click in the Python run window and choose *Change to Editor Directory*.

```
1  # AreaFormatted: Python 2 or 3 formated output, keyboard input, file input

   from numpy import *
   from sys import version
5  if int(version[0])>2:                     # Python 3 uses input, not raw_input
       raw_input=input
   name = raw_input( 'Key in your name: ')          # raw_input strings
   print("Hi ",name)
9  radius = eval(raw_input('Enter a radius: '))      # For numerical values
   print('you entered radius= %8.5f'%radius)              # formatted output
   print('Enter new name and r in file Name.dat')          # raw_input strings
   inpfile = open('Name.dat','r')                   # Read from file Name.dat
13 for line in inpfile:
       line = line.split()                   # Splits components of line
       name = line[0]                        # First entry in the list
       print(" Hi  %10s" %(name))            # print Hi + first entry
17     r = float(line[1])                     # convert string to float
       print(" r = %13.5f" %(r))             # convert to float & print
   inpfile.close()
   A = math.pi*r**2
21 print("Done, look in A.dat\n")
   outfile = open('A.dat','w')
   outfile.write('r=  %13.5f\n'%(r))
   outfile.write('A =  %13.5f\n'%(A))
25 outfile.close()
   print('r = %13.5f'%(r), ', A = %13.5f'%(A))              # Screen output
   print('\n Now example of integer input ')
   age=int(eval(raw_input ('Now key in your age as an integer: ')))
29 print("age: %4d years old, you don't look it!\n"%(age))
   print("Enter and return a character to finish")
   s = raw_input()
```

Listing 2.12   Directives.py Illustrates formatting via directives and escape characters.

```
   # Directives.py illustrates escape and formatting  characters
   import sys
3  print("hello \n")
   print("\t it's me")                               # tabulator
   b = 73
   print("decimal 73 as integer b = %d "%(b))  # for integer
7  print("as octal b = %o"%(b))                      # octal
```

```
   print("as hexadecimal b = %x "%(b))        # works hexadecimal
   print("learn \"Python\" ")                 # use of double quote symbol
   print("shows a backslash \\")              # use of \\
11 print('use of single \' quotes \' ')       # print single quotes
```

Listing 2.13   Limits.py Determines machine precision within a factor of 2. Note how we skip a line at the beginning of each class or method and how we align the closing brace vertically with its appropriate keyword (in italics)

```
   # Limits.py: determines approximate machine precision
2
   N = 10
   eps = 1.0
   for i in range(N):
6      eps = eps/2
       one_Plus_eps = 1.0 + eps
       print('eps = ', eps, ', one + eps = ', one_Plus_eps)
```

3

Errors and Uncertainties

> *To err is human, to forgive divine.*
>
> *— Alexander Pope*

Whether you are careful or not, errors and uncertainties are integral parts of a computation. In this chapter we examine some of the errors and uncertainties that may occur in computations. Although we do not keep repeating a mantra about watching for error, the lessons of this chapter apply to all other chapters as well.

3.1 Types of Errors

Some errors are the ones that humans inevitably make, but some are introduced by the computer. Computer errors arise because of the limited precision with which computers store numbers, or because algorithms or models are not perfect. Although it stifles creativity to keep thinking "error" when approaching a computation, it certainly is a waste of time and bad science to work with results that are meaningless ("garbage") because of errors.

Let's say that you have a program of high complexity. To gauge why errors should be of concern, imagine that your program has the logical flow

$$\text{start} \to U_1 \to U_2 \to \cdots \to U_n \to \text{end}, \tag{3.1}$$

where each unit U_i might be a statement or a step. If each unit has probability p of being correct, then the joint probability P of the whole program being correct is $P = p^n$. Let's also say we have a medium-sized program with $n = 1000$ steps and that the probability of each step being correct is almost one, $p \simeq 0.9993$. This means that you end up with $P \simeq \frac{1}{2}$, that is, a final answer that is as likely wrong as right (not a good way to build a bridge). The problem is that, as a scientist, you want a result that is correct—or at least in which the uncertainty is small and of known size, even if the code executes millions of steps.

Four general types of errors exist to plague your computations:

1. **Blunders or bad theory:** typographical errors entered with your program or data, running the wrong program, or having a fault in your reasoning (theory), using the wrong data file, and so on. (If your blunder count starts increasing, it may be time to go home or take a break.)

Computational Physics: Problem Solving with Python, Fourth Edition.
Rubin H. Landau, Manuel J. Páez, and Cristian C. Bordeianu.
© 2024 WILEY-VCH GmbH. Published 2024 by WILEY-VCH GmbH.

2. **Random errors:** imprecision caused by events such as fluctuations in electronics, cosmic rays, or someone pulling a plug. These may be rare, but you have no control over them and their likelihood increases with running time; while you may have confidence in a 20-second calculation, a week-long calculation may have to be run multiple times to check reproducibility.

3. **Approximation errors:** imprecision arising from simplifying the mathematics so that a problem can be solved on the computer. They include the replacement of infinite series by finite sums, infinitesimal intervals by finite ones, and variable functions by constants. For example,

$$\sin(x) = \sum_{n=1}^{\infty} \frac{(-1)^{n-1}x^{2n-1}}{(2n-1)!} \quad \text{(mathematically exact)},$$

$$\simeq \sum_{n=1}^{N} \frac{(-1)^{n-1}x^{2n-1}}{(2n-1)!} + \mathcal{E}(x, N) \quad \text{(algorithm)}. \tag{3.2}$$

Here $\mathcal{E}(x, N)$ is the approximation error, and it is the ignored series from $N+1$ to ∞. Since approximation error arises from the algorithm we use to approximate the mathematics, it is also called *algorithmic error*. For a good algorithm, the approximation error should decrease as N increases, and should vanish in the $N \to \infty$ limit. Specifically for (3.2), because the scale for N is set by the value of x, a small approximation error requires $N \gg x$. So if x and N are close in value, the approximation error will be large.

4. **Round-off errors:** imprecision arising from the finite number of digits used to store floating-point numbers. These "errors" are analogous to the uncertainty in the laboratory measurement of a physical quantity. The overall round-off error accumulates as the computer handles more numbers, that is, as the number of steps in a computation increases. This may cause some algorithms to become *unstable* with a concordant rapid increase in error. In some cases, round-off error may become the major component in your answer, leading to what computer experts call *garbage*.

   For example, if your computer kept four decimal places, then it will store $\frac{1}{3}$ as 0.3333 and $\frac{2}{3}$ as 0.6667, where the computer has "rounded off" the last digit in $\frac{2}{3}$. Accordingly, if we ask the computer to do as simple a calculation as $2\left(\frac{1}{3}\right) - \frac{2}{3}$, it would yield

$$2\left(\frac{1}{3}\right) - \frac{2}{3} = 0.6666 - 0.6667 = -0.0001 \neq 0. \tag{3.3}$$

So although the result may be small, it is not 0, and if we repeat this type of calculation millions of times, the final answer might not be small (small garbage begets large garbage).

When considering the precision of calculations it is good to recall our discussion in Chapter 2 of significant figures. For computational purposes, let us consider how the computer may store the floating-point number

$$a = 11223344556677889900 = 1.12233445566778899 \times 10^{19}. \tag{3.4}$$

Because the exponent is stored separately and is a small number, we may assume that it will be stored in full precision. In contrast, some of the digits of the mantissa may be truncated. In double precision, the mantissa of a will be stored in two words, the *most significant part* representing the decimal 1.12233, and the *least significant part* 44556677. The digits beyond 7 are lost. As we shall see soon, when we perform calculations with words of fixed length, it is inevitable that errors will be introduced (at least) into the least significant parts of the words.

3.1.1 Courting Disaster: Subtractive Cancelation

Calculations employing numbers that are stored approximately can only be expected to yield approximate answers. To demonstrate the effect of this type of uncertainty, we model the computer representation x_c of the exact number x as

$$x_c \simeq x(1 + \epsilon_x). \tag{3.5}$$

Here ϵ_x is the relative error in x_c, which we expect to be of a similar magnitude to the machine precision ϵ_m. If we apply this notation to the simple subtraction $a = b - c$, we obtain

$$a = b - c \Rightarrow a_c \simeq b_c - c_c \simeq b(1 + \epsilon_b) - c(1 + \epsilon_c)$$

$$\Rightarrow \frac{a_c}{a} \simeq 1 + \epsilon_b \frac{b}{a} - \frac{c}{a} \epsilon_c. \tag{3.6}$$

We see from (3.6) that the resulting error in a is essentially a weighted average of the errors in b and c, with no assurance that the last two terms will cancel. Of special importance here is the observation that the error in the answer a_c increases when we subtract two nearly equal numbers ($b \simeq c$) because a is then small, and we are subtracting off the most significant parts of both numbers and leaving the error-prone least-significant parts:

$$\frac{a_c}{a} \overset{\text{def}}{=} 1 + \epsilon_a \simeq 1 + \frac{b}{a}(\epsilon_b - \epsilon_c) \simeq 1 + \frac{b}{a} \max(|\epsilon_b|, |\epsilon_c|). \tag{3.7}$$

This shows that even if the relative errors in b and c cancel somewhat, they are multiplied by the large number b/a, which can significantly magnify the error. Because we cannot assume any sign for the errors, we must assume the worst.

Theorem If you subtract two large numbers and end up with a small one, the small one is less significant than the large numbers.

We have already seen an example of subtractive cancelation in the power series summation for $\sin x \simeq x - x^3/3! + \cdots$ for large x. A similar effect occurs for $e^{-x} \simeq 1 - x + x^2/2! - x^3/3! + \cdots$ for large x. Here the first few terms are large but of alternating sign, leading to an almost total cancelation in order to yield the final small result. In this case, subtractive cancelation can be eliminated by using the identity $e^{-x} = 1/e^x$ and then evaluating e^x, although round-off error will still remain.

3.1.2 Subtractive Cancelation Exercises

1) Remember back in high school when you learned that the quadratic equation

$$ax^2 + bx + c = 0 \tag{3.8}$$

has an analytic solution that can be written as either

$$x_{1,2} = \frac{-b \pm \sqrt{b^2 - 4ac}}{2a} \quad \text{or} \quad x'_{1,2} = \frac{-2c}{b \pm \sqrt{b^2 - 4ac}}. \tag{3.9}$$

Inspection of (3.9) indicates that subtractive cancelation (and consequently an increase in error) arises when $b^2 \gg 4ac$, as then the square root and its preceding term nearly cancel for one of the roots.

a) Write a program that calculates all solutions for arbitrary values of a, b, and c.

b) Investigate how errors in your computed answers become large as the subtractive cancelation increases, and relate this to the known machine precision. *Hint*: A good test case utilizes $a = 1$, $b = 1$, $c = 10^{-n}$, $n = 1, 2, 3, \ldots$.

2) As we have seen, subtractive cancelation occurs when summing a series with alternating signs. As another example, consider the finite sum

$$S_N^{(1)} = \sum_{n=1}^{2N} (-1)^n \frac{n}{n+1}. \tag{3.10}$$

If you sum the even and odd values of n separately, you get two sums:

$$S_N^{(2)} = -\sum_{n=1}^{N} \frac{2n-1}{2n} + \sum_{n=1}^{N} \frac{2n}{2n+1}. \tag{3.11}$$

All terms are positive in this form with just a single subtraction at the end of the calculation. Yet even this one subtraction and its resulting cancelation can be avoided by combining the series analytically to obtain

$$S_N^{(3)} = \sum_{n=1}^{N} \frac{1}{2n(2n+1)}. \tag{3.12}$$

Although all three summations $S^{(1)}$, $S^{(2)}$, and $S^{(3)}$ are mathematically equal, they may give different numerical results.

1) Write a double-precision program that calculates $S^{(1)}$, $S^{(2)}$, and $S^{(3)}$.

2) Assume $S^{(3)}$ to be the exact answer. Make a log–log plot of the relative error *versus* the number of terms, that is, of $\log_{10} |(S_N^{(1)} - S_N^{(3)})/S_N^{(3)}|$ *versus* $\log_{10}(N)$. Start with $N = 1$ and work up to $N = 1{,}000{,}000$. (Recall that $\log_{10} x = \ln x / \ln 10$.) The negative of the ordinate in this plot gives an approximate value for the number of significant figures.

3) See whether straight-line behavior for the error occurs in some regions of your plot. This indicates that the error is proportional to a power of N.

3) In spite of the power of your trusty computer, calculating the sum of even a simple series may require some thought and care. Consider the two series

$$S^{(up)} = \sum_{n=1}^{N} \frac{1}{n}, \qquad S^{(down)} = \sum_{n=N}^{1} \frac{1}{n}. \tag{3.13}$$

Both series are finite as long as N is finite, and when summed analytically both give the same answer. Nonetheless, because of round-off error, the numerical value of $S^{(up)}$ will not be precisely that of $S^{(down)}$.

a) Write a program to calculate $S^{(up)}$ and $S^{(down)}$ as functions of N.

b) Make a log–log plot of $(S^{(up)} - S^{(down)})/(|S^{(up)}| + |S^{(down)}|)$ *versus* N.

c) Observe the linear regime on your graph and explain why the downward sum is generally more precise.

3.1.3 Round-Off Errors

Let's start by seeing how error arises from a single division of the computer representations of two numbers:

$$a = \frac{b}{c} \Rightarrow a_c = \frac{b_c}{c_c} = \frac{b(1 + \epsilon_b)}{c(1 + \epsilon_c)},$$

$$\Rightarrow \frac{a_c}{a} = \frac{1 + \epsilon_b}{1 + \epsilon_c} \simeq (1 + \epsilon_b)(1 - \epsilon_c) \simeq 1 + \epsilon_b - \epsilon_c,$$

$$\Rightarrow \frac{a_c}{a} \simeq 1 + |\epsilon_b| + |\epsilon_c|. \tag{3.14}$$

Here we have ignored the very small ϵ^2 terms, and have added the absolute value of the errors since we can't assume that good fortune will lead to errors canceling each other. Because we add the errors in absolute value, this same rule holds for multiplication. Equation (3.14) is just the basic rule of error propagation from elementary laboratory work: You add the uncertainties in each quantity involved in an analysis to arrive at the overall uncertainty.

We can even generalize this model to estimate the error in the evaluation of a general function $f(x)$, that is, the difference in the value of the function evaluated at x and at x_c:

$$\mathcal{E} = \frac{f(x) - f(x_c)}{f(x)} \simeq \frac{df(x)/dx}{f(x)} (x - x_c). \tag{3.15}$$

So, for example,

$$f(x) = \sqrt{1 + x}, \qquad \frac{df}{dx} = \frac{1}{2} \frac{1}{\sqrt{1 + x}} = \frac{1}{4} f(x)(x - x_c) \tag{3.16}$$

$$\Rightarrow \mathcal{E} \simeq \frac{1}{2}\sqrt{1 + x}(x - x_c) = \frac{x - x_c}{2(1 + x)}. \tag{3.17}$$

If we evaluate this expression for $x = \pi/4$ and assume an error in the fourth place of x, we obtain a similar relative error of 1.5×10^{-4} in $\sqrt{1 + x}$.

3.1.4 Round-Off Error Accumulation

There is a useful model for approximating how round-off error accumulates in a calculation involving a large number of steps. As illustrated in Figure 3.1, we view the error in each step of a calculation as a literal "step" in a *random walk*, that is, a walk for which each step is in a

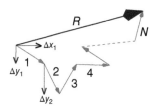

Figure 3.1 A schematic of the N steps in a random walk simulation that ends up a distance $R = \sqrt{N}$ from the origin. Notice how the Δx's for each step add vectorially.

random direction. As we will derive and simulate in Chapter 4, the total distance R covered in N steps of length r, is, on the average,

$$R \simeq \sqrt{N} \, r. \tag{3.18}$$

By analogy, the total relative error ϵ_{ro} arising after N calculational steps each with machine precision error ϵ_m is, on the average,

$$\epsilon_{ro} \simeq \sqrt{N} \, \epsilon_m. \tag{3.19}$$

If the round-off errors in a particular algorithm do not accumulate in a random manner, then a detailed analysis is needed to predict the dependence of the error on the number of steps N. In some cases, there may be no cancellation, and the error may increase as $N\epsilon_m$. Even worse, in some recursive algorithms, where the error generation is coherent, such as the upward recursion for spherical Bessel functions, there may be an $N!$ increase in error.

3.2 Experimental Error Investigation

Algorithms play a vital role in computational physics. Your **problem** is to take an algorithm and decide

1) Does it converge, and if so, how fast?
2) Even if it converges, are the answers precise?
3) How expensive (time-consuming) is the algorithm?

Your first thought might be "What a dumb problem! All algorithms converge if you let them run long enough. If you want more precision, then just let them run longer." Well, some algorithms may be asymptotic expansions that just approximate a function in certain regions of parameter space, and converge only up to a point. Yet even if a uniformly convergent power series is used as the algorithm, including more terms may decrease the algorithmic error but increase the round-off error. And because round-off errors eventually diverge to infinity, the best we can hope for is a "best" approximation. *Good algorithms are good not only because fewer steps take less time, but also because fewer steps produces less round-off error.*

Let's assume that an algorithm takes N steps to find a good answer. As a rule of thumb, the approximation (algorithmic) error decreases rapidly, often as some inverse power of the number of terms used:

$$\epsilon_{app} \simeq \frac{\alpha}{N^{\beta}}. \tag{3.20}$$

Here α and β are empirical constants that change for different algorithms and may be only approximately constant, and even then only as $N \to \infty$. The fact that the error *must* fall off for large N is just a statement that the algorithm works.

In contrast to algorithmic error, round-off error grows slowly and somewhat randomly with N. If the round-off errors in each step of the algorithm are not correlated, then we know from previous discussion that we can model the accumulation of error as a random walk with step size equal to the machine precision ϵ_m:

$$\epsilon_{ro} \simeq \sqrt{N} \epsilon_m. \tag{3.21}$$

This is the slow growth with N that we expect from round-off error. The total error in a computation is the sum of the two types of errors:

$$\epsilon_{tot} = \epsilon_{app} + \epsilon_{ro} \tag{3.22}$$

$$\epsilon_{tot} \simeq \frac{\alpha}{N^\beta} + \sqrt{N}\epsilon_m. \tag{3.23}$$

For small N we expect the first term to be the larger of the two, but as N grows it will be overcome by the ever-increasing round-off error.

As an example, in Figure 3.2, we present a log–log plot of the relative error in numerical integration using the Simpson integration rule (Chapter 5). We use the $\log_{10}$ of the relative error because its negative tells us the number of decimal places of precision obtained.[1] Let us assume $\mathcal{A}$ is the exact answer and $A(N)$ the computed answer. If

$$\frac{\mathcal{A} - A(N)}{\mathcal{A}} \simeq 10^{-9}, \quad \text{then} \quad \log_{10}\left|\frac{\mathcal{A} - A(N)}{\mathcal{A}}\right| \simeq -9. \tag{3.24}$$

We see in Figure 3.2 that the error does show a rapid decrease for small N, consistent with an inverse power law (3.20). In this region, the algorithm is converging. As N keeps increasing, the error starts to look somewhat erratic, with a slow increase on the average. In accordance with (3.22), in this region, round-off error has grown larger than the approximation error and will continue to grow for increasing N. Clearly then, the smallest total error will be obtained if we can stop the calculation at the minimum near 10^{-14}, that is, when $\epsilon_{approx} \simeq \epsilon_{ro}$.

In realistic calculations you would not know the exact answer; after all, if you did, then why would you bother with the computation? However, you may know the exact answer for a similar calculation, and you can use that similar calculation to perfect your numerical technique. Alternatively, now that you understand how the total error in a computation behaves, you should be able to look at a table or, better yet, a graph like

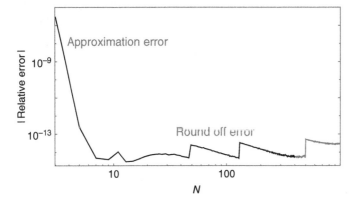

Figure 3.2 A log–log plot of relative error *versus* the number of points used for a numerical integration. The ordinate value of $\sim 10^{-14}$ at the minimum indicates that $\sim$14 decimal places of precision are obtained before round-off error begins to build up. Notice that while the round-off error does fluctuate indicating a statistical aspect of error accumulation, on the average it is increasing but more slowly than did the algorithm's error decrease.

---

1 Most computer languages use $\ln x = \log_e x$. Yet because $x = a^{\log_a x}$, we have $\log_{10} x = \ln x / \ln 10$.

Figure 3.2, of your answer and deduce the manner in which your algorithm is converging. Specifically, at some point, you should see that the mantissa of the answer changes only in the less significant digits, with that place moving further to the right of the decimal point as the calculation executes more steps. Eventually, however, as the number of steps becomes even larger, round-off error leads to a fluctuation in the less significant digits, with a gradual increase on the average. It is best to quit the calculation before this occurs.

Based upon this understanding, an approach to obtaining the best approximation is to deduce when your answer behaves like (3.22). To do that, we call $\mathcal{A}$ the exact answer and $A(N)$ the computed answer after N steps. We assume that for large enough values of N, the approximation converges as

$$A(N) \simeq \mathcal{A} + \frac{\alpha}{N^\beta}, \tag{3.25}$$

that is, the round-off error term in (3.22) is still small. We then run our computer program with $2N$ steps, which should give a better answer, and use that answer to eliminate the unknown $\mathcal{A}$:

$$A(N) - A(2N) \simeq \frac{\alpha}{N^\beta}. \tag{3.26}$$

To see if these assumptions are correct and determine what level of precision is possible for the best choice of N, plot $\log_{10}|[A(N) - A(2N)]/A(2N)|$ *versus* $\log_{10} N$, similar to what we have performed in Figure 3.2. If you obtain a rapid straight-line drop-off, then you know you are in the region of convergence and can deduce a value for β from the slope. As N gets larger, you should see the graph change from a straight-line decrease to a slow increase as round-off error begins to dominate. A good place to quit is before this. In any case, now you understand the error in your computation and therefore have a chance to control it.

As an example of how different kinds of errors enter into a computation, we assume we know the analytic form for the approximation and round-off errors:

$$\epsilon_{app} \simeq \frac{1}{N^2}, \qquad \epsilon_{ro} \simeq \sqrt{N}\epsilon_m, \tag{3.27}$$

$$\Rightarrow \quad \epsilon_{tot} = \epsilon_{approx} + \epsilon_{ro} \simeq \frac{1}{N^2} + \sqrt{N}\epsilon_m. \tag{3.28}$$

The total error is then a minimum when

$$\frac{d\epsilon_{tot}}{dN} = \frac{-2}{N^3} + \frac{1}{2}\frac{\epsilon_m}{\sqrt{N}} = 0, \tag{3.29}$$

$$\Rightarrow \quad N^{5/2} = \frac{4}{\epsilon_m}. \tag{3.30}$$

For a double-precision calculation ($\epsilon_m \simeq 10^{-15}$), the minimum total error occurs when

$$N^{5/2} \simeq \frac{4}{10^{-15}} \quad \Rightarrow \quad N \simeq 1099, \quad \Rightarrow \quad \epsilon_{tot} \simeq 4 \times 10^{-6}. \tag{3.31}$$

In this case most of the error is as a result of round-off and is not approximation error.

Seeing that the total error is mainly round-off error $\propto \sqrt{N}$, an obvious way to decrease the error is to use a smaller number of steps N. Let us assume we do this by finding another algorithm that converges more rapidly with N, for example, one with approximation error behaving like

$$\epsilon_{app} \simeq \frac{2}{N^4}. \tag{3.32}$$

The total error is now

$$\epsilon_{\text{tot}} = \epsilon_{\text{ro}} + \epsilon_{\text{app}} \simeq \frac{2}{N^4} + \sqrt{N}\epsilon_m. \tag{3.33}$$

The number of points for minimum error is found as before:

$$\frac{d\epsilon_{\text{tot}}}{dN} = 0 \Rightarrow N^{9/2} \Rightarrow N \simeq 67 \Rightarrow \epsilon_{\text{tot}} \simeq 9 \times 10^{-7}. \tag{3.34}$$

The error is now smaller by a factor of 4, with only 1/16 as many steps needed. Subtle are the ways of the computer. In this case, the better algorithm is quicker and, by using fewer steps, produces less round-off error.

Exercise   Estimate the error for a double-precision calculation.

3.3   Errors with Power Series

A classic numerical problem is the summation of a series to evaluate a function. As an example, consider the infinite series for $\sin x$:

$$\sin x = x - \frac{x^3}{3!} + \frac{x^5}{5!} - \frac{x^7}{7!} + \cdots \qquad \text{(exact)}. \tag{3.35}$$

Your **problem** is to use just this series to calculate $\sin x$ for $x < 2\pi$ and $x > 2\pi$, with an absolute error in each case of less than 1 part in 10^8. While in a mathematical sense an infinite series is exact and always converges, it is not an algorithm because computers can't sum an infinite number of terms. An algorithm would be the finite sum

$$\sin x \simeq \sum_{n=1}^{N} \frac{(-1)^{n-1}x^{2n-1}}{(2n-1)!} \qquad \text{(algorithm)}. \tag{3.36}$$

But how do we decide when to stop summing? (Do not even think of saying, "When the answer agrees with a table or with the built-in library function.") One approach would be to stop summing when the next term is smaller than the precision desired. Clearly then, if x is large this would require very large N. In fact, for very large x, one would have to go far out in the series before the terms even start to decrease, let alone the series as a whole converges.

We should also be wary of the algorithm (3.36) because it would have us calculate x^{2n-1} and then divide that by $(2n-1)!$. This is not good computation. On the one hand, both $(2n-1)!$ and x^{2n-1} can individually get very large and thereby cause overflows, despite the fact that their quotient may be small. On the other hand, powers and factorials are very expensive (time-consuming) to evaluate on the computer. Consequently, a better approach is to use a single multiplication to relate the next term in the series to the previous one:

$$\frac{(-1)^{n-1}x^{2n-1}}{(2n-1)!} = \frac{-x^2}{(2n-1)(2n-2)} \frac{(-1)^{n-2}x^{2n-3}}{(2n-3)!}$$

$$\Rightarrow \text{nth term} = \frac{-x^2}{(2n-1)(2n-2)} \times (n-1)\text{th term}. \tag{3.37}$$

While we might want to insure absolute accuracy for $\sin x$, that is not easy to do. What is easy to do is to assume that the error in the summation is approximately the last term summed (this assumes no round-off error). To obtain a relative error of 1 part in 10^8, we then would stop the calculation when

$$\left| \frac{n\text{th term}}{\text{sum}} \right| < 10^{-8}, \tag{3.38}$$

where "term" is the last term kept in the series (3.36) and "sum" is the accumulated sum of all the terms. In general, you are free to pick any tolerance level you desire, although if it is too close to, or smaller than, machine precision, your calculation may not be able to attain it. A pseudocode for performing the summation is

```
  term = x, sum = x, eps = 10^(-8)          # Initialize do
2   do term = -term*x*x/(2n-1)/(2*n-2);        # New wrt old
      sum = sum + term                          # Add term
      while abs(term/sum) > eps              # Break iteration
    end do
```

3.3.1 Implementation and Assessment

1) Write a program that implements this pseudocode for the indicated x values. Start with a tolerance of 10^{-8} as in (3.38). Present the results as a table with headings x N sum |sum−sin(x)|/sin(x), where sin(x) is the value obtained from the built-in function Math.sin(x) (you may assume that the built-in function is exact). The last column here is the relative error in your computation.

2) Show that for sufficiently small values of x, your algorithm converges (the changes in sum are smaller than your tolerance level) and that the series converges to the correct answer.

3) Compare the number of decimal places of precision obtained with that expected from (3.38).

4) Without using the identity $\sin(x + 2n\pi) = \sin(x)$, show that there is a range of somewhat large values of x for which the algorithm converges, but that it converges to the wrong answer.

5) Observe how significant subtractive cancelations occur when large terms are added together to give small answers. In particular, print out the near-perfect cancellation around $n \simeq x/2$.

6) Show that as you keep increasing x, you will reach a regime where the algorithm stops converging.

7) Now make use of the identity $\sin(x + 2n\pi) = \sin(x)$ to compute $\sin x$ for large x values where the series otherwise would diverge.

8) By progressively increasing x from 1 to 10, and then from 10 to 100, use your program to determine experimentally when the series starts to lose accuracy and when it no longer converges.

9) Make a series of graphs of the error *versus* N for different values of x. You should get curves similar to those in Figure 3.3.

10) Repeat the calculation using a "bad" version of the algorithm (one that calculates factorials) and compare the answers.

11) Set your tolerance level to a number smaller than machine precision and see how this affects your conclusions.

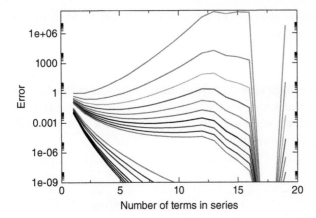

Figure 3.3 The error in the summation of the series for e^{-x} *versus N* for various *x* values. The values of *x* increase vertically for each curve. Note that a negative initial slope corresponds to decreasing algorithmic error with *N*, and that the dip indicates a rapid convergence followed by a rapid increase in error. (courtesy of J. Wiren.)

Note that because this series summation is such a simple, correlated process, the round-off error does not accumulate randomly as it might for a more complicated computation, and we do not obtain the error behavior (3.25). We will see the predicted error behavior when we examine integration in Chapter 5.

3.3.2 Error in Specular Reflection

For a perfectly reflecting surface, the basic law of optics tells us that the angle of incidence equals the angle of reflection (Figure 3.4 left. If no light is absorbed during a reflection, a light ray would continue to reflect endlessly (Figure 3.4 right). With an origin placed at the center of the circular mirror, we locate the ray by the angle θ. For an initial angle $\phi < \pi$, the angle increases by 2ϕ after each reflection:

$$\theta_{\text{new}} = \theta_{\text{old}} + 2\phi. \tag{3.39}$$

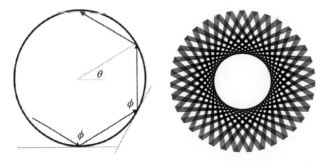

Figure 3.4 *Left*: Specular reflection within a circular mirror in which the incident angle equals the angle of reflection. *Right*: Infinite internal reflections between two circular mirrors.

Although this appears to indicate that θ increases endlessly, the addition or subtraction of 2π to θ does not change the location on the circle, and so if ϕ/π is a rational number,

$$\frac{\phi}{\pi} = \frac{n}{m}, \tag{3.40}$$

the ray will fall upon itself and form a geometric figure (Figure 3.4 right).

1) Determine the path followed by a light ray for a perfectly reflecting mirror.
2) Plot the light trajectories for a range of values for the initial angle ϕ.
3) Repeat the previous calculation using just four places of precision. You can do this by using the Python command `round`, for instance, `round(1.234567,4) = 1.234`. You should find that a significant relative error accumulates. As in large and complicated calculations with many steps and finite precision, this type of error increases as the number of calculational steps increases.

3.4 Errors in Bessel Functions

Accumulating round-off errors often limits the ability of a program to calculate accurately. Your **problem** is to compute the spherical Bessel and Neumann functions $j_l(x)$ and $n_l(x)$. These functions are, respectively, the regular/irregular (nonsingular/singular at the origin) solutions of the differential equation

$$x^2 f''(x) + 2x f'(x) + \left[x^2 - l(l+1)\right] f(x) = 0. \tag{3.41}$$

The spherical Bessel functions are related to the Bessel function of the first kind by $j_l(x) = \sqrt{\pi/2x}\, J_{n+1/2}(x)$. They occur in many physical problems, such as the expansion of a plane wave into spherical partial waves,

$$e^{i\mathbf{k}\cdot\mathbf{r}} = \sum_{l=0}^{\infty} i^l\, (2l+1) j_l(kr) P_l(\cos\theta). \tag{3.42}$$

Figure 3.5 shows what the first few j_l looks like, and Table 3.1 gives some explicit values. For the first two l values, the explicit forms are

$$j_0(x) = +\frac{\sin x}{x}, \qquad j_1(x) = +\frac{\sin x}{x^2} - \frac{\cos x}{x} \tag{3.43}$$

$$n_0(x) = -\frac{\cos x}{x}. \qquad n_1(x) = -\frac{\cos x}{x^2} - \frac{\sin x}{x}. \tag{3.44}$$

3.4.1 Numerical Recursion (Method)

The classic way to calculate $j_l(x)$ would be by summing its power series for small values of x/l and summing its asymptotic expansion for large x/l values. The approach we adopt here is based on the *recursion relations*

$$j_{l+1}(x) = \frac{2l+1}{x} j_l(x) - j_{l-1}(x), \qquad \text{(up)}, \tag{3.45}$$

$$j_{l-1}(x) = \frac{2l+1}{x} j_l(x) - j_{l+1}(x), \qquad \text{(down)}. \tag{3.46}$$

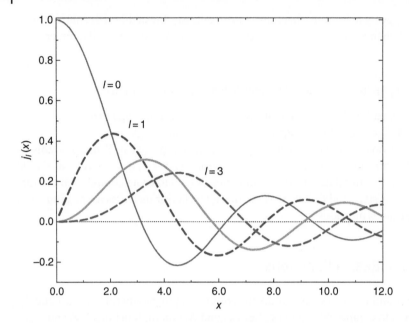

Figure 3.5 The first four spherical Bessel functions $j_l(x)$ as functions of x. Notice that for small x, the values for increasing l become progressively smaller.

Table 3.1 Approximate values for spherical Bessel functions (from Maple).

x	$j_3(x)$	$j_5(x)$	$j_8(x)$
0.1	$+9.518519719 \, 10^{-6}$	$+9.616310231 \, 10^{-10}$	$+2.901200102 \, 10^{-16}$
1	$+9.006581118 \, 10^{-3}$	$+9.256115862 \, 10^{-05}$	$+2.826498802 \, 10^{-08}$
10	$-3.949584498 \, 10^{-2}$	$-5.553451162 \, 10^{-02}$	$+1.255780236 \, 10^{-01}$

Equations (3.45) and (3.46) are the same mathematical relation, one written for upward recurrence from small to large l values, and the other for downward recurrence from large l to small l. We shall see that with just a few additions and multiplications, recurrence relations permit rapid, simple computation of the entire set of j_l values for fixed x and all l.

To recur upward in l for fixed x, we start with the known forms for j_0 and j_1 (3.43) and use (3.45). As you will prove for yourself, this upward recurrence usually seems to work at first, but then fails. The reason for the failure can be seen from the plots of $j_l(x)$ and $n_l(x)$ *versus* x (Figure 3.5). If we start at $x \simeq 2$ and $l = 0$, we see that as we recur j_l up to larger l values with (3.45), we are essentially taking the difference of two "large" functions to produce a "small" value for j_l. This process suffers from the dreaded subtractive cancelation that always reduces precision. As we continue recurring, we take the difference of two small functions, each with large errors, and produce a yet smaller function with a yet larger error. After a while, we are left with only round-off error (garbage).

To be more specific, let us call $j_l^{(c)}$ the numerical value we compute as an approximation for $j_l(x)$. Even if we start with pure j_l, after a short while the computer's lack of precision effectively mixes in a bit of $n_l(x)$:

$$j_l^{(c)} = j_l(x) + \epsilon n_l(x). \tag{3.47}$$

This is inevitable because both j_l and n_l satisfy the same differential equation and, on that account, the same recurrence relation. The admixture of n_l becomes a problem when the numerical value of $n_l(x)$ is much larger than that of $j_l(x)$ because even a minuscule amount of a very large number may be large.

The simple solution to this problem (*Miller's device*) is to use (3.46) for downward recursion of the j_l values starting at a large value $l = L$. This avoids subtractive cancelation by taking small values of $j_{l+1}(x)$ and $j_l(x)$ and producing a larger $j_{l-1}(x)$ by addition. While the error may still behave like a Neumann function, the actual magnitude of the error will *decrease* quickly as we move downward to smaller l values. In fact, if we start iterating downward with arbitrary values for $j_{L+1}^{(c)}$ and $j_L^{(c)}$, after a short while we will arrive at the correct l dependence for this value of x. Although the precise value of $j_0^{(c)}$ so obtained will not be correct because it depends upon the arbitrary values assumed for $j_{L+1}^{(c)}$ and $j_L^{(c)}$, the relative values will be accurate. The absolute values are fixed from the known value (3.43), $j_0(x) = \sin x / x$. Because the recurrence relation is a linear relation between the j_l values, we need only normalize all the computed values via

$$j_l^N(x) = j_l^c(x) \times \frac{j_0^{anal}(x)}{j_0^c(x)}. \tag{3.48}$$

Accordingly, after you have finished the downward recurrence, you obtain the final answer by normalizing all $j_l^{(c)}$ values based on the known value for j_0.

3.4.2 Implementation and Assessment: Recursion Relations

A program implementing recurrence relations is most easily written using subscripts. If you need to polish up on your skills with subscripts, you may want to study our program `Bessel.py` in Listing 3.1 before writing your own.

1) Write a program that uses both upward and downward recursion to calculate $j_l(x)$ for the first 25 l values for $x = 0.1$, 1, and 10.
2) Tune your program so that at least one method gives "good" values (meaning a relative error $\simeq 10^{-10}$). See Table 3.1 for some sample values.
3) Show the convergence and stability of your results.
4) Compare the upward and downward recursion methods, printing out $l, j_l^{(up)}, j_l^{(down)}$, and the relative difference $|j_l^{(up)} - j_l^{(down)}| / (|j_l^{(up)}| + |j_l^{(down)}|)$.
5) The errors in computation depend on x, and for certain values of x, both up and down recursions give similar answers. Explain the reason for this.

3.5 Code Listing

Listing 3.1 Bessel.py Determines spherical Bessel functions by downward recursion (you should modify this to also work by upward recursion).

```
# Bessel.py
from visual import *
from visual.graph import *

Xmax = 40.
Xmin = 0.25
step = 0.1                                    # Global class variables
order = 10; start = 50        # Plot j_order
graph1 = gdisplay(width = 500, height = 500, title = 'Sperical Bessel, \
    L = 1 (red), 10',xtitle = 'x', ytitle = 'j(x)',\
    xmin=Xmin,xmax=Xmax,ymin=-0.2,ymax=0.5)
funct1 = gcurve(color=color.red)
funct2 = gcurve(color=color.green)

def down (x, n, m):                          # Method down, recurs downward
    j = zeros( (start + 2), float)
    j[m + 1] = j[m] = 1.                      # Start with anything
    for k in range(m, 0, - 1):
        j[k - 1] = ( (2.*k + 1.)/x)*j[k]  -  j[k + 1]
    scale = (sin(x)/x)/j[0]                   # Scale solution to known j[0]
    return j[n] * scale

for x in arange(Xmin, Xmax, step):
    funct1.plot(pos = (x, down(x, order, start)))

for x in arange(Xmin, Xmax, step):
    funct2.plot(pos = (x, down(x,1, start)))
```

4

Monte Carlo Simulations

This chapter starts with a discussion of how computers generate numbers that appear random, but really aren't, and how we can test for that. We then explore how these pseudorandom numbers are used to incorporate the element of chance into simulations. We do this first by simulating a random walk, and then by simulating the spontaneous decay of an atom or nucleus. In Section 5.5, we show how to use these random numbers to evaluate integrals, and in Chapter 17, we investigate the use of random numbers to simulate thermal processes and the fluctuations inherent in quantum systems.

Some people are attracted to computing because of its deterministic nature; it's nice to have a place in one's life where nothing is left to chance. Barring machine errors or undefined variables, you get the same output every time you feed your program the same input. Nevertheless, many computer cycles are used for *Monte Carlo* calculations that at their very core include elements of chance. These are calculations in which random-like numbers generated by the computer are used to *simulate* natural random processes, such as thermal motion or radioactive decay, or to solve equations on the average. Indeed, much of computational physics' great achievements have come about from the ability of computers to solve previously intractable problems using these so-called Monte Carlo techniques.

4.1 Random Numbers

We define a *sequence* $r_1, r_2, \ldots$ as *random* if there are no correlations among the numbers. Yet being random does not mean that all the numbers in the sequence are equally likely to occur. If all the numbers in a sequence are equally likely to occur, then the sequence is called *uniform*, which doesn't necessarily mean that it is random. To illustrate, 1, 2, 3, 4, ... is uniform, but probably not random. Further, it is possible to have a sequence of numbers that, in some sense, are random but have very short-range correlations among themselves, for example,

$$r_1, (1 - r_1), r_2, (1 - r_2), r_3, (1 - r_3), \ldots \tag{4.1}$$

have short-range but not long-range correlations.

Computational Physics: Problem Solving with Python, Fourth Edition.
Rubin H. Landau, Manuel J. Páez, and Cristian C. Bordeianu.
© 2024 WILEY-VCH GmbH. Published 2024 by WILEY-VCH GmbH.

Mathematically, the likelihood of a number occurring is described by a distribution function $P(r)$, where $P(r)\,dr$ is the probability of finding r in the interval $[r, r + dr]$. A *uniform* distribution means that $P(r) =$ a constant. The standard random-number generator on computers generates uniform distributions between 0 and 1. In other words, the standard random-number generator outputs numbers in this interval, each with an equal probability, yet each independent of the previous numbers. As we shall see, numbers can also be more likely to occur in certain regions than other, yet still be random.

By their very nature, computers, being deterministic devices, cannot generate random numbers. Because computed random number must contain correlations, they are not truly random. Although it may be a bit of work, if we know a computed random number r_m and its preceding numbers, then it should be possible to figure out r_{m+1}. For this reason, computers are said to generate *pseudorandom numbers* (yet with our incurable laziness we won't bother saying "pseudo" all the time). While more sophisticated generators do a better job at hiding the correlations, experience shows that if you look hard enough, or use pseudorandom numbers long enough, you will be able to discern correlations. A primitive alternative to generating random numbers is to read in a table of truly random numbers generated by naturally random processes such as radioactive decay, or to connect the computer to an experimental device that measures random events. These alternatives are not ideal for production work, but have actually been used as a check in times of doubt.

4.1.1 Random Number Generation

The *linear congruent* or *power residue* method is the common way of generating a pseudorandom sequence of numbers $0 \le r_i \le M - 1$ over the interval $[0, M - 1]$. To obtain the next random number r_{i+1}, you multiply the present random number r_i by the constant a, add another constant c, take the *modulus* by M, and then keep just the fractional part (remainder):[1]

$$r_{i+1} \stackrel{\text{def}}{=} (a\,r_i + c)\,\text{mod}\,M = \text{remainder}\left(\frac{a\,r_i + c}{M}\right). \tag{4.2}$$

The value for r_1 (the *seed*) is frequently supplied by the user. The mod operator is usually built into the software, for example, in Python it's the percent sign %. *Remaindering* is essentially a bit-shift operation that ends up with the least significant part of the input number, and thereby counts on the randomness of round-off errors to generate a random sequence.

For example, $c = 1, a = 4, M = 9$, and $r_1 = 3$ produces the sequence

$$r_1 = 3, \tag{4.3}$$

$$r_2 = (4 \times 3 + 1)\text{mod}\,9 = 13\,\text{mod}\,9 = \text{rem}\,\tfrac{13}{9} = 4, \tag{4.4}$$

$$r_3 = (4 \times 4 + 1)\text{mod}\,9 = 17\,\text{mod}\,9 = \text{rem}\,\tfrac{17}{9} = 8, \tag{4.5}$$

$$r_4 = (4 \times 8 + 1)\text{mod}\,9 = 33\,\text{mod}\,9 = \text{rem}\,\tfrac{33}{9} = 6, \tag{4.6}$$

$$r_{5-10} = 7, 2, 0, 1, 5, 3. \tag{4.7}$$

---

1 You may obtain the same result for the modulus operation by subtracting M until any further subtractions would leave a negative number; what remains is the *remainder*.

We thus obtain a sequence of length $M = 9$, after which the entire sequence repeats. If we want numbers in the range $[0, 1]$, we divide the r's by $M = 9$:

$$0.333, \ 0.444, \ 0.889, \ 0.667, \ 0.778, \ 0.222, \ 0.000, \ 0.111, \ 0.555, \ 0.333. \quad (4.8)$$

This is still a sequence of length 9, but is no longer a sequence of integers. If random numbers in the range $[A, B]$ are needed, you only need to **scale**:

$$x_i = A + (B - A)r_i, \quad 0 \leq r_i \leq 1, \ \Rightarrow \ A \leq x_i \leq B. \quad (4.9)$$

As a rule of thumb: *Before using a random-number generator in your programs, you should check its range and that it produces numbers that "look" random* (we'll explain further). Although not a mathematical proof, you should always make a graphical display of your random numbers. Your visual cortex is quite refined at recognizing patterns and will tell you immediately if there is a pattern in your random numbers. For instance, Figure 4.1 shows generated sequences from "good" and "bad" generators. It is clear which is not random (although if you look hard enough at the random points, your mind may well pick out patterns there too).

The linear congruent method (4.2) produces integers in the range $[0, M - 1]$ and therefore becomes completely correlated if a particular integer comes up a second time (the whole cycle then repeats). In order to obtain a longer sequence, a and M should be large numbers, but not so large that the product ar_{i-1} overflows. On a computer using 48-bit integer arithmetic, the built-in random-number generator may use M values as large as $2^{48} \simeq 3 \times 10^{14}$. A 32-bit generator may use $M = 2^{31} \simeq 2 \times 10^9$. If your program uses approximately this many random numbers, you may need to reseed (start the sequence over again with a different initial value) during intermediate steps to avoid the cycle repeating.

Your computer probably has random-number generators that are better than the one you will compute with the power residue method. In Python we use `random.random()`, the Mesenna Twister generator. We recommend that you use the best one you can find rather than write your own. To initialize a random sequence, you need to plant a seed in it. In

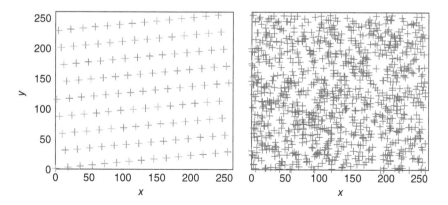

Figure 4.1 *Left*: A plot of successive random numbers $(x, y) = (r_i, r_{i+1})$ generated with a deliberately "bad" generator. *Right*: A plot generated with the built in random number generator. While the plot on the right is not proof that the distribution is random, the plot on the left is proof enough that the distribution is not random.

Python the statement `random.seed(None)` seeds the generator with the system time (see `Walk.py` in Listing 4.1).

$$M = 2^{48}, \ c = B \, (\text{base} \, 16) = 13 \, (\text{base} \, 8), \tag{4.10}$$

$$a = 5\text{DEECE66D} \, (\text{base} \, 16) = 273673163155 \, (\text{base} \, 8). \tag{4.11}$$

4.1.2 Computing a Random Sequence

For scientific work we recommend using an industrial-strength random-number generator. To see why, here we assess how *bad* a careless application of the power residue method can be.

1) Write a simple program to generate random numbers using the linear congruent method (4.2).
2) For pedagogical purposes, try the unwise choice: $(a, c, M, r_1) = (57, 1, 256, 10)$. Determine the *period*, that is, how many numbers are generated before the sequence repeats.
3) Take your pedagogical sequence of random numbers and look for correlations by observing clustering on a plot of successive pairs $(x_i, y_i) = (r_{2i-1}, r_{2i})$, $i = 1, 2, \ldots$. (Do *not* connect the points with lines.) You may "see" correlations (Figure 4.1), which means that you should not use this sequence for serious work.
4) Make your own version of Figure 4.2; that is, plot r_i versus i.
5) Test the built-in random-number generator on your computer for correlations by plotting the same pairs as above. (This should be good for serious work.)

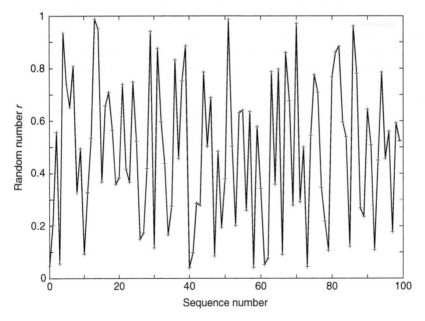

Figure 4.2 A plot of a uniform pseudorandom sequence r_i *versus i*. The points are connected to make it easier to follow the order. While this does not prove that a distribution is random, it at least shows the range of values and that there is fluctuation.

6) Test the linear congruent method again with reasonable constants like those in (4.10) and (4.11). Compare the scatterplot you obtain with that of the built-in random-number generator. (This should be good for semi-serious work.)

4.2   Simulating a Random Walk

Consider a perfume molecule released in the front of a classroom. Soon everyone can smell it. A molecule collides randomly with other molecules in the air and eventually reaches your nose despite the fact that you are hidden in the last row behind a newspaper. Your **problem** is to determine how many collisions, on the average, a perfume molecule makes in traveling a distance R. You are given the fact that a molecule travels an average (*root-mean-square*) distance r_{rms} between collisions.

There are a number of ways to simulate a random walk with (surprise, surprise) different assumptions leading to different behaviors. We will present a simple model for a 2D walk, and end up with a model for *normal diffusion*. The research literature is full of discussions of various versions of a random walk. For example, Browning motion corresponds to the limit in which the individual step lengths approach zero, and with no time delay between steps. Additional refinements include collisions within a moving medium (*abnormal diffusion*), including the velocities of the particles, or even pausing between steps. Models such as these are discussed in Chapter 14, *Fractals & Statistical Growth*.

In our random-walk simulation (Figure 4.3) an artificial *walker* takes sequential steps with the *direction* of each step *independent* of the direction of the previous step. We start at the origin and take N steps in the XY plane of *lengths* (not coordinates)

$$(\Delta x_1, \Delta y_1), (\Delta x_2, \Delta y_2), (\Delta x_3, \Delta y_3), \dots, (\Delta x_N, \Delta y_N). \tag{4.12}$$

Although each step may be in a different direction, the distances along each Cartesian axis just add algebraically. Accordingly, the radial distance R from the starting point after N steps is

$$\begin{aligned} R^2 &= (\Delta x_1 + \Delta x_2 + \cdots + \Delta x_N)^2 + (\Delta y_1 + \Delta y_2 + \cdots + \Delta y_N)^2 \\ &= \Delta x_1^2 + \Delta x_2^2 + \cdots + \Delta x_N^2 + 2\Delta x_1 \Delta x_2 + 2\Delta x_1 \Delta x_3 + 2\Delta x_2 \Delta x_1 + \cdots \\ &\quad + (x \to y). \end{aligned} \tag{4.13}$$

If the walk is random, the particle is equally likely to travel in any direction at each step. If we take the average of a large number of such random steps, all the cross terms in (4.13)

Figure 4.3 *Left*: A schematic of the N steps in a random walk simulation that end up a distance R from the origin. Notice how the Δx's for each step add vectorially. *Right*: A simulated walk in 3D from *Walk3D.py*.

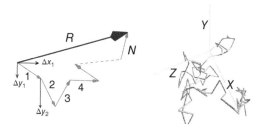

will vanish, and we will be left with

$$R_{rms}^2 = \langle R^2 \rangle \simeq \langle \Delta x_1^2 + \Delta x_2^2 + \cdots + \Delta x_N^2 + \Delta y_1^2 + \Delta y_2^2 + \cdots + \Delta y_N^2 \rangle$$
$$= \langle \Delta x_1^2 + \Delta y_1^2 \rangle + \langle \Delta x_2^2 + \Delta y_2^2 \rangle + \cdots$$
$$= N\langle r^2 \rangle = N r_{rms}^2,$$
$$\Rightarrow R_{rms} \simeq \sqrt{N} r_{rms}, \tag{4.14}$$

where $r_{rms} = \sqrt{\langle r^2 \rangle}$ is the *root-mean-square (RMS)* step size.

To summarize, if the walk is random, then we expect that after a large number of steps the average *vector* distance from the origin will vanish:

$$\langle \vec{R} \rangle = \langle x \rangle \vec{i} + \langle y \rangle \vec{j} \simeq 0. \tag{4.15}$$

Yet $R_{rms} = \sqrt{\langle R_i^2 \rangle}$ does not vanish. Equation (4.14) indicates that the average *scalar* distance from the origin is $\sqrt{N} r_{rms}$, where each step is of average length r_{rms}. In other words, the vector endpoint will be distributed uniformly in all quadrants, and so the displacement vector averages to zero, but the average length of that vector does not. For large N values, $\sqrt{N} r_{rms} \ll N r_{rms}$ (the value if all steps were in one direction on a straight line), but does not vanish. In our experience, computational simulations agree with this theory, but rarely perfectly, with the level of agreement depending upon the details of how the averages are taken and how the randomness is built into each step.

4.2.1 Random Walk Implementation

The program `walk.py` in Listing 4.1 is a sample random-walk simulation. Its key element is the random values for the x and y components of each step,

```
x += (random.random() - 0.5)*2.        # -1 =< x =< 1
y += (random.random() - 0.5)*2.        # -1 =< y =< 1
```

where here we have omitted the scaling factor that normalizes each step to length 1. When using your computer to simulate a random walk, you should expect to obtain (4.14) only as the average displacement, averaged over many trials, not necessarily as the answer for each trial. You final answer will depend on just how you take your random steps (Figure 4 4 right).

Start at the origin and take a 2D random walk with your computer:

1) To increase the amount of randomness, independently choose random values for $\Delta x'$ and $\Delta y'$ in the range $[-1, 1]$. Then normalize them so that each step is of unit length

$$\Delta x = \frac{1}{L} \Delta x', \ \Delta y = \frac{1}{L} \Delta y', \ L = \sqrt{\Delta x'^2 + \Delta y'^2}. \tag{4.16}$$

2) Use a plotting program to draw maps of several independent 2D random walks, each of 1000 steps. Based on your simulations, comment on whether the results look like what you would expect a random walk to look like.

3) If you have your walker taking N steps in a single trial, then conduct a total number $K \simeq \sqrt{N}$ of trials. Each trial should have N steps and start with a different seed.

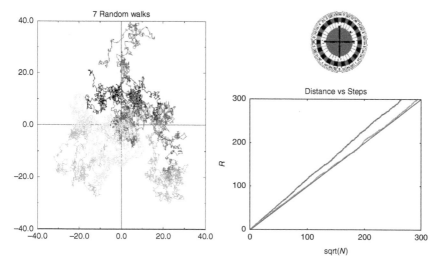

Figure 4.4 *Left*: The steps taken in seven 2D random walk simulations. *Right*: The distance covered in two walks of N steps using different schemes for including randomness. The theoretical prediction (4.14) is the straight line.

4) Calculate the mean square distance R^2 for each trial and then take the average of R^2 for all your K trials:

$$\langle R^2(N) \rangle = \frac{1}{K} \sum_{k=1}^{K} R_{(k)}^2(N). \tag{4.17}$$

5) Check the validity of the assumptions made in deriving the theoretical result (4.14) by checking how well

$$\frac{\langle \Delta x_i \Delta x_{j \neq i} \rangle}{R^2} \simeq \frac{\langle \Delta x_i \Delta y_j \rangle}{R^2} \simeq 0. \tag{4.18}$$

Do your checking for both a single (long) run and for the average over trials.

6) Plot the RMS distance, $R_{\text{mrs}} = \sqrt{\langle R^2(N) \rangle}$ as a function of $\sqrt{N}$. Values of N should start with a small number, where $R \simeq \sqrt{N}$ is not expected to be accurate, and end at a quite large value, where two or three places of accuracy should be expected on the average.

7) ⊙ Repeat the preceding and following analysis for a 3D walk as well.

4.2.2 Random Walks in a Brain

It has recently been realized that understanding the brain goes beyond just understanding the networks of neurons in it, to also understanding the effects of the fluid-filled extracellular spaces between the neurons [Nicholson, 2022].[2] This is important for understanding the molecular diffusion of radiographers, drugs, metabolites, and molecular signals within the brain.

---

2 Further discussion of the brain may be found in Chapter 11, *Neural Nets and Artificial Intelligence*.

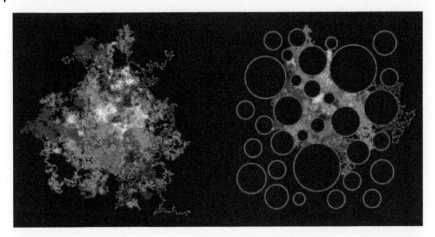

Figure 4.5 Fifty 2D random walk simulations exploring the diffusion of chemical probes within the brain from ([Nicholson, 2022] with permission from AIP publishing). The walks take 1500 equal-size steps with each walk assigned to one of six colors. *Left*: The walks with no impediments. *Right*: Circular impediments representing extracellular spaces that block out regions inaccessible to the walks.

Random-walk simulations, like the ones we have already examined, have provided understanding of diffusion within the brain. Specifically, the left of Figure 4.5 presents research results for 50 random walks of 1500 equal-size steps within a brain model, with each walk assigned to one of six colors Nicholson [2022]. Note the striking similarity of these walks to those we have shown on the left of Figure 4.4. On the right of Figure 4.5, we show the results of a model for diffusion in the brain that accounts for the extracellular spaces between neurons by randomly placing circular obstructions within the simulation volume.

1) Try to reproduce the simulation shown on the left of Figure 4.5 by recording and plotting 50 walks, with each walk assigned to one of six colors. Start the walks at the origin, use equal-sized steps, and restrict the simulation space to two dimensions.

2) As shown of the right of Figure 4.4, determine the average over all your walks of the RMS distance covered R_{rms}.

3) Take the same 2D space covered in your simulations, and now insert circular obstructions of varied sizes, similar to those on the right of Figure 4.5.

4) Yet again conduct and record 50 walks, with each walk assigned to one of six colors. Start the walks at the origin, use equal-sized steps, but include obstructions that *stop* the walks when they hit them.

5) Determine again the average over all your obstructed walks of the RMS distance covered R_{rms}. The obstructions should lead to a decreased R_{rms}.

6) Again conduct and record 50 walks, with each walk assigned to one of six colors. Start the walks at the origin, use equal-sized steps, but include obstructions that *repel*, but don't stop, the walks when they hit.

7) Again determine the average over all your obstructed walks of the RMS distance covered R_{rms}, and compare to the previous two results.

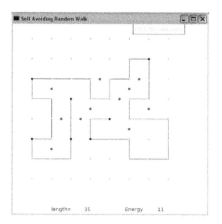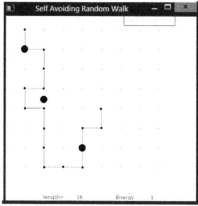

Figure 4.6 Two self-avoiding random walks that simulate protein chains with hydrophobic (H) monomers in large dots, and polar (P) monomers in small dots. The dark dots on the right indicate regions where two H monomers are not directly connected.

8) In Einstein's 1905 paper *Investigations on the Theory of the Browning Movement*, he proposed that the effective diffusion coefficient within a medium $D\frac{R^2_{rms}}{2dt}$, where d is the number of spatial dimensions (2 for a 2D simulation), and t is the average time for a walk. How much of an effect on D do the obstructions cause?
9) Repeat the problem for a three-dimensional volume.

4.2.3 Random Protein Folding

A protein is a large biological molecule made up of molecular chains (the residues of amino acids). These chains are formed from *monomers*, that is, molecules that bind chemically with other molecules. More specifically, the chains consist of non-polar hydrophobic (H) monomers that are repelled by water, and polar (P) monomers that are attracted by water. The actual structure of a protein results from a *folding process* in which random coils of chains rearrange themselves into a configuration of minimum energy. We want to model that process on the computer.

Although molecular dynamics (Chapter 18) may be used to simulate protein folding, it is much slower than Monte-Carlo techniques, and even then, it is hard to find the lowest energy states. Here we create a simple Monte-Carlo simulation in which you to take a random walk in a 2D square lattice [Yue *et al.*, 1995]. At the end of each step, you randomly choose an H or a P monomer and drop it on the lattice, with your choice weighted such that H monomers are more likely than P ones. The walk is restricted such that the only positions available after each step are the three neighboring sites, with the already occupied sites excluded (this is why this technique is known as a *self-avoiding random walk*).

The goal of the simulation is to find the lowest energy state of a sequence of H and P monomers with links of various lengths. It then may be compared to those in nature. Just how best to find such a state is an active research topic [Yue *et al.*, 1995]. The energy of a chain is defined as

$$E = -\epsilon f, \tag{4.19}$$

where ϵ is a positive constant and f is the number of H–H neighbor *not* connected directly (P–P and H–P bonds do not count at lowering the energy). So if the neighbor next to an H is another H, it lowers the energy, but if it is a P it does not lower the energy. We show a typical simulation result in Figure 4.6. Accordingly, for a given length of chain, we expect the natural state(s) of an H–P sequence to be those with the largest possible number f of H–H contacts. That is what we are looking for.

1) Modify the random walk program we have already developed so that it simulates a self-avoiding random walk. The key here is that the walk stops at a corner, or when there are no empty neighboring sites available.
2) Make a random choice as to whether the monomer is an H or a P, with a weighting such that there are more H's than P's.
3) Produce a visualization that shows the positions occupied by the monomers, with the H and P monomer indicated by different color dots. Our visualization, shown in Figure 4.6, is produced by the program `ProteinFold.py`, given in Listing 4.2.
4) After the walk ends, record the energy and length of the chain.
5) Run many folding simulations and save the outputs, categorized by length and energy.
6) Examine the state(s) of lowest energy for various chain lengths and compare the results to those from molecular dynamic simulations and actual protein structures (available on the Web).
7) Do you think this simple model has some merit?
8) ⊙ Extend the folding to 3D.

4.3 Spontaneous Decay

Your **problem** is to simulate the time dependence of the decay of a small number N of radioactive particles.[3] In particular, you are to determine the connection between exponential decay and *stochastic* decay (containing elements of chance). Realizing that exponential decay is a good model only when there are very large numbers of particles, the exponential model is no longer accurate as the number of decaying particles decreases, as it always does. Accordingly, our simulation should be closer to nature than is the exponential decay model (Figure 4.7). In fact, if you "listen" to the output of the decay simulation code, what you will hear sounds very much like a Geiger counter, an intuitively convincing demonstration of the realism of the simulation.

Spontaneous decay is a natural process in which a particle, with no external stimulation, decays into other particles. Although the probability of decay of any one particle in any one time interval is constant, just when it decays is random. Inasmuch as the presence, or decay, of any one particle does not influence the decay of any other particle, the probability of decay is not influenced by how long the particle has been around, or how many other particles are still around. In other words, the probability $\mathcal{P}$ of any one particle decaying per unit time interval is a constant, yet when that particle decays, it is gone forever. Of course,

---

3 Spontaneous decay is also discussed in Chapter 6, where we fit it to an exponential.

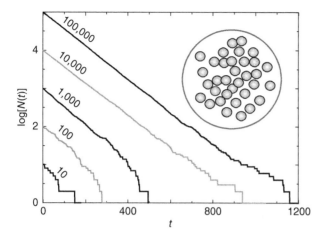

Figure 4.7 *Circle*: A sample containing *N* nuclei, each of which has the same probability of decaying per unit time, *Graphs*: Semilog plots of the number of nuclei versus time for five simulations with differing initial numbers of nuclei. Exponential decay would be a straight line without bumps, similar to the initial behavior for $N = 100,000$.

as the total number N of particles decays with time, so will the number that decay per unit time, but the probability of any one particle decaying in some time interval remains the same for as long as that particle exists.

4.3.1 Discrete Decay Model

Imagine having a sample containing $N(t)$ radioactive nuclei at time t (Figure 4.7 circle). Let ΔN be the number of particles that decay in some small time interval Δt. We convert the statement "the probability $\mathcal{P}$ of any one particle decaying per unit time is a constant" into the equation

$$\mathcal{P} = \frac{\Delta N(t)/N(t)}{\Delta t} = -\lambda, \tag{4.20}$$

$$\Rightarrow \frac{\Delta N(t)}{\Delta t} = -\lambda N(t), \tag{4.21}$$

where the constant λ is called the *decay rate* and the minus sign indicates a decreasing number. Because $N(t)$ decreases in time, the *activity* $\Delta N(t)/\Delta t$ (sometimes also called the decay rate) also decreases with time. In addition, because the total activity is proportional to the total number of particles present, it too is stochastic with an exponential-like decay in time. [Actually, because the number of decays $\Delta N(t)$ is proportional to the difference in random numbers, it tends to show even larger statistical fluctuations than does $N(t)$.]

Equation (4.21) is a *finite-difference equation* relating the experimental quantities $N(t)$, $\Delta N(t)$, and Δt. Although a difference equation cannot be integrated the way a differential equation can, it can be simulated numerically. Because the process is random, we cannot predict a single value for $\Delta N(t)$, although we can predict the average number of decays when observations are made on many identical systems of N decaying particles.

4.3.2 The Exponential Decay Approximation

When the number of particles $N \to \infty$ and the observation time interval $\Delta t \to 0$, the difference equation (4.21) becomes a differential equation, and we obtain the familiar exponential decay law:

$$\frac{\Delta N(t)}{\Delta t} \longrightarrow \frac{dN(t)}{dt} = -\lambda N(t). \tag{4.22}$$

This equation can be integrated to obtain the time dependencies of the total number of particles and of the total activity:

$$N(t) = N(0)e^{-\lambda t} = N(0)e^{-t/\tau}, \tag{4.23}$$

$$\frac{dN}{dt}(t) = -\lambda N(0)e^{-\lambda t} = \frac{dN}{dt}(0)\,e^{-\lambda t}. \tag{4.24}$$

In this limit we can identify the decay rate λ with the inverse lifetime:

$$\lambda = \frac{1}{\tau}. \tag{4.25}$$

We see from its derivation that exponential decay is a good description of nature for a large number of particles, that is, when $\Delta N/N \simeq 0$. However, in nature, $N(t)$ can be a small number, and in that case we have a statistical, as opposed to a continuous process. The basic law of nature (4.20) is always valid, but as we will see in the simulation, exponential decay (4.24) becomes less and less accurate as the number of particles gets smaller and smaller.

4.3.3 Discrete Decay Simulation

A program for simulating radioactive decay is surprisingly simple, but not without its subtleties. We increase time in discrete steps of Δt, and for each time interval we count the number of nuclei that have decayed during that Δt. The simulation quits when there are no nuclei left to decay. Such being the case, we have an outer loop over the time steps Δt, and an inner loop over the remaining nuclei for each time step. The pseudocode is simple (as is the code):

```
   input N, lambda
2  t=0
   while N > 0
      Delta = 0
      for i = 1..N
6        if (r_i < lambda) Delta = Delta + 1
      end for
      t = t +1
      N = N - Delta
10     Output t, Delta, N
   end while
```

When we pick a value for the decay rate $\lambda = 1/\tau$ to use in our simulation, we are setting the scale for times. For example, if the actual decay rate is $\lambda = 0.3 \times 10^6\,\text{s}^{-1}$, and if we decide to measure times in units of $10^{-6}\,\text{s}$, then we will choose random numbers $0 \le r_i \le 1$, which leads to λ values lying someplace near the middle of the range (e.g. $\lambda \simeq 0.3$). Alternatively, we can use a value of $\lambda = 0.3 \times 10^6 \, \text{s}^{-1}$ in our simulation and then scale the random numbers to the range $0 \leq r_i \leq 10^6$. However, unless you plan to compare your simulation to experimental data, you do not have to worry about the scale for time, but instead should focus on the physics behind the slopes and the derived functional dependencies.

`Decay.py` is our sample simulation of spontaneous decay. An extension of this program, `DecaySound.py`, in Listing 4.3, adds a beep each time an atom decays (unfortunately this works only with Windows). When we listen to the simulation it sounds like a Geiger counter, with its randomness and with a decay rate that decreases in time. This provides some rather convincing evidence of the realism of the simulation.

4.3.4 Decay Implementation and Visualization

Write a program to simulate radioactive decay using the simple program in Listing 4.3 as a guide. You should obtain results similar to those in Figure 4.7.

1) Plot the logarithm of the number left $\ln N(t)$ and the logarithm of the decay rate $\ln \Delta N(t)/\Delta t$ *versus* time. Note that the simulation measures time in steps of Δt (generation number).
2) Check that you obtain what looks like exponential decay when you start with large values for $N(0)$, but that the decay displays its stochastic nature for small $N(0)$ (large $N(0)$ values are also stochastic; they just don't look it at first).
3) Create two plots, one showing that the slopes of $N(t)$ *versus* t are *independent* of $N(0)$, and another showing that the slopes are proportional to the value for λ.
4) Create a plot showing that within expected statistical variations, $\ln N(t)$ and $\ln \Delta N(t)$ are proportional.
5) Explain in your own words how a process that is spontaneous and random at its very heart can lead to exponential decay.
6) How does your simulation show that the decay is exponential-like and not a power law such as $N = \beta t^{-\alpha}$?

4.4 Testing and Generating Random Distributions

Since the computer's random numbers are generated according to a definite rule, they must be correlated with each other. This can affect a simulation that assumes truly random events. Therefore it is wise to test a random-number generator to obtain a numerical measure of its uniformity and randomness before you stake your scientific reputation on it. In fact, some tests are simple enough for you to make it a habit to run them simultaneously with your simulation. In the examples to follow, we test for randomness and uniformity.

1) Probably the most obvious, but often neglected, test for randomness and uniformity is just to look at the numbers generated. For example, Table 4.1 presents some output from Python's `random` method. If you just look at these numbers you will know immediately that they all lie between 0 and 1, that they appear to differ from each other, and that there is no obvious pattern (like 0.3333).

Table 4.1 A table of a uniform, pseudo-random sequence r_i generated by Python's random method.

0.04689502438508175	0.20458779675039795	0.5571907470797255	0.05634336673593088
0.9360668645897467	0.7399399139194867	0.6504153029899553	0.8096333704183057
0.3251217462543319	0.49447037101884717	0.14307712613141128	0.32858127644188206
0.5351001685588616	0.9880354395691023	0.9518097953073953	0.36810077925659423
0.6572443815038911	0.7090768515455671	0.5636787474592884	0.3586277378006649
0.38336910654033807	0.7400223756022649	0.4162083381184535	0.3658031553038087
0.7484798900468111	0.522694331447043	0.14865628292663913	0.1741881539527136
0.41872631012020123	0.9410026890120488	0.1167044926271289	0.8759009012786472
0.5962535409033703	0.4382385414974941	0.166837081276193	0.27572940246034305
0.832243048236776	0.45757242791790875	0.7520281492540815	0.8861881031774513
0.04040867417284555	0.14690149294881334	0.2869627609844023	0.27915054491588953
0.7854419848382436	0.502978394047627	0.688866810791863	0.08510414855949322
0.48437643825285326	0.19479360033700366	0.3791230234714642	0.9867371389465821

2) As we have seen, a quick visual test (Figure 4.2) involves taking this same list and plotting it with r_i as ordinate and i as abscissa. Observe how there appears to be a uniform distribution between 0 and 1 and no particular correlation between points (although your eye and brain will try to recognize some kind of pattern).

3) As we have seen, an effective test for randomness is performed by making a scatterplot of $(x_i = r_{2i}, y_i = r_{2i+1})$ for many i values. If your points have noticeable regularity, the sequence is not random. If the points are random, they should uniformly fill a square with no discernible pattern (a cloud), as in Figure 4.1.

4) A simple test of uniformity evaluates the kth moment of a distribution:

$$\langle x^k \rangle = \frac{1}{N} \sum_{i=1}^{N} x_i^k. \tag{4.26}$$

If the numbers are distributed *uniformly*, then (4.26) is approximately the moment of the distribution function $P(x)$:

$$\frac{1}{N} \sum_{i=1}^{N} x_i^k \simeq \int_0^1 dx\, x^k P(x) \simeq \frac{1}{k+1} + O\left(\frac{1}{\sqrt{N}}\right). \tag{4.27}$$

If (4.27) holds for your generator, then you know that the distribution is uniform. If the deviation from (4.27) varies as $1/\sqrt{N}$, then you *also* know that the distribution is random because the $1/\sqrt{N}$ result derives from assuming randomness.

5) Another simple test determines the near-neighbor correlation in your random sequence by taking sums of products for small k:

$$C(k) = \frac{1}{N} \sum_{i=1}^{N} x_i x_{i+k}, \quad (k = 1, 2, \ldots). \tag{4.28}$$

If your random numbers x_i and x_{i+k} are distributed with the joint probability distribution $P(x_i, x_{i+k}) = 1$ and are independent and uniform, then (4.28) can be approximated as an integral:

$$\frac{1}{N}\sum_{i=1}^{N} x_i x_{i+k} \simeq \int_0^1 dx \int_0^1 dy \, xy \, P(x,y) = \int_0^1 dy \, xy = \frac{1}{4}. \tag{4.29}$$

If (4.29) holds for your random numbers, then you know that they are uniform and independent. If the deviation from (4.29) varies as $1/\sqrt{N}$, then you *also* know that the distribution is random.

6) Test your random-number generator with (4.27) for $k = 1, 3, 7$ and $N = 100, 10\,000,$ $100\,000$. In each case print out

$$\sqrt{N}\left|\frac{1}{N}\sum_{i=1}^{N} x_i^k - \frac{1}{k+1}\right| \tag{4.30}$$

to check that it is of order 1.

4.5 Code Listings

Listing 4.1   Walk.py Calls the random-number generator from the random package. Note that a different seed is needed to obtain a different sequence.

```
1  # Walk.py  Random walk with graph
   from visual import *
   from visual.graph import *
   import random

5
   random.seed(None)                    # Seed generator, None => system clock
   jmax = 20
   x   = 0.;          y = 0.                         # Start at origin
9  graph1 = gdisplay(width=500, height=500, title='Random Walk', xtitle='x',
                     ytitle='y')
   pts = gcurve(color = color.yellow)
   for i in range(0, jmax + 1):
13     pts.plot(pos = (x, y) )                       # Plot points
       x += (random.random() - 0.5)*2.               # -1 =< x =< 1
       y += (random.random() - 0.5)*2.               # -1 =< y =< 1
       pts.plot(pos = (x, y))
17     rate(100)
```

Listing 4.2   ProteinFold.py A self-avoiding random walk.

```
1  # ProteinFold.py: Self avoiding random walk
   # Stops in corners or  occupied neighbors
   # energy  =  -f|eps, f=1 if neighbour = H, f=0 if p
   # Yellow dot indicates unconnected neighbor

5
   from visual import *;  import random

   Maxx = 500;  Maxy = 500;   ran = 20; L = 100;  m = 100;  n = 100
9  size  = 8;   size2 = size*2;   nex = 0
   M = []; DD = []                  # Arrays for polymer & grid

   graph1 = display(width=Maxx, height=Maxy, title='Protein Folding',
13                   range=ran)
   positions = points(color=color.cyan, size = 2)
```

```
     def selectcol():              # Select atom's colors
17       hp = random.random()      # Select H or P
         if hp <= 0.7:
             col = (1,0,0)          # Hydrophobic color red
             r = 2
21       else:
             col = (1,1,1)          # Polar color white
             r = 1
         return col,r
25
     def findrest(m,length,fin,fjn):  # Check links energies
         ener = 0
         for t in range(m,length+1):   # Next link not considered
29           if DD[t][0]==fin and DD[t][1]==fjn and DD[t][2]==2:
                 ener = 1                  # Red unlinked neighbor
         return ener
33   def findenergy(length,DD):        # Finds energy of each link
         energy = 0
         for n in range (0,length+1):
             i = DD[n][0]
37           j = DD[n][1]
             cl = DD[n][2]
             if cl==1: pass    # if white
             else:             # red
41               if n < length+1:
                     imin = int(i-1)            # Check neighbor i-1,j
                     js = int(j)
                     if imin >= 0:
45                       e = findrest(n+2,length,imin,js) # Return energy 1
                         energy = energy + e
                         if e==1:    # Plot yellow dot at neighbour
                             xol = 4*(i-0.5)-size2
49                           yol = -4*j+size2
                             points(pos=(xol,yol),color=color.yellow, size=6)
                     ima = i+1
                     js = j
53                   if ima<=size-1:      # Check neighborr i+1,j
                         e = findrest(n+2,length,ima,js)
                         energy = energy+e
                         if e == 1:     # Plot yellow dot at neighbor
57                           xol = 4*(i+0.5)-size2
                             yol = -4*j+size2
                             points(pos=(xol,yol),color=color.yellow, size=6)
                     iss = i
61                   jma = j+1
                     if jma <= size-1:    # Check neighbor i,j+1
                         e = findrest(n+2,length,iss,jma)
                         energy = energy+e
65                       if e == 1:      # Plot yellow dot at neighbor
                             xol = 4*i-size2                # Start at middle
                             yol = -4*(j+0.5)+size2
                             points(pos=(xol,yol),color=color.yellow, size=6)
69                   iss = i
                     jmi = j-1
                     if jmi >= 0:         # Check neighbor i, j-1
                         e = findrest(n+2,length,iss,jmi)
73                       energy = energy +e
                         if e==1:        # Plot yellow dot at neighbour
                             xol = 4*i-size2                # Start at middle
                             yol = -4*(j-0.5)+size2
77                           points(pos=(xol,yol),color=color.yellow, size=6)
         return energy

     def grid():                          # Plot grid
81       for j in range(0,size):
             yp = -4*j+size2                    # World to screen coord
             for i in range (0,size):          # Horizontal row
                 xp = 4*i-size2
85               positions.append(pos = (xp,yp))
```

```
      grid ()
      length = 0
      while 1:                             # Adjust for desired number of walks
 89      pts2 = label(pos=(-5, -18), box=0)
         length = 0
         grid = zeros((size,size))
         D = zeros((L,m,n))
 93      DD = []
         i = size/2                        # Center of grid
         j = size/2
         xol = 4*i-size2
 97      yol = -4*j+size2
         col,c = selectcol()
         grid[i,j] = c                # Particle in center
         M = M+[points(pos=(xol,yol),color=col, size=6)] # Red point at center
101      print(" start        ")
         DD = DD+[[i,j,c]]
         while (i>0 and i<size-1 and j>0 and j<size-1 and (grid[i+1,j] == 0
                 or grid[i-1,j] == 0 or grid[i,j+1] == 0 or grid[i,j-1] == 0)):
105         r = random.random()
            if r < 0.25 :                # Probability 25%
                if grid[i+1,j]==0:  i += 1  # Step right if empty
            elif 0.25 < r and r < 0.5:        # Step left
109             if grid[i-1,j] == 0: i -= 1
            elif 0.50 < r and r < 0.75:          # Up
                if grid[i,j-1]==0:  j -= 1
            else :                      # Down
113             if grid[i,j+1]==0:  j+=1
            if grid[i,j] == 0:
                col,c = selectcol()
                grid[i,j] = 2                     # Occupy grid point
117             length += 1 # Increase length as occupied
                DD = DD+[[i,j,c]]
                xp = 4*i-size2
                yp = -4*j+size2
121             curve(pos=[(xol,yol),(xp,yp)])   # Connect last to new position
                M = M + [points(pos=(xp,yp), color=col,size=6)]
                xol = xp                       # Start new line
                yol = yp
125         while (j == (size-1) and i != 0 and i != (size-1)):   # Bottom row
                r1 = random.random()
                if r1 < 0.2:                    # Probability 20% move left
                    if grid[i-1,j] == 0:  i -= 1
129             elif r1 > 0.2 and r1 < 0.4:              # Probability 20% move right
                    if grid[i+1,j] == 0:  i += 1
                else:                      # Probability 60% move up
                    if grid[i,j-1] == 0:  j-=1
133             if grid[i,j] == 0:
                    col,c = selectcol()            # Increase length
                    grid[i,j] = 2                 # Grid point occupied
                    length += 1
137                 DD = DD + [[i,j,c]]
                    xp = 4*i - size2
                    yp = -4*j + size2
                    curve(pos=[(xol,yol),(xp,yp)])     # Line connecting new point
141                 M = M +[points(pos=(xp,yp), color=col,size=6)]
                    xol = xp
                    yol = yp      # Last row; Stop if corner or occupied neighbors
                if (i==0 or i==(size-1)) or (grid[i-1,size-1]!=0 and
                    grid[i+1,size-1]!=0):
145                 break
            while (j == 0 and i != 0 and i != (size-1)):   # First row
                r1 = random.random()
                if r1<0.2:
149                 if grid[i-1,j] == 0:  i -= 1
                elif r1>0.2 and r1<0.4:
                    if grid[i+1,j]==0:   i += 1
                else:
153                 if grid[i,j+1]==0:   j += 1
```

```python
            if grid[i,j]==0:
                col,c = selectcol()
                grid[i,j] = 2
157             length += 1
                DD = DD + [[i,j,c]]
                xp = 4*i - size2
                yp = -4*j + size2
161             curve(pos=[(xol,yol),(xp,yp)])
                M = M + [points(pos=(xp,yp), color=col,size=6)]
                xol = xp
                yol = yp
165         if i==(size-1) or i==0 or (grid[i-1,0]!=0 and grid[i+1,0]!=0):
                break
        while (i==0 and j !=0 and j !=(size-1)): # First column
            r1 = random.random()
169         if r1<0.2:
                if grid[i,j-1] == 0:  j -= 1
            elif r1 > 0.2 and r1 < 0.4:
                if grid[i,j+1] == 0:  j += 1
173         else:
                if grid[i+1,j] == 0:  i += 1
            if grid[i,j] == 0:
                col,c = selectcol()
177             grid[i,j] = c
                length += 1
                DD = DD+[[i,j,c]]
                xp = 4*i - size2
181             yp = -4*j + size2
                curve(pos=[(xol,yol),(xp,yp)])
                M = M +[points(pos=(xp,yp), color=col,size=6)]
                xol = xp
185             yol = yp
            if j==(size-1) or j==0 or (grid[0,j+1]!=0 and grid[0,j-1]!=0):
                break
        while (i==(size-1) and j !=0 and j !=(size-1)): # Last column
189         r1 = random.random()
            if r1 < 0.2:
                if grid[i,j-1] == 0: j -= 1
            elif r1 > 0.2 and r1 < 0.4:
193             if grid[i,j+1] == 0:  j += 1
            else:
                if grid[i-1,j] == 0:  i -= 1
            if grid[i,j] == 0:
197             col,c = selectcol()
                grid[i,j] = c
                length += 1
                col,c=selectcol()
201             DD = DD + [[i,j,c]]
                xp = 4*i - size2
                yp = -4*j + size2
                curve(pos=[(xol,yol),(xp,yp)])
205             M = M +[points(pos=(xp,yp), color=col,size=6)]
                xol = xp
                yol = yp
            if j--(size-1) or (grid[size-1,j+1]!=0 and grid[size-1,j-1]!=0):
209             break
    label(pos=(-10, -18), text='Length=', box=0)
    label(pos=(10,18,0), text='Click for new walk',color=color.red, display=graph1)
    pts2.text = '%4s' %length
213 label(pos=(5,-18,0), text='Energy',box=0)
    evalue=label(pos=(10, -18), box=0) # Energy
    evalue.text = '%4s' %findenergy(length,DD)        # Walk length walk
    print("energy is ",findenergy(length,DD))
217 print("dd")
    graph1.mouse.getclick()               # Detect mouse click
    for obj in graph1.objects:            # Start new walk
        if (obj is positions or obj is curve):       continue
221     obj.visible = 0 # Clear curve
```

Listing 4.3 DecaySound.py Simulates spontaneous decay in which a decay occurs if a random number is smaller than the decay parameter. The *winsound* package lets us play a beep each time there is a decay, and this leads to the sound of a Geiger counter.

```
1  # DecaySound.py spontaneous decay simulation

   from visual import *; from visual.graph import *; import random, winsound

5  lambda1 = 0.005                                    # Decay constant
   max = 80.;  time_max = 500;    seed = 68111
   number = nloop = max                               # Initial value
   graph1 = gdisplay(title ='Spontaneous Decay',xtitle='Time',\
9                    ytitle = 'Number')
   decayfunc = gcurve(color = color.green)
   for time in arange(0, time_max + 1):               # Time loop
       for atom in arange(1, number + 1 ):            # Decay loop
13         decay = random.random()
           if (decay < lambda1):
               nloop = nloop - 1                      # A decay
               winsound.Beep(600, 100)                # Sound beep
17     number = nloop
       decayfunc.plot( pos = (time, number) )
       rate(30)
```

5

Differentiation and Integration

> *We start this chapter with a short discussion of numerical differentiation, an important, if rather straight-forward, topic. We derive the algorithms for differentiation that will be used throughout the book. The majority of the chapter covers several algorithms for numerical integration, a basic tool of scientific computation. We end with a discussion of Monte Carlo integration techniques, which are fundamentally different from all the others.*

5.1 Differentiation Algorithms

Problem Figure 5.1 shows the trajectory of a projectile with air resistance. The dots indicate the times t at which measurements were made and tabulated. Your **problem** is to determine the projectile's velocity dy/dt as a function of time. Note that because there is realistic air resistance present, there is no analytic function to differentiate.

You probably did rather well in your first calculus course and feel competent at taking derivatives. However, you may never have taken derivatives of numerical data using the elementary definition

$$\frac{dy(t)}{dt} \stackrel{\text{def}}{=} \lim_{h \to 0} \frac{y(t+h) - y(t)}{h}. \tag{5.1}$$

In fact, even a computer will have problems with this kind of limit because it is wrought with subtractive cancellation; as h is made smaller, the computer's finite word length causes the numerator to fluctuate between 0 and the machine precision c_m, while as the denominator approaches zero, overflows will occurs.

5.1.1 Forward Difference

The most direct method for numerical differentiation starts by expanding a function in a Taylor series to obtain its value at a small step h away:

$$y(t+h) = y(t) + h\frac{dy(t)}{dt} + \frac{h^2}{2!}\frac{d^2y(t)}{dt^2} + \frac{h^3}{3!}\frac{dy^3(t)}{dt^3} + \cdots, \tag{5.2}$$

Computational Physics: Problem Solving with Python, Fourth Edition.
Rubin H. Landau, Manuel J. Páez, and Cristian C. Bordeianu.
© 2024 WILEY-VCH GmbH. Published 2024 by WILEY-VCH GmbH.

$$\frac{y(t+h) - y(t)}{h} = \frac{dy(t)}{dt} + \frac{h}{2!}\frac{d^2y(t)}{dt^2} + \frac{h^2}{3!}\frac{d^3y(t)}{dt^3} + \cdots. \tag{5.3}$$

If we ignore the h^2 terms in (5.3), we obtain the *forward-difference* algorithm for the derivative:

$$\left.\frac{dy(t)}{dt}\right|_{\text{fd}} \stackrel{\text{def}}{=} \frac{y(t+h) - y(t)}{h}. \tag{5.4}$$

An estimate of the error follows from substituting the Taylor series (5.2):

$$\left.\frac{dy(t)}{dt}\right|_{\text{fd}} \simeq \frac{dy(t)}{dt} - \frac{h}{2}\frac{dy^2(t)}{dt^2} + \cdots. \tag{5.5}$$

You can think of this approximation as using two points to represent the function by a straight line in the interval from x to $x + h$ (Figure 5.1 left). The approximation (5.4) has an error proportional to h (unless the heavens look down upon you kindly and makes y'' vanish). We can make the approximation error [the terms left off on the RHS of (5.4)] smaller by making h smaller, yet precision will be lost for too small an h when $y(t + h) \simeq y(t)$.

To try out this forward-difference algorithm, we'll take $y(t) = a + bt^2$. The exact derivative is $y' = 2bt$, while the computed derivative is

$$\left.\frac{dy(t)}{dt}\right|_{\text{fd}} \simeq \frac{y(t+h) - y(t)}{h} = 2bt + bh. \tag{5.6}$$

This clearly becomes a good approximation only for small $h \ll 1/b$.

5.1.2 Central Difference

An improved approximation to the derivative starts with the basic definition (5.1), or, geo-metrically, as shown on the right of Figure 5.1. Now, rather than making a single step of h forward, we form a *central difference* by stepping forward half a step and backward half a step:

$$\left.\frac{dy(t)}{dt}\right|_{\text{cd}} \equiv D_{\text{cd}}\, y(t) \stackrel{\text{def}}{=} \frac{y(t+h/2) - y(t-h/2)}{h}. \tag{5.7}$$

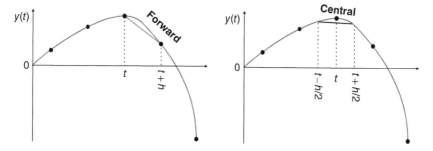

Figure 5.1 A trajectory of a projectile experiencing air resistance. *Left*: Forward-difference approximation (slanted line) and *Right*: central-difference approximation (horizontal line) for the numerical first derivative at time t. (A tangent to the curve at t would yield the correct derivative.) The central difference is seen to be more accurate than the forward difference.

We estimate the error in the central-difference algorithm by substituting the Taylor series for $y(t + h/2)$ and $y(t - h/2)$ into (5.7):

$$y\left(t + \frac{h}{2}\right) - y\left(t - \frac{h}{2}\right) \simeq \left[y(t) + \frac{h}{2}y'(t) + \frac{h^2}{8}y''(t) + \frac{h^3}{48}y'''(t) + \mathcal{O}(h^4)\right]$$

$$- \left[y(t) - \frac{h}{2}y'(t) + \frac{h^2}{8}y''(t) - \frac{h^3}{48}y'''(t) + \mathcal{O}(h^4)\right]$$

$$= hy'(t) + \frac{h^3}{24}y'''(t) + \mathcal{O}(h^5),$$

$$\Rightarrow \quad \left.\frac{dy(t)}{dt}\right|_{cd} \simeq y'(t) + \frac{1}{24}h^2 y'''(t) + \mathcal{O}(h^4). \tag{5.8}$$

The important difference between this central-difference algorithm and the forward difference one is that when $y(t - h/2)$ is subtracted from $y(t + h/2)$, all terms containing an even power of h in the two Taylor series cancel. This makes the central-difference algorithm accurate to order h^2 (h^3 before division by h), while the forward difference is accurate only to order h. If the $y(t)$ is smooth, that is, if $y'''h^2/24 \ll y''h/2$, then you can expect the error in central-difference algorithm to be smaller than with the forward difference algorithm.

If we now return to our parabola example (5.6), we will see that the central difference gives the exact derivative, independent of h:

$$\left.\frac{dy(t)}{dt}\right|_{cd} \simeq \frac{y(t + h/2) - y(t - h/2)}{h} = 2bt. \tag{5.9}$$

This is to be expected because the higher derivatives equal zero for a second-order polynomial.

5.2 Extrapolated Difference

Since a differentiation rule based on keeping a certain number of terms in a Taylor series also provides an expression for the error (the terms not included), we can reduce the theoretical error further by forming a combination of approximations whose summed errors extrapolate to zero. One such algorithm is the central-difference algorithm (5.7) using a half-step back and a half-step forward. A second algorithm is another central-difference approximation, but this time using quarter-steps:

$$\left.\frac{dy(t, h/2)}{dt}\right|_{cd} \overset{\text{def}}{=} \frac{y(t + h/4) - y(t - h/4)}{h/2} \tag{5.10}$$

$$\simeq y'(t) + \frac{h^2}{96}\frac{d^3 y(t)}{dt^3} + \cdots.$$

A combination of the two, called the *extended difference algorithm*, eliminates both the quadratic and linear terms:

$$\left.\frac{dy(t)}{dt}\right|_{ed} \overset{\text{def}}{=} \frac{4D_{cd}y(t, h/2) - D_{cd}y(t, h)}{3} \tag{5.11}$$

$$\simeq \frac{dy(t)}{dt} - \frac{h^4 y^{(5)}(t)}{4 \times 16 \times 120} + \cdots. \tag{5.12}$$

Equation (5.11) is the extended-difference algorithm and (5.12) is its error, with D_{cd} representing the central-difference algorithm. If $h = 0.4$ and $y^{(5)} \simeq 1$, then there will be only one place of round-off error and the truncation error will be approximately machine precision ϵ_m; this really is the best you can hope for.

When working with these, and similar higher-order methods, it is important to remember that while they may work as designed for well-behaved functions, they may fail badly for functions containing noise, as do data from computations or measurements. If noise is evident, it may be better to first smooth the data, or fit them with some analytic function using the techniques of Chapter 6, and then differentiate.

5.2.1 Second Derivatives

Let's say that you have measured the position $y(t)$ *versus* time for a particle (Figure 5.1). Your **problem** now is to determine the force on the particle. Newton's second law tells us that force and acceleration are linearly related:

$$F = ma = m\frac{d^2y}{dt^2}. \tag{5.13}$$

So by determining the derivative d^2y/dt^2 from the $y(t)$ values, we determine the force.

The concerns we have expressed about errors in first derivatives are even more of a concern for second derivatives, where additional subtractions may lead to additional cancellations. Let's look again at the central-difference method:

$$\left.\frac{dy(t)}{dt}\right|_{cd} \simeq \frac{y(t+h/2) - y(t-h/2)}{h}. \tag{5.14}$$

This algorithm gives the derivative at t by moving forward and backward from t by $h/2$. As our second derivative algorithm, we'll take the central difference of the first derivative:

$$\left.\frac{d^2y(t)}{dt^2}\right|_{cd} \simeq \frac{y'(t+h/2) - y'(t-h/2)}{h}$$

$$\simeq \frac{[y(t+h) - y(t)] - [y(t) - y(t-h)]}{h^2} \tag{5.15}$$

$$= \frac{y(t+h) + y(t-h) - 2y(t)}{h^2}. \tag{5.16}$$

As we did for first derivatives, we determine the second derivative at t by evaluating the function in the region surrounding t. Although the form (5.16) is more compact and requires fewer steps than (5.15), it may increase subtractive cancellation by first storing the "large" number $y(t+h) + y(t-h)$, and then subtracting another large number $2y(t)$ from it. We ask you to explore this difference as an exercise.

5.2.1.1 Assessment
Write a program to calculate the second derivative of $\cos t$ using the central-difference algorithms (5.15) and (5.16). Test it over four cycles. Start with $h \simeq \pi/10$ and keep reducing h until you reach machine precision. Is there any noticeable differences between (5.15) and (5.16)?

The approximation errors in numerical differentiation decrease with decreasing step size h. In turn, round-off errors increase with decreasing step size as you have to take more steps and do more calculations. Remember from our discussion in Chapter 3 that the best approximation occurs for an h that minimizes the sum of application and round-off errors $\epsilon_{app} + \epsilon_{ro}$, and that occurs when $\epsilon_{ro} \simeq \epsilon_{app}$.

We have already estimated the approximation error in numerical differentiation rules by using the Taylor series expansion of $y(x + h)$. The approximation error with the forward-difference algorithm (5.4) is $\mathcal{O}(h)$, while that with the central-difference algorithm (5.8) is $\mathcal{O}(h^2)$:

$$\epsilon_{app}^{fd} \simeq \frac{y'' h}{2}, \qquad \epsilon_{app}^{cd} \simeq \frac{y''' h^2}{24}. \tag{5.17}$$

To obtain a rough estimate of the round-off error, we observe that differentiation essentially subtracts the value of a function at argument x from that of the same function at argument $x + h$, and then divides the difference by h: $y' \simeq [y(t + h) - y(t)]/h$. As h is made continually smaller, we eventually reach the round-off error limit where $y(t + h)$ and $y(t)$ differ by just machine precision ϵ_m:

$$\epsilon_{ro} \simeq \frac{\epsilon_m}{h}. \tag{5.18}$$

Consequently, round-off and approximation errors become equal when

$$\epsilon_{ro} \simeq \epsilon_{app}, \tag{5.19}$$

$$\frac{\epsilon_m}{h} \simeq \epsilon_{app}^{fd} = \frac{y^{(2)} h}{2}, \qquad \frac{\epsilon_m}{h} \simeq \epsilon_{app}^{cd} = \frac{y^{(3)} h^2}{24}, \tag{5.20}$$

$$\Rightarrow \quad h_{fd}^2 = \frac{2\epsilon_m}{y^{(2)}}, \qquad \Rightarrow \quad h_{cd}^3 = \frac{24\epsilon_m}{y^{(3)}}. \tag{5.21}$$

We take $y' \simeq y^{(2)} \simeq y^{(3)}$ (which may be crude in general, although not bad for e^t or $\cos t$) and assume double precision, $\epsilon_m \simeq 10^{-15}$:

$$h_{fd} \simeq 4 \times 10^{-8}, \qquad h_{cd} \simeq 3 \times 10^{-5}, \tag{5.22}$$

$$\Rightarrow \quad \epsilon_{fd} \simeq \frac{\epsilon_m}{h_{fd}} \simeq 3 \times 10^{-8}, \quad \Rightarrow \quad \epsilon_{cd} \simeq \frac{\epsilon_m}{h_{cd}} \simeq 3 \times 10^{-11}. \tag{5.23}$$

This may seem contradictory because the better algorithm leads to a larger h value. It is not. The ability to use a larger h means that the error in the central-difference method is about 1000 times smaller than the error in the forward-difference method.

The programming for numerical differentiation is simple:

```
FD = ( y(t+h) - y(t) ) /h;                  // Forward diff
CD = ( y(t+h/2) - y(t-h/2) ) /h;            // Central diff
ED = (8*(y(t+h/4)-y(t-h/4)) - (y(t+h/2)-y(t-h/2)))/3/h; //extra
```

1) Use forward-, central-, and extrapolated-difference algorithms to differentiate the functions $\cos t$ and e^t at $t = 0.1, 1.0$, and 100.
   a) Print out the derivative and its relative error $\mathcal{E}$ as functions of h. Reduce the step size h until the error equals machine precision $h \simeq \epsilon_m$.
   b) Plot $\log_{10} |\mathcal{E}|$ *versus* $\log_{10} h$ and check whether the number of decimal places obtained agrees with the estimates in the text.

c) See if you can identify regions where algorithmic (series truncation) error dominates at large h and round-off error at small h in your plot. Do the slopes agree with our model's predictions?

5.3 Integration Algorithms

Problem *Integrate a Spectrum* An experiment measured $dN(t)/dt$, the number of particles entering a counter per unit time. Your **problem** is to integrate this spectrum to obtain the number of particles that entered the counter in the first second:

$$N(1) = \int_0^1 \frac{dN(t)}{dt}\, dt. \tag{5.24}$$

5.3.1 Box Counting

The integration of a function may require some cleverness to do analytically, but is relatively straightforward on a computer. An ancient way to perform numerical integration is to take a piece of graph paper and count the number of boxes or *quadrilaterals* lying below a curve of the integrand. For this reason, numerical integration is also called *numerical quadrature*, even when it becomes more sophisticated than simple box counting.

The Riemann definition of an integral is the limit of the sum over boxes as the width h of the box approaches zero (Figure 5.2):

$$\int_a^b f(x)\, dx = \lim_{h \to 0} \left[h \sum_{i=1}^{(b-a)/h} f(x_i) \right]. \tag{5.25}$$

The numerical integral of a function $f(x)$ is approximated as the equivalent of a finite sum over boxes of height $f(x)$ and width w_i:

$$\int_a^b f(x)\, dx \simeq \sum_{i=1}^{N} f(x_i) w_i. \tag{5.26}$$

This is similar to the Riemann definition (5.25), except that there is no limit to an infinitesimal box size. Equation (5.26) is the standard form for all integration algorithms; the function $f(x)$ is evaluated at N points in the interval $[a, b]$, and the function values $f_i \equiv f(x_i)$ are summed with each term in the sum weighted by w_i. While, in general, the sum in (5.26) gives the exact integral only when $N \to \infty$, it may be exact for finite N if the

Figure 5.2 The integral $\int_a^b f(x)\, dx$ is the area under the graph of $f(x)$ from a to b. Here we break up the area into four regions of equal widths h and five integration points.

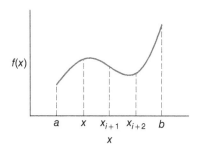

integrand is a polynomial. The different integration algorithms (also called Newton-Coates formulas) amount to different ways of choosing the points x_i and weights w_i. Generally, the precision increases as N gets larger, at least until round-off error becomes significant. Since the "best" integration rule depends on the specific behavior of $f(x)$, there is no universally best rule. In fact, some of the automated integration schemes found in subroutine libraries and computational environments switch from one method to another, as well as change the methods for different intervals, until they find ones that work well for each interval.

In general, you should not attempt a numerical integration of an integrand that contains a singularity without first somehow removing the singularity. You may be able to do this very simply by breaking the interval down into several subintervals so the singularity is at an endpoint where an integration point is not placed, or by a change of variable; for example:

$$\int_{-1}^{1} |x| f(x)\, dx = \int_{-1}^{0} f(-x)\, dx + \int_{0}^{1} f(x)\, dx, \tag{5.27}$$

$$\int_{0}^{1} x^{1/3}\, dx = \int_{0}^{1} 3y^3\, dy, \qquad (y \overset{\text{def}}{=} x^{1/3}), \tag{5.28}$$

$$\int_{0}^{1} \frac{f(x)\, dx}{\sqrt{1-x^2}} = 2 \int_{0}^{1} \frac{f(1-y^2)\, dy}{\sqrt{2-y^2}}, \qquad (y^2 \overset{\text{def}}{=} 1 - x). \tag{5.29}$$

Likewise, if your integrand has a very slow variation in some region, you can speed up the integration by changing to a variable that compresses that region and places few points there, or divides up the interval and performs several integrations. Conversely, if your integrand has a very rapid variation in some region, you may want to change to variables that expand that region to ensure that no oscillations are missed.

5.3.2 Trapezoid Rule

The trapezoid and Simpson's integration rules both use evenly spaced values of x (Figure 5.3). They use N points x_i, $i = 1, N$, evenly spaced a distance h apart throughout the integration region $[a, b]$, and *include the endpoints* in the integration region. This means that there are $(N - 1)$ intervals, each of length h:

$$h = \frac{b-a}{N-1}, \qquad x_i = a + (i-1)h, \qquad i = 1, N, \tag{5.30}$$

where we start our counting at $i = 1$. The trapezoid rule takes each integration interval i, and constructs a trapezoid of width h in it (Figure 5.3). This approximates $f(x)$ by a straight line in each interval i, and uses the average height $(f_i + f_{i+1})/2$ as the value for f. The area of each such trapezoid is

$$\int_{x_i}^{x_i+h} f(x)\, dx \simeq \frac{h(f_i + f_{i+1})}{2} = \tfrac{1}{2} h f_i + \tfrac{1}{2} h f_{i+1}. \tag{5.31}$$

In terms of our standard integration formula (5.26), the "rule" in (5.31) is for $N = 2$ points with weights $w_i \equiv \tfrac{1}{2}$ (Table 5.1).

In order to apply the trapezoid rule to the entire region $[a, b]$, we add the contributions from each subinterval:

$$\int_{a}^{b} f(x)\, dx \simeq \tfrac{h}{2} f_1 + h f_2 + h f_3 + \cdots + h f_{N-1} + \tfrac{h}{2} f_N. \tag{5.32}$$

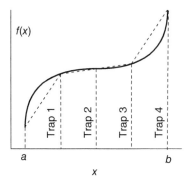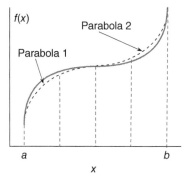

Figure 5.3 Different shapes used to approximate the areas under the curve. *Left*: Straight-line sections used for the trapezoid rule. *Right*: Two parabolas used in Simpson's rule.

Table 5.1 Elementary weights for uniform-step integration rules.

Name	Degree	Elementary weights
Trapezoid	1	$(1,1)\frac{h}{2}$
Simpson's	2	$(1,4,1)\frac{h}{3}$
$\frac{3}{8}$	3	$(1,3,3,1)\frac{3}{8}h$
Mile	4	$(14,64,24,64,14)\frac{h}{45}$

You will notice that because the internal points are counted twice (at the end of one interval and at the beginning of the next), they have weights of $h/2 + h/2 = h$, whereas the endpoints are counted just once, and on that account have weights of only $h/2$. In terms of our standard integration rule (5.61), we have

$$w_i = \left\{ \frac{h}{2}, h, \dots, h, \frac{h}{2} \right\} \qquad \text{(Trapezoid rule)}. \tag{5.33}$$

In Listing 5.1, we provide a simple implementation of the trapezoid rule.

5.3.3 Simpson's Rule

Simpson's rule approximates the integrand $f(x)$ by a parabola within each equally spaced interval (Figure 5.3 right):

$$f(x) \simeq \alpha x^2 + \beta x + \gamma. \tag{5.34}$$

The area under the parabola for each interval is

$$\int_{x_i}^{x_i+h} (\alpha x^2 + \beta x + \gamma)\, dx = \left. \frac{\alpha x^3}{3} + \frac{\beta x^2}{2} + \gamma x \right|_{x_i}^{x_i+h}. \tag{5.35}$$

In order to relate the parameters α, β, and γ to the function, we consider an interval from -1 to $+1$, in which case

$$\int_{-1}^{1} (\alpha x^2 + \beta x + \gamma)\, dx = \frac{2\alpha}{3} + 2\gamma. \tag{5.36}$$

But we notice that

$$f(-1) = \alpha - \beta + \gamma, \quad f(0) = \gamma, \quad f(1) = \alpha + \beta + \gamma, \tag{5.37}$$

$$\Rightarrow \quad \alpha = \frac{f(1) + f(-1)}{2} - f(0), \quad \beta = \frac{f(1) - f(-1)}{2}, \quad \gamma = f(0). \tag{5.38}$$

In this way, we can express the integral as the weighted sum over the values of the function at three points:

$$\int_{-1}^{1} (\alpha x^2 + \beta x + \gamma)\, dx = \frac{f(-1)}{3} + \frac{4f(0)}{3} + \frac{f(1)}{3}. \tag{5.39}$$

Seeing that three values of the function are needed, we apply this result to our problem by evaluating the integral over two adjacent intervals, in which case we evaluate the function at the two endpoints and in the middle (Table 5.1):

$$\int_{x_i-h}^{x_i+h} f(x)\, dx = \int_{x_i}^{x_i+h} f(x)\, dx + \int_{x_i-h}^{x_i} f(x)\, dx$$

$$\simeq \frac{h}{3} f_{i-1} + \frac{4h}{3} f_i + \frac{h}{3} f_{i+1}. \tag{5.40}$$

Take note: Simpson's rule requires the elementary integration to be over *pairs* of intervals, which, in turn, requires that the *total number of intervals be even or that the number of points N be odd*. In order to apply Simpson's rule to the entire interval, we add up the contributions from each pair of subintervals, counting all but the first and last endpoints twice:

$$\int_{a}^{b} f(x)dx \simeq \frac{h}{3} f_1 + \frac{4h}{3} f_2 + \frac{2h}{3} f_3 + \frac{4h}{3} f_4 + \cdots + \frac{4h}{3} f_{N-1} + \frac{h}{3} f_N. \tag{5.41}$$

In terms of our standard integration rule (5.26), we have

$$w_i = \left\{ \frac{h}{3}, \frac{4h}{3}, \frac{2h}{3}, \frac{4h}{3}, \ldots, \frac{4h}{3}, \frac{h}{3} \right\} \quad \text{(Simpson's rule)}. \tag{5.42}$$

The sum of these weights provides a useful check on your integration:

$$\sum_{i=1}^{N} w_i = (N-1)h. \tag{5.43}$$

Remember, the number of points N must be odd for Simpson's rule.

5.3.4 Simple Integration Error Estimates

In general, you should choose an integration rule that gives an accurate answer using the least number of integration points. We obtain a crude estimate of the *approximation* or *algorithmic error* $\mathcal{E}$ for the equal-spacing rules and their relative error ϵ, by expanding $f(x)$ in a Taylor series around the midpoint of the integration interval. We then multiply that error by the number of intervals N to estimate the error for the entire region $[a, b]$. For the trapezoid and Simpson's rules this yields

$$\mathcal{E}_t = O\left(\tfrac{[b-a]^3}{N^2}\right) f^{(2)}, \quad \mathcal{E}_s = O\left(\tfrac{[b-a]^5}{N^4}\right) f^{(4)}, \quad \epsilon_{t,s} = \tfrac{\mathcal{E}_{t,s}}{f}, \tag{5.44}$$

where ϵ is a measure of the relative error. We see that the third-derivative term in Simpson's rule cancels (much like the central-difference method does in differentiation). Equations (5.44) are illuminating in showing how increasing the sophistication of an integration rule leads to an error that decreases with a higher inverse power of N, yet is also proportional to higher derivatives of f. Consequently, for small intervals and functions $f(x)$ with well-behaved derivatives, Simpson's rule should converge more rapidly and be more accurate than the trapezoid rule.

To model the round-off error in integration, we assume that after N steps the *relative* round-off error is random and of the form

$$\epsilon_{ro} \simeq \sqrt{N}\epsilon_m, \tag{5.45}$$

where ϵ_m is the machine precision, $\epsilon \sim 10^{-7}$ for single precision, and $\epsilon \sim 10^{-15}$ for double precision. Inasmuch as most scientific computations are performed with doubles, we will assume double precision. We want to determine an N that minimizes the total error, that is, the sum of the approximation and round-off errors:

$$\epsilon_{tot} \simeq \epsilon_{ro} + \epsilon_{app}. \tag{5.46}$$

This occurs, approximately, when the two errors are of equal magnitude, which we approximate even further by assuming that the two errors are equal:

$$\epsilon_{ro} = \epsilon_{app} = \frac{\mathcal{E}_{trap,simp}}{f}. \tag{5.47}$$

To continue the search for optimum N for a general function f, we set the scale of function size and the lengths by assuming

$$\frac{f^{(n)}}{f} \simeq 1, \quad b-a=1 \quad \Rightarrow \quad h = \frac{1}{N}. \tag{5.48}$$

The estimate (5.47), when applied to the **trapezoid rule**, yields

$$\sqrt{N}\epsilon_m \simeq \frac{f^{(2)}(b-a)^3}{fN^2} = \frac{1}{N^2}, \tag{5.49}$$

$$\Rightarrow \quad N \simeq \frac{1}{(\epsilon_m)^{2/5}} = \left(\frac{1}{10^{-15}}\right)^{2/5} = 10^6, \tag{5.50}$$

$$\Rightarrow \epsilon_{ro} \simeq \sqrt{N}\epsilon_m = 10^{-12}. \tag{5.51}$$

The estimate (5.47), when applied to **Simpson's rule**, yields

$$\sqrt{N}\epsilon_m = \frac{f^{(4)}(b-a)^5}{fN^4} = \frac{1}{N^4}, \tag{5.52}$$

$$\Rightarrow N = \frac{1}{(\epsilon_m)^{2/9}} = \left(\frac{1}{10^{-15}}\right)^{2/9} = 2154, \tag{5.53}$$

$$\Rightarrow \epsilon_{ro} \simeq \sqrt{N}\epsilon_m = 5\times10^{-14}. \tag{5.54}$$

These results are illuminating in that they show how:

- Simpson's rule requires fewer points and has less error than the trapezoid rule.
- It is possible to obtain an error close to machine precision with Simpson's rule (and with other higher-order integration algorithms).
- Obtaining the *best* numerical approximation to an integral is not achieved by letting $N \rightarrow \infty$, but with a relatively small $N \leq 1000$. Larger N only gives you more round-off errors.

5.3.5 Higher-Order Algorithms

As in numerical differentiation, we can use the known functional dependence of the error on interval size h to reduce the integration error. For simple algorithms like the trapezoid and Simpson's rules, we have the analytic estimates (5.47), while for others you may have to experiment to determine an approximate h dependence. To illustrate, if $A(h)$ and $A(h/2)$ are the values of the integral determined for intervals h and $h/2$, respectively, and if we assume that the numerical evaluation of the integral has an error whose expansion has a leading error term proportional to h^2,

$$A(h) \simeq \int_a^b f(x)\,dx + \alpha h^2 + \beta h^4 + \cdots, \tag{5.55}$$

then

$$A\left(\tfrac{h}{2}\right) \simeq \int_a^b f(x)\,dx + \tfrac{\alpha h^2}{4} + \tfrac{\beta h^4}{16} + \cdots. \tag{5.56}$$

Consequently, we can make the h^2 term in the error vanish by computing the integral as the combination

$$A \simeq \tfrac{4}{3} A\left(\tfrac{h}{2}\right) - \tfrac{1}{3} A(h) \simeq \int_a^b f(x)\,dx - \tfrac{\beta h^4}{4} + \cdots. \tag{5.57}$$

Clearly this particular trick (Romberg's extrapolation) works only if the h^2 term dominates the error.

In Table 5.1, we have given the weights for several equal-interval rules. We see that the Simpson's rule uses two intervals, the three-eighths rule uses three, and the Milne rule four.[1] Do remember, these are single-interval rules, and when strung together to obtain a rule for the entire integration range, the points that end one interval and begin the next are counted twice. You can easily determine the number of elementary intervals integrated over, and check whether you and we have written the weights right, by summing just the weights for any rule. The sum gives the integral of $f(x) = 1$ and must equal h times the number of intervals (which in turn equals $b - a$):

$$\sum_{i=1}^{N} w_i = h \times N_{\text{intervals}} = b - a. \tag{5.58}$$

---

1 There is, not coincidentally, a Mile Computer Center at Oregon State University, although there no longer is a central computer there

5.4 Gaussian Quadrature

It is often useful to rewrite the basic integration formula (5.26) with a weighting function $W(x)$ separate from the integrand:

$$\int_a^b f(x)\, dx \equiv \int_a^b W(x)g(x)\, dx \simeq \sum_{i=1}^N w_i g(x_i). \tag{5.59}$$

In the Gaussian quadrature approach to integration, the N points and weights in (5.59) are chosen to make the integration exact if $g(x)$ were a $(2N-1)$-degree polynomial. To obtain this incredible optimization, the points x_i end up having a specific distribution over $[a, b]$. In general, if $g(x)$ is smooth, or can be made smooth by factoring out some $W(x)$ (see Table 5.2), Gaussian quadrature will produce higher accuracy than the trapezoid and Simpson's rules for the same number of points. Sometimes the integrand may not be smooth because it has different behaviors in different regions, in which case you could integrate each region separately, and then add the results. In fact, some "smart" integration subroutines decide for themselves how many intervals to use and which rule to use in each.

All the rules indicated in Table 5.2 are a form of Gaussian quadrature following the general form (5.59). We can see that in one case the weighting function is an exponential, in another a Gaussian, and in several cases, there is an integrable singularity. In contrast to the equally spaced rules, there is never an integration point at the extremes of the intervals (a or b), with differing N values leading to completely differing sets of points and weights.

The derivation of the Gaussian points will be outlined below, but we point out here that for ordinary Gaussian (Gauss–Legendre) integration, the points y_i turn out to be the N zeros of the Legendre polynomials, with the weights related to the derivatives,

$$P_N(y_i) = 0, \qquad w_i = \frac{2}{(1 - y_i^2)[P_N'(y_i)]^2}. \tag{5.60}$$

Programs to generate these points and weights are standard in mathematical function libraries, are found in tables Abramowitz and Stegun [1972], or can be computed, as we do in our `gauss.py` program, which also scales the points to span a specified region. As a check that your program's points are correct, you may want to compare them to this four-point set:

$\pm y_i$	w_i
0.33998 10435 84856	0.65214 51548 62546
0.86113 63115 94053	0.34785 48451 37454

Table 5.2 Types of Gaussian integration rules.

Integral	Name	Integral	Name
$\int_{-1}^1 f(y)\, dy$	Gauss	$\int_{-1}^1 \frac{F(y)}{\sqrt{1-y^2}}\, dy$	Gauss–Chebyshev
$\int_{-\infty}^\infty e^{-y^2} F(y)\, dy$	Gauss–Hermite	$\int_0^\infty e^{-y} F(y)\, dy$	Gauss–Laguerre
$\int_0^\infty \frac{e^{-y}}{\sqrt{y}} F(y)\, dy$	Associated Gauss–Laguerre		

5.4.1 Mapping Gaussian Points

Our standard integration rule (5.26) for the general interval $[a, b]$ is

$$\int_a^b f(x)\, dx \simeq \sum_{i=1}^N f(x_i) w_i. \tag{5.61}$$

With Gaussian points and weights, the y interval $-1 < y_i \leq 1$ must be *mapped* onto the x interval $a \leq x \leq b$. Here are some mappings we have found useful in our work. In all cases, (y_i, w_i') are the elementary Gaussian points and weights for the interval $[-1, 1]$, and we want to scale the x with various ranges.

1) $[-1, 1] \rightarrow [a, b]$ uniformly, $(a+b)/2 = $ **midpoint**:

$$x_i = \frac{b+a}{2} + \frac{b-a}{2} y_i, \quad w_i = \frac{b-a}{2} w_i', \tag{5.62}$$

$$\Rightarrow \int_a^b f(x)\, dx = \frac{b-a}{2} \int_{-1}^1 f[x(y)]\, dy. \tag{5.63}$$

2) $[0 \rightarrow \infty]$, $a = $ **midpoint**:

$$x_i = a\frac{1+y_i}{1-y_i}, \quad w_i = \frac{2a}{(1-y_i)^2} w_i'. \tag{5.64}$$

3) $[-\infty \rightarrow \infty]$, **scale set by** a:

$$x_i = a\frac{y_i}{1-y_i^2}, \quad w_i = \frac{a(1+y_i^2)}{(1-y_i^2)^2} w_i'. \tag{5.65}$$

4) $[a \rightarrow \infty]$, $a + 2b = $ **midpoint**:

$$x_i = \frac{a+2b+ay_i}{1-y_i}, \quad w_i = \frac{2(b+a)}{(1-y_i)^2} w_i'. \tag{5.66}$$

5) $[0 \rightarrow b]$, $ab/(b+a) = $ **midpoint**:

$$x_i = \frac{ba(1+y_i)}{b+a-(b-a)y_i}, \quad w_i = \frac{2ab^2}{(b+a-(b-a)y_i)^2} w_i'. \tag{5.67}$$

As you can see, even if your integration range extends out to infinity, there will be points at large, but not infinite x values. As you keep increasing the number of integration points N, the last x_i gets larger, but always remains finite.

5.4.2 Gaussian Quadrature Derivation ⊙

We want to perform a numerical integration with N integration points:

$$\int_{-1}^{+1} f(x)\, dx = \sum_{i=1}^N w_i f(x_i), \tag{5.68}$$

where $f(x)$ is a polynomial of degree $(2N - 1)$ or less. The unique property of Gaussian quadrature is that (5.68) will be exact, as long as we ignore the effect of round-off error. Determining the x_i's and w_i's require some knowledge of special functions and some cleverness [Hildebrand, 1956]. The knowledge needed is the two properties of Legendre polynomials $P_N(x)$ of order N:

1) $P_N(x)$ is orthogonal to every polynomial of order less than N.
2) $P_N(x)$ has N real roots in the interval $-1 \leq x \leq 1$.

We define a new polynomial of degree equal to or less than N obtained by dividing the integrand $f(x)$ by the Legendre polynomial $P_N(x)$:

$$q(x) \overset{\text{def}}{=} \frac{f(x)}{P_N(x)}, \tag{5.69}$$

$$\Rightarrow \quad f(x) = q(x)P_N(x) + r(x). \tag{5.70}$$

Here $r(x)$ is an (unknown) polynomial of degree N or less, which we will not need to determine. If we now substitute (5.70) into (5.68), and use the fact that P_N is orthogonal to every polynomial of degree less than or equal to N, only the second, $r(x)$, term remains:

$$\int_{-1}^{+1} f(x)\, dx = \int_{-1}^{+1} q(x)P_N(x)\, dx + \int_{-1}^{+1} r(x)\, dx = \int_{-1}^{+1} r(x)\, dx. \tag{5.71}$$

Yet because $r(x)$ is a polynomial of degree N or less, we can use a standard N point rule to evaluate the integral exactly.

Now that we know it is possible to integrate a $(2N - 1)$ or less degree polynomial with just N points, we display some cleverness to determine just what those points will be. We substitute (5.70) into (5.68) and note that

$$\int_{-1}^{+1} f(x)\, dx = \sum_{i=1}^{N} w_i\, q(x_i) P_N(x_i) + \sum_{i=1}^{N} w_i\, r(x_i) = \sum_{i=1}^{N} w_i\, r(x_i). \tag{5.72}$$

The cleverness is realizing that if we choose the N integration points to be the zeros (roots) of the Legendre polynomial $P_N(x)$, then the first term on the RHS of (5.72) will vanish because $P_N(x_i) = 0$ for each x_i:

$$\int_{-1}^{+1} f(x)\, dx = \sum_{i=1}^{N} w_i\, r(x_i). \tag{5.73}$$

This is our proof that the N integration points over the interval $(-1, 1)$ are the N zeros of the Legendre polynomial $P_N(x)$. As indicated in (5.60), the weights are related to the derivative of the Legendre polynomials evaluated at the roots of the polynomial. We leave the derivation of the weights to Hildebrand [1956].

5.5 Monte Carlo Integrations

Imagine   yourself as a farmer walking to your furthermost field to add some algae-eating fish to a pond having an algae explosion. You get there only to read the instruction label on the fish container and discover that you need to know the area of the pond in order to determine the correct number of fish to add. Your **problem** is to measure the area of this irregularly shaped pond with just the materials at hand [Gould *et al.*, 2006].

It is hard to believe that Monte Carlo techniques can be used to evaluate integrals. After all, we do not want to gamble on the values! While it is true that other methods are

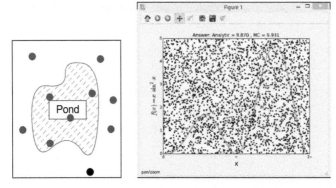

Figure 5.4 *Left*: Throwing stones into a pond as a technique for measuring its area. The ratio of "hits" to total number of stones thrown equals the ratio of the area of the pond to that of the box. *Right*: The evaluation of an integral via a Monte Carlo (stone-throwing) technique based on the ratio of areas.

preferable for single and double integrals, it turns out that Monte Carlo techniques are best when the dimensionality of integrations gets large! For our pond problem, we will use a *sampling* technique (Figure 5.4):

1) Walk off a box that completely encloses the pond, and remove any pebbles lying on the ground within the box.
2) Measure the lengths of the sides in natural units like your *feet*. This lets you calculate the area of the enclosing box A_{box}.
3) Grab a bunch of pebbles, count their number, and then throw them up in the air in random directions.
4) Count the number of splashes in the pond N_{pond} and the number of pebbles lying on the ground within your box N_{box}.
5) Assuming that you threw the pebbles uniformly and randomly, the number of pebbles falling into the pond should be proportional to the area of the pond A_{pond}. You determine that area from the simple ratio

$$\frac{N_{pond}}{N_{pond} + N_{box}} = \frac{A_{pond}}{A_{box}} \quad \Rightarrow \quad A_{pond} = \frac{N_{pond}}{N_{pond} + N_{box}} A_{box}. \tag{5.74}$$

5.5.1 Stone Throwing Implementation

Use sampling (Figure 5.4) to perform a 2D integration and thereby determine π:

1) Imagine a circular pond enclosed in a square of side $2\,(r = 1)$.
2) We know the analytic answer that the area of a circle $\oint dA = \pi$.
3) Generate a sequence of random numbers $-1 \le r_i \le +1$.
4) For $i = 1$ to N, pick $(x_i, y_i) = (r_{2i-1}, r_{2i})$.
5) If $x_i^2 + y_i^2 < 1$, let $N_{pond} = N_{pond} + 1$; otherwise let $N_{box} = N_{box} + 1$.
6) Use (5.74) to calculate the area, and in this way π.
7) Increase N until you get π to three significant figures (we don't ask much – that's only slide-rule accuracy).

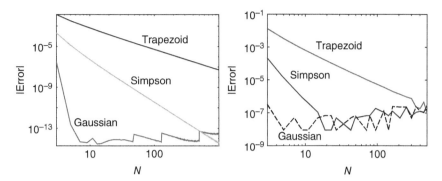

Figure 5.5 Log–log plots of the error in the integration of exponential decay using the trapezoid rule, Simpson's rule, and Gaussian quadrature *versus* the number of integration points N. Approximately, 15 decimal places of precision are attainable with double precision (*left*), and 7 places with single precision (*right*). The algorithms are seen to stop converging when round-off error (the fluctuating and increasing part near the bottom) starts to dominate.

Listing 5.3 provides our code vonNeuman.py that performs a Monte Carlo integration via stone throwing.

5.5.2 Integration Error Investigation

1) Write a double-precision program to integrate e^{-t} from 0 to 1 numerically using the trapezoid rule, the Simpson's rule, Gaussian quadrature, and Monte Carlo (MC) integration. In this case, there is an analytic answer with which to compare:

$$\frac{dN(t)}{dt} = e^{-t} \quad \Rightarrow \quad N(1) = \int_0^1 e^{-t}\, dt = 1 - e^{-1}. \tag{5.75}$$

2) Compute the relative error $\epsilon = |(\text{numerical-exact})/\text{exact}|$ in each case. Present your data in the tabular form

N	ϵ_T	ϵ_S	ϵ_G
2	$\cdots$	$\cdots$	$\cdots$
$\vdots$			
10	$\cdots$	$\cdots$	$\cdots$

with spaces or tabs separating the fields. Try N values of 2, 10, 20, 40, 80, 160, …. (*Hint:* Even numbers may not be the assumption of every rule.)

3) Make a $\log_{10} - \log_{10}$ plot of the relative error ϵ *versus* N, as in Figure 5.5. You should observe

$$\epsilon \simeq CN^\alpha \quad \Rightarrow \quad \log \epsilon = \alpha \log N + \text{constant}. \tag{5.76}$$

If your graph is similar to straight line, this means that error obeys a power law. The ordinate on your plot will be the negative of the number of decimal places of precision in your calculation.

4) Use your plot or table to estimate the power-law dependence of the error ϵ on the number of points N, and to determine the number of decimal places of precision in your calculation. Do this for both the trapezoid and Simpson's rules, and in both the algorithmic and round-off error regimes. (Note that it may be hard to make N large enough to reach the round-off error regime for the trapezoid rule because the approximation error is so large.)

In Listing 5.2, we give a sample program that performs an integration with Gaussian points. The method gauss generates the points and weights and may be useful in other applications as well.

5.6 Mean Value and N–D Integration

The standard Monte Carlo technique for integration is based on the *mean value theorem* (presumably familiar from elementary calculus):

$$I = \int_a^b dx\, f(x) = (b-a)\langle f\rangle. \tag{5.77}$$

The theorem states the obvious if you think of integrals as areas: The value of the integral of some function $f(x)$ between a and b equals the length of the interval $(b-a)$ times the mean value of the function over that interval $\langle f\rangle$ (Figure 5.6). The Monte Carlo integration algorithm simply uses random points to evaluate the mean in (5.77). With a sequence $a \le x_i \le b$ of N uniform random numbers, we determine the *sample mean* by *sampling* the function $f(x)$ at N points:

$$\langle f\rangle \simeq \frac{1}{N}\sum_{i=1}^N f(x_i). \tag{5.78}$$

This gives us the very simple integration rule:

$$\int_a^b dx\, f(x) \simeq (b-a)\frac{1}{N}\sum_{i=1}^N f(x_i) = (b-a)\langle f\rangle. \tag{5.79}$$

Equation (5.79) can be thought of as our standard algorithm for integration (5.26) with the points x_i chosen randomly, and with uniform weights $w_i = (b-a)/N$. Seeing that no attempt has been made to obtain an optimal answer for a given value of N, this does not seem like it would be an efficient means to evaluate integrals, but you must admit it is simple. If we let the number of samples of $f(x)$ approach infinity, $N \to \infty$, or if we keep the number of samples finite and take the average of infinitely many runs, the laws of statistics

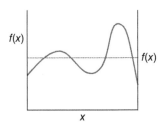

Figure 5.6 The area under the curve $f(x)$ is the same as that under the horizontal line whose height $y = \langle f\rangle$.

assure us that (5.79) will approach the correct answer, at least if there were no round-off errors.

For readers who are familiar with statistics, we remind you that the uncertainty in the value obtained for the integral I after N samples of $f(x)$ is measured by the standard deviation σ_I. If σ_f is the standard deviation of the integrand f in the sampling, then for normal distributions we have

$$\sigma_I \simeq \frac{1}{\sqrt{N}} \, \sigma_f. \tag{5.80}$$

So for large N, the error in the value obtained for the integral should decrease as $1/\sqrt{N}$.

Let's say that we want to calculate some properties of a small atom such as magnesium with 12 electrons. To do that we need to integrate the atomic wave functions over the three coordinates for each of 12 electrons. This amounts to a $3 \times 12 = 36$-D integral. If we use 64 points for each integration, this requires about $64^{36} \simeq 10^{65}$ evaluations of the integrand. If the computer were fast and could evaluate the integrand a million times per second, this would take about 10^{59} seconds, which is significantly longer than the age of the universe ($\sim 10^{17}$ seconds).

5.6.1 10-D MC Error Investigation

When we perform a multidimensional integration, the relative error in the Monte Carlo technique, being statistical, decreases as $1/\sqrt{N}$. This is valid even if the N points are distributed over D dimensions. In contrast, when we use these same N points to perform a D-dimensional integration as D separate 1D integrals using a rule such as Simpson's, we use N/D points for each integration. For fixed N, this means that the number of points used for each integration decreases as the number of dimensions D increases, and so the error in each integration *increases* with D. Furthermore, the total error will be approximately N times the error in each integral. If you put these trends together and do the analysis for a particular integration rule, you will find that for a dimension $D \simeq 3$–4, the error in Monte Carlo integration is approximately equal to that of conventional schemes. For larger values of D, the Monte Carlo method is more accurate!

5.6.2 Implementation: 10-D Monte Carlo Integration

Your **problem** is to find a way to perform multidimensional integrations so that you live long enough to savor the results. Specifically, evaluate the 10D integral

$$I = \int_0^1 dx_1 \int_0^1 dx_2 \cdots \int_0^1 dx_{10} \left(x_1 + x_2 + \cdots + x_{10} \right)^2. \tag{5.81}$$

Check your numerical answer against the analytic one, $\frac{155}{6}$.

It is easy to generalize mean value integration to many dimensions by picking random points in a multidimensional space. For example, in 2D:

$$\int_a^b dx \int_c^d dy \, f(x,y) \simeq (b-a)(d-c) \frac{1}{N} \sum_i^N f(\mathbf{x}_i) = (b-a)(d-c) \langle f \rangle. \tag{5.82}$$

Use a built-in random-number generator to perform the 10D Monte Carlo integration in (5.81).

1) Conduct 16 trials and take the average as your answer.
2) Try sample sizes of $N = 2, 4, 8, \ldots, 8192$.
3) Plot the relative error *versus* $1/\sqrt{N}$, and see if a linear behavior occurs.
4) What is your estimate for the accuracy of the integration?
5) Show that for a dimension $D \simeq 3$–4, the error in multidimensional Monte Carlo integration is approximately equal to that of conventional schemes, and that for larger values of D, the Monte Carlo method is more accurate.

5.7 MC Variance Reduction

It is common in many physical applications to integrate a function with an approximately Gaussian dependence on x. The rapid falloff of the integrand means that our Monte Carlo integration technique would require an incredibly large number of points to obtain even modest accuracy. Your **problem** is to make Monte Carlo integration more efficient for rapidly varying integrands.

If the function being integrated never differs much from its average value, then the standard Monte Carlo mean value method (5.79) should work well with a large, but manageable, number of points. Yet for a function with a large *variance* (i.e., one that is not "flat"), many of the evaluations of the function may occur for x values at which the function is very small, and thus makes a very small contribution to the final value of the integral; so it's basically a waste of time to expend much effort in regions where the integrand is very small. The efficiency of the integration can be improved by mapping the function f into a different function g that has a smaller variance over the interval. We indicate two methods here and refer you to Press *et al.* [2007] and Koonin [1986] for more details.

The first method is a *variance reduction* in which we devise a flatter function over which to integrate. Suppose we construct a function $g(x)$ with the following properties on $[a, b]$:

$$|f(x) - g(x)| \le \epsilon, \qquad \int_a^b dx\, g(x) = J. \tag{5.83}$$

We now evaluate the integral of the difference $f(x) - g(x)$ and add the result to J:

$$\int_a^b dx\, f(x) = \int_a^b dx\, [f(x) - g(x)] + J. \tag{5.84}$$

If we are clever enough to find a simple $g(x)$ that makes the variance of $f(x) - g(x)$ less than that of $f(x)$, we can obtain even more accurate answers.

5.8 Importance Sampling and von Neumann Rejection

A second method for improving Monte Carlo integration is *importance sampling*, so called because it samples the integrand in the most important regions. It derives from the identity

$$I = \int_a^b dx\, f(x) = \int_a^b dx\, w(x) \frac{f(x)}{w(x)}. \tag{5.85}$$

Figure 5.7 The von Neumann rejection technique for generating random points with weight $W(x)$. A random point is accepted if it lies below the curve of $W(x)$ and rejected if it lies above. This generates a random distribution weighted by whatever $W(x)$ function is plotted.

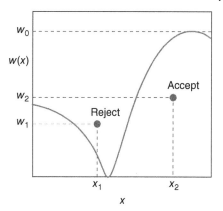

If we use a *probability distribution* for our random numbers that incorporates $w(x)$, the integral can be approximated as

$$I = \left\langle \frac{f}{w} \right\rangle \simeq \frac{1}{N} \sum_{i=1}^{N} \frac{f(x_i)}{w(x_i)}. \tag{5.86}$$

The improvement arising from (5.86) is that with a judicious choice of weighting function $w(x) \propto f(x)$, we can make $f(x)/w(x)$ more constant and thus easier to integrate accurately.

A simple and ingenious method for generating random points with a probability distribution $w(x)$ was deduced by von Neumann. This method is essentially the same as the rejection or sampling method used to guess the area of a pond, only now the pond has been replaced by the weighting function $w(x)$, and the arbitrary box around the lake by the arbitrary constant w_0. Imagine a graph of $w(x)$ *versus* x (Figure 5.7). Walk off your box by placing the line $w = w_0$ on the graph, with the only condition being $w_0 \geq w(x)$. We next "throw stones" at this graph and count only those splashes that fall into the $w(x)$ pond. That is, we generate uniform distributions in x and $y \equiv W$ with the maximum y value equal to the width of the box w_0:

$$(x_i, W_i) = (r_{2i-1}, w_0 r_{2i}). \tag{5.87}$$

We then reject all x_i that does not fall into the pond:

$$\text{if } W_i < w(x_i), \quad \text{accept,} \quad \text{if } W_i > w(x_i), \quad \text{reject.} \tag{5.88}$$

The x_i values so accepted will be weighted by $w(x)$ (Figure 5.7), with the largest number of acceptances occurring where $w(x)$ is large, in this case for midrange x. In Chapter 17, we apply a variation of the rejection technique known as the *Metropolis algorithm*.

5.9 Code Listings

Listing 5.1 TrapMethods.py Integrates a function $f(y)$ with the trapezoid rule. Note that the step size h depends upon the size of interval here and that the weights at the ends and middle of the intervals differ.

```
1  # TrapMethods.py: trapezoid integration , a<x<b, N pts , N-1 intervals

   from numpy import *
```

```
 5  def func(x):
        return 5*(sin(8*x))**2*exp(-x*x)-13*cos(3*x)

    def trapezoid(A,B,N):
 9      h = (B - A)/(N - 1)                          # step size
        sum = (func(A)+func(B))/2                    # (1st + last)/2
        for i in range(1, N-1):
            sum += func(A+i*h)
13      return h*sum
    A = 0.5
    B = 2.3
    N = 1200
17  print(trapezoid(A,B,N-1))
```

Listing 5.2   IntegGauss.py Integrates the function $f(x)$ via Gaussian quadrature. The points and weights are generated in the method `gauss`, which will be the same for other applications as well. Note that the level of desired precision is set by the parameter `eps`, which should be set by the user, as should the value for `job`, which controls the mapping of the points onto arbitrary intervals [they are generated in $(-1, 1)$].

```
 1  # IntegGauss.py: Gaussian quadrature generator of pts & wts

    from numpy import *
    from sys import version
 5
    max_in = 11                          # Numb intervals
    vmin = 0.; vmax = 1.                 # Int ranges
    ME = 2.7182818284590452354E0         # Euler's const
 9  w = zeros( (2001), float )
    x = zeros( (2001), float )

    def f(x):                            # The integrand
13      return (exp( - x) )
    def gauss(npts, job, a, b, x, w):
        m = i = j = t = t1 = pp = p1 = p2 = p3 = 0.
        eps = 3.E-14    # Accuracy: ******ADJUST THIS*******!
17      m = int((npts + 1)/2 )
        for i in range(1, m + 1):
            t = cos(math.pi*(float(i) - 0.25)/(float(npts) + 0.5) )
            t1 = 1
21          while( (abs(t - t1) ) >= eps):
                p1 = 1. ;   p2 = 0.
                for j in range(1, npts + 1):
                    p3 = p2;    p2 = p1
25                  p1 = ((2.*float(j)-1)*t*p2 - (float(j)-1.)*p3)/(float(j))
                pp = npts*(t*p1 - p2)/(t*t - 1.)
                t1 = t; t = t1   - p1/pp
            x[i - 1] = - t;   x[npts - i] = t
29          w[i - 1] = 2./( (1. - t*t)*pp*pp)
            w[npts - i] = w[i - 1]
        if (job == 0):
            for i in range(0, npts):
33              x[i] = x[i]*(b - a)/2. + (b + a)/2.
                w[i] = w[i]*(b - a)/2.
        if (job == 1):
            for i in range(0, npts):
37              xi  = x[i]
                x[i] = a*b*(1. + xi) / (b + a - (b - a)*xi)
                w[i] = w[i]*2.*a*b*b/( (b + a - (b-a)*xi)*(b + a - (b-a)*xi))
        if (job == 2):
41          for i in range(0, npts):
                xi = x[i]
                x[i] = (b*xi + b + a + a) / (1. - xi)
                w[i] = w[i]*2.*(a + b)/( (1. - xi)*(1. - xi) )
45  def gaussint (no, min, max):
        quadra = 0.
```

```
        gauss (no, 0, min, max, x, w)          # Returns pts & wts
        for n in  range(0, no):
49          quadra   += f(x[n]) * w[n]          # Calculate integral
        return (quadra)
    for i in range(3, max\_in + 1, 2):
        result = gaussint(i, vmin, vmax)
53      print (" i ", i, " err ", abs(result - 1 + 1/ME))
    print ("Enter and return any character to quit")
```

Listing 5.3   vonNeuman.py Performs a Monte Carlo integration via stone throwing.

```
#  vonNeuman: Monte-Carlo integration via stone throwing

import random
4 from visual.graph import *

N        = 100   # points to plot the function
graph    = display(width=500,height=500,title='vonNeumann Rejection Int')
8 xsinx   = curve(x=list(range(0,N)), color=color.yellow, radius=0.5)
pts      = label(pos=(-60, -60), text='points=', box=0)          # Labels
pts2     = label(pos=(-30, -60), box=0)
inside   = label(pos=(30,-60), text='accepted=', box=0)
12 inside2 = label(pos=(60,-60), box=0)
arealbl  = label(pos=(-65,60), text='area=', box=0)
arealbl2 = label(pos=(-35,60), box=0)
areanal  = label(pos=(30,60), text='analytical=', box=0)
16 zero    = label(pos=(-85,-48), text='0', box=0)
five     = label(pos=(-85,50), text='5', box=0)
twopi    = label(pos=(90,-48), text='2pi', box=0)

20 def fx (x):   return x*sin(x)*sin(x)                          # Integrand

def plotfunc():                                   # Plot function
    incr = 2.0*pi/N
24  for i in range(0,N):
        xx       = i*incr
        xsinx.x[i] = ((80.0/pi)*xx-80)
        xsinx.y[i] = 20*fx(xx)-50
28  box          = curve(pos=[(-80,-50), (-80,50), (80,50),
                     (80,-50), (-80,-50)], color=color.white)     # box

plotfunc()                             # Box area = h x w =5*2pi
32 j            = 0
Npts          = 3001                      # Pts inside box
analyt        = (pi)**2                   # Analytical integral
areanal.text = 'analytical=%8.5f'%analyt
36 genpts       = points(size=2)
for i in range(1,Npts):                   # points inside box
    rate(500)                                  #  slow process
    x = 2.0*pi*random.random()
40  y = 5*random.random()
    xp = x*80.0/pi-80
    yp = 20.0*y-50
    pts2.text = '%4s' %i
44  if y *<*= fx(x):                             # Below curve
        j += 1
        genpts.append(pos=(xp,yp), color=color.cyan)
        inside2.text='%4s'%j
48  else:   genpts.append(pos=(xp,yp), color=color.green)
    boxarea = 2.0*pi*5.0
    area = boxarea*j/(Npts-1)
    arealbl2.text = '%8.5f'%area
```

6

Trial-and-Error Searching and Data Fitting

This chapter adds some more tools to our computational toolbox. First, we examine ways to solve equations via a trial-and-error search. In Chapter 8 we will combine trial-and-error searching with the solution of ordinary differential equations to solve the general quantum eigenvalue problem. The second half of this chapter examines the fitting of curves to data. There we examine interpolating within a table of numbers, and least-squares fitting of a function to data, the latter often requiring a search.

6.1 Quantum Bound States I

Many computational techniques use well-defined algorithms leading to definite outcomes. In contrast, some techniques use trial-and-error algorithms in which internal decisions are made as to what steps to follow, and in which a number of solutions may be tried before one is settled upon, or not. (We already did some of this when we summed a power series until the terms became small.) Writing this type of program is usually challenging because we must foresee a number of possible outcomes, with the chance of failure always present.

Probably the most standard problem in quantum mechanics,[1] is to solve for the energies of a particle of mass m bound within a 1D square well of radius a:

$$V(x) = \begin{cases} -V_0, & \text{for } |x| \leq a, \\ 0, & \text{for } |x| \geq a. \end{cases} \tag{6.1}$$

As shown in quantum mechanics texts [Gottfried and Yan, 2004], the energies of the bound states $E = -E_B < 0$ within this well are solutions of the transcendental equations

$$\sqrt{10 - E_B} \, \tan\left(\sqrt{10 - E_B}\right) = \sqrt{E_B} \quad \text{(even)}, \tag{6.2}$$

$$\sqrt{10 - E_B} \, \cotan\left(\sqrt{10 - E_B}\right) = \sqrt{E_B} \quad \text{(odd)}, \tag{6.3}$$

where even and odd refer to the symmetry of the wave function. Here we have chosen units such that $\hbar = 1$, $2m = 1$, $a = 1$, and $V_0 = 10$.

---

1 We solve this same problem in Section 13.1 using an approach that is applicable to almost any potential, and which also provides the wave functions. The approach here is specialized to the eigenenergies of a square well.

Computational Physics: Problem Solving with Python, Fourth Edition.
Rubin H. Landau, Manuel J. Páez, and Cristian C. Bordeianu.
© 2024 WILEY-VCH GmbH. Published 2024 by WILEY-VCH GmbH.

Your **problem** is to

1) Find several bound-state energies E_B for even wave functions (6.2).
2) Explore how making the potential deeper, say, by changing the 10 to a 20 or a 30, affects the number of bound states and their energies.

6.2 Bisection Search

Trial-and-error root finding looks for a value of x for which

$$f(x) \simeq 0, \tag{6.4}$$

where we follow the convention of moving what might, otherwise, be on the right-hand-side (RHS) of an equation to the left-hand side (LHS) in order to leave just a 0 on the RHS. The search procedure starts with a guessed value for x, substitutes that guess into $f(x)$ (the "trial"), and then sees how different the LHS is from zero (the "error"). The algorithm then changes x based on the error, and tries out the new guess in $f(x)$. The procedure continues until $f(x) \simeq 0$ to some desired level of precision, or until the changes in x are insignificant, or when the search seems endless.

The most elementary trial-and-error technique is the *bisection algorithm*. It is reliable but slow. If you know some interval in which $f(x)$ changes sign, then the bisection algorithm will always converge to the root by finding progressively smaller and smaller intervals within which the zero lies. Other techniques, such as the Newton–Raphson method we describe next, may converge more quickly, but if the initial guess does not get you close to the zero, it may become unstable and move off into the wilderness.

The basis of the bisection algorithm is shown in Figure 6.1. We start with two values of x, x_-, and x_+, between which we know a zero occurs. (You can determine these by making a graph or by stepping through different x values and looking for a sign change.) To be specific, let us say that $f(x)$ is negative at x_- and positive at x_+:

$$f(x_-) < 0, \quad f(x_+) > 0. \tag{6.5}$$

(Note that it may well be that $x_- > x_+$ if the function changes from positive to negative as x increases.) Thus we start with the interval $x_+ \leq x \leq x_-$, within which we know a zero occurs. As you can see in `Bisection.py` in Listing 6.1, the algorithm then picks a new x value

Figure 6.1 A graphical representation of the steps involved in solving for a zero of $f(x)$ using the bisection algorithm. The bisection algorithm takes the midpoint of the interval as the new guess for x, with each step reducing the interval size by one-half. Four steps are shown here.

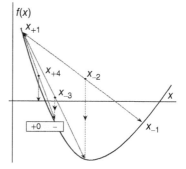

equal to the midpoint of the interval, and then sets a new interval as the half of the previous interval in which the sign changed:

```
x = ( plus + minus ) / 2
if ( f(x) f(plus) > 0 ) plus = x
else minus = x
```

This process continues until the value of $f(x)$ is less than a predefined level of precision, or until a predefined (large) number of subdivisions occurs.

The example in Figure 6.1 shows the first interval extending from $x_- = x_{+1}$ to $x_+ = x_{-1}$. We then bisect that interval at x, and because $f(x) < 0$ at the midpoint, we set $x_- = x_{-2} = x$ and label it x_{-2} to indicate the second step. We then use $x_{+2} = x_{+1}$ and x_{-2} as the next interval and continue the process. We see that only x_- changes for the first three steps in this example, but that for the fourth step x_+ finally changes. The changes then become too small for us to show.

6.2.1 Bisection Exercises

1) The first step in implementing any search algorithm is to get an idea of what your function looks like. For the present problem you do this by making a plot or a table of $f(E) = \sqrt{10 - E_B} \tan(\sqrt{10 - E_B}) - \sqrt{E_B}$ versus E_B. Note from your plot some approximate values at which $f(E_B) = 0$. Your program should be able to find more exact values for these zeros.
2) Write a program that implements the bisection algorithm and uses it to find some solutions of (6.2).
3) *Warning*: Seeing that the tan function has singularities, some care is suggested. In fact, your graphics program may not function accurately near these singularities. One cure is to use a different, but equivalent, form of the equation. Show that an equivalent form of (6.2) is

$$\sqrt{E} \cot(\sqrt{10 - E}) - \sqrt{10 - E} = 0. \tag{6.6}$$

4) Make a second plot of (6.6), which also has singularities but at different places. Use this plot to choose some x values that bracket the zeros.
5) After you have found a solution, evaluate $f(E_B)$ and thus determine the precision of your solution.
6) Compare the roots you find with those given by Maple or Mathematica.

6.3 Newton–Raphson Search

The Newton–Raphson algorithm can find roots of the $f(x) = 0$ more quickly than the bisection method. As we see graphically in Figure 6.2, this algorithm is the equivalent of drawing a straight line $f(x) \simeq mx + b$ tangent to the curve at an x value for which $f(x) \simeq 0$, and then using the intercept of the line with the x-axis at $x = -b/m$ as an improved guess for the root. If the "curve" were actually a straight line, the answer would be exact; otherwise, it is a good approximation if the guess is close

Figure 6.2 A graphical representation of the steps involved in solving for a zero of $f(x)$ using the Newton–Raphson method. The Newton–Raphson method takes the new guess as the zero of the line tangent to $f(x)$ at the old guess. Two guesses are shown.

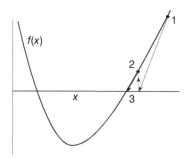

enough to the root for $f(x)$ to be nearly linear. The process continues until some set level of precision is reached or until too many guesses fail to find a root. If a guess is in a region where $f(x)$ is nearly linear (Figure 6.2), then the convergence is very rapid.

The analytic formulation of the Newton–Raphson algorithm starts with an old guess x_0, and expresses a new guess x as the old guess plus a correction Δx:

$$x_0 = \text{old guess}, \quad \Delta x = \text{unknown correction} \tag{6.7}$$

$$\Rightarrow x = x_0 + \Delta x. \tag{6.8}$$

We next expand the known function $f(x)$ in a Taylor series around x_0, and keep only the linear term in the expansion:

$$f(x = x_0 + \Delta x) \simeq f(x_0) + \left.\frac{df}{dx}\right|_{x_0} \Delta x. \tag{6.9}$$

We determine the correction Δx by calculating the point at which this linear approximation to $f(x)$ crosses the x-axis:

$$f(x_0) + \left.\frac{df}{dx}\right|_{x_0} \Delta x = 0, \tag{6.10}$$

$$\Rightarrow \Delta x = -\frac{f(x_0)}{df/dx|_{x_0}}. \tag{6.11}$$

The procedure is repeated, starting at the improved x, until some set level of precision is obtained.

The Newton–Raphson algorithm (6.11) requires evaluation of the derivative df/dx at each value of x_0. In many cases, you may have an analytic expression for the derivative and can build it into the algorithm. However, especially for more complicated problems, it is simple enough to just use a numerical forward-difference approximation to the derivative:

$$\frac{df}{dx} \simeq \frac{f(x + \delta x) - f(x)}{\delta x}, \tag{6.12}$$

where δx is some small change in x that you chose [different from the Δ used for searching in (6.11)]. While a central-difference approximation for the derivative would be more accurate, it would require additional evaluation of the f's, and once you find a zero, it does not matter how you got there. In Listing 6.2, we give a program NewtonCD.py that implements the search with the central difference derivative.

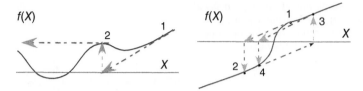

Figure 6.3 Two examples of how the Newton–Raphson algorithm may fail if the initial guess is not in the region where $f(x)$ can be approximated by a straight line. *Left*: A guess lands at a local extremum (minimum/maximum), that is, a place where the derivative vanishes, and so the next guess ends up at $x = \infty$. *Right*: The search has fallen into an infinite loop. The technique known as "backtracking" could eliminate this problem.

6.3.1 Search + Backtracking

Two examples of possible problems with the Newton–Raphson algorithm are shown in Figure 6.3. On the left, we see a case where the search takes us to an x value where the function has a local extremum (minimum or maximum), that is, where $df/dx = 0$. Because $\Delta x = -f/(df/dx)$, this leads to a horizontal tangent (division by zero), and so the next guess is $x = \infty$, from where it is hard to return. When this happens, you need to start your search with a different guess, and pray that you do not fall into this trap again. In cases where the correction is very large, but maybe not infinite, you may want to try backtracking (described below), and hope that by taking a smaller step you will not get into as much trouble.

In Figure 6.3 right we see a case where a search falls into an infinite loop surrounding the zero, without ever getting there. A solution to this problem is *backtracking*. As the name implies, in cases where the new guess $x_0 + \Delta x$ leads to an increase in the magnitude of the function, $|f(x_0 + \Delta x)|^2 > |f(x_0)|^2$, you can backtrack somewhat and try a smaller guess, say, $x_0 + \Delta x/2$. If the magnitude of f still increases, then you just need to backtrack some more, say, by trying $x_0 + \Delta x/4$ as your next guess, and so forth. Because you know that the tangent line leads to a local decrease in $|f|$, eventually an acceptable small enough step should be found.

The problem in both these cases is that the initial guesses were not close enough to the regions where $f(x)$ is approximately linear. So again, a good plot or table may help produce a good first guess. Alternatively, you may want to start your search with the bisection algorithm, and then switch to the faster Newton–Raphson algorithm when you get closer to the zero.

Exercise

1) Use the Newton–Raphson algorithm to find some energies E_B that are solutions of (6.2). Compare this solution with the one found with the bisection algorithm.

2) Again, notice that the 10 in this equation is proportional to the strength of the potential that causes the binding. See if making the potential deeper, say, by changing the 10 to a 20 or a 30, produces more or deeper bound states. (Note that in contrast to the bisection algorithm, your initial guess must be closer to the answer for the Newton–Raphson algorithm to work.)

3) Modify your algorithm to include backtracking and then try it out on some difficult cases.

4) Evaluate $f(E_B)$ and thus determine directly the precision of your solution.

6.4 Magnetization Search

Problem Determine $M(T)$ the magnetization as a function of temperature for simple magnetic materials.

A collection of N spin-1/2 particles each with magnetic moment μ is at temperature T. The collection has an external magnetic field B applied to it, and comes to equilibrium with N_L particles in the lower energy state (spins aligned with the magnetic field), and with N_U particles in the upper energy state (spins opposed to the magnetic field). The Boltzmann distribution law tells us that the relative probability of a state with energy E is proportional to $\exp(-E/(k_B T))$, where k_B is Boltzmann's constant. For a dipole with moment μ, its energy in a magnetic field is given by the dot product $E = -\mu \cdot \mathbf{B}$. Accordingly, spin-up particle have lower energy in a magnetic field than spin-down particles, and thus are more probable.

Applying the Boltzmann distribution to our spin problem, we have that the number of particles in the lower energy level (spin up) is

$$N_L = N \frac{e^{\mu B/(k_B T)}}{e^{\mu B/(k_B T)} + e^{-\mu B/(k_B T)}}, \tag{6.13}$$

while the number of particles in the upper energy level (spin down) is

$$N_U = N \frac{e^{-\mu B/(k_B T)}}{e^{\mu B/(k_B T)} + e^{-\mu B/(k_B T)}}. \tag{6.14}$$

As discussed in [Kittel, 2018], we now assume that the molecular magnetic field $B = \lambda M$ is much larger than the applied magnetic field, and so replace B by the molecular field. This permits us to eliminate B from the preceding equations. The *magnetization* $M(T)$ is given by the individual magnetic moment μ times the net number of particles pointing in the direction of the magnetic field:

$$M(T) = \mu \times (N_L - N_U) \tag{6.15}$$

$$= N\mu \tanh\left(\frac{\lambda \mu M(T)}{k_B T}\right). \tag{6.16}$$

Note that this definition appears to make sense because as the temperature approaches zero, all spins will be aligned along the direction of B, and so $M(T = 0) = N\mu$.

$M(T)$ via Searching Equation (6.16) relates the magnetization and the temperature. However, it is not really a solution to our problem because M appears on the LHS of the equation as well as within the hyperbolic function on the RHS. Generally, a *transcendental equation* of this sort does not have an analytic solution that would give M as a function of the temperature T. But by sort of working backward we can find a numerical solution. To do that, we first express (6.16) in terms of the reduced magnetization m, the reduced

temperature t, and the Curie temperature T_c:

$$m(t) = \tanh\left(\frac{m(t)}{t}\right), \tag{6.17}$$

$$m(T) = \frac{M(T)}{N\mu}, \qquad t = \frac{T}{T_c}, \qquad T_c = \frac{N\mu^2\lambda}{k_B}. \tag{6.18}$$

While it is no easier to find an analytic solution to (6.17) than it was to (6.16), the simpler form of (6.17) makes the programming easier.

One approach to a trial-and-error solution is to define a function

$$f(m,t) = m - \tanh\left(\frac{m(t)}{t}\right), \tag{6.19}$$

and then, for a variety of fixed $t = t_i$ values, search for those m values at which $f(m, t_i) = 0$. (One could just as well fix the value of m to m_j and search for the value of t for which $f(m_j, t) = 0$; once you have a solution, you have a solution.) Each zero so found provides a single value of $m(t_i)$. A plot or a table of these values for a range of t_i values then provides the best we can do for the desired solution $m(t)$.

Figure 6.4 shows three plots of $f(m, t)$ as a function of the reduced magnetization m, each plot for a different value of the reduced temperature. As you can see, other than the uninteresting solution at $m = 0$, there is only one solution (a zero) and it's near $m = 1$ for $t = 0.5$. There is no solution at the other temperatures.

1) Find the root of (6.19) to six significant figures for $t = 0.5$ using the bisection algorithm.
2) Find the root of (6.19) to six significant figures for $t = 0.5$ using the Newton–Raphson algorithm.
3) Compare the time it takes to find the solutions for the bisection and Newton–Raphson algorithms.

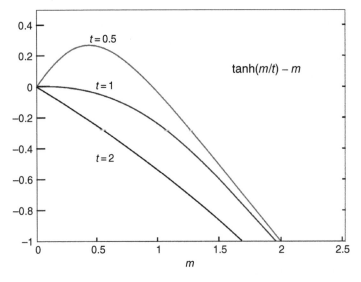

Figure 6.4 A function of the reduced magnetism m at three reduced temperatures t. A zero of this function determines the value of the magnetism at a particular value of t.

4) Construct a plot of the reduced magnetization $m(t)$ as a function of the reduced temperature t.

6.5 Data Fitting

Data fitting is an art worthy of serious study by all scientists [Bevington and Robinson, 2003]. In the sections to follow we just scratch the surface by examining how to interpolate within a table of numbers and how to do a least-squares fit to data. We also show how to go about making a least-squares fit to nonlinear functions using some of the search techniques and subroutine libraries.

Problem   The cross sections measured for the resonant scattering of neutrons from a nucleus are given in Table 6.1. Your problem is to determine values for the cross sections at energy values lying between those in the table.

You can solve this problem in a number of ways. The simplest is to numerically *interpolate* between the values of the experimental $f(E_i)$ given in Table 6.1. This is direct and easy, but does not account for there being experimental noise in the data. A more appropriate solution (discussed in Section 6.7) is to find the *best fit* of a theoretical function to the data. We start with what we believe to be the "correct" theoretical description of the data,

$$f(E) = \frac{f_r}{(E - E_r)^2 + \Gamma^2/4}, \tag{6.20}$$

where f_r, E_r, and Γ are unknown parameters. We then adjust the parameters to obtain the best fit. This is a best fit in a statistical sense, but in fact may not pass through all (or any) of the data points. For an easy, yet effective, introduction to statistical data analysis, we recommend Bevington and Robinson [2003].

These two techniques of interpolation and least-squares fitting are powerful tools that let you treat tables of numbers as if they were analytic functions, and sometimes let you deduce statistically meaningful constants or conclusions from measurements. In general, you can view data fitting as *global* or *local*. In global fits, a single function of x is used to represent the entire set of numbers in a table such as Table 6.1. While it may be spiritually satisfying to find a single function that passes through all the data points, if that function is not the correct function for describing the data, the fit may show nonphysical behavior (such as large oscillations) between the data points. The rule of thumb is that if you must interpolate, keep it local and view global interpolations with a critical eye.

Table 6.1   Experimental values for a scattering cross section ($f(E)$ in the theory), each with absolute error $\pm\sigma_i$, as a function of energy (x_i in the theory).

i	1	2	3	4	5	6	7	8	9
E_i (MeV)	0	25	50	75	100	125	150	175	200
$g(E_i)$ (MB)	10.6	16.0	45.0	83.5	52.8	19.9	10.8	8.25	4.7
Error (MB)	9.34	17.9	41.5	85.5	51.5	21.5	10.8	6.29	4.14

Consider Table 6.1 as ordered data. We call the independent variable x and its tabulated values $x_i (i = 1, 2, ...)$, and assume that the dependent variable is the function $g(x)$, with tabulated values $g_i = g(x_i)$. We assume that $g(x)$ can be approximated as an $(n-1)$th-degree polynomial in each interval i:

$$g_i(x) \simeq a_0 + a_1 x + a_2 x^2 + \cdots + a_{n-1} x^{n-1}. \tag{6.21}$$

Seeing that our fit is local, we do not assume that one $g(x)$ can fit all the data in the table, but instead use a different polynomial, that is, a different set of a_i values, for each interval. Each polynomial will be of low degree, and multiple polynomials will be needed to span the entire table. If some care is taken, the set of polynomials so obtained will behave well enough to be used in further calculations without introducing much unwanted noise or discontinuities in $g(x)$ or its derivatives.

The classic interpolation formula was created by Lagrange. He figured out a closed-form expression that directly fits the $(n-1)$ order polynomial (6.21) to n values of the function $g(x)$ evaluated at the points x_i. The formula for each interval is written as the sum of polynomials:

$$g(x) \simeq g_1 \lambda_1(x) + g_2 \lambda_2(x) + \cdots + g_n \lambda_n(x), \tag{6.22}$$

$$\lambda_i(x) = \prod_{j(\neq i)=1}^{n} \frac{x - x_j}{x_i - x_j} = \frac{x - x_1}{x_i - x_1} \frac{x - x_2}{x_i - x_2} \cdots \frac{x - x_n}{x_i - x_n}. \tag{6.23}$$

For three points, (6.22) provides a second-degree polynomial, while for eight points it gives a seventh-degree polynomial. For example, assume we are given the points and function values

$$x_{1-4} = (0, 1, 2, 4) \qquad g_{1-4} = (-12, -12, -24, -60). \tag{6.24}$$

With four points, the Lagrange formula determines a third-order polynomial that reproduces each of the tabulated values:

$$g(x) = \frac{(x-1)(x-2)(x-4)}{(0-1)(0-2)(0-4)}(-12) + \frac{x(x-2)(x-4)}{(1-0)(1-2)(1-4)}(-12)$$

$$+ \frac{x(x-1)(x-4)}{(2-0)(2-1)(2-4)}(-24) + \frac{x(x-1)(x-2)}{(4-0)(4-1)(4-2)}(-60),$$

$$\Rightarrow g(x) = x^3 - 9x^2 + 8x - 12. \tag{6.25}$$

As a check we see that

$$g(4) = 4^3 - 9(4^2) + 32 - 12 = -60, \qquad g(0.5) = -10.125. \tag{6.26}$$

If the data contain little noise, this polynomial can be used with some confidence within the range of the data, but with risk beyond the range of the data.

Notice that Lagrange interpolation makes no restriction that the points x_i be evenly spaced. Usually, the Lagrange fit is made to only a small region of the table with a small value of n, despite the fact that the formula works perfectly well for fitting a high-degree polynomial to the entire table. The difference between the value of the polynomial evaluated at some x and that of the actual function can be shown to be the *remainder*

$$R_n \simeq \frac{(x - x_1)(x - x_2) \cdots (x - x_n)}{n!} g^{(n)}(\zeta), \tag{6.27}$$

where ζ lies somewhere in the interpolation interval. What is significant here is that we see that if significant high derivatives exist in $g(x)$, then the remainder can be very large. For example, a table of noisy data would have significantly high derivatives.

6.5.1 Lagrange Fitting

Consider the experimental neutron scattering data in Table 6.1. The expected theoretical functional form that describes these data is (6.20), and our empirical fits to these data are shown in Figure 6.5.

1) Write a subroutine to perform an n-point Lagrange interpolation using (6.22). Treat n as an arbitrary input parameter. (You may also do this exercise with the spline fits discussed in Section 6.5.2.)
2) Use the Lagrange interpolation formula to fit the entire experimental spectrum with one polynomial. (This means that you must fit all nine data points with an 8th degree polynomial.) Then use this fit to plot the cross section in steps of 5 MeV.
3) Use your graph to deduce the resonance energy E_r (your peak position) and Γ (the full width at half-maximum). Compare your results with those predicted by a theorist friend, $(E_r, \Gamma) = (78, 55)$ MeV.
4) A more realistic use of Lagrange interpolation is for local interpolation with a small number of points, such as three. Interpolate the preceding cross-sectional data in 5-MeV steps using three-point Lagrange interpolation for each interval. (Note that the end intervals may be special cases.)
5) We deliberately have not discussed *extrapolation* of data because it can lead to serious *systematic* errors; the answer you get may well depend more on the function you assume than on the data you input. Add some adventure to your life and use the programs you have written to extrapolate to values outside Table 6.1. Compare your results to the theoretical Breit–Wigner shape (6.20).

This example shows how easy it is to go wrong with a high-degree-polynomial fit to data with errors. Although the polynomial is guaranteed to pass through all the data points, the representation of the function away from these points can be quite unrealistic. Using a low-order interpolation formula, say, $n = 2$ or 3, in each interval usually eliminates the wild oscillations, but may not have any theoretical justification. If these local fits are matched together carefully, as we discuss in the following section on cubic spline interpolation, then a rather continuous curve results. Nonetheless, you must recall that if the data contain errors, a curve that actually passes through them may lead you astray. We discuss how to do this properly with least-square fitting in Section 6.7.

6.5.2 Cubic Spline Interpolation

If you have followed our suggestions and tried to interpolate the resonant cross section with Lagrange interpolation, then you saw that fitting parabolas (three-point interpolation) within a table may avoid the erroneous and possibly catastrophic deviations of a high-order formula. (A two-point interpolation, which connects the points with straight lines, may not lead you far astray, but it is rarely pleasing to the eye or precise.) A sophisticated variation

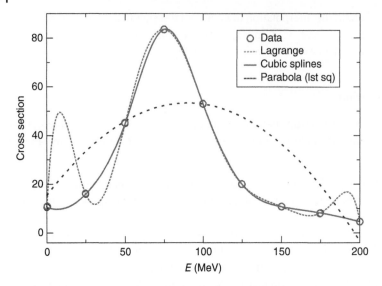

Figure 6.5 Three fits to data. *Dashed*: Lagrange interpolation using an 8th degree polynomial; *Short dashes*: cubic splines fit; *Long dashed*: Least-squares parabola fit.

of an $n = 4$ interpolation, known as *cubic splines*, often leads to surprisingly smooth and eye-pleasing fits. In this approach (Figure 6.5), cubic polynomials are fit to the function in each interval, with the additional constraint that the first and second derivatives of the cubics be continuous from one interval to the next. This continuity of slope and curvature makes the spline fit particularly eye-pleasing. The analytic approach is analogous to using a flexible spline drafting tool (a lead wire within a rubber sheath), from which the method draws its name.

The series of cubic polynomials obtained by spline-fitting a table of data can be integrated and differentiated, and is guaranteed to have well-behaved derivatives. The existence of meaningful derivatives is an important consideration. As a case in point, if the interpolated function is a potential, you can take the derivative to obtain the force. The complexity of simultaneously matching polynomials and their derivatives over all the interpolation points leads to many simultaneous linear equations to be solved. This makes splines unattractive for hand calculations, yet easy for computers and, not surprisingly, popular in both calculations and computer drawing programs. To illustrate, the smooth solid curve in Figure 6.5 is a spline fit.

The basic approximation of splines is the representation of the function $g(x)$ in the subinterval $[x_i, x_{i+1}]$ with a cubic polynomial:

$$g(x) \simeq g_i(x), \quad \text{for } x_i \leq x \leq x_{i+1}, \tag{6.28}$$

$$g_i(x) = g_i + g_i'(x - x_i) + \frac{1}{2}g_i''(x - x_i)^2 + \frac{1}{6}g_i'''(x - x_i)^3. \tag{6.29}$$

This representation makes it clear that the coefficients in the polynomial equal the values of $g(x)$ and its first, second, and third derivatives at the tabulated points x_i. Derivatives beyond the third vanish for a cubic. The computational chore is to determine these derivatives in terms of the N tabulated g_i values. The matching of g_i at the *nodes* that connect one interval to the next provides the equations

$$g_i(x_{i+1}) = g_{i+1}(x_{i+1}), \quad i = 1, N - 1. \tag{6.30}$$

The matching of the first *and* second derivatives at each interval's boundaries provides the equations

$$g'_{i-1}(x_i) = g'_i(x_i), \quad g''_{i-1}(x_i) = g''_i(x_i). \tag{6.31}$$

The additional equations needed to determine all constants are obtained by matching the third derivatives at adjacent nodes. Values for the third derivatives are found by approximating them in terms of the second derivatives:

$$g'''_i \simeq \frac{g''_{i+1} - g''_i}{x_{i+1} - x_i}. \tag{6.32}$$

As discussed in Chapter 5, a central-difference approximation would be more accurate than a forward-difference approximation, yet (6.32) keeps the equations simpler.

It is straightforward, although complicated, to solve for all the parameters in (6.29). We leave that to the references Thompson [1992] and Press *et al.* [2007]. We can see, however, that matching at the boundaries of the intervals results in only $(N - 2)$ linear equations for N unknowns. Further input is required. It usually is taken to be the boundary conditions at the endpoints $a = x_1$ and $b = x_N$, specifically, the second derivatives there $g''(a)$ and $g''(b)$. There are several ways to determine these second derivatives:

Natural spline: Set $g''(a) = g''(b) = 0$; that is, permit the function to have a slope at the endpoints but no curvature. This is "natural" because the derivative vanishes for the flexible spline drafting tool (its ends being unconstrained).

Input values for g' at the boundaries: The computer uses $g'(a)$ to approximate $g''(a)$. If you do not know the first derivatives, you can calculate them numerically from the table of g_i values.

Input values for g'' at the boundaries: Knowing values is of course better than approximating them, but it requires the user to input information. If the values of g'' are not known, they can be approximated by applying a forward-difference approximation to the tabulated values:

$$g''(x) \simeq \frac{[g(x_3) - g(x_2)]/[x_3 - x_2] - [g(x_2) - g(x_1)]/[x_2 - x_1]}{[x_3 - x_1]/2}. \tag{6.33}$$

6.5.3 Cubic Spline Quadrature

A powerful integration scheme is to fit an integrand with splines, and then integrate the cubic polynomials analytically. If the integrand $g(x)$ is known only at its tabulated values, then this is about as good an integration scheme as is possible; if you have the ability to calculate the function directly for arbitrary x values, then Gaussian quadrature may be preferable. We know that the spline fit to g in each interval is the cubic (6.29)

$$g(x) \simeq g_i + g'_i(x - x_i) + \frac{1}{2}g''_i(x - x_i)^2 + \frac{1}{6}g'''_i(x - x_i)^3. \tag{6.34}$$

It is easy to integrate this to obtain the integral of g for this interval and then to sum over all intervals:

$$\int_{x_i}^{x_{i+1}} g(x)\,dx \simeq \left(g_i x + \frac{1}{2}g_i'x^2 + \frac{1}{6}g_i''x^3 + \frac{1}{24}g_i'''x^4\right)\Bigg|_{x_i}^{x_{i+1}}, \tag{6.35}$$

$$\int_{x_j}^{x_k} g(x)\,dx = \sum_{i=j}^{k} \left(g_i x + \frac{1}{2}g_i'x_i^2 + \frac{1}{6}g_i''x^3 + \frac{1}{24}g_i'''x^4\right)\Bigg|_{x_i}^{x_{i+1}}. \tag{6.36}$$

Making the intervals smaller does not necessarily increase precision, as subtractive cancellations in (6.35) may get large.

Spline Fit of Cross Section (Implementation) Fitting a series of cubics to data is a little complicated to program yourself, so we recommend using a library routine. We have adapted the `splint.c` and the `spline.c` functions from Press *et al.* [2007] to produce the `SplineInteract.py` program shown in Listing 6.3.

Your **problem** for this section is to carry out the assessment in Section 6.5.1 using cubic spline interpolation rather than Lagrange interpolation.

6.6 Fitting Exponential Decay

Figure 6.6 presents actual experimental data on the number of decays ΔN of the π meson as a function of time [Stetz *et al.*, 1973]. Notice that the time has been "binned" into intervals $\Delta t = 10$-ns, and that the smooth curve is the theoretical exponential decay expected if there were a very large numbers of pions (which there's not). Your problem is to deduce the lifetime τ of the π meson from these data (the tabulated lifetime is 2.6×10^{-8} seconds).

Assume that we start with N_0 particles at time $t = 0$ that can decay to other particles.[2] If we wait a short time Δt, then a small number ΔN of the particles will decay *spontaneously*, that is, with no external influences. This decay is a stochastic process, which means that an

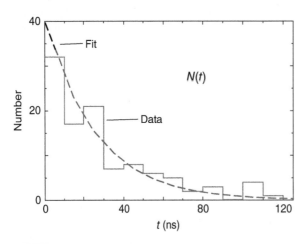

Figure 6.6 A reproduction of the experimental measurement of Stetz *et al.* [1973] giving the number of decays of π mesons as a function of time since their creation. Measurements were made during time intervals (box sizes) of 10-inch width. The dashed curve is the result of a linear least-square fit to the log $N(t)$.

---

2 Spontaneous decay is discussed further and simulated in Section 4.3.

element of chance helps determine just when a decay will occur, and so no two experiments are expected to give exactly the same results. The basic law of nature for spontaneous decay is that the number of decays ΔN in a time interval Δt is proportional to the number of particles $N(t)$ present at that time and to the time interval

$$\Delta N(t) = -\frac{1}{\tau}N(t)\Delta t \quad \Rightarrow \quad \frac{\Delta N(t)}{\Delta t} = -\lambda N(t). \tag{6.37}$$

Here $\tau = 1/\lambda$ is the *lifetime* of the particle, with λ a rate parameter. The actual decay *rate* is given by the second equation in (6.37). If the number of decays ΔN is very small compared to the number of particles N, and if we look at vanishingly small time intervals, then the difference equation (6.37) becomes the differential equation

$$\frac{dN(t)}{dt} \simeq -\lambda N(t) = \frac{1}{\tau}N(t). \tag{6.38}$$

This differential equation has an exponential solution for the number as well as for the decay rate:

$$N(t) = N_0 e^{-t/\tau}, \quad \frac{dN(t)}{dt} = -\frac{N_0}{\tau}e^{-t/\tau} = \frac{dN(0)}{dt}e^{-t/\tau}. \tag{6.39}$$

Equation (6.39) is the theoretical formula we wish to fit to the data in Figure 6.6. The output of such a fit is a best-fit value for the lifetime τ.

6.7 Least-Squares Fitting

Books have been written and careers have been spent discussing what is meant by a "good fit" to experimental data. We cannot do justice to the subject here and refer the reader to Bevington and Robinson [2003], Press *et al.* [2007], and Thompson [1992]. However, we will emphasize three points:

1) If the data being fit contain errors, then the "best fit" in a statistical sense should not pass through all the data points.
2) If the theory is not an appropriate one for the data (e.g., the parabola in Figure 6.5), then its best fit to the data may not be a good fit at all. This is good, for this is how we know that the theory is not appropriate.
3) Only for the simplest case of a linear least-squares fit can we write down a closed-form solution to evaluate and obtain the best fit. More realistic problems are usually solved by *trial-and-error* search procedures, sometimes using sophisticated subroutine libraries. In Section 6.8, we show how to conduct such a nonlinear search using familiar tools.

Imagine that you have measured N_D data values of the independent variable y as a function of the dependent variable x:

$$(x_i, y_i \pm \sigma_i), \quad i = 1, N_D, \tag{6.40}$$

where $\pm\sigma_i$ is the experimental uncertainty in the ith value of y. (For simplicity we assume that all the errors σ_i occur in the dependent variable, although this is hardly ever true [Thompson, 1992]). For our problem, y is the number of decays as a function of time, and x_i is the times. Our goal is to determine how well a mathematical function $y = g(x)$ (also called a *theory* or a *model*) can describe these data. Additionally, if the theory contains some parameters or constants, our goal is also to determine the best values for these parameters. We assume that the theory function $g(x)$ contains, in addition to the functional dependence on x, an additional dependence upon M_P parameters $\{a_1, a_2, \ldots, a_{M_P}\}$. Notice that the *parameters* $\{a_m\}$ are not *variables*, in the sense of numbers read from a meter, but rather are parts of the theoretical model, such as the size of a box, the mass of a particle, or the depth of a potential well. For the exponential decay function (6.39), the parameters are the lifetime τ and the initial decay rate $\frac{dN}{dt}(0)$. We indicate this as

$$g(x) = g(x; \{a_1, a_2, \ldots, a_{M_P}\}) = g(x; \{a_m\}), \tag{6.41}$$

where the a_i's are parameters and x the independent variable.

We use the chi-square, χ^2, measure as a gauge of how well a theoretical function g reproduces data [Bevington and Robinson, 2003]:

$$\chi^2 \stackrel{\text{def}}{=} \sum_{i=1}^{N_D} \left(\frac{y_i - g(x_i; \{a_m\})}{\sigma_i} \right)^2, \tag{6.42}$$

where the sum is over the N_D experimental points $(x_i, y_i \pm \sigma_i)$. The definition (6.42) is such that smaller values of χ^2 are better fits, with $\chi^2 = 0$ occurring if the theoretical curve went through the center of every data point. Notice also that the $1/\sigma_i^2$ weighting means that measurements with larger errors contribute less to χ^2.[3] *Least-squares fitting* refers to adjusting the parameters in the theory until a minimum in χ^2 is found, that is, finding a curve that produces the least value for the summed squares of the deviations of the data from the function $g(x)$. In general, this is the best fit possible and the best way to determine the parameters in a theory. The M_P parameters $\{a_m, m = 1, M_P\}$ that make χ^2 an extremum are found by solving the M_P equations:

$$\frac{\partial \chi^2}{\partial a_m} = 0, \quad \Rightarrow \quad \sum_{i=1}^{N_D} \frac{[y_i - g(x_i)]}{\sigma_i^2} \frac{\partial g(x_i)}{\partial a_m} = 0, \quad (m = 1, M_P). \tag{6.43}$$

Often, the function $g(x; \{a_m\})$ has a sufficiently complicated dependence on the a_m values for (6.43) to produce M_P simultaneous nonlinear equations in the a_m values. In these cases, solutions are found by a trial-and-error search through the M_P-dimensional parameter space, as we do in Section 6.8. To be safe, when such a search is completed, you should check that the minimum χ^2 you found is *global* and not *local*. One way to do that is to repeat the search for a whole grid of starting values, and if different minima are found, to pick the one with the lowest χ^2.

6.7.1 Least-Squares Implementation

When the deviations from theory are as a result of random errors, and when these errors are described by a Gaussian distribution, there are some useful rules of thumb to remember. You know that your fit is good if the value of χ^2 calculated via the definition (6.42) is approximately equal to the number of degrees of freedom $\chi^2 \simeq N_D - M_P$, where N_D is the number of data points and M_P is the number of parameters in the theoretical function. If your χ^2 is much less than $N_D - M_P$, it doesn't mean that you have a "great" theory or really precise

---

3 If you are not given the errors, you can guess them on the basis of the apparent deviation of the data from a smooth curve, or you can weigh all points equally by setting $\sigma_i \equiv 1$ and continue with the fitting.

measurements; instead, you probably have too many parameters, or have assigned errors (σ_i values) that are too large. In fact, too small a χ^2 may indicate that you are fitting the random scatter in the data rather than missing approximately one-third of the error bars, as expected if the errors are random. If your χ^2 is significantly greater than $N_D - M_P$, the theory may not be good, you may have significantly underestimated your errors, or you may have errors that are not random.

The M_P simultaneous equations (6.43) can be simplified considerably if the functions $g(x; \{a_m\})$ depend *linearly* on the parameter values a_i, e.g.,

$$g\left(x; \{a_1, a_2\}\right) = a_1 + a_2 x. \tag{6.44}$$

In this case (also known as *linear regression*), as shown in Figure 6.7, there are $M_P = 2$ parameters, the slope a_2, and the y intercept a_1. Notice that while there are only two parameters to determine, there still may be an arbitrary number N_D of data points to fit. Remember, a unique solution is not possible unless the number of data points is equal to or greater than the number of parameters. For this linear case, there are just two derivatives,

$$\frac{\partial g(x_i)}{\partial a_1} = 1, \qquad \frac{\partial g(x_i)}{\partial a_2} = x_i, \tag{6.45}$$

and after substitution, the χ^2 minimization equations (6.43) can be solved:

$$a_1 = \frac{S_{xx}S_y - S_x S_{xy}}{\Delta}, \qquad\qquad a_2 = \frac{SS_{xy} - S_x S_y}{\Delta}, \tag{6.46}$$

$$S = \sum_{i=1}^{N_D} \frac{1}{\sigma_i^2}, \qquad S_x = \sum_{i=1}^{N_D} \frac{x_i}{\sigma_i^2}, \qquad S_y = \sum_{i=1}^{N_D} \frac{y_i}{\sigma_i^2}, \tag{6.47}$$

$$S_{xx} = \sum_{i=1}^{N_D} \frac{x_i^2}{\sigma_i^2}, \qquad S_{xy} = \sum_{i=1}^{N_D} \frac{x_i y_i}{\sigma_i^2}, \qquad \Delta = SS_{xx} - S_x^2. \tag{6.48}$$

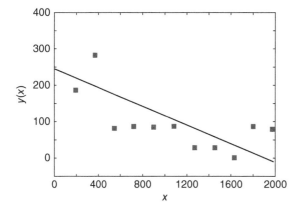

Figure 6.7 A linear least-squares best fit of a straight line to data. The deviation of theory from experiment is greater than would be expected from statistics, which means that a straight line is not a good theory to describe these data.

Statistics also gives you an expression for the *variance* or uncertainty in the deduced parameters:

$$\sigma_{a_1}^2 = \frac{S_{xx}}{\Delta}, \qquad \sigma_{a_2}^2 = \frac{S}{\Delta}. \tag{6.49}$$

These are measures of the uncertainties in the values of the fitted parameters arising from the uncertainties σ_i in the measured y_i values. A measure of the dependence of the parameters on each other is given by the *correlation coefficient*:

$$\rho(a_1, a_2) = \frac{\text{cov}(a_1, a_2)}{\sigma_{a_1}\sigma_{a_2}}, \qquad \text{cov}(a_1, a_2) = \frac{-S_x}{\Delta}. \tag{6.50}$$

Here $\text{cov}(a_1, a_2)$ is the *covariance* of a_1 and a_2, and vanishes if a_1 and a_2 are independent. The correlation coefficient $\rho(a_1, a_2)$ lies in the range $-1 \leq \rho \leq 1$, with a positive ρ indicating that the errors in a_1 and a_2 are likely to have the same sign, and a negative ρ indicating opposite signs.

The preceding analytic solutions for the parameters are of the form found in statistics books, but are not optimal for numerical calculations because subtractive cancellation can decrease the accuracy of the answers. A rearrangement of the equations can decrease this type of error [Thompson 1992]:

$$a_1 = \bar{y} - a_2\bar{x}, \quad a_2 = \frac{S_{xy}}{S_{xx}}, \quad \bar{x} = \frac{1}{N}\sum_{i=1}^{N_d} x_i, \quad \bar{y} = \frac{1}{N}\sum_{i=1}^{N_d} y_i$$

$$S_{xy} = \sum_{i=1}^{N_d} \frac{(x_i - \bar{x})(y_i - \bar{y})}{\sigma_i^2}, \quad S_{xx} = \sum_{i=1}^{N_d} \frac{(x_i - \bar{x})^2}{\sigma_i^2}. \tag{6.51}$$

In Fit.py in Listing 6.4, we give a program that fits a parabola to some data. You can use it as a model for fitting a line to data, although you can also use our closed-form expressions for a straight-line fit.

6.7.2 Linear Quadratic Fit

As indicated earlier, as long as the function being fitted depends *linearly* on the unknown parameters a_i, the condition of minimum χ^2 leads to a set of simultaneous linear equations

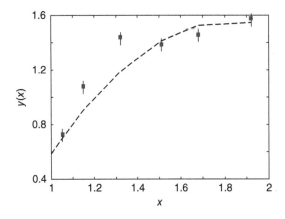

Figure 6.8 A linear least-squares best fit of a parabola to data. Here we see that the fit misses approximately one-third of the points, as expected from the statistics for a good fit.

for the a's that can be solved by hand, or on the computer using matrix techniques. To illustrate, suppose we want to fit the quadratic polynomial

$$g(x) = a_1 + a_2 x + a_3 x^2 \qquad (6.52)$$

to the experimental measurements $(x_i, y_i, i = 1, N_D)$ shown in Figure 6.8. Because this $g(x)$ is linear in the parameters a_i, the fit is still linear, even though x is raised to the second power. [However, if we tried to a fit a function of the form $g(x) = (a_1 + a_2 x) \exp(-a_3 x)$ to the data, then we would not be able to make a linear fit because there is not a linear dependence on a_3.]

The best fit of this quadratic to the data is obtained by applying the minimum χ^2 condition (6.43) for $M_p = 3$ parameters and N_D (still arbitrary) data points. Equation (6.43) leads to the three simultaneous equations for a_1, a_2, and a_3:

$$\sum_{i=1}^{N_D} \frac{[y_i - g(x_i)]}{\sigma_i^2} \frac{\partial g(x_i)}{\partial a_1} = 0, \qquad \frac{\partial g}{\partial a_1} = 1, \qquad (6.53)$$

$$\sum_{i=1}^{N_D} \frac{[y_i - g(x_i)]}{\sigma_i^2} \frac{\partial g(x_i)}{\partial a_2} = 0, \qquad \frac{\partial g}{\partial a_2} = x, \qquad (6.54)$$

$$\sum_{i=1}^{N_D} \frac{[y_i - g(x_i)]}{\sigma_i^2} \frac{\partial g(x_i)}{\partial a_3} = 0, \qquad \frac{\partial g}{\partial a_3} = x^2. \qquad (6.55)$$

Note: Because the derivatives are independent of the parameters (the a's), the a dependence arises only from the terms in square brackets in the sums, and because those terms have a linear dependence on the a's, these equations are linear in the a's.

Exercise   Show that after some rearrangement, (6.53)–(6.55) can be written as

$$S a_1 + S_x a_2 + S_{xx} a_3 = S_y, \qquad (6.56)$$

$$S_x a_1 + S_{xx} a_2 + S_{xxx} a_3 = S_{xy},$$

$$S_{xx} a_1 + S_{xxx} a_2 + S_{xxxx} a_3 = S_{xxy}.$$

Here the definitions of the S's are simple extensions of those used in (6.46)–(6.48) and are programmed in `Fit.py` in Listing 6.4. After placing the three unknown parameters into a vector x and the known three terms in (6.56) into a vector $\vec{b}$, these equations assume the matrix form:

$$A\vec{x} = \vec{b}, \qquad (6.57)$$

$$A = \begin{bmatrix} S & S_x & S_{xx} \\ S_x & S_{xx} & S_{xxx} \\ S_{xx} & S_{xxx} & S_{xxxx} \end{bmatrix}, \quad \vec{x} = \begin{bmatrix} a_1 \\ a_2 \\ a_3 \end{bmatrix}, \quad \vec{b} = \begin{bmatrix} S_y \\ S_{xy} \\ S_{xxy} \end{bmatrix}.$$

The solution for the parameter vector $\vec{x}$ is obtained by solving the matrix equations. Although for 3×3 matrices we can write out the solution in closed form, for larger problems the numerical solution requires a matrix computation.

6.7.2.1 Linear Quadratic Fit Assessment

1) Fit the quadratic (6.52) to the following data sets [given as $(x_1, y_1), (x_2, y_2), \ldots$]. In each case indicate the values found for the *as*, the number of degrees of freedom, *and* the value of χ^2.
   a) $(0, 1)$
   b) $(0, 1), (1, 3)$
   c) $(0, 1), (1, 3), (2, 7)$
   d) $(0, 1), (1, 3), (2, 7), (3, 15)$
2) Find a fit to the last set of data to the function $y = Ae^{-bx^2}$.
   Hint: A judicious change of variables will permit you to convert this to a linear fit. Does a minimum χ^2 still have meaning here?

6.8 Nonlinear Fit to a Resonance

Recall how earlier in this chapter we interpolated the values in Table 6.1 in order to obtain the experimental cross section σ as a function of energy. Although we did not use it, we also gave the theory describing these data, namely, the Breit–Wigner resonance formula (6.20):

$$f(E) = \frac{f_r}{(E - E_r)^2 + \Gamma^2/4}. \tag{6.58}$$

Your **problem** is to determine what values for the parameters E_r, f_r, and Γ in (6.58) provide the best fit to the data in Table 6.1.

Since (6.58) is not a linear function of the parameters (E_r, f_r, Γ), the three equations that result from minimizing χ^2 are not linear equations, and so cannot be solved by the techniques of *linear* algebra. However, in our study of the masses on a string problem we show how to use the Newton–Raphson algorithm to search for solutions of simultaneous nonlinear equations. That technique involved expansion of the equations about the previous guess to obtain a set of linear equations, and then solving the linear equations with the matrix libraries. We now use this same combination of fitting, trial-and-error searching, and matrix algebra to conduct a nonlinear least-squares fit of (6.58) to the data in Table 6.1.

Recall that the condition for a best fit is to find values of the M_P parameters a_m in the theory $g(x, a_m)$ that minimize $\chi^2 = \sum_i [(y_i - g_i)/\sigma_i]^2$. This leads to the M_P equations (6.43) to solve

$$\sum_{i=1}^{N_D} \frac{[y_i - g(x_i)]}{\sigma_i^2} \frac{\partial g(x_i)}{\partial a_m} = 0, \qquad (m = 1, M_P). \tag{6.59}$$

To find the form of these equations appropriate to our problem, we rewrite our theory function (6.58) in the notation of (6.59):

$$a_1 = f_r, \quad a_2 = E_R, \quad a_3 = \Gamma^2/4, \ x = E, \tag{6.60}$$

$$\Rightarrow g(x) = \frac{a_1}{(x - a_2)^2 + a_3}. \tag{6.61}$$

The three derivatives required in (6.59) are then

$$\frac{\partial g}{\partial a_1} = \frac{1}{(x - a_2)^2 + a_3}, \quad \frac{\partial g}{\partial a_2} = \frac{-2a_1(x - a_2)}{[(x - a_2)^2 + a_3]^2}, \quad \frac{\partial g}{\partial a_3} = \frac{-a_1}{[(x - a_2)^2 + a_3]^2}.$$

Substitution of these derivatives into the best-fit condition (6.59) yields three simultaneous equations in a_1, a_2, and a_3 that we need to solve in order to fit the $N_D = 9$ data points (x_i, y_i) in Table 6.1:

$$\sum_{i=1}^{9} \frac{y_i - g(x_i, a)}{(x_i - a_2)^2 + a_3} = 0, \qquad \sum_{i=1}^{9} \frac{y_i - g(x_i, a)}{[(x_i - a_2)^2 + a_3]^2} = 0,$$

$$\sum_{i=1}^{9} \frac{\{y_i - g(x_i, a)\}(x_i - a_2)}{[(x_i - a_2)^2 + a_3]^2} = 0. \tag{6.62}$$

Even without the substitution of (6.58) for $g(x, a)$, it is clear that these three equations depend on the a's in a nonlinear fashion. That's okay because in Section 6.3 we derived the N-dimensional Newton–Raphson search for the roots of

$$f_i(a_1, a_2, \ldots, a_N) = 0, \ i = 1, N, \tag{6.63}$$

where we have made the change of variable $y_i \to a_i$ for the present problem. We use that same formalism here for the $N = 3$ equations (6.62) by writing them as

$$f_1(a_1, a_2, a_3) = \sum_{i=1}^{9} \frac{y_i - g(x_i, a)}{(x_i - a_2)^2 + a_3} = 0, \tag{6.64}$$

$$f_2(a_1, a_2, a_3) = \sum_{i=1}^{9} \frac{\{y_i - g(x_i, a)\}(x_i - a_2)}{[(x_i - a_2)^2 + a_3]^2} = 0, \tag{6.65}$$

$$f_3(a_1, a_2, a_3) = \sum_{i=1}^{9} \frac{y_i - g(x_i, a)}{[(x_i - a_2)^2 + a_3]^2} = 0. \tag{6.66}$$

Because $f_r \equiv a_1$ is the peak value of the cross section, $E_R \equiv a_2$ is the energy at which the peak occurs, and $\Gamma = 2\sqrt{a_3}$ is the full width of the peak at half-maximum, good guesses for the a's can be extracted from a graph of the data. To obtain the nine derivatives of the three f's with respect to the three unknown a's, we use two nested loops over i and j, along with the forward-difference approximation for the derivative

$$\frac{\partial f_i}{\partial a_j} \simeq \frac{f_i(a_j + \Delta a_j) - f_i(a_j)}{\Delta a_j}, \tag{6.67}$$

where Δa_j corresponds to a small, say $\leq 1\%$, change in the parameter value.

Nonlinear Fit Exercise Use the Newton–Raphson algorithm as outlined in Section 6.8 to conduct a nonlinear search for the best-fit parameters of the Breit–Wigner theory (6.58) to the data in Table 6.1. Compare the deduced values of (f_r, E_R, Γ) to that obtained by inspection of the graph.

6.9 Code Listings

Listing 6.1 The **Bisection.py** code is a simple implementation of the bisection algorithm for finding a zero of a function, in this case $2\cos x - x$.

```
1  # Bisection.py: zero of f(x) via Bisection algorithm within [a,b]

   from vpython import *
   eps = 1e-3;  Nmax = 100;  a = 0.0; b = 7.0      # Precision, [a,b]
5
   def f(x): return 2*cos(x) - x               # Your function here

   def Bisection(Xminus, Xplus, Nmax, eps):         # Do not change
9      for it in range(0, Nmax):
           x = (Xplus + Xminus)/2.
           print(" it =", it, " x = ", x, " f(x) =", f(x))
           if (f(Xplus)*f(x) > 0.): Xplus = x      # Change x+ to x
13         else: Xminus = x                         # Change x- to x
           if (abs(f(x) ) <= eps):                  # Converged?
               print("\n Root found with precision eps = ", eps)
               break
17         if it == Nmax-1: print ("\n No root after N iterations\n")
       return x

   root = Bisection(a, b, Nmax, eps)
21 print(" Root =", root)
```

Listing 6.2 **NewtonCD.py** Uses the Newton–Raphson method to search for a zero of the function $f(x)$. A central-difference approximation is used to determine fd/dx.

```
1  # NewtonCD.py     Newton Search with central difference

   from math import cos

5  x = 1111.;  dx = 3.e-4; eps = 0.002; Nmax = 100;       # Parameters

   def f(x):  return 2*cos(x) - x # Function

9  for it in range(0, Nmax + 1):
       F = f(x)
       if (abs(F) *<*= eps):                            # Converged?
           print("\n Root found, f(root) =", F, ", eps = " , eps)
13         break
       print("Iteration # = ", it, " x = ", x, " f(x) = ", F)
       df = (f(x+dx/2) -  f(x-dx/2))/dx                 # Central diff
       dx = - F/df
17     x   += dx                                        # New guess
```

Listing 6.3 **SplineInteract.py** Performs a cubic spline fit to data with interactive control.

```
1  # SplineInteract.py  Spline fit with slide to control number of points

   from visual import *;                    from visual.graph import *;
   from visual.graph import gdisplay, gcurve
5  from visual.controls import slider, controls, toggle

   x = array([0., 0.12, 0.25, 0.37, 0.5, 0.62, 0.75, 0.87, 0.99])   # input
   y = array([10.6, 16.0, 45.0, 83.5, 52.8, 19.9, 10.8, 8.25, 4.7])
9  n = 9;   np = 15

   # Initialize
   y2 = zeros( (n), float); u = zeros( (n), float)
```

```
13  graph1 = gdisplay(x=0,y=0,width=500, height=500,
                        title='Spline Fit', xtitle='x', ytitle='y')
    funct1 = gdots(color = color.yellow)
    funct2 = gdots(color = color.red)
17  graph1.visible = 0

    def update():                               # Nfit = 30 = output
        Nfit = int(control.value)
21      for i in range(0, n):                   # Spread out points
            funct1.plot(pos = (x[i], y[i]) )
            funct1.plot(pos = (1.01*x[i], 1.01*y[i]) )
            funct1.plot(pos = (.99*x[i], .99*y[i]) )
25          yp1 = (y[1]-y[0]) / (x[1]-x[0]) - (y[2]-y[1])/ \
                    (x[2]-x[1])+(y[2]-y[0])/(x[2]-x[0])
            ypn = (y[n-1] - y[n-2])/(x[n-1] - x[n-2]) - (y[n-2]-y[n-3])/(x[n-2]-x[n-3]) +
                    (y[n-1]-y[n-3])/(x[n-1]-x[n-3])
            if (yp1 > 0.99e30):  y2[0] = 0.;  u[0] = 0.
29          else:
                y2[0] = - 0.5
                u[0] = (3./(x[1] - x[0]) )*( (y[1] - y[0])/(x[1] - x[0]) - yp1)
            for i in range(1, n - 1):                    # Decomp loop
33              sig = (x[i] - x[i - 1])/(x[i + 1] - x[i - 1])
                p = sig*y2[i - 1] + 2.
                y2[i] = (sig - 1.)/p
                u[i] = (y[i+1]-y[i])/(x[i+1]-x[i]) - (y[i]-y[i-1])/(x[i]-x[i-1])
37              u[i] = (6.*u[i]/(x[i + 1] - x[i - 1]) - sig*u[i - 1])/p
            if (ypn > 0.99e30):  qn = un = 0.            # Test for natural
            else:
                qn = 0.5;
41              un = (3/(x[n-1]-x[n-2]))*(ypn - (y[n-1]-y[n-2])/(x[n-1]-x[n-2]))
            y2[n - 1] = (un - qn*u[n - 2])/(qn*y2[n - 2] + 1.)
            for k in range(n - 2, 1, - 1):
                y2[k] = y2[k]*y2[k + 1] + u[k]
45          for i in range(1, Nfit + 2):                 # Begin fit
                xout = x[0] + (x[n - 1] - x[0])*(i - 1)/(Nfit)
                klo = 0;      khi = n - 1                 # Bisection algor
                while (khi - klo >1):
49                  k = (khi + klo) >> 1
                    if (x[k] > xout): khi = k
                    else: klo = k
                h = x[khi] - x[klo]
53                  if (x[k] > xout):  khi = k
                    else: klo = k
                h = x[khi] - x[klo]
                a = (x[khi] - xout)/h
57              b = (xout - x[klo])/h
                yout = a*y[klo] + b*y[khi] +
                        ((a*a*a-a)*y2[klo]+(b*b*b-b)*y2[khi])*h*h/6
                funct2.plot(pos = (xout, yout) )
                    c = controls(x=500,y=0,width=200,height=200)      # Control via slider
61  control = slider(pos=(-50,50,0), min = 2, max = 100, action = update)
    toggle(pos = (0, 35, - 5), text1 = "Number of points", height = 0)
    control.value = 2
    update()
65
    while 1:
        c.interact()
        rate(50)                                          # update < 10/sec
69      funct2.visible = 0
```

Listing 6.4 Fit.py Performs a least-squares fit of a parabola to data using the NumPy linage package to solve the set of linear equations $S\vec{a} = \vec{s}$.

```
1  # Fit.py: Linear least square fit via matrix solution

   import pylab as p
   from numpy import*; from numpy.linalg import inv, solve
5
```

```
   Nd = 7
   A = zeros( (3,3), float );   bvec = zeros((3,1), float)          # Initialize
   ss= sx = sxx = sy = sxxx = sxxxx = sxy = sxy = sxxy = 0.
 9 x = array([1., 1.1, 1.24, 1.35, 1.451, 1.5, 1.92])              # x values
   y = array([0.52, 0.8, 0.7, 1.8, 2.9, 2.9, 3.6])                 # y values
   sig = array([0.1, 0.1, 0.2, 0.3, 0.2, 0.1, 0.1])               # Error bars
   xRange = arange(1.0, 2.0, 0.1)                                  # For plots
13 p.plot(x, y, 'bo')                                              # Blue data
   p.errorbar(x,y,sig)
   p.title('Least Square Fit of Parabola to Blue Data')
   p.xlabel('x');  p.ylabel('y');   p.grid(True)                  # Plot grid
17
   for i in range(0, Nd):
           sig2 = sig[i] * sig[i]
           ss  += 1. / sig2;    sx   += x[i]/sig2;        sy    += y[i]/sig2
21         rhl  = x[i] * x[i];  sxx  += rhl/sig2;   sxxy  += rhl * y[i]/sig2
           sxy += x[i]*y[i]/sig2; sxxx +=rhl*x[i]/sig2; sxxxx +=rhl*rhl/sig2
   A   = array([ [ss,sx,sxx], [sx,sxx,sxxx], [sxx,sxxx,sxxxx] ])
   bvec = array([sy, sxy, sxxy])
25 xvec = multiply(inv(A), bvec)                                 # Invert matrix
   print('\n x via Inverse A\n', xvec, '\n' )
   xvec = solve(A, bvec)                                          # Solve via elimination
   print('\n x via Elimination \n', xvec, '\n Fit to Parabola\n')
29 print('y(x) = a0 + a1 x + a2 x^2\n a0 =', x[0],'a1 =', x[1], 'a2 =', x[2])
   print('\n i   xi      yi      yfit  ')
   for i in range(0, Nd):
       s = xvec[0] + xvec[1]*x[i] + xvec[2]*x[i]*x[i]
33     print(" %d %5.3f  %5.3f  %8.7f" %(i, x[i], y[i], s))
   # red line is the fit, red dots the fits at y[i]m
   curve  = xvec[0] + xvec[1]*xRange + xvec[2]*xRange**2
   points = xvec[0] + xvec[1]*x + xvec[2]*x**2
37 p.plot(xRange, curve,'r', x, points, 'ro')
   p.show()
```

7

Matrix Computing and N–D Searching

This chapter discusses how to compute with matrices, and, in particular, the use of the Python matrix and linear algebra packages. The chapter ends with a discussion of how to speed up large matrix computations.

7.1 Masses on a String and N–D Searching

Problem Two masses with weights $(W_1, W_2) = (10, 20)$ are connected by three pieces of string with lengths $(L_1, L_2, L_3) = (3, 4, 4)$, and hung from a horizontal bar of length $L = 8$ (Figure 7.1). Find the angles assumed by the strings and the tensions exerted by the strings.

In spite of the fact that this is a simple problem requiring no more than first-year physics to formulate, the coupled transcendental equations that result are just about impossible to solve analytically.[1] We approach it as a matrix problem combined with a trial-and-error search.

We start with the geometric constraints that the horizontal length of the structure is L, and that the strings begin and end at the same height (Figure 7.1):

$$L_1 \cos \theta_1 + L_2 \cos \theta_2 + L_3 \cos \theta_3 = L, \tag{7.1}$$

$$L_1 \sin \theta_1 + L_2 \sin \theta_2 - L_3 \sin \theta_3 = 0, \tag{7.2}$$

$$\sin^2 \theta_1 + \cos^2 \theta_1 = 1, \tag{7.3}$$

$$\sin^2 \theta_2 + \cos^2 \theta_2 = 1, \tag{7.4}$$

$$\sin^2 \theta_3 + \cos^2 \theta_3 = 1. \tag{7.5}$$

Observe that since we treat $\sin \theta$ and $\cos \theta$ as independent variables, we have included three trigonometric identities as independent equations. The basics physics (Figure 7.2) says that because there are no accelerations, the sum of the forces in the horizontal and vertical

1 Almost impossible anyway, as L. Molar has supplied me with an analytic solution.

Computational Physics: Problem Solving with Python, Fourth Edition.
Rubin H. Landau, Manuel J. Páez, and Cristian C. Bordeianu.
© 2024 WILEY-VCH GmbH. Published 2024 by WILEY-VCH GmbH.

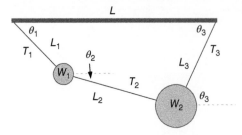

Figure 7.1 Two masses with weights (W_1, W_2) are connected by three pieces of string of lengths (L_1, L_2, L_3), and hung from a horizontal bar of length L. The lengths are all known, but the angles and the tensions in the strings are to be determined.

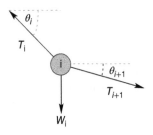

Figure 7.2 A free-body diagram for one weight in equilibrium. Balancing the forces in the x and y directions for all weights leads to the equations of static equilibrium.

directions must equal zero:

$$T_1 \sin \theta_1 - T_2 \sin \theta_2 - W_1 = 0, \tag{7.6}$$

$$T_1 \cos \theta_1 - T_2 \cos \theta_2 = 0, \tag{7.7}$$

$$T_2 \sin \theta_2 + T_3 \sin \theta_3 - W_2 = 0, \tag{7.8}$$

$$T_2 \cos \theta_2 - T_3 \cos \theta_3 = 0. \tag{7.9}$$

Here W_i is the weight of mass i and T_i is the tension in string i. Note that because we do not have a rigid structure, we cannot assume an equilibrium of torques.

Equations (7.1)–(7.9) are nine simultaneous, *nonlinear* equations, which being nonlinear cannot be solved with *linear* algebra. However, you can extend the Newton–Raphson algorithm to multiple equations, and *search* for a solution. To that effect, we rename the nine unknown angles and tensions as the subscripted variable y_i, and place all of the variables into a vector:

$$\mathbf{y} = \begin{bmatrix} x_1 \\ x_2 \\ x_3 \\ x_4 \\ x_5 \\ x_6 \\ x_7 \\ x_8 \\ x_9 \end{bmatrix} = \begin{bmatrix} \sin \theta_1 \\ \sin \theta_2 \\ \sin \theta_3 \\ \cos \theta_1 \\ \cos \theta_2 \\ \cos \theta_3 \\ T_1 \\ T_2 \\ T_3 \end{bmatrix}. \tag{7.10}$$

The nine equations to be solved are written in a general form with zeros on the right-hand sides (RHS), and also placed in a vector:

$$f_i(x_1, x_2, \ldots, x_N) = 0, \quad i = 1, N, \tag{7.11}$$

$$\mathbf{f}(\mathbf{y}) = \begin{bmatrix} f_1(\mathbf{y}) \\ f_2(\mathbf{y}) \\ f_3(\mathbf{y}) \\ f_4(\mathbf{y}) \\ f_5(\mathbf{y}) \\ f_6(\mathbf{y}) \\ f_7(\mathbf{y}) \\ f_8(\mathbf{y}) \\ f_9(\mathbf{y}) \end{bmatrix} = \begin{bmatrix} 3x_4 + 4x_5 + 4x_6 - 8 \\ 3x_1 + 4x_2 - 4x_3 \\ x_7 x_1 - x_8 x_2 - 10 \\ x_7 x_4 - x_8 x_5 \\ x_8 x_2 + x_9 x_3 - 20 \\ x_8 x_5 - x_9 x_6 \\ x_1^2 + x_4^2 - 1 \\ x_2^2 + x_5^2 - 1 \\ x_3^2 + x_6^2 - 1 \end{bmatrix} = \mathbf{0}. \tag{7.12}$$

In words, we are looking for set of nine x_i values for which all nine f_i's vanish simultaneously. Although these equations are not very complicated (the physics after all is elementary), the terms quadratic in x make them nonlinear.

The search procedure guesses a solution, expands the nonlinear equations and keeps just the linear terms, solves the linear equations, and makes a better guess based on how close the previous guess was to making $\mathbf{f} = 0$. The search starts with the approximate solution at any one stage called the set x_i, and assumes that there are (yet unknown) corrections Δx_i for which

$$f_i(x_1 + \Delta x_1, x_2 + \Delta x_2, \ldots, x_9 + \Delta x_9) = 0, \ i = 1, 9. \tag{7.13}$$

We solve for the approximate Δx_i's by assuming that our previous solution is close enough to the actual one for two terms in the Taylor series to be accurate:

$$f_i(x_1 + \Delta x_1, .., x_9 + \Delta x_9) \simeq f_i(x_1, .., x_9) + \sum_{j=1}^{9} \frac{\partial f_i}{\partial x_j} \Delta x_j = 0, \ i = 1, 9.$$

We now have a set of nine linear equations in the nine unknowns Δx_i, which we express as a single matrix equation

$$f_1 + \partial f_1/\partial x_1 \, \Delta x_1 + \partial f_1/\partial x_2 \, \Delta x_2 + \cdots + \partial f_1/\partial x_9 \, \Delta x_9 = 0,$$
$$f_2 + \partial f_2/\partial x_1 \, \Delta x_1 + \partial f_2/\partial x_2 \, \Delta x_2 + \cdots + \partial f_2/\partial x_9 \, \Delta x_9 = 0,$$
$$\ddots$$
$$f_9 + \partial f_9/\partial x_1 \, \Delta x_1 + \partial f_9/\partial x_2 \, \Delta x_2 + \cdots + \partial f_9/\partial x_9 \, \Delta x_9 = 0,$$

or

$$\begin{bmatrix} f_1 \\ f_2 \\ \ddots \\ f_9 \end{bmatrix} + \begin{bmatrix} \partial f_1/\partial x_1 & \partial f_1/\partial x_2 & \cdots & \partial f_1/\partial x_9 \\ \partial f_2/\partial x_1 & \partial f_2/\partial x_2 & \cdots & \partial f_2/\partial x_9 \\ & \ddots & & \\ \partial f_9/\partial x_1 & \partial f_9/\partial x_2 & \cdots & \partial f_9/\partial x_9 \end{bmatrix} \begin{bmatrix} \Delta x_1 \\ \Delta x_2 \\ \ddots \\ \Delta x_9 \end{bmatrix} = 0. \tag{7.14}$$

Note now that the derivatives and the f's are all evaluated at known values of the x_i's, so that only the vector of the Δx_i values is unknown. We write this equation in matrix notation as

$$\mathbf{f} + F' \Delta \mathbf{x} = 0, \quad \Rightarrow \quad F' \Delta \mathbf{x} = -\mathbf{f}, \tag{7.15}$$

$$\Delta \mathbf{x} = \begin{bmatrix} \Delta x_1 \\ \Delta x_2 \\ \vdots \\ \Delta x_9 \end{bmatrix}, \quad \mathbf{f} = \begin{bmatrix} f_1 \\ f_2 \\ \vdots \\ f_9 \end{bmatrix}, \quad F' = \begin{bmatrix} \partial f_1/\partial x_1 & \cdots & \partial f_1/\partial x_9 \\ \partial f_2/\partial x_1 & \cdots & \partial f_2/\partial x_9 \\ & \ddots & \\ \partial f_9/\partial x_1 & \cdots & \partial f_9/\partial x_9 \end{bmatrix},$$

where we use bold characters to denote the vector and matrix nature of the equations.

The equation $F' \Delta \mathbf{x} = -\mathbf{f}$ is in the standard form for the solution of a linear equation (often written $A\mathbf{x} = \mathbf{b}$), where $\Delta \mathbf{x}$ is the vector of unknowns and $\mathbf{b} = -\mathbf{f}$. Matrix equations are solved using the techniques of linear algebra, which we will discuss shortly. In a formal sense, the solution of (7.15) is obtained by multiplying both sides of the equation by the inverse of the F' matrix:

$$\Delta \mathbf{x} = -F'^{-1} \mathbf{f}, \tag{7.16}$$

where the inverse must exist if there is to be a unique solution. Although we are dealing with matrices now, this solution is identical in form to that of the 1D problem, equation (6.11), $\Delta x = -(1/f')f$. The abstract notation for matrices is seen to reveal the simplicity that lies within.

As we have noted for the single-equation Newton–Raphson method in Section 6.3, even if we can derive analytic expressions for the derivatives $\partial f_i/\partial x_j$, there are $9 \times 9 = 81$ such derivatives for this (small) problem, and entering them all would be both time-consuming and error-prone. In contrast, it is straightforward to program up a forward-difference approximation for the derivatives,

$$\frac{\partial f_i}{\partial x_j} \simeq \frac{f_i(x_j + \Delta x_j) - f_i(x_j)}{\Delta x_j}, \tag{7.17}$$

where, for partial derivatives, each individual x_j is varied independently, and the δx_j are arbitrary small changes. While a central-difference approximation for the derivative would be more accurate, it would also require more evaluations of the f's, and once we find a solution it does not matter how accurate our algorithm for the derivative was.

As also discussed for the 1D Newton–Raphson method (Section 6.3.1), the method can fail if the initial guess is not close enough to the zeros of all nine f's. The *backtracking* technique (using fractional Δx guesses) may be applied here as well, in the present case, progressively decreasing the corrections Δx_i until $|f|^2 = |f_1|^2 + |f_2|^2 + \cdots + |f_N|^2$ becomes acceptably small.

7.2 Matrix Generalities

Often a physical theory is easier to understand when expressed more abstractly with matrices, an example being the rotation of solid bodies with the inertia tensor. It should not be surprising then that scientific computing often involve matrices. This is a good thing since computers are good at continued repetition of simple instructions, and that is just what matrix manipulations involve. And as physical systems get more realistic and more complex, the equations used to describe them often involves the manipulations of large matrices, which are no harder to program up than small ones.

We strongly recommend the use of the powerful and robust linear algebra libraries that have been perfected over decades. Their programs are usually an order of magnitude, or more, faster than the elementary methods found in linear algebra texts,[2] are designed to minimize round-off error, and are often "robust," that is, have a high chance of working well for a broad class of problems. An additional value of library routines is that you can often run the same program either on a desktop machine or on a parallel supercomputer, with matrix routines automatically adapting to the local architecture.

The most basic matrix problem is the system of linear equations:

$$A\mathbf{x} = \mathbf{b}, \tag{7.18}$$

where A is a known $N \times N$ matrix, $\mathbf{x}$ is an unknown vector of length N, and $\mathbf{b}$ is a known vector of length N. The obvious way to solve this equation is to multiply both sides by the inverse of A:

$$\mathbf{x} = A^{-1}\mathbf{b}. \tag{7.19}$$

Both the direct solution (7.18) as it stands, and the determination of a matrix's inverse are standards in subroutine libraries. The direct solution via Gaussian elimination or lower–upper (LU) decomposition tends to be faster, but sometime you may want the inverse for other purposes.

If you have to solve the matrix equation

$$A\mathbf{x} = \lambda\mathbf{x}, \tag{7.20}$$

with $\mathbf{x}$ an unknown vector and λ an unknown parameter, then the solution (7.19) will not be of much help because the RHS contains the unknowns λ and $\mathbf{x}$. Equation (7.20) is the *eigenvalue problem*, and its solutions exist for only certain, if any, values of λ. To find the solution, we use the identity matrix I to rewrite (7.20) as

$$[A - \lambda I]\mathbf{x} = 0. \tag{7.21}$$

We see that multiplication of (7.21) by $[A - \lambda I]^{-1}$ yields the *trivial solution*

$$\mathbf{x} = 0 \quad \text{(trivial solution)}. \tag{7.22}$$

While the trivial solution is a bona fide solution, it is nonetheless trivial. A more interesting solution results from the condition that forbids us from multiplying both sides of (7.21) by $[A - \lambda I]^{-1}$, namely, the nonexistence of the inverse. If you recall that Crammer's rule for the inverse requires division by $\det[A - \lambda I]$, it is clear that the inverse fails to exist (and in this way eigenvalues *do* exist) when

$$\det[A - \lambda I] = 0. \tag{7.23}$$

The λ values that satisfy this *secular equation* are the eigenvalues of (7.20). If you are interested in only the eigenvalues for (7.20), you should look for a matrix routine that

---

2 Although we prize the book Press *et al.* [2007] and what it has accomplished, we cannot recommend taking matrix subroutines from it. They are neither optimized nor documented for easy, stand-alone use, whereas the subroutine libraries recommended in this chapter are.

solves (7.23). First you need a subroutine to calculate the determinant of a matrix, and then a search routine to zero in on the solution of (7.23). Such routines are available in libraries.

Many programming bugs arise from the improper use of arrays.[3] This may be as a result of the extensive use of matrices in scientific computing, or to the complexity of keeping track of indices and dimensions. In any case, here are some rules of thumb to observe:

Tests: *Always* test a library routine on a small problem whose answer you know (such as the tests in Section 7.4). Then you'll know if you are supplying it with the right arguments and if you have all the links working.

Paging: Operating systems store data and variables in fixed-length, contiguous blocks of memory called *pages* that are treated as single entries in a page table. If there is not enough room in RAM for a page, then the entire page gets stored in *virtual memory*, which means it gets placed on a slow disk. This is called *paging*. If your program deals with matrices that take up lots of memory, it may be near the memory limit at which paging occurs, and then even a slight increase in the matrix's size may lead to an order-of-magnitude increase in execution time. (There are similar issues with computer caches, which are small amounts of superfast memory, such as 1–8 MB, that feeds the CPU.)

Computers are finite: Unless you are careful, your matrices may use up so much memory that your computation will slow down significantly, especially if it starts to use virtual memory. As a case in point, let's say that you performing some matrix calculations involving 4-D matrices, with each index having a *physical dimension* of 200, for example, A[200] [200] [200] [200]. A single array of $(200)^4$ 64-byte words occupies $\simeq 16$ GB of memory.

Processing time: Matrix operations such as inversion require on the order of N^3 steps for a square matrix of dimension N. Therefore, doubling the dimensions of a 2D square matrix (as happens when the number of integration steps is doubled) leads to an *eightfold* increase in processing time.

Matrix storage: While we think of matrices as multidimensional blocks of stored numbers, the computer stores them as linear strings. For instance, a matrix a[3,3] in Python, is stored in *row-major order*:

$$a_{0,0} \; a_{0,1} \; a_{0,2} \; a_{1,0} \; a_{1,1} \; a_{12} \; a_{2,0} \; a_{2,1} \; a_{2,2} \; \dots .$$

This differs from Fortran, where subscripts usually start at 1, and where the storage is in *column-major order*:

$$a_{1,1} \; a_{2,1} \; a_{3,1} \; a_{1,2} \; a_{2,2} \; a_{3,2} \; a_{1,3} \; a_{2,3} \; a_{3,3} \; \dots .$$

It is important to keep this storage scheme in mind since some routines assume linear storage, and sometimes Python and Fortran programs get intermixed.

Minimizing stride: *Stride*, the amount of memory skipped in order to get to the next element needed in a calculation, should be minimized. For instance, summing the diagonal elements of a matrix to form the trace

$$\operatorname{Tr} A = \sum_{i=1}^{N} a(i, i) \tag{7.24}$$

---

3 Even a vector $V(N)$ is called an "array," albeit a 1D one.

involves large stride because the diagonal elements are stored far apart for large N. However, the sum

$$b(i) = a(i) + a(i+1) \tag{7.25}$$

has stride 1 because adjacent elements of a are accessed. The basic rule for accessing indexed variables is

- Keep the stride low, preferably at 1, which in practice means:
- Vary the rightmost index first on Python and C arrays.

Accessing matrices: Sometimes your effort at elegant programming may be very inefficient. For example, it may be elegant to put all your data in one matrix with many indices, such as $V_{N,M,k,k',Z,A}$, but it may require the computer to make large *strides* as you go through thousands of k and k', values. A more efficient approach might be to break up the data into several matrices, each with fewer indices, such as $V_{N,M}$, $U_{k,k'}$, and $W_{Z,A}$.

7.3 Matrices in Python

7.3.1 Lists as Arrays

A *list* is Python's built-in sequence of numbers or objects. Although called a "list," it is similar to what other computer languages call an "array." It may be easier for you to think of a Python list as a container that holds a bunch of items in a definite order. (Soon we will describe the higher-level, and recommended, *array* data type available with the *NumPy* package.) In this section, we review some of Python's native *list* features.

Python interprets a sequence of ordered items, $L = l_0, l_1, \ldots, l_{N-1}$, as a *list* and represents it with a single symbol L:

```
>>> L = [1, 2, 3]              # Create list
>>> L[0]                       # Print element 0 (first)
1                              # Python output          4
>>> L                          # Print entire list
[1, 2, 3]                      # Output
>>> L[0] = 5                   # Change element 0
>>> L                                                   8
[5, 2, 3]
>>> len(L)                     # Length of list
3
>>> for items in L: print items   # for loop over items  12
5
2
3
```

Observe that square brackets with comma separators such as [1, 2, 3] are used for lists, and that a square bracket is also used to indicate the index for a list item, as in line 2 (L[0]). Lists contain sequences of arbitrary objects that are *mutable* or changeable. As we see in line 7 in the L command, an entire list can be referenced as a single object, in this case to obtain its printout.

Python also has a built-in type of list, known as a *tuple*, with elements that are not mutable. Tuples are indicated by round parenthesis (.., .., .), with individual elements still referenced by square brackets:

```
>>> T = (1, 2, 3, 4)                    # Create tuple
>>> T[3]                                # Print element 3
4
>>> T
    (1, 2, 3, 4)                        # Print entire tuple
>>> T[0] = 5                            # Attempt to change element 0
    Traceable (most recent call last):
    T[0] = 5
    Error: 'tuple' object does not support item assignment
```

Note the error message that arises when we try to change an element of a tuple.

Many languages require you to specify the size of an array before you can start storing objects in it. In contrast, Python lists are *dynamic*, which means that their sizes adjust as needed. In addition, while a list is essentially one dimensional, a compound list can be created by having the individual elements themselves as lists:

```
>>> L = [[1,2], [3,4], [5,6]]           # A list of lists
>>> L
    [[1, 2], [3, 4], [5, 6]]
>>> L[0]                                # The first element
    [1, 2]
```

Here are some more list operations:

Operation	Effect	Operation	Effect
L = [1, 2, 3, 4]	Form list	L1 + L2	Concatenate lists
L[i]	*i*th element	len(L)	Length of list *L*
i in L	True if *i* in *L*	L[i:j]	Slice from *i* to *j*
for i in L	Iteration index	L.append(x)	Append *x* to end of *L*
L.count(x)	Number of *x*'s in *L*	L.index(x)	Location of 1st *x* in *L*
L.remove(x)	Remove 1st *x* in *L*	L.reverse()	Reverse elements in *L*
L.sort()	Order elements in *L*		

7.3.2 NumPy Matrices

Although we have just described Python's basic `array` data type, it is rather limited and we suggest using NumPy arrays, which converts Python lists into arrays. In order to use NumPy, you must import *NumPy* into your programs, as we show here running our program `Matrix.py` from a shell (the >>>):

```
>>> from numpy import *                  # Import NumPy package
>>> vector1 = array([1, 2, 3, 4, 5])     # Fill 1D array
>>> print('vector1 =',vector1)           # Print array (parens if Python 3)
vector1 =  [1 2 3 4 5]                    # Output
>>> vector2 = vector1 + vector1          # Add 2 vectors
>>> print('vector2=',vector2)            # Print vector2
vector2= [ 2  4  6  8 10]                 # Output
>>> vector2 = 3 * vector1                # Multi array by scalar
>>> print ('3 * vector1 = ', vector2)    # Print vector
3 * vector1 =  [ 3  6  9 12 15]           # Output
>>> matrix1 = array(([0,1],[1,3]))       # An array of arrays
>>> print(matrix1)                       # Print matrix1
[[0 1]
 [1 3]]
```

```
>>> print ('vector1.shape= ',vector1.shape)                              15
vector1.shape =  (5)
>>> print  (matrix1 * matrix1)               # Matrix multiply          17
  [[0  1]
   [1  9]]                                                               19
```

We see here that we have initialized an array object, have added two 1D array objects together, and have printed out the result. Likewise, we see that multiplying an array by a constant does, in fact, multiply each element by that constant (line 8). We then construct a "matrix" as a 1D array of two 1D arrays, and when we print it out, we note that it does indeed look like a matrix. However, when we multiply this matrix by itself, the result is not the $\begin{bmatrix} 1 & 3 \\ 3 & 10 \end{bmatrix}$, that one normally expects from matrix multiplication. So if you need actual mathematical matrices, then you need to use NumPy!

Now we give some examples of the use of NumPy, but do refer the reader to the NumPy Tutorial [NumPy, 2023] and to the articles in *Computing in Science & Engineering* [CiSE, 2015] for more information. To start, we note that a NumPy array can hold up to 32 dimensions (32 indices), but each element must be of the same type (a *uniform* array). The elements are not restricted to just floating-point numbers or integers, but can be any object, as long as all elements are of this same type. (Compound objects may be useful, for example, for storing parts of data sets.) There are various ways to create arrays, with square brackets [...] used for indexing in all cases. We start with a Python list (tuples work as well) and create an array from it:

```
>>> from NumPy  import *                                                 1
>>> a = array( [1, 2, 3, 4] )          # Array from a list
>>> a                                  # Check with print               3
array([1, 2, 3, 4])
```

Notice that it is essential to have the square brackets within the round parentheses because the square brackets produce the list object while the round parentheses indicate a function argument. Note too that because the data in our original list were all integers, the created array is one of 32-bit integer data types, which we can check by affixing the `type` method:

```
>>> a.dtype
type('int32')                                                            2
```

If we had started with floating-point numbers, or a mix of floats and int's, we would have ended up with floating-point arrays:

```
>>> b = array([1.2, 2.3, 3.4])
>>> b                                                                    2
array([ 1.2,   2.3,   3.4])
>>> b.dtype                                                              4
type('float64')
```

When describing NumPy arrays, the number of "dimensions," `dim`, means the number of indices, which as we said can be as high as 32. What might be called the "size" or "dimensions" of a matrix in mathematics is called the *shape* of a NumPy array.

Furthermore, NumPy does have a `size` method that returns the total number of elements. Because Python's lists and tuples are all one dimensional, if we want an array of a particular shape, we can attain that by affixing the `reshape` method when we create the array. Where Python has a `range` function to generate a sequence of numbers, NumPy has an `arrange` function that creates an array, rather than a list. Here we use it and then reshape the 1D array into a 3 × 4 array:

```
>>> import numpy as np                                              1
>>> np.arange(12)                      # List of 12 int's in 1D array
array([ 0,  1,  2,  3,  4,  5,  6,  7,  8,  9, 10, 11])             3
>>> np.arange(12).reshape((3,4))       # Create, shape to 3x4 array
array([[ 0,  1,  2,  3],                                           5
       [ 4,  5,  6,  7],
       [ 8,  9, 10, 11]])                                          7
>>> a = np.arange(12).reshape((3,4))          # Give array a name
>>> a                                                              9
array([[ 0,  1,  2,  3],
       [ 4,  5,  6,  7],                                          11
       [ 8,  9, 10, 11]])
>>> a.shape                                    # Shape = ?         13
(3L, 4L)
>>> a.ndim                                     # Dimension?       15
2
>>> a.size                      # Size of a (number of elements)? 17
12
```

Note that here we have imported NumPy as the object `np`, and then affixed the `arrange` and `reshape` methods to this object. We then checked the shape of `a`, and found it to have three rows and four columns of long integers (Python 3 may just say `int`). Note too, as we see on line 9, NumPy uses parentheses () to indicate the shape of an array, and so `(3L, 4L)` indicates an array with three rows and four columns of long int's.

Now that we have shapes on our minds, we should note that NumPy offers a number of ways to change shapes. For example, we can transpose an array with the `.T` method, or `reshape` into a vector:

```
>>> from numpy import *
>>> a = arrange(12).reshape((3,4))              # Give array a name    2
>>> a
array([[ 0,  1,  2,  3],                                               4
       [ 4,  5,  6,  7],
       [ 8,  9, 10, 11]])                                              6
>>> a.T                                          # Transpose
array([[ 0,  4,  8],                                                   8
       [ 1,  5,  9],
       [ 2,  6, 10],                                                  10
       [ 3,  7, 11]])
>>> b = a.reshape( (1,12) )                       # Form vector length 12  12
>>> b
array([[ 0,  1,  2,  3,  4,  5,  6,  7,  8,  9, 10, 11]])              14
```

And again, `(1,12)` indicates an array with one row and 12 columns. Yet another handy way to take a matrix and extract just what you want from it is to use Python's *slice* operator `start:stop:step:` to take a slice out of an array:

```
>>> a
array([[ 0,  1,  2,  3],                                               2
       [ 4,  5,  6,  7],
       [ 8,  9, 10, 11]])                                              4
```

```
>>> a[:2, :]                              # First 2 rows
array([[0, 1, 2, 3],
       [4, 5, 6, 7]])
>>> a[:,1:3]                              # Columns 1-3
array([[ 1,  2],
       [ 5,  6],
       [ 9, 10]])
```

Note here how Python indices start counting from 0, and so 1:3 means indices 0, 1, 2 (without the 3). Slicing can be very useful in speeding up programs by picking out and placing in memory just the specific data elements from a large data set that need to be processed. This avoids the time-consuming jumping through large segments of memory, as well as excessive reading from disk.

Finally, we remind you that while all elements in a NumPy array must be of the same data type, that data type can be compound. For example, an array of arrays:

```
>>> from numpy import *
>>> M = array( [ (10, 20), (30,40), (50, 60) ] )     # Array of 3 arrays
>>> M
array([[10, 20],
       [30, 40],
       [50, 60]])
>>> M.shape
(3L, 2L)
>>> M.size
6
>>> M.dtype
type('int32')
```

Furthermore, an array can be composed of complex numbers by specifying the `complex` data type as an option on the `array` command. NumPy then uses the j symbol for the imaginary number i:

```
>>> c = array( [ [1,complex(2,2)], [complex(3,2),4] ], dtype=complex )
>>> c
array([[ 1.+0.j,  2.+2.j],
       [ 3.+2.j,  4.+0.j]])
```

In Section 7.3.3, we discuss using true mathematical matrices with NumPy, which is one use of an array object. Here we note that if you wanted the familiar matrix product from two arrays, you would use the `dot` function, whereas `*` is used for an element-by-element (direct) product:

```
>>> matrix1= array( [[0,1], [1,3]])
>>> matrix1
array([[0, 1],
       [1, 3]])
>>> print ( dot(matrix1,matrix1) )       # Matrix or dot product
[[ 1  3]
 [ 3 10]]
>>> print (matrix1 * matrix1)            # Element-by-element product
[[0 1]
 [1 9]]
```

NumPy is actually optimized to work well with arrays, and in part this is because arrays are handled and processed much as if they were simple, scalar variables.[4] For example, here is another example of *slicing*, a technique that is also used in ordinary Python with lists and tuples, in which two indices separated by a colon indicate a range:

```
from visual import *
2 stuff = zeros(10, float)
t = arrange(4)
stuff[3:7] = \sqrt(t+1)
```

Here we start by creating the NumPy array `stuff` of floats, all of whose 10 elements are initialized to zero. Then we create the array `t` containing the four elements [0, 1, 2, 3] by assigning 4 variables uniformly in the range 0–4 (the "a" in `arange` creates floating-point variables, `range` creates integers). Next, we use a slice to assign [sqrt(0+1), sqrt(1+1), sqrt(2+1), sqrt(3+1)] = [1, 1.414, 1.732, 2] to the middle elements of the `stuff` array. Note that the NumPy version of the `sqrt` function, one of many universal function (*functions*) supported by NumPy, has the amazing property of automatically outputting an array whose length is that of its argument, in this case, the array `t`. In general, much of the power of NumPy comes from its *broadcasting* operation, an operation in which values are assigned to multiple elements via a single assignment statement. Broadcasting permits Python to *vectorize* array operations, which means that the same operation can be performed on different array elements in parallel (or nearly so). Broadcasting also speeds up processing because array operations use C instead of Python, and with a minimum of array copies being made. Here is a simple aspect of broadcasting:

```
w = zeros(100, float)
w = 23.7
```

The first line creates the NumPy array `w`, and the second line "broadcasts" the value 23.7 to all elements in the array. There are many possible array operations in NumPy and various rules pertaining to them; we recommend that the serious user explore the extensive NumPy documentation for additional information.

7.3.3 NumPy Linear Algebra Library

The array objects of NumPy are not the same as mathematical matrices. Fortunately, there is NumPy's LinearAlgebra package that treats 2D arrays as mathematical matrices, and also provides a simple interface to the powerful *linear algebra package* (LAPACK) linear algebra library. As we keep saying, there is much to be gained in speed and reliability from using these libraries rather than writing your own matrix routines.

Our first example from linear algebra is the standard matrix equation

$$A\mathbf{x} = \mathbf{b}, \tag{7.26}$$

where we have used a bold character to represent a 1D matrix (a vector). Equation (7.26) describes a set of linear equations with $\mathbf{x}$ an unknown vector and A a known matrix. Now we take A to be a 3×3, $\mathbf{b}$ to be 3×1, and let the program figure out that $\mathbf{x}$ must be 3×1.[5]

---

4 We thank Bruce Sherwood for helpful comments on these points.

We start by importing all the packages, by inputting a matrix and a vector, and by printing out *A* and **x**:

```
    >>> from numpy import *
  2 >>> from numpy.linalg import*
    >>> A = array( [ [1,2,3], [22,32,42], [55,66,100] ] )   # Array of arrays
    >>> print ('A =', A)
    A = [[   1    2    3]
  6      [ 22  32  42]
         [ 55  66 100]]
    >>> b = array([1,2,3])
    >>> print ('b =', b)
 10 b = [1 2 3]
```

Seeing that we have the matrices A and **b**, we can go ahead and solve $A\mathbf{x} = \mathbf{b}$ using NumPy's `solve` command, and then test how close $A\mathbf{x} - \mathbf{b}$ is to a zero vector:

```
    >>> from numpy.linalg import solve
  2 >>> x = solve(A, b)                                 # Finds solution
    >>> print ('x =', x)
    x = [ -1.4057971   -0.1884058    0.92753623]       # The solution
    >>> print ('Residual =', dot(A, x) - b)            # LHS-RHS
  6
    Residual = [4.44089210e-16   0.00000000e+00  -3.55271368e-15]
```

This is really quite impressive. We have solved the entire set of linear equations (by elimination) with just the single command `solve`, performed a matrix multiplication with the single command `dot`, did a matrix subtraction with the usual operator, and are left with a residual essentially equal to machine precision.

Although there are more efficient numerical approaches, another way to solve

$$A\mathbf{x} = \mathbf{b} \tag{7.27}$$

is to calculate the inverse A^{-1}, and then multiply both sides of the equation by the inverse, yielding

$$\mathbf{x} = A^{-1}\mathbf{b}. \tag{7.28}$$

```
    >>> from numpy.linalg import in                                             1
    >>>   dot(in(A), A)                         # Test inverse
                                                                                3
    array([[  1.00000000e+00,  -1.33226763e-15,  -1.77635684e-15],
           [  8.88178420e-16,   1.00000000e+00,   0.00000000e+00],             5
           [ -4.44089210e-16,   4.44089210e-16,   1.00000000e+00]])
    >>> print ('x =', multiply(in(A), b))                                       7
    x = [-1.4057971   -0.1884058    0.92753623]          # Solution
    >>> print ('Residual =', dot(A, x) - b)                                     9

    Residual = [  4.44089210e-16   0.00000000e+00  -3.55271368e-15]            11
```

Here we first tested that `in(A)` is in fact the inverse of `A` by seeing if `A` times `in(A)` equals the identity matrix. Then we used the inverse to solve the matrix equation directly, and got the same answer as before and an error at the level of machine precision as before.

---

5 Don't be bothered by the fact that although we think of these vectors as 3×1, they sometimes get printed out as 1×3; think of all the trees being saved!

Our second example comes from finding the principal-axes of a cube, and requires us to find a coordinate system in which the inertia tensor is diagonal. This entails solving the eigenvalue problem,

$$I\omega = \lambda\omega, \tag{7.29}$$

where I is the inertia matrix (tensor), ω is an unknown eigenvector, and λ is an unknown eigenvalue. The program `Eigen.py` solves for the eigenvalues and vectors, and shows how easy it is to deal with matrices. Here is an interpretive version:

```
>>> from numpy import*                                                          1
>>> from numpy.linalg import big
>>> I = array( [[2./3,-1./4], [-1./4,2./3]] )                                   3
>>> print('I =\n', I)
I =                                                                             5
 [[ 0.66666667 -0.25       ]
  [-0.25        0.66666667]]                                                     7
>>> Es, evectors = big(A)                    # Solves eigenvalue problem
>>> print('Eigenvalues =', Es, '\n Eigenvector Matrix =\n', evectors)           9
Eigenvalues =   [ 0.91666667  0.41666667]
 Eigenvector Matrix =                                                          11
 [[ 0.70710678  0.70710678]
  [-0.70710678  0.70710678]]                                                   13
>>> vec = array([ evectors[0, 0], evectors[1, 0] ] )
>>> LHS = dot(I, vec)                         # Matrix x vector                 15
>>> RHS = Es[0]*vec                           # Scalar multi
>>> print('LHS - RHS =', LHS-RHS)             # Test for zero                   17
LHS - RHS = [  1.11022302e-16  -1.11022302e-16]
```

We see how, after setting up the array `I` on line 3, we solved for its eigenvalues and eigenvectors with the single statement `Es, evectors = big(I)` on line 8. We then extracted the first eigenvector on line 14, and use it, along with the first eigenvalue, to check that (7.29) is in fact satisfied to machine precision.

Well, we think by now you have some idea of the power of NumPy. In Table 7.1 we give some more NumPy operators.

7.4 Exercise: Tests Before Use

Before you direct the computer to go off crunching numbers on a million elements of some matrix, it's a good idea to try out your procedures on a small matrix, especially one for which you know the right answer. In this way, it will take you only a short time to realize how hard it is to get the calling procedure perfectly right! Here are some exercises.

1) Find the numerical inverse of $A = \begin{bmatrix} +4 & -2 & +1 \\ +3 & +6 & -4 \\ +2 & +1 & +8 \end{bmatrix}$

   a) As a general check, applicable even if you do not know the analytic answer, check your inverse in both directions; that is, check that $AA^{-1} = A^{-1}A = I$, and note the number of decimal places to which this is true. This also gives you some idea of the precision of your calculation.

Table 7.1 The operators of NumPy and their effects.

Operator	Effect	Operator	Effect
dot(a, b[,out])	Dot product arrays	vdot(a, b)	Dot product
inner(a, b)	Inner product arrays	outer(a, b)	Outer product
tensordot(a, b)	Tensor dot product	einsum()	Einstein sum
linalg.matrix_power(M, n)	Matrix to power n	kron(a, b)	Kronecker product
linalg.cholesky(a)	Cholesky decomp	linalg.qr(a)	QR factorization
linalg.svd(a)	Singular val decomp	linalg.eig(a)	Eigenproblem
linalg.eigh(a)	Hermitian eigen	linalg.eigvals(a)	General eigen
linalg.eigvalsh(a)	Hermitian eigenvals	linalg.norm(x)	Matrix norm
linalg.cond(x)	Condition number	linalg.det(a)	Determinant
linalg.slogdet(a)	Sign and log(det)	trace(a)	Diagnol sum
linalg.solve(a, b)	Solve equation	linalg.tensorsolve(a, b)	Solve $ax = b$
linalg.lstsq(a, b)	Least-squares solve	linalg.inv(a)	Inverse
linalg.pinv(a)	Penrose inverse	linalg.tensorinv(a)	Inverse N–D array

b) Determine the number of decimal places of agreement there is between your numer-
ical inverse and the analytic result: $A^{-1} = \dfrac{1}{263}\begin{bmatrix} +52 & +17 & +2 \\ -32 & +30 & +19 \\ -9 & -8 & +30 \end{bmatrix}$. Is this similar to the
error in AA^{-1}?

2) Consider the same matrix A as before, here being used to describe three simultaneous
linear equations, $Ax = b$, or explicitly,

$$\begin{bmatrix} a_{00} & a_{01} & a_{02} \\ a_{10} & a_{11} & a_{12} \\ a_{20} & a_{21} & a_{22} \end{bmatrix}\begin{bmatrix} x_0 \\ x_1 \\ x_2 \end{bmatrix} = \begin{bmatrix} b_0 \\ b_1 \\ b_2 \end{bmatrix}. \tag{7.30}$$

Now the vector **b** on the RHS is assumed known, and the problem is to solve for the
vector **x**. Use an appropriate subroutine to solve these equations for the three different
x vectors appropriate to these three different **b** values on the RHS:

$$b_1 = \begin{bmatrix} +12 \\ -25 \\ +32 \end{bmatrix}, \quad b_2 = \begin{bmatrix} +4 \\ -10 \\ +22 \end{bmatrix}, \quad b_3 = \begin{bmatrix} +20 \\ -30 \\ +40 \end{bmatrix}.$$

The solutions should be

$$x_1 = \begin{bmatrix} +1 \\ -2 \\ +4 \end{bmatrix}, \quad x_2 = \begin{bmatrix} +0.312 \\ -0.038 \\ +2.677 \end{bmatrix}, \quad x_3 = \begin{bmatrix} +2.319 \\ -2.965 \\ +4.790 \end{bmatrix}. \tag{7.31}$$

3) Consider the matrix $A = \begin{bmatrix} \alpha & \beta \\ -\beta & \alpha \end{bmatrix}$, where you are free to use any values you want for α
and β. Use a numerical eigenvalue solver to show that the eigenvalues and eigenvectors

are the complex conjugates

$$\mathbf{x}_{1,2} = \begin{bmatrix} +1 \\ \mp i \end{bmatrix}, \quad \lambda_{1,2} = \alpha \mp i\beta. \tag{7.32}$$

4) Use your eigenvalue solver to find the eigenvalues of the matrix

$$A = \begin{bmatrix} -2 & +2 & -3 \\ +2 & +1 & -6 \\ -1 & -2 & +0 \end{bmatrix}. \tag{7.33}$$

a) Verify that you obtain the eigenvalues $\lambda_1 = 5$, $\lambda_2 = \lambda_3 = -3$. Beware, double roots can cause problems. In particular, there is a uniqueness issue with their eigenvectors because any combination of these eigenvectors is also an eigenvector.

b) Verify that the eigenvector for $\lambda_1 = 5$ is proportional to

$$\mathbf{x}_1 = \frac{1}{\sqrt{6}}\begin{bmatrix} -1 \\ -2 \\ +1 \end{bmatrix}. \tag{7.34}$$

c) The eigenvalue -3 corresponds to a double root. This means that the corresponding eigenvectors are degenerate, which in turn means that they are not unique. Two linearly independent ones are

$$\mathbf{x}_2 = \frac{1}{\sqrt{5}}\begin{bmatrix} -2 \\ +1 \\ +0 \end{bmatrix}, \quad \mathbf{x}_3 = \frac{1}{\sqrt{10}}\begin{bmatrix} 3 \\ 0 \\ 1 \end{bmatrix}. \tag{7.35}$$

In this case, it's not clear what your eigenvalue solver will give for the eigenvectors. Try to find a relationship between your computed eigenvectors with the eigenvalue -3 and these two linearly independent ones.

5) Imagine that your model of some physical system results in $N = 100$ coupled linear equations in N unknowns:

$$a_{00}y_0 + a_{01}y_1 + \cdots + a_{0(N-1)}y_{N-1} = b_0,$$
$$a_{10}y_0 + a_{11}y_1 + \cdots + a_{1(N-1)}y_{N-1} = b_1,$$
$$\cdots$$
$$a_{(N-1)0}y_0 + a_{(N-1)1}y_1 + \cdots + a_{(N-1)(N-1)}y_{N-1} = b_{N-1}.$$

In many cases the a and b values are known, so your exercise is to solve for all the x values, taking a as the *Hilbert* matrix and $\mathbf{b}$ as its first column:

$$[a_{ij}] = a = \begin{bmatrix} \frac{1}{i+j-1} \end{bmatrix} = \begin{bmatrix} 1 & \frac{1}{2} & \frac{1}{3} & \frac{1}{4} & \cdots & \frac{1}{100} \\ \frac{1}{2} & \frac{1}{3} & \frac{1}{4} & \frac{1}{5} & \cdots & \frac{1}{101} \\ \ddots & & & & & \\ \frac{1}{100} & \frac{1}{101} & \cdots & & \cdots & \frac{1}{199} \end{bmatrix}, \tag{7.36}$$

$$[b_i] = \mathbf{b} = \begin{bmatrix} \frac{1}{i} \end{bmatrix} = \begin{bmatrix} 1 \\ 1/2 \\ 1/3 \\ \ddots \\ 1/100 \end{bmatrix}. \tag{7.37}$$

Compare to the analytic solution

$$
\begin{bmatrix} y_1 \\ y_2 \\ \ddots \\ y_N \end{bmatrix} = \begin{bmatrix} 1 \\ 0 \\ \ddots \\ 0 \end{bmatrix}.
\tag{7.38}
$$

7.5 Solution to String Problem

In Section 7.1 we set up the solution to our two masses on a string problem as a matrix problem. Now we have the matrix tools needed to solve it. Your **problem** is to check out the physical reasonableness of the solution for a variety of weights and lengths. You should check that the deduced tensions are positive and that the deduced angles correspond to a physical geometry (e.g., with a sketch). Inasmuch as this is a realistic problem, we know that the sine and cosine functions must be less than 1 in magnitude and that the tensions should be similar in magnitude to the weights of the spheres. Our solution NewtonNDanimate.py, which is given in Listing 7.1, shows graphically the steps in the search.

1) See at what point your initial guess for the angles of the strings gets so bad that the computer is unable to find a physical solution.
2) A possible problem with the formalism we have just laid out is that by incorporating the identity $\sin^2 \theta_i + \cos^2 \theta_i = 1$ into the equations, we may be discarding some information about the sign of $\sin \theta$ or $\cos \theta$. If you look at Figure 7.1, you can observe that for some values of the weights and lengths, θ_2 may turn out to be negative, yet $\cos \theta$ should remain positive. We can build this condition into our equations by replacing $f_7 - f_9$ with f's based on the form

$$
f_7 = x_4 - \sqrt{1 - x_1^2}, \quad f_8 = x_5 - \sqrt{1 - x_2^2}, \quad f_9 = x_6 - \sqrt{1 - x_3^2}.
\tag{7.39}
$$

See if this makes any difference in the solutions obtained.
3) ⊙ Solve the similar three-mass problem. The approach is the same, but the number of equations is larger.

7.6 Spin States and Hyperfine Structure

The energy levels of hydrogen exhibit a *fine structure* splitting arising from the coupling of the electron's spin to its orbital angular momentum. (Or, you can think of this as the couplings of magnetic moments.) In addition, these finely split levels exhibit a smaller *hyperfine* splitting arising from the coupling of the electron's spin to the proton's spin. In Gaussian CGS units, the magnetic moment of a particle of charge q is related to its spin S by

$$
\mu = g \frac{q}{2m} S,
\tag{7.40}
$$

where g is the particle's g factor and m its mass. An electron has

$$q = -e, \quad S = \frac{\hbar}{2}\sigma, \quad g \simeq -2, \Rightarrow \mu_e \simeq (-2)\frac{-e}{2m_e}\frac{\sigma}{2} = \mu_B\sigma, \tag{7.41}$$

$$\mu_B = \frac{e\hbar}{2m_e} = 5.05082 \times 10^{-27} \text{ J/T}, \tag{7.42}$$

where μ_B is the electron's Bohr magneton. Because the proton's mass is ~ 2000 times larger than the electron's mass, the proton's Bohr magneton and magnetic interaction is ~ 2000 times smaller than the electron's:

$$\mu_B|_p = \frac{-e\hbar}{2m_p} = -\frac{m_e}{m_p}\mu_B|_e = -\frac{1}{1836.15}\mu_B. \tag{7.43}$$

Even though the electron's and the proton's spins (internal degrees of freedom) exist in different spaces, they are both spin 1/2 particles, and so both can be (separately) represented by the Pauli matrices:

$$\sigma = \sigma_x\hat{e}_x + \sigma_y\hat{e}_y + \sigma_z\hat{e}_z, \tag{7.44}$$

$$\sigma_x = \begin{bmatrix} 0 & 1 \\ 1 & 0 \end{bmatrix}, \quad \sigma_y = \begin{bmatrix} 0 & -i \\ i & 0 \end{bmatrix}, \quad \sigma_z = \begin{bmatrix} 1 & 0 \\ 0 & -1 \end{bmatrix}. \tag{7.45}$$

In terms of the Pauli matrices, the electron–proton interaction is

$$V = W\sigma_e \cdot \sigma_p = W(\sigma_x^e\sigma_x^p + \sigma_y^e\sigma_y^p + \sigma_z^e\sigma_z^p). \tag{7.46}$$

The spin 1/2 states for the electron and the proton, each, can be either up or down:

$$|\alpha\rangle = |\uparrow\rangle = \begin{bmatrix} 1 \\ 0 \end{bmatrix}, \quad |\beta\rangle = |\downarrow\rangle = \begin{bmatrix} 0 \\ 1 \end{bmatrix}. \tag{7.47}$$

1) Verify, that if both the electron and the proton start off in spin-up states,

$$|\psi\rangle = |\alpha^e\alpha^p\rangle, \tag{7.48}$$

then the interaction (7.46) produces the mixed state

$$V|\psi\rangle = W\sigma^e \cdot \sigma^p |\alpha^e\alpha^p\rangle = W(\sigma_x^e\sigma_x^p + \sigma_y^e\sigma_y^p + \sigma_z^e\sigma_z^p)|\alpha^e\alpha^p\rangle \tag{7.49}$$

$$= |\beta^e\beta^p\rangle + i|\beta^e\beta^p\rangle + |\alpha^e\alpha^p\rangle. \tag{7.50}$$

2) Show that the interaction matrix for the $|\alpha^e\alpha^p\rangle$ state is

$$\langle\alpha^e\alpha^p| V |\alpha^e\alpha^p\rangle = \begin{bmatrix} W & 0 & 0 & 0 \\ 0 & -W & 2W & 0 \\ 0 & 2W & -W & 0 \\ 0 & 0 & 0 & W \end{bmatrix}. \tag{7.51}$$

3) Use a symbolic manipulation program to show that the eigenvalues of V are:

$$-3W(\text{multiplicity 3, triplet state}), \quad W(\text{multiplicity 1, singlet state}), \tag{7.52}$$

where the triplet state refers to $|S = 1, m_S = \pm 1, 0\rangle$, and the singlet state to $|S = 0, m_S = 0\rangle$. Our program Hyperfine.py is given in Listing 7.2.

4) Evaluate the numerical value for the hyperfine splitting of the 1S state Bransden and Joachain [1991]:

$$v = \hbar \Delta E = \frac{4W}{\hbar}. \tag{7.53}$$

Compare this to the value measured by Bailey and Townsend [1921]:

$$v = 1420.405751800 \pm 0.000000028 \text{ Hz (measured)}. \tag{7.54}$$

In addition to being one of the most accurately measured quantities in physics, you should find that (7.54) agrees with theory.

7.7 Speeding Up Matrix Computing ⊙

Programs written in Fortran and C tend to be faster than those written in Python because the former are compiled languages (the entire program is processed in one fell swoop), while Python is interpreted line by line, although sometimes compiled. However, the NumPy linear algebra routines are mainly written in C and C++, and are fast. In any case, here we give some techniques to help you speed up your large matrix computations.

7.7.1 Vectorization

A powerful feature of NumPy is its high-level *vectorization*. This is a simple and automatic process in which a single operation acts on an entire array, as opposed to each element individually, and leads to order-of-magnitude speedups. Our examples will use small matrices, and so while the relative speedup will be significant, the absolute savings of time will still be small. However, if you were dealing with very large matrices, and especially doing it often in a program, then speeding up your program may be worth the effort.

Here is our code TuneNumPy.py that compares the speed of a calculation using a *for* loop to evaluate a function for each of 100,000 elements in an array, versus the speed using NumPy's vectored evaluation of that function for an array object:

```
   # TuneNumpy.py: Comparison of NumPy op versus for loop
 2 from dateline import dateline
   import numpy as np

   def f(x):   return x**2-3*x + 4
 6 x = np.arange(1e5)              # An array of 100,000 integers
   for j in range(0, 3):          # Repeat comparison three time
       t1 = datetime.now()
       y = [f(i) for i in x]      # The for loop
10     t2 = datetime.now()
       print (' For for loop, t2-t1 =', t2-t1)
       t1 = datetime.now()
       y = f(x)                   # Vectored evaluation
14     t2 = datetime.now()
       print (' For vector function, t2-t1 =', t2-t1)
   Output:
   For for loop,        t2-t1 = 0:00:00.384000
18 For vector function , t2-t1 = 0:00:00.009000
```

Recall that we defined *stride* as the amount of memory skipped in order to get to the next element needed in a calculation. It is important to have your program minimize stride in order to avoid jumping through memory to find a needed value. For example, for a 1000×1000 array, the computer moves one word to get to the next column, but 1000 words to get to the next row. Clearly better to do a column-by-column calculation than a row-by-row one. To see this in action, we enter a 3×3 array of integers using NumPy's `arange` to create a 1D array. We then reshape it into a 3×3 array, and determine the strides for rows and columns calls:

```
>>> from numpy import *
>>> A = arange(0,90,10)                                              2
>>> A
array([ 0, 10, 20, 30, 40, 50, 60, 70, 80])                         4
>>> A = A.reshape((3,3))
>>> A                                                                6
array([[ 0, 10, 20],
       [30, 40, 50],
       [60, 70, 80]])                                               8
>>> A.strides                                                       10
(12, 4)
```

Line 11 tells us that it takes 12 bytes (3 words) to get to the same position in the next row, but only 4 bytes (one word) to get to the same position in the next column. It's clearly cheaper to go from column to column than row to row.

An easy way to cut down on memory jumping is to use Python's *slice* operator that extracts just the desired part of a list (like taking a "slice" through the center of a jelly doughnut):

```
ListName[StartIndex:StopBeforeIndex:Step].
```

The convention is that if no argument is given, then the slice starts at 0 and stops at the end of the list. For example:

```
>>> A = arange(0,90,10).reshape((3,3))                              1
>>> A
array([[ 0, 10, 20],                                                3
       [30, 40, 50],
       [60, 70, 80]])                                               5
>>> A[:2,:]              # First two rows (start at 2, go to end)
array([[ 0, 10, 20],                                                7
       [30, 40, 50]])
>>> A[:,1:3]            # Columns 1-3 (start at 1, end at 4)         9
array([[10, 20],
       [40, 50],                                                   11
       [70, 80]])
>>> A[::2,:]            # Every second row                         13
array([[ 0, 10, 20],
       [60, 70, 80]])                                              15
```

This is called *view-based indexing*, with the indexed notation returning a new array object that *points* to the address of the original data, as opposed to storing the values of the new array (think "pointers" in C). For instance, you can optimize a calculation of forward and central difference derivatives quite elegantly:

```
>>> x = arange(0,20,2)                                              1
>>> x
array([ 0, 2, 4, 6, 8, 10, 12, 14, 16, 18])                        3
>>> y = x**2
>>> y                                                               5
```

```
array([  0,    4,   16,   36,   64, 100, 144, 196, 256, 324], dtype=int32)
>>> dy_dx = ((y[1:]-y[:1])/(x[1:]-x[:-1]))              # Forward difference       7
>>> dy_dx
array([  2.,     8.,    18.,    32.,    50.,    72.,    98.,   128.,   162.])      9
>>> dy_dx_c = ((y[2:]-y[:-2])/(x[2:]-x[:-2]))           # Central difference
>>> dy_dx_c                                                                        11
array([  4.,    8.,   12.,   16.,   20.,   24.,   28.,   32.])
```

We note that the values of the derivatives are different because forward difference is evaluated at the start of the interval while central difference at the center.

7.7.2 Speedup Exercises

1) Timing an operation

```
import time
start = time.time()
print("hello")
end = time.time()          4
print(end - start)
```

2) Run the two simple codes listed below, timing how long each takes. Note that although each has the same number of arithmetic operations, one takes significantly more time because it makes large jumps through memory.

Sequential column references

```
for j = 1, 999999;
    x(j) = m(1,j)                    // Sequential column reference
```

Sequential row references

```
for j = 1, 999999;
    x(j) = m(j,1)                    // Sequential row reference        2
```

3) Test the effect of stride on your machine by comparing the time it takes to run these two programs. Run for increasing column size idiom and compare the times for loop *A versus* those for loop *B*. Loop *A* steps through the matrix vec in column order, while loop *B* steps through in row order. Both loops take us through all the elements of the matrix, but the stride is different.

Loop A bad (large) stride

```
Dimension vec(N, M)                    // Stride 1 fetch (f90)
      for j = 1, M;                                                      2
        for i=1, N;    Ansi = Ansi + vec(i,j)*vec(i,j)
```

Loop B good (small) stride

```
Dimension vec(N, M)                    // Stride dim fetch (f90)         1
      for i = 1, N;
        for j=1, M;    Ansi = Ansi + vec(i,j)*vec(i,j)
```

4) The penultimate example of memory usage is large-matrix multiplication:

$$[C] = [A] \times [B] \quad \Rightarrow \quad c_{ij} = \sum_{k=1}^{N} a_{ik} \times b_{kj}. \tag{7.55}$$

Test the effect of stride on your machine by comparing the time it takes to run these two programs. Run for increasing column size N

GOOD Python (min stride)

```
1  for i = 1, N; {                                    // Row
      for j = 1, N; {                                 // Column
         c(i,j) = 0.0                                 // Initialize
      for k = 1, N; {                                 
5        c(i,j) = c(i,j) + a(i,k)*b(k,j) }}}          // Accumulate
```

BAD Python (max stride)

```
   for j = 1, N; {                                    // Initialization
      for i = 1, N; {
3        c(i,j) = 0.0 }
      for k = 1, N; {
         for i = 1, N; {c(i,j) = c(i,j) + a(i,k)*b(k,j) }}}
```

5) Use NumPy's vectorized function evaluation to determine the speedup in the matrix multiplication [A][B], where the matrices contain at least 10^5 floating-point numbers. Compare the direct multiplication to application of the elementary rule for each element:

$$[BA]_{ij} = \sum_{k} a_{ik} b_{kj}. \tag{7.56}$$

6) Determine the speedup obtained by using Python stripping to reduce stride in evaluating the forward-difference and central-difference derivatives over an array of at least 10^5 floating-point numbers.

7.8 Code Listing

Listing 7.1 The code **NewtonNDanimate.py** shows the step-by-step search for solution of the two-mass-on-a-string problem via a Newton–Raphson search.

```
# NewtonNDanimate.py:              MultiDimension Newton Search

3  from visual import *
   from numpy.linalg import solve
   from visual.graph import *

7  scene = display(x=0,y=0,width=500,height=500,
                    title='String and masses configuration')
   tempe = curve(x=range(0,500),color=color.black)

11 n = 9
   eps = 1e-3
   deriv = zeros( (n, n), float)
   f = zeros( (n), float)
15 x = array([0.5, 0.5, 0.5, 0.5, 0.5, 0.5, 0.5, 1., 1., 1.])
```

```
 def plotconfig():
      for obj in scene.objects:
19        obj.visible=0                    # Erase previous configuration
      L1 = 3.0
      L2 = 4.0
      L3 = 4.0
23    xa = L1*x[3]                          # L1*cos(th1)
      ya = L1*x[0]                          # L1 sin(th1)
      xb = xa+L2*x[4]                       # L1*cos(th1)+L2*cos(th2)
      yb = ya+L2*x[1]                       # L1*sin(th1)+L2*sen(th2)
27    xc = xb+L3*x[5]                       # L1*cos(th1)+L2*cos(th2)+L3*cos(th3)
      yc = yb-L3*x[2]                       # L1*sin(th1)+L2*sen(th2)-L3*sin(th3)
      mx = 100.0                            # for linear coordinate transformation
      bx = -500.0                           # from 0=*<* x =*<*10
31    my = -100.0                           # to    -500 =*<*x_window=>500
      by = 400.0                            # same transformation for y
      xap = mx*xa+bx                        # to keep aspect ratio
      yap = my*ya+by
35    ball1 = sphere(pos=(xap,yap), color=color.cyan,radius=15)
      xbp = mx*xb+bx
      ybp = my*yb+by
      ball2 = sphere(pos=(xbp,ybp), color=color.cyan,radius=25)
39    xcp = mx*xc+bx
      ycp = my*yc+by
      x0 = mx*0+bx
      y0 = my*0+by
43    line1 = curve(pos=[(x0,y0),(xap,yap)], color=color.yellow,radius=4)
      line2 = curve(pos=[(xap,yap),(xbp,ybp)], color=color.yellow,radius=4)
      line3 = curve(pos=[(xbp,ybp),(xcp,ycp)], color=color.yellow,radius=4)
      topline = curve(pos=[(x0,y0),(xcp,ycp)], color=color.red,radius=4)
47
 def F(x, f):                              # F function
      f[0] = 3*x[3]   +   4*x[4]   +   4*x[5]   -   8.0
      f[1] = 3*x[0]   +   4*x[1]   -   4*x[2]
51    f[2] = x[6]*x[0]   -   x[7]*x[1]   -   10.0
      f[3] = x[6]*x[3]   -   x[7]*x[4]
      f[4] = x[7]*x[1]   +   x[8]*x[2]   -   20.0
      f[5] = x[7]*x[4]   -   x[8]*x[5]
55    f[6] = pow(x[0], 2)   +   pow(x[3], 2)   -   1.0
      f[7] = pow(x[1], 2)   +   pow(x[4], 2)   -   1.0
      f[8] = pow(x[2], 2)   +   pow(x[5], 2)   -   1.0

59 def dFi_dXj(x, deriv, n):               # Derivatives
      h = 1e-4
      for j in range(0, n):
          temp = x[j]
63        x[j] = x[j] + h/2.
          F(x, f)
          for i in range(0, n): deriv[i, j] = f[i]
          x[j] = temp
67    for j in range(0, n):
          temp = x[j]
          x[j] = x[j] - h/2.
          F(x, f)
71        for i in range(0, n): deriv[i, j] = (deriv[i, j] - f[i])/h
          x[j] = temp

 for it in range(1, 100):
75    rate(1)                              # 1 second between graphs
      F(x, f)
      dFi_dXj(x, deriv, n)
      B = array([[-f[0]], [-f[1]], [-f[2]], [-f[3]], [-f[4]], [-f[5]],\
79    [-f[6]], [-f[7]], [-f[8]]])
      sol = solve(deriv, B)
      dx = take(sol, (0, ), 1)             # First column of sol
      for i in range(0, n):
83       x[i]  = x[i]  + dx[i]
      plotconfig()
      errX = errF = errXi = 0.0
      for i in range(0, n):
```

```
87          if ( x[i] != 0.): errXi = abs(dx[i]/x[i])
            else:   errXi = abs(dx[i])
            if ( errXi > errX): errX = errXi
            if ( abs(f[i]) > errF ):  errF = abs(f[i])
91          if ( (errX <= eps) and (errF <= eps) ): break

    print('Number of iterations = ', it , "\n Final Solution:")
    for i in range(0, n):
95          print('x[', i, '] = ', x[i])
```

Listing 7.2   Hyperfine.py Hyperfine splitting in H using symbolic package SymPy.

```
    # Hyperfine.py: Hydrogen hyperfine structure using Sympy

3   from sympy import *
    import numpy as np, matplotlib.pyplot as plt

    W, mue, mup, B = symbols('W mu_e mu_p B')           # Symbols & Hamiltonian
7   H  = Matrix([[W,0,0,0],[0,-W,2*W,0],[0,2*W,-W,0],[0,0,0,W]])
    Hmag = Matrix([[ -(mue+mup)*B,0,0,0],[0,-(mue-mup)*B,0,0],[0,0,-(-mue+mup)*B,0],
          [0,0,0,(mue+mup)*B]])                       # H with external B
    print ("\n Hyperfine Hamiltonian H =", H )
11  print ("\n Eigenvalues and multiplicities of H =",H.eigenvals() )
    print ("\n Hmag =", Hmag)
    Htot = H + Hmag                                      # Hamiltonian + pertubation
    print ("\n Htot = H + Hmag =", Htot)
15  print ("\n Eigenvalues of matrix HB")
    e1, e2, e3, e4 = Htot.eigenvals()                    # 4 eigenvalues
    print (" e1 = ", e1, "\n e2 = ", e2, "\n e3 = ", e3, "\n e4 = ", e4)
    print ("\n After substitute mu_e = 1, and mu_p = 0 in eigenvalues")
19  print (" e1 = ",e1.subs([(mue,1),(mup,0)]),"\n e2       =
          ",e2.subs([(mue,1),(mup,0)]))
    print (" e3 = ",e3.subs([(mue,1),(mup,0)]),"\n e4 = ",e4.subs([(mue,1),(mup,0)]))
    b = np.arange(0,4,0.1)
    E = 1
23  E4 = -E + np.sqrt(b**2 +4*E**2)
    E3 = E - b
    E2 = E + b
    E1 = -E - np.sqrt(b**2 +4*E**2)
27  plt.figure()
    plt.plot(b,E1, label='E1'); plt.plot(b,E2, label='E2')
    plt.plot(b,E3, label='E3');  plt.plot(b,E4, label='E4')
    plt.legend(); plt.text(-0.4,1, 'E')
31  plt.xlabel(' Magnetic Field B')
    plt.title('Hyperfine Splitting of H Atom 1S Level')
    plt.show()
```

8

Differential Equations and Nonlinear Oscillations

In this chapter we develop numerical methods for solving ordinary differential equations, and focus on applying those tools to nonlinear systems. We start with simple systems that have analytic solutions, and use them to test various differential-equation solvers. We then let the oscillations become large so that nonlinear effects are important, and investigate nonlinear resonances and beating. In Chapter 16, Continuous Nonlinear Dynamics, we make a related study of the realistic pendulum and its chaotic behavior.

8.1 Nonlinear Oscillators

Figure 8.1 shows a mass m attached to a spring that exerts a restoring force toward the origin, as well as a hand that exerts a time-dependent external force on the mass. The restoring force exerted by the spring is nonlinear.

Problem Solve for the motion of the mass as a function of time for an arbitrary restoring force. You may assume the motion is constrained to one dimension.

This is a classical mechanics problem and so Newton's second law provides us with the equation of motion

$$F_k(x) + F_{ext}(x, t) = m\frac{d^2x}{dt^2}, \tag{8.1}$$

where $F_k(x)$ is an arbitrary restoring force exerted by the spring and $F_{ext}(x, t)$ is the external force. Because we are not told just how the spring departs from being linear, we'll just try out some different spring models. As our first model, we'll look at a potential that is linear for small displacements x, but becomes nonlinear for large x values:

$$V(x) \simeq \frac{1}{2}kx^2\left(1 - \frac{2}{3}\alpha x\right), \tag{8.2}$$

$$\Rightarrow \quad F_k(x) = -\frac{dV(x)}{dx} = -kx(1 - \alpha x) \tag{8.3}$$

$$\Rightarrow \quad m\frac{d^2x}{dt^2} = -kx(1 - \alpha x), \tag{8.4}$$

Computational Physics: Problem Solving with Python, Fourth Edition.
Rubin H. Landau, Manuel J. Páez, and Cristian C. Bordeianu.

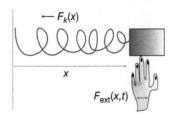

Figure 8.1 A mass m (the block) attached to a spring with restoring force $F_k(x)$ as well as driven by an external time-dependent driving force (the hand).

where we have omitted the time-dependent external force. Equation (8.4) is the second-order ordinary differential equation (ODE) we need to solve. If $\alpha x \ll 1$, we should have essentially harmonic motion, but as $x \to 1/\alpha$ the anharmonic effects should increase.

We can understand the basic physics of this model by looking at the curves on the left in Figure 8.2. As long as $x < 1/\alpha$, there will be a *restoring force* and the motion will be periodic (repeated exactly and indefinitely in time), though it may not be harmonic. If the amplitude of oscillation is large, there will be an asymmetry in the motion to the right and left of the equilibrium position. And if $x > 1/\alpha$, the force will become repulsive and the mass will be pushed away from the origin.

As a second model of a nonlinear oscillator, we assume that the spring's potential function is proportional to some arbitrary *even* power p of x:

$$V(x) = \frac{1}{p}kx^p, \quad (p \text{ even}). \tag{8.5}$$

We require an even p to ensure that the force,

$$F_k(x) = -\frac{dV(x)}{dx} = -kx^{p-1}, \tag{8.6}$$

contains an odd power of p, which guarantees that it is a *restoring* force for positive and negative x values. We display some characteristics of this potential on the right in Figure 8.2. We see that $p = 2$ is the harmonic oscillator and that $p = 6$ is nearly a square well with the mass moving almost freely until it hits the wall at $x \simeq \pm 1$. Regardless of the p value, the motion will be periodic, but it will be harmonic only for $p = 2$. Newton's law

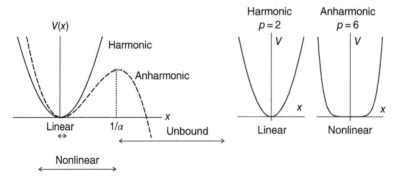

Figure 8.2 *Left*: The potentials of an harmonic oscillator (solid curve) and of an anharmonic oscillator (dashed curve). If the amplitude becomes too large for the anharmonic oscillator, the motion becomes unbound. *Right*: The shapes of the potential energy function $V(x) \propto |x|^p$ for $p = 2$ and $p = 6$. The "linear" and "nonlinear" labels refer to the restoring force derived from these potentials.

(8.1) gives the second-order ODE we need to solve:

$$m\frac{d^2x}{dt^2} = F_{\text{ext}}(x, t) - kx^{p-1}. \tag{8.7}$$

8.2 ODE Review

The background material in this section is presented to avoid confusion over semantics. The well-versed reader may want to skim or skip it.

8.2.1 Order

A general form for a *first-order* differential equation is

$$\frac{dy}{dt} = f(t, y), \tag{8.8}$$

where the "order" refers to the degree of the derivative on the LHS. The derivative or force function $f(t, y)$ on the RHS, is arbitrary. For instance, even if $f(t, y)$ is a nasty function of y and t such as

$$\frac{dy}{dt} = -3t^2 y + t^9 + y^7, \tag{8.9}$$

this is still a first-order differential equation. A general form for a *second-order* differential equation is

$$\frac{d^2y}{dt^2} + \lambda\frac{dy}{dt} = f\left(t, \frac{dy}{dt}, y\right). \tag{8.10}$$

The derivative function f on the RHS is arbitrary and may involve any power of the first derivative as well. To illustrate,

$$\frac{d^2y}{dt^2} + \lambda\frac{dy}{dt} = -3t^2\left(\frac{dy}{dt}\right)^4 + t^9 y(t) \tag{8.11}$$

is a second-order differential equation, as in Newton's law (8.1).

In the differential equations (8.8) and (8.10), the time t is the *independent* variable and the position y is the *dependent* variable. This means that we are free to vary the time at which we want a solution, but not the value of the position y at that time. Note that we often use the symbol y or Y for the dependent variable, but that this is just a symbol which may refer to other variables. For example, in some applications, we use y to describe a position instead of t.

8.2.2 Ordinary and Partial

Equations such as (8.1) and (8.8) are ODEs because they contain only *one* independent variable, in these cases t. In contrast, an equation such as the Schrödinger equation,

$$i\hbar\frac{\partial\psi(\mathbf{x}, t)}{\partial t} = -\frac{\hbar^2}{2m}\left[\frac{\partial^2\psi}{\partial x^2} + \frac{\partial^2\psi}{\partial y^2} + \frac{\partial^2\psi}{\partial z^2}\right] + V(\mathbf{x})\psi(\mathbf{x}, t), \tag{8.12}$$

contains four independent variables, and this makes it a *partial differential equation* (PDE). The partial derivative symbol ∂ is used to indicate that the dependent variable ψ depends simultaneously on several independent variables. In the early parts of this book, we limit ourselves to ordinary differential equations, yet in Chapters 20–27, we'll examine a variety of PDEs.

8.2.3 Linear and Nonlinear

Part of the strength of computational science is that we are no longer limited to solving linear equations. A *linear equation* is one in which only the first power of y or $d^n y/d^n t$ appears; a *nonlinear* equation may contain higher powers. For example,

$$\frac{dy}{dt} = g^3(t)y(t) \quad \text{(linear)}, \qquad \frac{dy}{dt} = \lambda y(t) - \lambda^2 y^2(t) \quad \text{(nonlinear)}. \tag{8.13}$$

An important property of linear equations is the *law of linear superposition* that lets us add different solutions together to form new ones. As a case in point, if $A(t)$ and $B(t)$ are solutions of the linear equation in (8.13), then

$$y(t) = \alpha A(t) + \beta B(t) \tag{8.14}$$

is also a solution for arbitrary values of the constants α and β. In contrast, even if we were clever enough to guess that the solution of the nonlinear equation in (8.13) is

$$y(t) = \frac{a}{1 + be^{-\lambda t}}, \tag{8.15}$$

(which we invite you to verify), this wouldn't work if we tried to obtain a more general solution by adding together two such solutions:

$$y_1(t) = \frac{a}{1 + be^{-\lambda t}} + \frac{a'}{1 + b'e^{-\lambda t}} \tag{8.16}$$

(which you we invite you to verify).

8.2.4 Initial and Boundary Conditions

The general solution of a first-order differential equation contains one arbitrary constant. The general solution of a second-order differential equation contains two such constants, and so forth. For any specific problem, these constants are usually determined by the *initial conditions*. For a first-order equation the sole initial condition may be the position $y(t)$ at some time. For a second-order equation, the two initial conditions may be the position and velocity at some time. Regardless of how powerful the hardware and software that you utilize, mathematics remains valid, and so you must know the initial conditions in order to obtain a unique solution to a differential equation.

In addition to the initial conditions, it is possible to further restrict the solutions of differential equations. One such way is by *boundary conditions* that constrain the solution to have fixed values at the boundaries of the solution space. In Chapter 13, we discuss how to extend the techniques of this chapter to boundary-value problems.

8.3 Dynamic Form of ODEs

A standard form for ODEs, which has proven to be useful in both numerical analysis [Press *et al.*, 2007] and classical dynamics [Scheck, 2010; Tabor, 1989; José and Salatan, 1998], is to express ODEs of *any order* as N simultaneous first-order ODEs in the N unknowns, y^i, $i = 0, N - 1$:

$$\frac{dy^{(0)}}{dt} = f^{(0)}(t, \{y^{(i)}\}),$$ (8.17)

$$\frac{dy^{(1)}}{dt} = f^{(1)}(t, \{y^{(i)}\})$$ (8.18)

$$\vdots \qquad \vdots$$

$$\frac{dy^{(N-1)}}{dt} = f^{(N-1)}(t, \{y^{(i)}\}).$$ (8.19)

Note, f can contain an explicit dependence on any or all of the $y^{(i)}$s, but not explicitly on a derivative $dy^{(i)}/dt$. These equations can be expressed more succinctly by use of the N-dimensional vectors (indicated here in **boldface**) **y** and **f**:

$$d\mathbf{y}(t)/dt = \mathbf{f}(t, \mathbf{y}),$$ (8.20)

$$\mathbf{y} = \begin{bmatrix} y^{(0)}(t) \\ y^{(1)}(t) \\ \ddots \\ y^{(N-1)}(t) \end{bmatrix}, \qquad \mathbf{f} = \begin{bmatrix} f^{(0)}(t, \mathbf{y}) \\ f^{(1)}(t, \mathbf{y}) \\ \ddots \\ f^{(N-1)}(t, \mathbf{y}) \end{bmatrix}.$$ (8.21)

The utility of such compact notation is that we can study the properties of the ODEs, as well as develop algorithms to solve them, by dealing with the single equation (8.20), without having to worry about individual $y^{(i)}$'s. To see how this works in practice, let's convert Newton's law

$$\frac{d^2x}{dt^2} = \frac{1}{m} F\left(t, x, \frac{dx}{dt}\right),$$ (8.22)

to this standard form. The rule is that the RHS may *not* contain any explicit derivatives, although individual components of $y^{(i)}$ may represent derivatives. To pull this off, we define the position x as the first dependent variable $y^{(0)}$, and the velocity dx/dt as the second dependent variable $y^{(1)}$:

$$y^{(0)}(t) \overset{\text{def}}{=} x(t), \qquad y^{(1)}(t) \overset{\text{def}}{=} \frac{dx}{dt} = \frac{dy^{(0)}(t)}{dt}.$$ (8.23)

The second-order ODE (8.22) now becomes two simultaneous first-order ODEs:

$$\frac{dy^{(0)}}{dt} = y^{(1)}(t), \qquad \frac{dy^{(1)}}{dt} = \frac{1}{m} F(t, y^{(0)}, y^{(1)}).$$ (8.24)

This expresses the acceleration [the second derivative in (8.22)] as the first derivative of the velocity $y^{(1)}$. These equations are now in the standard form (8.20), with the derivative or force function **f** having the two components

$$f^{(0)} = y^{(1)}(t), \qquad f^{(1)} = \frac{1}{m} F(t, y^{(0)}, y^{(1)}),$$ (8.25)

where F may be an explicit function of time as well as of position and velocity. To be even more specific, applying these definitions to our spring problem (8.7), we obtain the coupled first-order equations

$$\frac{dy^{(0)}}{dt} = y^{(1)}(t), \qquad \frac{dy^{(1)}}{dt} = \frac{1}{m}\left[F_{\text{ext}}(x, t) - ky^{(0)}(t)^{p-1}\right],$$ (8.26)

where $y^{(0)}(t)$ is the position of the mass at time t and $y^{(1)}(t)$ is its velocity. In the standard form, the components of the force function and the initial conditions are

$$f^{(0)}(t,\mathbf{y}) = y^{(1)}(t), \qquad\qquad f^{(1)}(t,\mathbf{y}) = \frac{1}{m}\left[F_{\text{ext}}(x,t) - k(y^{(0)})^{p-1}\right],$$

$$y^{(0)}(0) = x_0, \qquad\qquad y^{(1)}(0) = v_0. \tag{8.27}$$

8.4 ODE Algorithms

The classic way to solve an ODE is shown in Figure 8.3. One starts with the known initial value of the dependent variable, $y_0 \equiv y(t = 0)$, and then uses the derivative function $f(t,y)$ to advance the initial value one small step h forward in time to produce $y(t = h) \equiv y_1$. Once you can do that, you can solve the ODE for all t values by just continuing to step to larger times, one small h at a time.[1] Error is always a concern when integrating differential equations because derivatives require small differences, and small differences are prone to subtractive cancellations and round-off error accumulation. In addition, because this stepping procedure is a continuous extrapolation of the initial conditions, with each step building on a previous extrapolation, this is somewhat like a castle built on sand; in contrast to interpolation, there are no tabulated values on which to anchor your solution. It is simplest if the time steps used throughout the integration remain constant in size, and that is mostly what we shall do. Industrial-strength algorithms, such as the one we discuss in Section 8.4.2, adapt the step size by making h larger in regions where y varies slowly (this speeds up the integration and cuts down on round-off error), and making h smaller in regions where y varies rapidly.

8.4.1 Euler's Rule

Euler's rule (Figure 8.4) is the simplest algorithm for integrating the differential equation (8.8) by one step. It is just an application of the forward-difference algorithm for the derivative:

$$\frac{d\mathbf{y}(t)}{dt} \simeq \frac{\mathbf{y}(t_{n+1}) - \mathbf{y}(t_n)}{h} = \mathbf{f}(t,\mathbf{y}), \tag{8.28}$$

$$\Rightarrow \quad \mathbf{y}_{n+1} \simeq \mathbf{y}_n + h\mathbf{f}(t_n,\mathbf{y}_n), \tag{8.29}$$

where $y_n \overset{\text{def}}{=} y(t_n)$ is the value of y at time t_n. We know from our discussion of differentiation that the error in the forward-difference algorithm is $\mathcal{O}(h^2)$, and so then this too is the error in Euler's rule.

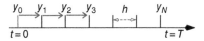

Figure 8.3 A sequence of uniform steps of length h taken in solving a differential equation. The solution starts at time $t = 0$, and is integrated in steps of h until $t = T$.

---

1 To avoid confusion, notice that $y^{(n)}$ is the nth component of the y vector, while y_n is the value of y after n time steps. Yes, there is a price to pay for elegance in notation.

Figure 8.4 Euler's algorithm for integration of a differential equation one step forward in time. This linear extrapolation with the slope evaluated at the initial point is seen to lead to an error Δ.

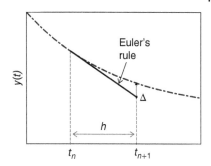

To indicate the simplicity of this algorithm, we apply it to our oscillator problem (8.4) for the first time step:

$$y_1^{(0)} = x_0 + v_0 h, \quad y_1^{(1)} = v_0 + h\frac{1}{m}\left[F_{ext}(t=0) + F_k(t=0)\right]. \tag{8.30}$$

Compare these to the projectile equations familiar from first-year physics,

$$x = x_0 + v_0 h + \tfrac{1}{2}ah^2, \quad v = v_0 + ah. \tag{8.31}$$

We see that with Euler's rule, the acceleration does not contribute to the change in distance (no h^2 term), yet it does contribute to the change in velocity (and so will contribute belatedly to the distance in the next time step). This is clearly a simple algorithm that requires very small h values to obtain precision. Yet using small values for h increases the number of steps and the accumulation of round-off error, which may lead to instability.[2] Whereas we do not recommend Euler's algorithm for general use, it is commonly used to start off more precise algorithms.

8.4.2 Runge–Kutta Rule

Although no one algorithm is good for solving all ODEs, the fourth-order Runge–Kutta algorithm, `rk4`, or its extension with adaptive step size, `rk45`, comes close. In spite of `rk4` being our recommended standard, we derive the simpler `rk2` here, and just state the result for `rk4`.

The Runge–Kutta algorithm for integrating a differential equation is based upon the formal (exact) integral of our differential equation:

$$\frac{dy}{dt} = f(t,y) \Rightarrow y(t) = \int f(t,y)\, dt \tag{8.32}$$

$$\Rightarrow y_{n+1} = y_n + \int_{t_n}^{t_{n+1}} f(t,y)\, dt. \tag{8.33}$$

To derive the second-order Runge–Kutta algorithm `rk2` (Figure 8.5 and `rk2.py`), we expand $f(t,y)$ in a Taylor series about the *midpoint* of the integration interval and retain

---

2 Instability is often a problem when you integrate a $y(t)$ that decreases as the integration proceeds, analogous to upward recursion of spherical Bessel functions. In this case, and if you have a linear ODE, you are best off integrating *inward* from large times to small times and then scaling the answer to agree with the initial conditions.

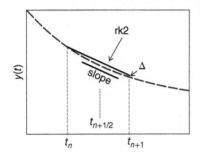

Figure 8.5 The rk2 algorithm for integration of a differential equation uses a slope (bold line segment) evaluated at the interval's midpoint, and is seen to lead to a smaller error than Euler's algorithm in Figure 8.4.

two terms in the expansion:

$$f(t,y) \simeq f(t_{n+1/2}, y_{n+1/2}) + (t - t_{n+1/2})\frac{df}{dt}(t_{n+1/2}) + \mathcal{O}(h^2). \tag{8.34}$$

Since $(t - t_{n+1/2})$ raised to any odd power is equally positive and negative over the interval $t_n \le t \le t_{n+1}$, the integral of the $(t - t_{n+1/2})$ term in (8.34) vanishes and we obtain the **rk2 algorithm**:

$$\int_{t_n}^{t_{n+1}} f(t,y)\, dt \simeq f(t_{n+1/2}, y_{n+1/2})h + \mathcal{O}(h^3), \tag{8.35}$$

$$\Rightarrow \quad y_{n+1} \simeq y_n + hf(t_{n+1/2}, y_{n+1/2}) + \mathcal{O}(h^3). \tag{8.36}$$

We see that while **rk2** contains the same number of terms as Euler's rule, it obtains a higher level of precision by taking advantage of the cancellation of the $\mathcal{O}(h)$ terms. The price for improved precision is having to evaluate the derivative function and the solution y at the middle of the time interval, $t = t_n + h/2$. And there's the rub, for we do not know the value of $y_{n+1/2}$ and cannot use this algorithm to determine it. The way out of this quandary is to use Euler's algorithm to determine $y_{n+1/2}$:

$$y_{n+1/2} \simeq y_n + \frac{1}{2}h\frac{dy}{dt} = y_n + \frac{1}{2}hf(t_n, y_n). \tag{8.37}$$

Putting the pieces all together gives the complete **rk2** algorithm:

$$\mathbf{y}_{n+1} \simeq \mathbf{y}_n + \mathbf{k}_2, \qquad (\texttt{rk2}) \tag{8.38}$$

$$\mathbf{k}_2 = h\mathbf{f}\left(t_n + \frac{h}{2}, \mathbf{y}_n + \frac{\mathbf{k}_1}{2}\right), \quad \mathbf{k}_1 = h\mathbf{f}(t_n, \mathbf{y}_n), \tag{8.39}$$

where we use boldface to indicate the vector nature of y and f. We see that the known derivative function **f** is evaluated at the ends and the midpoint of the interval, but only the (known) initial value of the dependent variable **y** is required. This makes the algorithm self-starting.

As an example of the use of **rk2**, we apply it to our spring problem:

$$y_1^{(0)} = y_0^{(0)} + hf^{(0)}\left(\frac{h}{2}, y_0^{(0)} + k_1\right) \tag{8.40}$$

$$\simeq x_0 + h\left[v_0 + \frac{h}{2}F_k(0)\right], \tag{8.41}$$

$$y_1^{(1)} = y_0^{(1)} + hf^{(1)} \left[\left(\tfrac{h}{2}, y_0 + \tfrac{h}{2}f(0), y_0 \right) \right] \tag{8.42}$$

$$\simeq v_0 + \frac{h}{m} \left[F_{\text{ext}} \left(\tfrac{h}{2} \right) + F_k \left(y_0^{(1)} + \tfrac{k_1}{2} \right) \right]. \tag{8.43}$$

These equations say that the position $y^{(0)}$ changes because of the initial velocity and force, while the velocity $y^{(1)}$ changes because of the external force at $t = h/2$ and the internal force at two intermediate positions. We see that the position $y^{(0)}$ now has an h^2 time dependence, which at last brings us up to the level of first-year physics.

The fourth-order Runge–Kutta method `rk4.py` (Listing 8.1) obtains $\mathcal{O}(h^4)$ precision by approximating y as a Taylor series up to order h^2 (a parabola) at the midpoint of the interval, which again leads to cancellation of lower-order error. All in all, `rk4` provides an excellent balance of power, precision, and programming simplicity. With rk4 there are four intermediate slopes, and these are approximated with the Euler algorithm:

$$\mathbf{y}_{n+1} = \mathbf{y}_n + \tfrac{1}{6}(\mathbf{k}_1 + 2\mathbf{k}_2 + 2\mathbf{k}_3 + \mathbf{k}_4), \tag{8.44}$$

$$\mathbf{k}_1 = h\mathbf{f}(t_n, \mathbf{y}_n), \qquad\qquad \mathbf{k}_2 = h\mathbf{f}\left(t_n + \tfrac{h}{2}, \mathbf{y}_n + \tfrac{\mathbf{k}_1}{2}\right),$$

$$\mathbf{k}_3 = h\mathbf{f}\left(t_n + \tfrac{h}{2}, \mathbf{y}_n + \tfrac{\mathbf{k}_2}{2}\right), \qquad\qquad \mathbf{k}_4 = h\mathbf{f}(t_n + h, \mathbf{y}_n + \mathbf{k}_3).$$

This provides an improved approximation to $f(t, y)$ near the midpoint. Although `rk4` is computationally more expensive than the Euler method, its precision is much better, and sometimes is made up by the ability to use larger step sizes h.

A variation of `rk4`, known as the Runge–Kutta–Fehling method [Mathews, 2002], or `rk45`, varies the step size while doing the integration with the hope of obtaining better precision and maybe better speed. Our implementation, `rk45.py`, is given in Listing 8.2. It automatically doubles the step size and tests to see how an estimate of the error changes. If the error is still within acceptable bounds, the algorithm will continue to use the larger step size and thus speed up the computation; if the error is too large, the algorithm will decrease the step size until an acceptable error is found. As a consequence of the extra information obtained in the testing, the algorithm does obtain $\mathcal{O}(h^5)$ precision, but sometimes at the expense of extra computing time. Whether that extra time is recovered by being able to use a larger step size depends upon the application.

8.4.3 Adams-Bashful-Moulton Predictor-Corrector Rule

Another approach for obtaining high precision in an ODE algorithm uses the solution from two previous steps, y_{n-2} and y_{n-1}, in addition to y_n, to predict y_{n+1}. (The Euler and `rk` methods use just one previous step.) Many of these methods tend to be like a Newton's search method; we start with a guess or *prediction* for the next step, and then use an algorithm, such as `rk4`, to check on the prediction and thereby obtain a *correction*. As with `rk45`, one can use the correction as a measure of the error and then adjust the step size to obtain improved precision [Press *et al.*, 2007]. For those readers who may want to explore such methods, **ABM.py** in Listing 8.3 gives our implementation of the *Adams-Bashful-Moulton* predictor-corrector scheme.

8.4.4 Assessment: rk2 *versus* rk4 *versus* rk45

While you are free to do as you please, unless you are very careful, we recommend that you do *not* write your own `rk4` or `rk45` methods. You will be using this algorithm for some high-precision work, and unless you get every fraction and method call just right, your code may appear to work well, but still not give all the precision that you could obtain. And so we give you `rk4.py`, and `rk45.py` codes to use. However, we do recommend that you write your own `rk2`, as doing so will make it clearer as to how the Runge–Kutta methods work, but without all the pain and danger of `rk4`.

1) Write your own `rk2` method, with the derivative function $f(t,x)$ a separate method.
2) Use your `rk2` to solve the equation of motion (8.7) or (8.26). Plot both the position $x(t)$ and velocity dx/dt as functions of time.
3) Once your ODE solver is running, do a number of things to check that it is working well and that you know what h values to use:
   a) Adjust the parameters in your potential so that it corresponds to a pure harmonic oscillator (set $p = 2$ or $\alpha = 0$). For an oscillator initially at rest, we have an analytic result with which to compare:
   $$x(t) = A \sin(\omega_0 t), \quad v = \omega_0 A \cos(\omega_0 t), \quad \omega_0 = \sqrt{k/m}. \tag{8.45}$$
   b) Pick values of k and m such that the period $T = 2\pi/\omega$ is a nice number with which to work (something like $T = 1$).
   c) Start with a step size $h \simeq T/5$ and make h smaller until the solution looks smooth, has a period that remains constant for a large number of cycles, and agrees with the analytic result. Always try to start with a large h so that you can see a bad solution turn good.
   d) Make sure that you have exactly the same initial conditions for the analytic and numerical solutions (zero displacement, nonzero velocity), and then plot the two together. It is good if you cannot tell them apart, yet that is not much of a test since it only ensures approximately two places of agreement.
   e) Try different initial velocities and verify that a *harmonic* oscillator is *isochronous*, that is, that its period does *not* change as the amplitude varies.
4) Now that you know you can get a good solution of an ODE with `rk2`, compare the solutions obtained with the `rk2`, `rk4`, and `rk45` solvers.
5) Make a table of comparisons similar to Table 8.1, where we compare `rk4` and `rk45` for the two equations
   $$2yy'' + y^2 - y'^2 = 0, \tag{8.46}$$
   $$y'' + 6y^5 = 0, \tag{8.47}$$
   with initial conditions $[y(0), y'(0)] = [1,1]$. Although nonlinear, (8.46) does have the analytic solution,[3] $y(t) = 1 + \sin t$. Equation (8.47) corresponds to our standard potential (8.5), with $p = 6$. Although we have not tuned `rk45`, Listing 8.2 shows that by setting

---

3 Be warned, the `rk` procedures may be inaccurate for this equation if integrated through the point $y(t) = 0$, as then the equation becomes $y'^2 = 0$, which is problematic.

Table 8.1 Comparison of ODE solvers for different equations.

Eqn. no.	Method	Initial h	No. of flops	Time (ms)	Relative error
(8.46)	rk4	0.01	1000	5.2	2.2×10^{-8}
	rk45	1.00	72	1.5	1.8×10^{-8}
(8.47)	rk4	0.01	227	8.9	1.8×10^{-8}
	rk45	0.1	3143	36.7	5.7×10^{-11}

Figure 8.6 The logarithm of the relative error in the solution of an ODE obtained with rk4 using a differing number N of time steps over a fixed time interval. The logarithm approximately equals the negative of the number of places of precision. Increasing the number of steps used for a fixed interval is seen to lead to smaller errors.

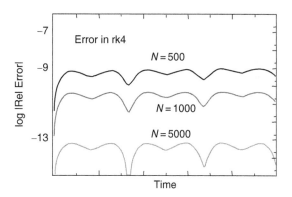

its tolerance parameter to a small enough number, rk45 will obtain better precision than rk4 (Figure 8.6), but that it requires ~10 times more floating-point operations and takes ~5 times longer. For (8.46), we obtained increased precision in less time.

8.5 Solution for Nonlinear Oscillations

Use your rk4 program to study anharmonic oscillations by trying powers in the range $p = 2$–12 for potential (8.5), or anharmonic strengths in the range $0 \leq \alpha x \leq 2$ for potential (8.2). Do *not* include any explicit time-dependent forces yet. Note that for large values of p, the forces and accelerations get large near the turning points, and so you may need a smaller step size h than that used for the harmonic oscillator.

1) Check that the solution remains periodic with constant amplitude and period for all initial conditions regardless of how nonlinear you make the force. In addition, check that the maximum speed occurs at $x = 0$ and zero velocity at the maximum $|x|$'s, the latter being a consequence of energy conservation.
2) Verify that nonharmonic oscillators are *nonisochronous*, that is, that vibrations with different amplitudes have different periods (Figure 8.7).
3) Explain why the shapes of the oscillations change for different p's or α's.
4) Devise an algorithm to determine the period T of the oscillation by recording times at which the mass passes through the origin. Note that because the motion may be asymmetric, you must record at least *three* times to deduce the period.
5) Construct a graph of the deduced period as a function of initial amplitude.

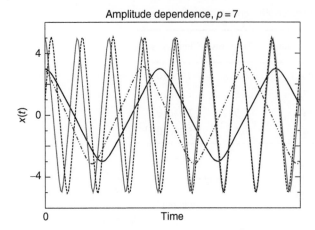

Figure 8.7 The position versus time for oscillations within the potential $V \propto x^7$ for four different initial amplitudes. Each is seen to have a different period.

6) Verify that the motion is oscillatory, but not harmonic, as the energy approaches $k/6\alpha^2$, or for $p > 6$.

7) Verify that for the anharmonic oscillator with $E = k/6\alpha^2$, the motion separates from oscillatory to translational. See how close you can get to this *separatrix* where a single oscillation takes an infinite time. (There is no separatrix for the power-law potential.)

8.5.1 Precision Assessment via E Conservation

We have not explicitly built energy conservation into our ODE solvers. Nonetheless, unless you have explicitly included a frictional force, it follows mathematically from the equations of motion that energy must be a constant for all values of p or α. That being the case, the constancy of energy is a demanding test of the numerics.

1) Plot the potential energy $PE(t) = V[x(t)]$, the kinetic energy $KE(t) = mv^2(t)/2$, and the total energy $E(t) = KE(t) + PE(t)$, for 50 periods. Comment on the correlation between $PE(t)$ and $KE(t)$ and how it depends on the potential parameters.

2) Check the long-term *stability* of your solution by plotting

$$-\log_{10}\left|\frac{E(t) - E(t = 0)}{E(t = 0)}\right| \simeq \text{number of places of precision} \qquad (8.48)$$

for a large number of periods (Figure 8.6). Because $E(t)$ should be independent of time, the numerator is the absolute error in your solution, and when divided by $E(0)$, becomes the relative error (say 10^{-11}). If you cannot achieve 11 or more places, then you need to decrease the value of h or debug.

3) Because a particle bound by a large-p oscillator is essentially "free" most of the time, you should observe that the average of its kinetic energy over time exceeds its average potential energy. This is actually the physics behind the Virial theorem for a power-law potential [Marion and Thornton, 2019]:

$$\langle KE \rangle = \frac{p}{2}\langle PE \rangle. \qquad (8.49)$$

Verify that your solution satisfies the Virial theorem. (Those readers who have worked on the perturbed oscillator problem can use this relation to deduce an effective p value, which should be between 2 and 3.)

8.6 Extensions: Nonlinear Resonances, Beats, Friction

Problem   So far our oscillations have been rather simple. We have ignored friction and have assumed that there are no external forces (hands) influencing the system's natural oscillations. Determine the following:

1) How the oscillations change when friction is included.
2) How the resonances and beats of nonlinear oscillators differ from those of linear oscillators.
3) How introducing friction affects resonances.

8.6.1   Friction

The world is full of friction, and not all of it is bad. While friction makes it harder to pedal a bike through the wind, it also lets you walk on ice, and generally adds stability to dynamical systems. The simplest models for frictional force are called *static*, *kinetic*, and *viscous* friction:

$$F_f^{(\text{static})} \leq -\mu_s N, \qquad F_f^{(\text{kinetic})} = -\mu_k N \frac{v}{|v|}, \qquad F_f^{(\text{viscous})} = -bv. \qquad (8.50)$$

Here N is the *normal force* on the object under consideration, μ and b are parameters, and v is the velocity. This model for static friction is appropriate for objects at rest, while the model for kinetic friction is appropriate for an object sliding on a dry surface. If the surface is lubricated, or if the object is moving through a viscous medium, then a frictional force proportional to some power of the velocity is a better model.[4]

1) Extend your harmonic oscillator code to include the three types of friction in (8.50), and observe how the motion differs for each.
2) *Hint*: For the simulation with static plus kinetic friction, each time the oscillator has $v = 0$, you need to check that the restoring force exceeds the static force of friction. If not, the oscillation must end at that instant. Check that your simulation terminates at nonzero x values.
3) For your simulations with viscous friction, investigate the qualitative changes that occur for increasing b values:

Under damped:	$b < 2m\omega_0$	Oscillate within decaying envelope
Critically damped:	$b = 2m\omega_0$	Nonconciliatory, finite decay time
Over damped:	$b > 2m\omega_0$	Nonconciliatory, infinite decay time

8.6.2   Resonances and Beats

Stable physical systems will oscillate if displaced slightly from their rest positions. The frequency ω_0 with which a stable system executes small oscillations about its rest positions

---

4 The effect of air resistance on projectile motion is studied in Section 13.4.

is called its *natural frequency*. If an external sinusoidal force is applied to this system, and if the frequency of the external force is equal to the natural frequency ω_0, then a *resonance* may occur in which the system absorbs energy from the external force and the amplitude of oscillation increases with time. If the oscillation and the driving force remain in phase over time, the amplitude of oscillation will increase continuously, unless there is some mechanism, such as friction or nonlinearities, to limit the growth. If the frequency of the driving force is close to, but not exactly equal to, the natural frequency of the system, then a related phenomena, known as *beating*, may occur. In beating there is interference between the natural oscillation and the external force. If the frequency of the external driving force is very close to the natural frequency, then the resulting motion,

$$x \simeq x_0 \sin \omega t + x_0 \sin \omega_0 t = \left(2x_0 \cos \frac{\omega - \omega_0}{2}t\right) \sin \frac{\omega + \omega_0}{2}t, \tag{8.51}$$

resembles the natural oscillation of the system at the average frequency $\frac{\omega+\omega_0}{2}$, yet with an amplitude $2x_0 \cos \frac{\omega-\omega_0}{2}t$ that varies slowly with a *beat frequency* $\frac{\omega - \omega_0}{2}$.

8.6.3 Time-Dependent Forces

To extend our simulation to include an external force,

$$F_{ext}(t) = F_0 \sin \omega t, \tag{8.52}$$

we need to include a time dependence in the force function $\mathbf{f}(t, \mathbf{y})$ of our ODE solver.

1) Add the sinusoidal time-dependent external force (8.52) to the space-dependent restoring force in your program (do not include friction yet).
2) Start with a very large value for the magnitude of the driving force F_0. This should lead to *mode locking* (the 500-pound-gorilla effect), where the system is overwhelmed by the driving force and, after the transients die out, the system oscillates in phase with the driver regardless of the driver's frequency.
3) Now lower F_0 until it is close to the magnitude of the natural restoring force of the system. You need to have this near equality for beating to occur.
4) Verify that the beat frequency for the harmonic oscillator (the number of variations in intensity per unit time) equals the frequency difference $(\omega - \omega_0)/2\pi$ in cycles per second, where $\omega \simeq \omega_0$.
5) Once you have a value for F_0 matched well with your system, make a series of runs in which you progressively increase the frequency of the driving force for the frequency range $\omega_0/10 \le \omega \le 10\omega_0$.
6) Make of plot of the maximum amplitude of oscillation *versus* the driver's ω.
7) Explore what happens when you make a nonlinear system resonate. If the nonlinear system is close to being harmonic, you should get beating in place of the blowup that occurs for the linear system. Beating occurs because the natural frequency changes as the amplitude increases, and thus the natural and forced oscillations fall out of phase. Yet once out of phase, the external force stops feeding energy into the system, and so the amplitude decreases, and with the decrease in amplitude, the frequency of the oscillator returns to its natural frequency, the driver and oscillator get back in phase, and the entire cycle repeats.

8) Investigate now how the inclusion of viscous friction modifies the curve of amplitude *versus* driver frequency. You should find that friction broadens the curve.

9) Explain how the character of the resonance changes as the exponent p in the potential $V(x) = k|x|^p/p$ is made larger and larger. At large p, the mass effectively "hits" the wall and falls out of phase with the driver, and so the driver is less effective at pumping energy into the system.

8.7 Code Listings

Listing 8.1 rk4.py solves an ODE with the RHS given by the method f( ) using rk4. The method f( ) is separate from the algorithm.

```
# rk4.py 4th order Runge Kutta application wi built in rk4

from visual.graph import *

#    Initialization
a = 0.
b = 10.
n = 100
ydumb = zeros((2), float);    y = zeros((2), float)
fReturn = zeros((2), float);  k1 = zeros((2), float)
k2 = zeros((2), float);       k3 = zeros((2), float)
k4 = zeros((2), float)
y[0] = 3.;    y[1] = -5.
t = a;        h = (b-a)/n;

def f( t, y):                       # Force function
    fReturn[0] = y[1]
    fReturn[1] = -100.*y[0]-2.*y[1] + 10.*sin(3.*t)
    return fReturn

graph1 = gdisplay(x=0,y=0, width = 400, height = 400, title = 'RK4',
            xtitle = 't', ytitle = 'Y[0]',xmin=0,xmax=10,ymin=-2,ymax=3)
funct1 = gcurve(color = color.yellow)
graph2 = gdisplay(x=400,y=0, width = 400, height = 400, title = 'RK4',
            xtitle = 't', ytitle = 'Y[1]',xmin=0,xmax=10,ymin=-25,ymax=18)
funct2 = gcurve(color = color.red)

def rk4(t,h,n):
    k1 = [0]*(n)
    k2 = [0]*(n)
    k3 = [0]*(n)
    k4 = [0]*(n)
    fR = [0]*(n)
    ydumb = [0]*(n)
    fR = f(t, y)                     # Returns RHS's
    for i in range(0, n):
        k1[i] = h*fR[i]
    for i in range(0, n):
        ydumb[i] = y[i] + k1[i]/2.
    k2 = h*f(t+h/2., ydumb)
    for i in range(0, n):
        ydumb[i] = y[i] + k2[i]/2.
    k3 = h*f(t+h/2., ydumb)
    for i in range(0, n):
        ydumb[i] = y[i] + k3[i]
    k4 = h*f(t+h, ydumb)
    for i in range(0, 2):
        y[i] = y[i] + (k1[i] + 2.*(k2[i] + k3[i]) + k4[i])/6.
    return y

while (t < b):                       # Time loop
    if ((t + h) > b):
```

```
53        h = b − t                    # Last step
       y = rk4(t,h,2)
       t = t + h
       rate(30)
57     funct1.plot(pos = (t, y[0]) )
       funct2.plot(pos = (t, y[1]) )
```

Listing 8.2   rk45.py solves an ODE with the RHS given by the method f( ) using rk4 with adaptive step size.

```
# rk45.py           Adaptive step size Runge Kutta

from visual.graph import *

a = 0.;  b = 10.                      # Error tolerance, endpoints
Tol = 1.0E−8
ydumb = zeros( (2), float)            # Initialize
y = zeros( (2), float)
fReturn = zeros( (2), float)
err = zeros( (2), float)
k1 = zeros( (2), float)
k2 = zeros( (2), float)
k3 = zeros( (2), float)
k4 = zeros( (2), float)
k5 = zeros( (2), float)
k6 = zeros( (2), float)
n = 20
y[0] = 1. ;    y[1] = 0.

h = (b − a)/n;    t = a;    j = 0
hmin = h/64;    hmax = h*64        # Min and max step sizes
flops = 0;    Eexact = 0. ;    error = 0.
sum = 0.

def f( t, y, fReturn ):                       # Force function
    fReturn[0] = y[1]
    fReturn[1] =    − 6.*pow(y[0], 5.)

graph1 = gdisplay( width = 600, height = 600, title = 'RK 45',
                   xtitle = 't', ytitle = 'Y[0]')
funct1 = gcurve(color = color.blue)
graph2 = gdisplay( width = 500, height = 500, title = 'RK45',
                   xtitle = 't', ytitle = 'Y[1]')
funct2 = gcurve(color = color.red)
funct1.plot(pos = (t, y[0]) )
funct2.plot(pos = (t, y[1]) )

while (t < b):                            # Loop over time
    funct1.plot(pos = (t, y[0]) )
    funct2.plot(pos = (t, y[1]) )
    if ( (t + h) > b ):
        h = b − t                          # Last step
    f(t, y, fReturn)              # Evaluate f, return in fReturn
    k1[0] = h*fReturn[0];         k1[1] = h*fReturn[1]
    for i in range(0, 2):
        ydumb[i] = y[i]  +  k1[i]/4
    f(t + h/4, ydumb, fReturn)
    k2[0] = h*fReturn[0];         k2[1] = h*fReturn[1]
    for i in range(0, 2):
        ydumb[i] = y[i] + 3*k1[i]/32  + 9*k2[i]/32
    f(t + 3*h/8, ydumb, fReturn)
    k3[0] = h*fReturn[0];   k3[1] = h*fReturn[1]
    for i in range(0, 2):
        ydumb[i] = y[i]  +  1932*k1[i]/2197 − 7200*k2[i]/2197.  + 7296*k3[i]/2197
    f(t + 12*h/13, ydumb, fReturn)
    k4[0] = h*fReturn[0]; k4[1] = h*fReturn[1]
    for i in range(0, 2):
```

```
                ydumb[i] = y[i] + 439*k1[i]/216  - 8*k2[i] +  3680*k3[i]/513  -
                    845*k4[i]/4104
           f(t  +  h, ydumb, fReturn)
60         k5[0] = h*fReturn[0]; k5[1] = h*fReturn[1]
           for i in range(0, 2):
                ydumb[i] = y[i]  - 8*k1[i]/27  + 2*k2[i] - 3544*k3[i]/2565  +
                    1859*k4[i]/4104  - 11*k5[i]/40
           f(t  +  h/2, ydumb, fReturn)
64         k6[0] = h*fReturn[0]; k6[1] = h*fReturn[1];
           for i in range(0, 2):
                err[i] = abs( k1[i]/360  -  128*k3[i]/4275  - 2197*k4[i]/75240 +
                    k5[i]/50.  + 2*k6[i]/55)
           if ( err[0] < Tol or err[1] < Tol or h <=  2*hmin ):  # Accept step
68             for i in range(0, 2):
                    y[i] = y[i]  +  25*k1[i]/216.  + 1408*k3[i]/2565.  + 2197*k4[i]/4104.
                        - k5[i]/5.
               t = t  +  h
               j = j  + 1
72         if ( err[0] == 0 or err[1] == 0 ):
               s = 0                                      # Trap division by 0
           else:
               s = 0.84*pow(Tol*h/err[0], 0.25)           # Reduce step
76         if ( s  < 0.75 and h > 2*hmin ):
               h /= 2.                                    # Increase step
           else:
               if ( s > 1.5 and 2* h  < hmax ):
80                  h *= 2.
           flops = flops  + 1
           E = pow(y[0], 6.)  +  0.5*y[1]*y[1]
           Eexact = 1.
84         error = abs( (E - Eexact)/Eexact )
           sum  += error
    print(" <error>=  ", sum/flops, ", flops = ", flops)
```

Listing 8.3 **ABM.py** solves an ODE with the RHS given by the method f( ) using the ABC predictor-corrector algorithm.

```
# ABM.py:   Adams BM method to integrate ODE
# Solves y' = (t - y)/2,    with y[0] = 1 over [0, 3]

4  from vpython import *

   numgr = graph(x=0, y=0, width=600, height=300, xmin=0.0, xmax = 3.0,
           title="Numerical Solution", xtitle='t', ytitle='y', ymax=2., ymin=0)
8  numsol = gcurve(color=color.red)
   exactgr = graph(x=0, y=300, width=600, height=300, title="Exact solution",
           xtitle='t', ytitle='y', xmax=3.0, xmin=0.0, ymax=2.0, ymin=0)

12 exsol =gcurve (color = color.cyan)
   n = 24                                   # N steps > 3
   A = 0; B = 3.
   t =[0]*500;     y =[0]*500;      yy=[0]*4
16
   def f(t, y):                             # RHS F function
       return  (t - y)/2.0

20 def rk4(t, yy, h1):
       for i in range(0, 3):
           t  = h1 * i
           k0 = h1 * f(t, y[i])
24         k1 = h1 * f(t + h1/2., yy[i] + k0/2.)
           k2 = h1 * f(t + h1/2., yy[i] + k1/2.)
           k3 = h1 * f(t + h1, yy[i] + k2 )
           yy[i + 1] = yy[i]  +  (1./6.) * (k0  + 2.*k1  + 2.*k2 + k3)
28         print(i,yy[ i])
       return yy[3]

   def ABM(a,b,N):
```

```
32  # Compute 3 additional starting values using rk
        h = (b-a) / N                                # step
        t[0] = a;     y[0] = 1.00;      F0  = f(t[0], y[0])
        for k in range(1, 4):
36          t[k] = a  +  k * h
        y[1]  = rk4(t[1], y, h)                       # 1st step
        y[2]  = rk4(t[2], y, h)                       # 2nd step
        y[3]  = rk4(t[3], y, h)                       # 3rd step
40      F1 = f(t[1], y[1])
        F2 = f(t[2], y[2])
        F3 = f(t[3], y[3])
        h2 = h/24.
44      for k in range(3, N):                                    # Predictor
            p = y[k]  +  h2*(-9.*F0  +   37.*F1 - 59.*F2 + 55.*F3)
            t[k + 1] = a + h*(k+1)                    # Next abscissa
            F4 = f(t[k+1], p)
48          y[k+1] = y[k] + h2*(F1-5.*F2 + 19.*F3 + 9.*F4)   # Corrector
            F0 = F1                                    # Update values
            F1 = F2
            F2 = F3
52          F3 = f(t[k + 1], y[k + 1])
        return t,y

    t, y = ABM(A,B,n)
56  for k in range(0, n+1):
        numsol.plot( t[k], y[k] )
        exsol.plot( t[k], 3.*exp(-t[k]/2.) -2. + t[k])
```

Part II

Data Science

9

Fourier Analyses

This chapter discusses Fourier series and Fourier transforms. When implemented as algorithms, both become the Discrete Fourier Transform (DFT), or its fast cousin, the Fast Fourier Transform (FFT). In Chapter 14, we discuss the Short-Time Fourier Transform, and in Chapter 12, we derive the Quantum Fourier Transform, the quantum computing version of the DFT.

9.1 Fourier Series

Consider again a particle oscillating either in the nonharmonic potential of (8.5):

$$V(x) = \frac{1}{p}k|x|^p, \qquad p \neq 2, \tag{9.1}$$

or in the perturbed harmonic oscillator potential (8.2),

$$V(x) = \frac{1}{2}kx^2\left(1 - \frac{2}{3}\alpha x\right). \tag{9.2}$$

While free oscillations in these potentials are always periodic, they are not truly sinusoidal. Your **problem** is to take the solution of one of these nonlinear oscillators and expand it in a Fourier series:

$$y(t) = b_0 \sin \omega_0 t + b_1 \sin 2\omega_0 t + \cdots. \tag{9.3}$$

For example, if your oscillator is sufficiently nonlinear to behave like the sawtooth function (Figure 9.1 left), then the Fourier spectrum you obtain should be similar to that shown on the right in Figure 9.1.

In general, when we undertake such a spectral analysis we want to analyze the steady-state behavior of a system. This means that we have to wait for the initial transients to die out. It is easy to identify just what the initial transient is for linear systems, but may be less apparent for nonlinear systems in which the "steady state" jumps among a number of configurations. In the latter case, we could construct different Fourier spectra at different times, as is done with the *Short-Time Fourier Transform* to be discussed in Chapter 10.

Part of our interest in nonlinear oscillations arises from their lack of study in traditional physics courses, where just (approximate) linear oscillations are often studied. If the force

Computational Physics: Problem Solving with Python, Fourth Edition.
Rubin H. Landau, Manuel J. Páez, and Cristian C. Bordeianu.
© 2024 WILEY-VCH GmbH. Published 2024 by WILEY-VCH GmbH.

on a particle is always toward its equilibrium position (a restoring force), then the resulting motion will be *periodic*, but not necessarily *harmonic*. A good example is the motion in the highly anharmonic potential, such as (9.1) with $p \simeq 10$, that produces an $x(t)$ looking like a series of pyramids; this motion is periodic but not harmonic.

Our approach is in contrast to the traditional one in which the *fundamental* oscillation is determined analytically, and the higher-frequency *overtones* are determined by perturbation theory [Landau and Lifshitz, 1976]. We start with the full solution, and decompose it into *harmonics* and *overtones*. When we speak of fundamentals, overtones, and harmonics, we speak of solutions to the linear *boundary-value problem*, for example, of waves on a plucked violin string. In this latter case, and when given the correct conditions (and enough musical skill), it is possible to excite individual harmonics, or sums of them, from the series (9.3).

You may recall from classical mechanics that the general solution for a vibrating system can be expressed as the sum of the *normal* modes of that system. These expansions are possible only if we have *linear operators* and, consequently, the *principle of superposition*: If $y_1(t)$ and $y_2(t)$ are solutions of some linear equation, then $\alpha_1 y_1(t) + \alpha_2 y_2(t)$ is also a solution. The principle of linear superposition does not hold when we solve nonlinear problems. Nevertheless, it is always possible to expand a *periodic* solution of a *nonlinear* problem in terms of trigonometric functions. This is a consequence of *Fourier's theorem* being applicable to any single-valued periodic function with only a finite number of discontinuities. We assume we know the period T, that is, that

$$y(t + T) = y(t). \tag{9.4}$$

This tells us the *true frequency* ω:

$$\omega \equiv \omega_1 = \frac{2\pi}{T}. \tag{9.5}$$

Any periodic function (often designated as the *signal*) can be expanded as a series of harmonic functions with frequencies that are multiples of the true frequency:

$$y(t) = \frac{a_0}{2} + \sum_{n=1}^{\infty} \left(a_n \cos n\omega t + b_n \sin n\omega t \right). \tag{9.6}$$

This equation represents the signal $y(t)$ as the simultaneous sum of pure tones of frequency $n\omega$. The coefficients a_n and b_n measure of the amount of $\cos n\omega t$ and $\sin n\omega t$ present in $y(t)$, respectively. The intensity or *power* at each frequency is proportional to $a_n^2 + b_n^2$.

The Fourier series (9.6) is a "best fit," in the least squares sense of Chapter 6, to a number of measurements of the signal. This means that the series converges to the *average* behavior of the signal, but misses the signal at discontinuities (at which points it converges to the mean), or at sharp corners (where it overshoots). A general function $y(t)$ may contain an infinite number of Fourier components, although low-accuracy reproduction is usually possible with a small number of components.

The coefficients a_n and b_n in (9.6) are determined by the standard techniques of orthogonal function expansion. To find them, multiply both sides of (9.6) by $\cos n\omega t$ or $\sin n\omega t$, integrate over one period, and project a single a_n or b_n:

$$\binom{a_n}{b_n} = \frac{2}{T} \int_0^T dt \binom{\cos n\omega t}{\sin n\omega t} y(t), \qquad \omega \stackrel{\text{def}}{=} \frac{2\pi}{T}. \tag{9.7}$$

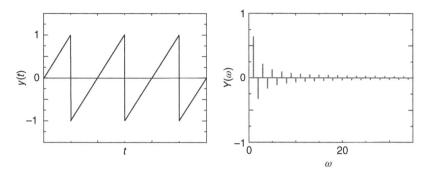

Figure 9.1 *Left*: A periodic sawtooth function. *Right*: The Fourier spectrum of frequencies contained in this function.

As seen in (Figure 9.1 right), the b_n's decrease in magnitude as the frequency increases, and can enter with a positive or negative sign, the negative sign indicating relative phase.

Awareness of the *symmetry* of the function $y(t)$ may eliminate the need to evaluate all the expansion coefficients. For example,

- a_0 is twice the average value of y: $a_0 = 2 \langle y(t) \rangle$.
- For an *odd function*, that is, one for which $y(-t) = -y(t)$, all a_n coefficients equal 0, and only half of the integration range is needed to determine b_n:

$$b_n = \frac{4}{T} \int_0^{T/2} dt\, y(t) \sin n\omega t. \tag{9.8}$$

However, if there is no input signal for $t < 0$, we do not have a truly odd function, and so small values of a_n may occur.
- For an *even function*, that is, one for which $y(-t) = y(t)$, all b_n coefficient equal 0, and only half the integration range is needed to determine a_n:

$$a_n = \frac{4}{T} \int_0^{T/2} dt\, y(t) \cos n\omega t. \tag{9.9}$$

9.1.1 Sawtooth and Half-Wave Functions

The sawtooth function (Figure 9.1 left) is described mathematically as

$$y(t) = \begin{cases} \frac{t}{T/2}, & \text{for } 0 \le t \le \frac{T}{2}, \\ \frac{t-T}{T/2}, & \text{for } \frac{T}{2} \le t \le T. \end{cases} \tag{9.10}$$

It is clearly periodic, nonharmonic, and discontinuous. Yet it is also odd and so can be represented more simply by shifting the signal to the left:

$$y(t) = \frac{t}{T/2}, \qquad -\frac{T}{2} \le t \le \frac{T}{2}. \tag{9.11}$$

Although the general shape of this function can be reproduced with only a few terms of the Fourier components, many components are needed to reproduce the sharp corners. As the function is odd, the Fourier series is a sine series, and (9.7) determines the b_n values:

$$b_n = \frac{2}{T} \int_{-T/2}^{+T/2} dt \, \sin n\omega t \frac{t}{T/2} = \frac{2}{n\pi}(-1)^{n+1},$$ (9.12)

$$\Rightarrow y(t) = \frac{2}{\pi}\left[\sin \omega t - \frac{1}{2}\sin 2\omega t + \frac{1}{3}\sin 3\omega t - \cdots\right].$$ (9.13)

The half-wave function

$$y(t) = \begin{cases} \sin \omega t, & \text{for } 0 < t < T/2, \\ 0, & \text{for } T/2 < t < T, \end{cases}$$ (9.14)

is periodic, nonharmonic (the upper half of a sine wave), and continuous, but with discontinuous derivatives. Because it lacks the sharp corners of the sawtooth function, it is easier to reproduce with a finite Fourier series. Equation (9.7) determines

$$a_n = \begin{cases} \dfrac{-2}{\pi(n^2-1)}, & n \text{ even or } 0, \\ 0, & n \text{ odd}, \end{cases} \qquad b_n = \begin{cases} \frac{1}{2}, & n = 1, \\ 0, & n \neq 1, \end{cases}$$

$$\Rightarrow y(t) = \frac{1}{2}\sin \omega t + \frac{1}{\pi} - \frac{2}{3\pi}\cos 2\omega t - \frac{2}{15\pi}\cos 4\omega t +.$$ (9.15)

9.1.2 Exercises: Fourier Series Summations

Hint: The program **FourierMatplot.py** written by Oscar Estrepe performs a Fourier analysis of a sawtooth function and produces the visualization shown on the right of Figure 9.1. You may want to use this program to help with this exercise.

1) **Sawtooth function**: Sum the Fourier series for the *sawtooth function* up to order $N = 2, 4, 10, 20$, and plot the results over two periods.
   (a) Check that in each case the series gives the mean value of the function *at* the points of discontinuity.
   (b) Check that in each case the series *overshoots* by about 9% the value of the function on either side of the discontinuity (the *Gibbs phenomenon*).
2) **Half-wave function**: Sum the Fourier series for the *half-wave function* up to order $N = 2, 4, 10, 20$, and plot the results over two periods. (The series converges quite well, doesn't it?)

9.2 Fourier Transforms

Although a Fourier *series* is the right tool for approximating or analyzing *periodic* functions, the Fourier *transform* or *integral* is the right tool for analyzing nonperiodic functions. We proceed from the series to the transform by imagining a system described by a continuum of "fundamental" frequencies, namely, *wave packets*.[1] While the difference between

---

1 We have chosen time and frequency as the conjugate variables here, but it could be otherwise, such as position x and wave vector k.

series and transforms may appear clear mathematically, when we approximate the Fourier integral as a finite sum, the two become equivalent.

By analogy with (9.6), we now imagine our function or signal $y(t)$ expressed in terms of a continuous series of harmonics (*inverse Fourier transform*):

$$y(t) = \int_{-\infty}^{+\infty} d\omega \, Y(\omega) \frac{e^{i\omega t}}{\sqrt{2\pi}}, \tag{9.16}$$

where for compactness we use a complex exponential function.[2] The expansion amplitude $Y(\omega)$ is analogous to the Fourier coefficients (a_n, b_n), and is called the *Fourier transform* of $y(t)$. The integral (9.16) is the inverse transform because it converts the transform to the signal. The *Fourier transform* converts the signal $y(t)$ to its transform $Y(\omega)$:

$$Y(\omega) = \int_{-\infty}^{+\infty} dt \, \frac{e^{-i\omega t}}{\sqrt{2\pi}} y(t). \tag{9.17}$$

The $1/\sqrt{2\pi}$ factor in both these integrals is a common normalization in quantum mechanics, but may not be in engineering, where only a single $1/2\pi$ factor is sometimes used. Likewise, the signs in the exponents are also conventions that do not matter as long as you maintain consistency.

If $y(t)$ is the measured response of a system (signal) as a function of time, then $Y(\omega)$ is the *spectral function* that measures the amount of frequency ω present in the signal. In many cases, it turns out that $Y(\omega)$ is a complex function with both positive and negative values, and with powers-of-ten variation in magnitude. Accordingly, it is customary to eliminate some of the extreme variations of $Y(\omega)$ by making a semilog plot of the squared modulus $|Y(\omega)|^2$ *versus* ω. This is called a *power spectrum* and provides an immediate view of the amount of power or strength in each component.

If the Fourier transform and its inverse are consistent with each other, we should be able to substitute (9.16) into (9.17) and obtain an identity:

$$Y(\omega) = \int_{-\infty}^{+\infty} dt \, \frac{e^{-i\omega t}}{\sqrt{2\pi}} \int_{-\infty}^{+\infty} d\omega' \, \frac{e^{i\omega' t}}{\sqrt{2\pi}} Y(\omega') \tag{9.18}$$

$$= \int_{-\infty}^{+\infty} d\omega' \left\{ \int_{-\infty}^{+\infty} dt \, \frac{e^{i(\omega'-\omega)t}}{2\pi} \right\} Y(\omega'). \tag{9.19}$$

For this to be an identity, the term in braces must be the *Dirac delta function*:

$$\int_{-\infty}^{+\infty} dt \, e^{i(\omega'-\omega)t} = 2\pi\delta(\omega' - \omega). \tag{9.20}$$

While the delta function is one of the most common and useful functions in theoretical physics, it is not well behaved in a mathematical sense, and misbehaves terribly in a computational sense. While it is possible to create numerical approximations to $\delta(\omega' - \omega)$, they may well be borderline pathological. It is certainly better for you to do the delta function part of an integration analytically.

---

2 Recall that $\exp(i\omega t) = \cos \omega t + i \sin \omega t$, and with the law of linear superposition this means that the real part of y gives the cosine series, and the imaginary part the sine series.

9.3 Discrete Fourier Transforms

If $y(t)$ or $Y(\omega)$ is known analytically or numerically, the integral (9.16) and (9.17) can be evaluated using the integration techniques studied earlier. In practice, the signal $y(t)$ is measured at just a finite number N of times t. The resultant *DFT* is an approximation, both because the signal is not known for all times, and because we integrate numerically [Briggs and Henson, 1995]. Once we have a discrete set of (approximate) transform values, they can be used to reconstruct the signal for any value of the time. In this way, the DFT can be thought of as a technique for interpolating, compressing, and extrapolating a signal.

We assume that the signal $y(t)$ is sampled at $(N + 1)$ discrete times (N time intervals), with a constant spacing $\Delta t = h$ between times:

$$y_k \stackrel{\text{def}}{=} y(t_k), \qquad k = 0, 1, 2, \ldots, N, \tag{9.21}$$

$$t_k \stackrel{\text{def}}{=} kh, \qquad h = \Delta t. \tag{9.22}$$

In other words, we measure $y(t)$ once every hth of a second for a total time of T. This correspondingly defines the signal's period T and the *sampling rate* s:

$$T \stackrel{\text{def}}{=} Nh, \qquad s = \frac{N}{T} = \frac{1}{h}. \tag{9.23}$$

Regardless of the true periodicity of the signal, when we choose a period T over which to sample the signal, the mathematics will inevitably produce a $y(t)$ that is periodic with period T,

$$y(t + T) = y(t). \tag{9.24}$$

We recognize this periodicity, and ensure that there are only N independent measurements used in the transform, by defining the first and last y's to be equal:

$$y_0 = y_N. \tag{9.25}$$

If we are analyzing a truly periodic function, then the N points should span one complete period, but not more. This guarantees their independence. Unless we make further assumptions, the N independent data $y(t_k)$ can determine no more than N independent transform values $Y(\omega_k)$.

The time interval T (which should be the period for periodic functions) is the largest time over which we measure the variation of $y(t)$. Consequently, it determines the lowest frequency contained in our Fourier representation of $y(t)$,

$$\omega_1 = \frac{2\pi}{T}. \tag{9.26}$$

The full range of frequencies in the spectrum ω_n is determined by the number of samples taken, and by the total sampling time $T = Nh$ as

$$\omega_n = n\omega_1 = n\frac{2\pi}{Nh}, \quad n = 0, 1, \ldots, N. \tag{9.27}$$

Here $\omega_0 = 0$ corresponds to the zero-frequency or *DC component* of the transform, that is, the part of the signal that does not oscillate.

The DFT algorithm follows from two approximations. First, we evaluate the integral (9.17) from time 0 to time T, over which the signal is measured, and not from $-\infty$ to $+\infty$.

Second, the trapezoid rule is used for the integration[3]:

$$Y(\omega_n) \stackrel{\text{def}}{=} \int_{-\infty}^{+\infty} dt\, \frac{e^{-i\omega_n t}}{\sqrt{2\pi}} y(t) \simeq \int_0^T dt\, \frac{e^{-i\omega_n t}}{\sqrt{2\pi}} y(t), \tag{9.28}$$

$$\simeq \sum_{k=1}^N h\, y(t_k) \frac{e^{-i\omega_n t_k}}{\sqrt{2\pi}} = h \sum_{k=1}^N y_k \frac{e^{-2\pi i k n/N}}{\sqrt{2\pi}}. \tag{9.29}$$

To keep the final notation more symmetric, the step size h is factored from the transform Y and a discrete function Y_n is defined:

$$Y_n \stackrel{\text{def}}{=} \frac{1}{h} Y(\omega_n) = \sum_{k=1}^N y_k \frac{e^{-2\pi i k n/N}}{\sqrt{2\pi}}, \quad n = 0, 1 \ldots, N. \tag{9.30}$$

With this same care in accounting, and with $d\omega \to 2\pi/Nh$, we invert the Y_n's:

$$y(t) \stackrel{\text{def}}{=} \int_{-\infty}^{+\infty} d\omega\, \frac{e^{i\omega t}}{\sqrt{2\pi}} Y(\omega), \tag{9.31}$$

$$\Rightarrow y(t) \simeq \sum_{n=1}^N \frac{2\pi}{Nh} \frac{e^{i\omega_n t}}{\sqrt{2\pi}} Y(\omega_n). \tag{9.32}$$

Once we know the N values of the transform, we can use (9.32) to evaluate $y(t)$ for any time t. There is nothing illegal about evaluating Y_n and y_k for arbitrarily large values of n and k, yet there is also nothing to be gained; because the trigonometric functions are periodic, we just get the old answers:

$$y(t_{k+N}) = y([k+N]h) = y(t_k), \tag{9.33}$$

$$Y(\omega_{n+N}) = Y([n+N]\omega_1) = Y(\omega_n). \tag{9.34}$$

Another way of stating this is to observe that none of the equations change if we replace $\omega_n t$ by $\omega_n t + 2\pi n$. There are still just N-independent output numbers for N independent inputs, with the transform and the reconstituted signal periodic.

We see from (9.27) that the larger we make the time $T = Nh$ over which we sample the function, the smaller will be the frequency steps or resolution.[4] Accordingly, if you want a smooth frequency spectrum, you will need to have a smaller frequency step $2\pi/T$, which means longer observation time T. While the best approach would be to measure the input signal for all times, in practice a measured signal $y(t)$ is often extended in time ("padded") by adding zeros for times beyond the last measured signal; this increases the value of T artificially and may lead to spurious conclusions. Although one may not think of padding as adding new information to the analysis, it does build in the assumption that the signal has no existence at times after the last measurement.

While periodicity is expected for a Fourier *series*, it is somewhat surprising for a Fourier *integral*, which has been touted as the right tool for nonperiodic functions. Clearly, if we input values of the signal for longer lengths of time, then the inherent period becomes longer, and if the repeat period T is very long, it may be of little consequence for times

---

3 The alert reader may be wondering what has happened to the $h/2$ with which the trapezoid rule weights the initial and final points. Actually, they are there, but because we have set $y_0 \equiv y_N$, two $h/2$ terms have been added to produce one h term.

4 See also Section 9.3.1 where we discuss the related phenomenon of aliasing.

short compared to the period. If $y(t)$ is actually periodic with period Nh, then the DFT is an excellent way of obtaining the Fourier series. If the input function is not periodic, then the DFT can be a bad approximation near the endpoints of the time interval, as the function will repeat there; likewise for the lowest frequencies.

The DFT and its inverse can be written in a concise and insightful way, and be evaluated efficiently, by introducing a complex variable Z for the exponential and then raising Z to various powers:

$$y_k = \frac{\sqrt{2\pi}}{N} \sum_{n=1}^{N} Z^{-nk} Y_n, \qquad Z = e^{-2\pi i/N}, \tag{9.35}$$

$$Y_n = \frac{1}{\sqrt{2\pi}} \sum_{k=1}^{N} Z^{nk} y_k, \qquad Z^{nk} \equiv [Z^n]^k. \tag{9.36}$$

With this formulation the computer needs to compute only powers of Z. We give our DFT code in Listing 9.1. If your preference is to avoid complex numbers, we can rewrite (9.35) in terms of separate real and imaginary parts by applying Euler's theorem with $\theta \overset{\text{def}}{=} 2\pi/N$:

$$Z = e^{-i\theta}, \quad \Rightarrow \quad Z^{\pm nk} = e^{\mp i nk\theta} = \cos nk\theta \mp i \sin nk\theta, \tag{9.37}$$

$$\Rightarrow \quad Y_n = \frac{1}{\sqrt{2\pi}} \sum_{k=1}^{N} \left[\cos(nk\theta) \mathbf{Re}\, y_k + \sin(nk\theta)\, \mathbf{Im}\, y_k \right.$$
$$\left. + \mathbf{i}(\cos(nk\theta)\, \mathbf{Im}\, y_k - \sin(nk\theta)\mathbf{Re}\, y_k) \right], \tag{9.38}$$

$$y_k = \frac{\sqrt{2\pi}}{N} \sum_{n=1}^{N} \left[\cos(nk\theta)\, \mathbf{Re}\, Y_n - \sin(nk\theta)\mathbf{Im}\, Y_n \right.$$
$$\left. + \mathbf{i}(\cos(nk\theta)\mathbf{Im}\, Y_n + \sin(nk\theta)\, \mathbf{Re}\, Y_n) \right]. \tag{9.39}$$

Readers new to DFTs are often surprised when they apply these equations to practical situations and end up with transforms Y having imaginary parts, despite the fact that the signal y is real. Equation (9.38) should make it clear that a real signal ($\mathbf{Im}\, y_k \equiv 0$) will yield an imaginary transform unless $\sum_{k=1}^{N} \sin(nk\theta)\, \mathbf{Re}\, y_k = 0$. This occurs only if $y(t)$ is an *even* function over $-\infty \le t \le +\infty$ *and* we integrate exactly. Because neither condition holds, the DFTs of real, even functions may have small imaginary parts. This is not as a result of an error in programming, and in fact yields a measure of the approximation error in the entire procedure.

The computation time for a DFT can be reduced even further by use of the *FFT* algorithm, as discussed in Section 9.5. An examination of (9.35) shows that the DFT is evaluated as a matrix multiplication of a vector of length N containing the Z values, by a vector of length N of y value. The time for this DFT scales like N^2, while the time for the FFT algorithm scales as $N\log_2 N$. Although this may not seem like much of a difference, for $N = 10^{2-3}$, the difference of 10^{3-5} is the difference between a minute and a week. For this reason, it is the FFT that is often used for on-line spectrum analysis.

9.3.1 Aliasing

The sampling of a signal by DFT for only a finite number of times and large Δt, limits the accuracy of the deduced high-frequency components present in the signal. Obviously,

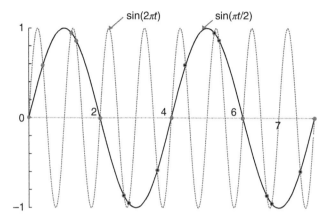

Figure 9.2 A plot of the functions $\sin(\pi t/2)$ and $\sin(2\pi t)$. If the sampling rate is not high enough, these signals may appear indistinguishable in a Fourier decomposition. If the sample rate is too low, and if both signals are present in a sample, the deduced low-frequency components may be contaminated by the higher-frequency ones.

good information about very high frequencies requires sampling the signal with small time steps so that all the wiggles can be included. While a poor deduction of the high-frequency components may be tolerable, if all we care about are the low-frequency ones, the inaccurate high-frequency components may contaminate the deduced low-frequency ones. This effect is called *aliasing* and is the cause of the Moiré pattern distortion in digital images.

As an example, consider Figure 9.2 showing the two functions $\sin(\pi t/2)$ and $\sin(2\pi t)$ for $0 \le t \le 8$, with their points of overlap in bold. If we were unfortunate enough to sample a signal containing these functions at the times $t = 0, 2, 4, 6, 8$, then we would measure $y \equiv 0$ and assume that there was no signal at all! However, if we were unfortunate enough to measure the signal at the filled dots in Figure 9.2, where $\sin(\pi t/2) = \sin(2\pi t)$, specifically, $t = 0, \frac{12}{10}, \frac{4}{3}, \dots$, then our Fourier analysis would completely miss the high-frequency components. In DFT jargon, we would say that the high-frequency component has been *aliased* by the low-frequency component. In other cases, some high-frequency values may be included in our sampling of the signal, but our sampling rate may not be high enough to include enough of them to separate the high-frequency component properly. In this case, some high-frequency signals would be included spuriously as part of the low-frequency spectrum, and this would lead to spurious low-frequency oscillations when the signal is synthesized from its Fourier components.

More precisely, aliasing occurs when a signal containing frequency f is sampled at a rate of $s = N/T$ measurements per unit time, with $s \le f/2$. In this case, the frequencies f and $f - 2s$ yield the same DFT, and we would not be able to determine that there are two frequencies present. That being the case, to avoid aliasing we want no frequencies $f > s/2$ to be present in our input signal. This is known as the *Nyquist criterion*. In practice, some applications avoid the effects of aliasing by filtering out the high frequencies from the signal, and then analyzing only the remaining low-frequency part. (The low-frequency *sinc filter* discussed in Section 9.4.4 is often used for this purpose.) Although filtering eliminates some high-frequency information, it lessens the distortion of the low-frequency components, and so may lead to improved reproduction of the signal.

If accurate values for the high frequencies are required, then you will need to increase the sampling rate s by increasing the number N of samples taken within the fixed sampling time $T = Nh$. By keeping the sampling time constant and increasing the number of samples taken, we make the time step h smaller and pick up the higher frequencies. By increasing the number N of frequencies that you compute, you move the previous higher-frequency components closer to the middle of the spectrum, and thus away from the error-prone ends.

If we increase the total time sampling time $T = Nh$ and keep h the same, then the sampling rate $s = N/T = 1/h$ remains the same. Since $\omega_1 = 2\pi/T$, this makes ω_1 smaller, which means we have more low frequencies recorded and a smoother frequency spectrum. And as we said, this is often carried out, after the fact, by padding the end of the data set with zeros.

Exercise

1) The sampling of a signal by DFT for only a finite number of times not only limits the accuracy of the deduced high-frequency components, but also contaminates the deduced low-frequency components (*aliasing*). Consider the two functions $\sin(\pi t/2)$ and $\sin(2\pi t)$ for $0 \leq t \leq 8$.
   (a) Make graphs of both functions on the same plot.
   (b) Perform a DFT on both functions.
   (c) Sample at times $t = 0, 2, 4, 6, 8, \ldots$ and draw conclusions.
   (d) Sample at times $t = 0, 12/10, 4/3, \ldots$ and draw conclusions about the high-frequency components (*Hint*: They may be *aliased* by the low-frequency components).
   (e) The *Nyquist criterion* states that when a signal containing frequency f is sampled at a rate of $s = N/T$, measurements per unit time, with $s \leq f/2$, then aliasing occurs. Verify specifically that the frequencies f and $f - 2s$ yield the same DFT.
2) Perform a Fourier analysis of the chirp signal $y(t) = \sin(60t^2)$. As seen in Figure 10.5, this signal is not truly periodic, and is better analyzed with methods soon to be discussed.

9.3.2 Assessments

Simple analytic input: It is always good to do simple checks before examining more complex problems, even if you are using a package's Fourier tool.
1) Sample the even signal

$$y(t) = 3\cos(\omega t) + 2\cos(3\omega t) + \cos(5\omega t). \tag{9.40}$$

   (a) Decompose this into its components.
   (b) Check that the components are essentially real and in the ratio $3:2:1$ (or $9:4:1$ for the power spectrum).
   (c) Verify that the frequencies have the expected values (not just ratios).
   (d) Verify the components sum up to give the input signal.
   (e) Experiment on the separate effects of picking different values of the step size h and of enlarging the measurement period $T = Nh$.
2) Sample the odd signal

$$y(t) = \sin(\omega t) + 2\sin(3\omega t) + 3\sin(5\omega t). \tag{9.41}$$

Decompose this into its components, and then check that they are essentially imaginary and in the ratio $1:2:3$ (or $1:4:9$ if a power spectrum is plotted). Check that they sum up to give the input signal.

3) Sample the mixed-symmetry signal

$$y(t) = 5\sin(\omega t) + 2\cos(3\omega t) + \sin(5\omega t). \qquad (9.42)$$

Decompose this into its components, and then check that they are in the ratio $5:2:1$ (or $25:4:1$ if a power spectrum is plotted). Check that they sum up to give the input signal.

4) Sample the signal

$$y(t) = 5 + 10\sin(t+2).$$

Compare and explain the results obtained by sampling (a) without the 5, (b) as given but without the 2, and (c) without the 5 and the 2.

5) In our discussion of aliasing, we examined Figure 9.2 showing the functions $\sin(\pi t/2)$ and $\sin(2\pi t)$. Sample the function

$$y(t) = \sin(\tfrac{\pi}{2}t) + \sin(2\pi t) \qquad (9.43)$$

and explore how aliasing occurs. Explicitly, we know that the true transform contains peaks at $\omega = \pi/2$ and $\omega = 2\pi$. Sample the signal at a rate that leads to aliasing, as well as at a higher sampling rate at which there is no aliasing. Compare the resulting DFTs in each case and check if your conclusions agree with the Nyquist criterion.

Highly nonlinear oscillator: Recall the numerical solution for oscillations of a spring with power $p = 12$ [see (9.1)]. Decompose the solution into a Fourier series and determine the number of higher harmonics that contribute at least 10%; for example, determine the n for which $|b_n/b_1| < 0.1$. Check that resuming the components reproduces the signal.

Nonlinearly perturbed oscillator: Remember the harmonic oscillator with a nonlinear perturbation (8.2):

$$V(x) = \tfrac{1}{2}kx^2 \left(1 - \tfrac{2}{3}\alpha x\right), \qquad F(x) = -kx(1 - \alpha x). \qquad (9.44)$$

For very small amplitudes of oscillation ($x \ll 1/\alpha$), the solution $x(t)$ essentially should be only the first term of a Fourier series. (*Warning*: The ω you use in your series must correspond to the *true* frequency of the system, not the ω_0 of small oscillations.

1) We want the signal to contain "approximately 10% nonlinearity." This being the case, fix your value of α so that $\alpha x_{max} \simeq 10\%$, where x_{max} is the maximum amplitude of oscillation. For the rest of the problem, keep the value of α fixed.

2) Decompose your numerical solution into a discrete Fourier spectrum.

3) Plot a graph of the percentage of importance of the first *two*, non-DC Fourier components as a function of the initial displacement for $0 < x_0 < 1/2\alpha$. You should find that higher harmonics are more important as the amplitude increases. Because both even and odd components are present, Y_n should be complex. Because a 10% effect in amplitude becomes a 1% effect in power, make sure that you make a semilog plot of the power spectrum.

4) As always, check that resummations of your transforms reproduce the signal.

9.3.3 Transforming Nonperiodic Functions

Consider an electron initially localized around $x = 5$. A model to describe this "localized" electron is a Gaussian multiplying a plane wave:

$$\psi(x, t = 0) = \exp\left[-\frac{1}{2}\left(\frac{x-5}{\sigma_0}\right)^2\right] e^{ik_0 x}, \tag{9.45}$$

where we use natural units in which $\hbar = 1$. This wave packet is not an eigenstate of the momentum operator $p = id/dx$, but rather contains a spread of momenta. Your **problem** is to evaluate the Fourier transform,

$$\psi(p) = \int_{-\infty}^{+\infty} dx \frac{e^{ipx}}{\sqrt{2\pi}} \psi(x, 0), \tag{9.46}$$

as a way of determining the momenta components in (9.45).

9.4 Noise Filtering

In the process of solving this problem, we examine two simple approaches: the use of auto-correlation functions and the use of filters. Both approaches find wide applications in science, with our discussion not doing the subjects justice. We will see filters again in the discussion of wavelets in Chapter 10.

You measure a signal $y(t)$ that obviously contains noise. Your **problem** is to determine the frequencies that would be present in the spectrum of the signal if there were no noise. Of course, once you have a Fourier transform from which the noise has been removed, you can transform it to obtain a noise-free signal $s(t)$.

9.4.1 Noise Reduction via Autocorrelation

We assume that the measured signal is the sum of the true signal $s(t)$, which we wish to determine, plus some unwelcome *noise* $n(t)$:

$$y(t) = s(t) + n(t), \tag{9.47}$$

One approach at removing the noise relies on the fact that noise is usually random, and thus should not be correlated with the signal. Yet what do we mean when we say that two functions are not correlated? Well, if the two tend to oscillate with their nodes and peaks in much the same places, then the two functions are clearly correlated. An analytic measure of the correlation of two arbitrary functions $y(t)$ and $x(t)$ is the *correlation function*

$$c(\tau) = \int_{-\infty}^{+\infty} dt\, y(t)\, x(t + \tau) \equiv \int_{-\infty}^{+\infty} dt\, y(t - \tau)\, x(t), \tag{9.48}$$

Here τ, the *lag time*, is a variable, and we assume that the average values of the functions have been subtracted off, so that they oscillate around zero. Even if the two signals have different magnitudes, if they have similar time dependencies, except for one lagging or leading the other, then for certain values of τ, the integrand in (9.48) will be positive for all values of t. For those values of τ, the two signals interfere constructively and produce a large value for the correlation function. In contrast, if both functions oscillate independently, regardless of the value of τ, then it is just as likely for the integrand to be positive as to be negative, in which case the two signals interfere destructively and produce a small value for the integral.

Before we apply the correlation function to our problem, let us study some of its properties. We use (9.16) to express c, y, and x in terms of their Fourier transforms:

$$c(\tau) = \int_{-\infty}^{+\infty} d\omega'' \, C(\omega'') \frac{e^{i\omega'' t}}{\sqrt{2\pi}}, \qquad y(t) = \int_{-\infty}^{+\infty} d\omega \, Y(\omega) \frac{e^{-i\omega t}}{\sqrt{2\pi}},$$

$$x(t+\tau) = \int_{-\infty}^{+\infty} d\omega' \, X(\omega') \frac{e^{+i\omega t}}{\sqrt{2\pi}}. \tag{9.49}$$

Seeing that ω, ω', and ω'' are dummy variables, other names may be used for them without changing the results. When we substitute these representations into the definition (9.48) of the correlation function, and assume that the resulting integrals converge well enough to be rearranged, we obtain

$$\int_{-\infty}^{+\infty} d\omega \, C(\omega) e^{i\omega t} = \int_{-\infty}^{+\infty} \frac{d\omega}{2\pi} \int_{-\infty}^{+\infty} d\omega' \, Y(\omega) X(\omega') e^{i\omega \tau} 2\pi \delta(\omega' - \omega)$$

$$= \int_{-\infty}^{+\infty} d\omega \, Y(\omega) X(\omega) e^{i\omega \tau},$$

$$\Rightarrow C(\omega) = \sqrt{2\pi} \, Y(\omega) X(\omega), \tag{9.50}$$

where the last line follows because ω'' and ω are equivalent dummy variables. Equation (9.50) says that the Fourier transform of the correlation function of two signals is proportional to the product of their transforms. (We shall see a related convolution theorem for filters.)

A special case of the correlation function $c(\tau)$ is the *autocorrelation function* $A(\tau)$ that measures the correlation of a time signal with itself:

$$A(\tau) \stackrel{\text{def}}{=} \int_{-\infty}^{+\infty} dt \, y(t) \, y(t+\tau) \equiv \int_{-\infty}^{+\infty} dt \, y(t) y(t-\tau). \tag{9.51}$$

This function is computed by taking a signal $y(t)$ that has been measured over some time period, and then averaging it over time using, $y(t+\tau)$ as a weighting function. This process is called *folding*, or *convoluting*, a function onto itself (as might be done with dough). To see how this folding removes noise from a signal, we go back to the measured signal (9.47), which was the sum of pure signal plus noise $s(t) + n(t)$. As an example, on the upper left in Figure 9.3, we show a signal that was constructed by adding random noise to a smooth signal. When we compute the autocorrelation function for this signal, we obtain a function (upper right in Figure 9.3) that looks like a broadened, smoothed version of the signal $y(t)$.

We can understand how the noise is removed by taking the Fourier transform of $s(t) + n(t)$ to obtain a simple sum of transforms:

$$Y(\omega) = S(\omega) + N(\omega), \tag{9.52}$$

$$\left\{ \begin{array}{c} S(\omega) \\ N(\omega) \end{array} \right\} = \int_{-\infty}^{+\infty} dt \left\{ \begin{array}{c} s(t) \\ n(t) \end{array} \right\} \frac{e^{-i\omega t}}{\sqrt{2\pi}}. \tag{9.53}$$

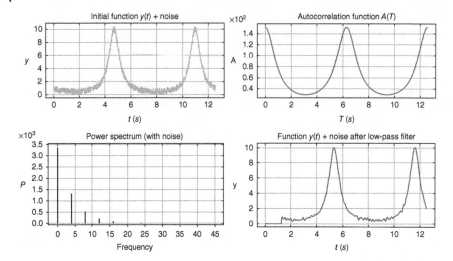

Figure 9.3 *From bottom left to right*: A function that is a signal plus noise $s(t) + n(t)$; the autocorrelation function versus time deduced by processing this signal; the power spectrum obtained from autocorrelation function; the signal plus noise after passage through a lowpass filter.

Because the autocorrelation function (9.51) for $y(t) = s(t) + n(t)$ involves the second power of y, is not a linear function, that is, $A_y \neq A_s + A_n$, but instead,

$$A_y(\tau) = \int_{-\infty}^{+\infty} dt\, [s(t)s(t + \tau) + s(t)n(t + \tau) + n(t)n(t + \tau)]. \tag{9.54}$$

If we assume that the noise $n(t)$ in the measured signal is truly random, then it should average to zero over long times and be uncorrelated at times t and $t + \tau$. This being the case, both integrals involving the noise vanish, and so

$$A_y(\tau) \simeq \int_{-\infty}^{+\infty} dt\, s(t)\, s(t + \tau) = A_s(\tau). \tag{9.55}$$

Thus, the part of the noise that is random tends to be averaged out of the autocorrelation function, and we are left with an approximation of the autocorrelation function of the pure signal.

So how does this help us? Application of (9.50) with $Y(\omega) = X(\omega) = S(\omega)$ tells us that the Fourier transform $A(\omega)$ of the autocorrelation function is proportional to $|S(\omega)|^2$:

$$A(\omega) = \sqrt{2\pi}\, |S(\omega)|^2. \tag{9.56}$$

The function $|S(\omega)|^2$ is the *power spectrum* of the pure signal. Thus, evaluation of the auto-correlation function of the noisy signal gives us the pure signal's power spectrum, which is often all that we need to know. For example, in Figure 9.3 we see a noisy signal (lower left, the autocorrelation function (lower right, which clearly is smoother than the signal, and finally, the deduced power spectrum (upper left). Notice that the broadband high-frequency components characteristic of noise are absent from the power spectrum.

You can easily modify the sample program `DFTcomplex.py` in Listing 9.1 or `DFTreal.py` in Listing 9.2 sample program `DFTcomplex.py` in Listing 9.1 to compute the autocorrelation function and then the power spectrum $A(\tau)$. The program `NoiseSincFilter.py` does just that.

9.4.2 Autocorrelation Function Exercises

1) Imagine that you have sampled the pure signal

$$s(t) = \frac{1}{1 - 0.9 \sin t}.$$

(9.57)

Although there is just a single sine function in the denominator, there is an infinite number of overtones as you can see from the expansion

$$s(t) \simeq 1 + 0.9 \sin t + (0.9 \sin t)^2 + (0.9 \sin t)^3 + \cdots.$$

(9.58)

(a) Compute the DFT $S(\omega)$. Make sure not to sample just one period, but also to cover the entire period. Also make sure to sample at enough times (fine scale) to obtain good sensitivity to the high-frequency components.

(b) Make a semilog plot of the power spectrum $|S(\omega)|^2$.

(c) Take your input signal $s(t)$ and compute its autocorrelation function $A(\tau)$ for a full range of τ values (an analytic solution is okay too).

(d) Compute the power spectrum indirectly by performing a DFT on the autocorrelation function. Compare your results to the spectrum obtained by computing $|S(\omega)|^2$ directly.

2) Add some random noise to the signal using a random number generator:

$$y(t_i) = s(t_i) + \alpha(2r_i - 1), \quad 0 \le r_i \le 1,$$

(9.59)

where α is an adjustable parameter and r_i are random numbers. Try several values of α, from small ones that just add some fuzz to the signal to large ones that nearly hide the signal.

(a) Plot your noisy data, their Fourier transform, and their power spectrum obtained directly from the transform with noise.

(b) Compute the autocorrelation function $A(\tau)$ and its Fourier transform $A(\omega)$.

(c) Compare the DFT of $A(\tau)$ to the true power spectrum. Comment on the effectiveness of reducing noise by use of the autocorrelation function.

(d) For what value of α do you essentially lose all the information in the input?

9.4.3 Filtering with Transforms

A filter (Figure 9.4) is a device that converts an input signal $f(t)$ to an output signal $g(t)$, with some specific property for $g(t)$. More specifically, an *analog filter* is defined as integration over an input function [Hartmann, 1998]:

$$g(t) = \int_{-\infty}^{+\infty} d\tau f(\tau) h(t - \tau) \stackrel{\text{def}}{=} f(t) * h(t).$$

(9.60)

Here the asterisk $*$ indicates a *convolution*, which we have already seen in the discussion of the autocorrelation function. The function $h(t)$ is the *unit response* or *transfer function* of

Figure 9.4 An input signal *f(t)* passes through a filter *h* that outputs the function *g(t)*.

the filter; it is the response of the filter to a unit impulse:

$$h(t) = \int_{-\infty}^{+\infty} d\tau\, \delta(\tau)\, h(t - \tau). \tag{9.61}$$

Equation (9.60) states that the output $g(t)$ of a filter equals the input $f(t)$ convoluted with the transfer function $h(t - \tau)$. Because the argument of the response function is delayed by a time τ relative to that of the signal in the integral (9.60), τ is called the *lag time*. While the integration is over all times, the response of a good detector usually peaks around zero time. In any case, the response must equal zero for $\tau > t$ because events in the future cannot affect the present (causality).

The *convolution theorem* states that the Fourier transform of the convolution $g(t)$ is proportional to the product of the transforms of $f(t)$ and $h(t)$:

$$G(\omega) = \sqrt{2\pi}\, F(\omega)\, H(\omega). \tag{9.62}$$

The theorem results from expressing the functions in (9.60) by their transforms, and using the resulting Dirac delta function to evaluate an integral (essentially what we did in our discussion of the correlation function).

Filtering, as we have defined it, is a linear process involving just the first powers of the signal f. This means that the output at one frequency is proportional to the input at that frequency. The constant of proportionality between the two may change with frequency, and thus suppress specific frequencies relative to others, but that constant remains fixed in time. Since the law of linear superposition is valid for linear filters, if the input to a filter is the sum of various functions, then the transform of the output will be the sum of the functions' Fourier transforms.

Filters that remove or decrease high-frequency components more than they do low-frequency ones, are called *lowpass* filters. Those that filter out the low frequencies are called *highpass filters*. A simple lowpass filter is the *RC* circuit on the left of Figure 9.5 that produces the transfer function

$$H(\omega) = \frac{1}{1 + i\omega\tau} = \frac{1 - i\omega\tau}{1 + \omega^2\tau^2}, \tag{9.63}$$

where $\tau = RC$ is the time constant. The ω^2 in the denominator leads to a decrease in the response at high frequencies and therefore makes this a lowpass filter (the $i\omega$ affects only the phase). A simple highpass filter is the *RC* circuit on the right in Figure 9.5 that produces the transfer function

$$H(\omega) = \frac{i\omega\tau}{1 + i\omega\tau} = \frac{i\omega\tau + \omega^2\tau^2}{1 + \omega^2\tau^2}. \tag{9.64}$$

We see that $H = 1$ at large ω, yet vanishes as $\omega \to 0$; as expected for a highpass filter.

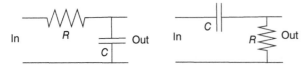

Figure 9.5 *Left*: An *CR* circuit arranged as a lowpass filter. *Right*: An *CR* circuit arranged as a highpass filter.

Figure 9.6 A delay-line filter in which the signal at different times is scaled by different amounts c_i.

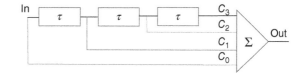

Filters composed of resistors and capacitors are fine for analog signal processing, but for digital processing we want a *digital filter* that has a specific response function for each frequency range. A physical model for a digital filter may be constructed from a delay line with taps at various spacing along the line (Figure 9.6) [Hartmann, 1998]. The signal read from tap n is just the input signal delayed by time $n\tau$, where the delay time τ is a characteristic of the particular filter. The output from each tap is described by the transfer function $\delta(t - n\tau)$, possibly with scaling factor c_n. As represented by the triangle on the right in Figure 9.6, the signals from all taps are ultimately summed together to form the total response function:

$$h(t) = \sum_{n=0}^{N} c_n \, \delta(t - n\tau). \tag{9.65}$$

In the frequency domain, the Fourier transform of a delta function is an exponential, and so (9.65) results in the transfer function

$$H(\omega) = \sum_{n=0}^{N} c_n \, e^{-i\,n\omega\tau}, \tag{9.66}$$

where the exponential indicates the phase shift from each tap.

If a digital filter is given a continuous time signal $f(t)$ as input, its output will be the discrete sum

$$g(t) = \int_{-\infty}^{+\infty} dt' \, f(t') \sum_{n=0}^{N} c_n \, \delta(t - t' - n\tau) = \sum_{n=0}^{N} c_n \, f(t - n\tau). \tag{9.67}$$

And of course, if the signal's input is a discrete sum, its output will remain a discrete sum. In either case, we see that knowledge of the filter coefficients c_i provides us with all we need to know about a digital filter. If we look back at our work on the DFT in Section 9.3, we can view a digital filter (9.67) as a Fourier transform in which we use an N-point approximation to the Fourier integral. The c_n's then contain both the integration weights and the values of the response function at the integration points. Accordingly, the transform can be viewed as a filter of the signal into specific frequencies.

9.4.4 Digital Filters: Windowed Sinc Filters ⊙

A popular way to separate the bands of frequencies in a signal is with a *windowed sinc filter* [Smith, 1999]. This filter is based on the observation that an ideal *lowpass* filter passes all frequencies below a cutoff frequency ω_c, and blocks all frequencies above this frequency. And because there tends to be more noise at high frequencies than at low frequencies, removing the high frequencies tends to remove more noise than signal, although some signal is inevitably lost. One use for windowed sinc filters is in reducing aliasing in DFTs by removing the high-frequency component of a signal before determining its Fourier components. The graph on the lower right in Figure 9.1 was obtained by passing our noisy signal through a sinc filter (using the program `NoiseSincFilter.py`).

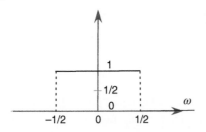

Figure 9.7 The rectangle function rect(ω) that is constant for a finite frequency interval. The Fourier transform of this function is sinc(t).

If both positive and negative frequencies are included, an ideal low-frequency filter will look like the rectangular pulse in frequency space:

$$H(\omega, \omega_c) = \text{rect}\left(\frac{\omega}{2\omega_c}\right), \qquad \text{rect}(\omega) = \begin{cases} 1, & \text{if } |\omega| \leq \frac{1}{2}, \\ 0, & \text{otherwise.} \end{cases} \qquad (9.68)$$

Here rect(ω) is the rectangular function (Figure 9.7). Although maybe not obvious, a rectangular pulse in the frequency domain has a Fourier transform that is proportional to the *sinc function* in the time domain [Smith, 1991]:

$$\int_{-\infty}^{+\infty} d\omega\, e^{-i\omega t} \text{rect}(\omega) = \text{sinc}\left(\frac{t}{2}\right) \stackrel{\text{def}}{=} n\frac{\sin(\pi t/2)}{\pi t/2}, \qquad (9.69)$$

where the π's are sometimes omitted. Consequently, we can filter out the high-frequency components of a signal by convoluting it with $\sin(\omega_c t)/(\omega_c t)$, a technique also known as the *Nyquist–Shannon* interpolation formula. In terms of discrete transforms, the time-domain representation of the sinc filter is simply

$$h[i] = \frac{\sin(\omega_c i)}{i\pi}. \qquad (9.70)$$

Because all frequencies below the cutoff frequency ω_c are passed with unit amplitude, while all higher frequencies are blocked, we can see the importance of a sinc filter.

In practice, there are a number of problems in using the sinc function as the filter. First, as formulated, the filter is *noncausal*; that is, there are coefficients at negative times, which violates causality because we do not start measuring the signal until $t = 0$. Second, in order to produce a perfect rectangular response, we would have to sample the signal at an infinite number of times. In practice, we sample at $(M + 1)$ points (M even) placed symmetrically around the main lobe of $\sin(\pi t)/\pi t$, and then shift times to purely positive values:

$$h[i] = \frac{\sin[2\pi\omega_c(i - M/2)]}{i - M/2}, \qquad 0 \leq t \leq M. \qquad (9.71)$$

As might be expected, a penalty is incurred for making the filter discrete, instead of the ideal rectangular response, we obtain some *Gibbs overshoot*, with rounded corners and oscillations beyond the corner.

There are two ways to reduce the departures from the ideal filter. The first is to increase the length of times over which the signal is sampled, which inevitably leads to longer compute times. The other way is to smooth out the truncation of the sinc function by multiplying it with a smoothly tapered curve, like the *Hamming window function*:

$$w[i] = 0.54 - 0.46 \cos\left(\frac{2\pi i}{M}\right). \qquad (9.72)$$

In this way the filter's kernel becomes

$$h[i] = \frac{\sin[2\pi\omega_c(i - M/2)]}{i - M/2}\left[0.54 - 0.46 \cos\left(\frac{2\pi i}{M}\right)\right]. \qquad (9.73)$$

The cutoff frequency ω_c should be a fraction of the sampling rate. The time length M determines the *bandwidth* over which the filter changes from 1 to 0.

Exercise Repeat the exercise that added random noise to a known signal, this time using the sinc filter to reduce the noise. See how small you can make the relative strength of the signal, and still be able to separate it from the noise.

9.5 Fast Fourier Transform ⊙

We have seen in (9.35) that a DFT can be written in the compact form

$$Y_n = \frac{1}{\sqrt{2\pi}} \sum_{k=1}^{N} Z^{nk} y_k, \quad Z = e^{-2\pi i/N}, \quad n = 0, 1, \dots, N-1. \tag{9.74}$$

Even if the signal elements y_k to be transformed are real, Z is complex, and therefore we must process both real and imaginary parts when computing transforms. Because both n and k range over N integer values, the $(Z^n)^k y_k$ multiplications in (9.74) require some N^2 multiplications and additions of complex numbers. As N gets large, as happens in realistic applications, this geometric increase in the number of steps slows down the computation.

In 1965, Cooley and Turkey discovered an algorithm[5] that reduces the number of operations necessary to perform a DFT from N^2 to roughly $N\log_2 N$ [Cooley and Tukey, 1965; Donnelly and Rust, 2005]. Although this may not seem like such a big difference, it represents a 100-fold speedup for 1000 data points, which changes a full day of processing into 15 min of work. Due to its widespread use (including cell phones), the FFT algorithm is considered one of the 10 most important algorithms of all time.

The idea behind the FFT is to utilize the periodicity inherent in the definition of the DFT (9.74) to reduce the total number of computational steps. Essentially, the algorithm divides the input data into two equal groups and transforms only one group, which requires $\sim (N/2)^2$ multiplications. It then divides the remaining (untransformed) group of data in half and transforms them, continuing the process until all the data have been transformed. The total number of multiplications required with this approach is approximately $N\log_2 N$.

Specifically, the FFTs time economy arises from the computationally expensive complex factor $Z^{nk}[= [(Z)^n]^k]$ having values that are repeated as the integers n and k vary sequentially. For instance, for $N = 8$,

$$Y_0 = Z^0 y_0 + Z^0 y_1 + Z^0 y_2 + Z^0 y_3 + Z^0 y_4 + Z^0 y_5 + Z^0 y_6 + Z^0 y_7,$$
$$Y_1 = Z^0 y_0 + Z^1 y_1 + Z^2 y_2 + Z^3 y_3 + Z^4 y_4 + Z^5 y_5 + Z^6 y_6 + Z^7 y_7,$$
$$Y_2 = Z^0 y_0 + Z^2 y_1 + Z^4 y_2 + Z^6 y_3 + Z^8 y_4 + Z^{10} y_5 + Z^{12} y_6 + Z^{14} y_7,$$
$$Y_3 = Z^0 y_0 + Z^3 y_1 + Z^6 y_2 + Z^9 y_3 + Z^{12} y_4 + Z^{15} y_5 + Z^{18} y_6 + Z^{21} y_7,$$
$$Y_4 = Z^0 y_0 + Z^4 y_1 + Z^8 y_2 + Z^{12} y_3 + Z^{16} y_4 + Z^{20} y_5 + Z^{24} y_6 + Z^{28} y_7,$$
$$Y_5 = Z^0 y_0 + Z^5 y_1 + Z^{10} y_2 + Z^{15} y_3 + Z^{20} y_4 + Z^{25} y_5 + Z^{30} y_6 + Z^{35} y_7,$$
$$Y_6 = Z^0 y_0 + Z^6 y_1 + Z^{12} y_2 + Z^{18} y_3 + Z^{24} y_4 + Z^{30} y_5 + Z^{36} y_6 + Z^{42} y_7,$$
$$Y_7 = Z^0 y_0 + Z^7 y_1 + Z^{14} y_2 + Z^{21} y_3 + Z^{28} y_4 + Z^{35} y_5 + Z^{42} y_6 + Z^{49} y_7,$$

---

[5] Actually, this algorithm has been discovered a number of times, for instance, in 1942 by Dandelion and Lancers Danielson and Lanczos [1942], as well as much earlier by Gauss.

where we include $Z^0 (\equiv 1)$ for clarity. When we actually evaluate these powers of Z, we find only four independent values:

$$Z^0 = \exp(0) = +1, \qquad\qquad Z^1 = \exp\left(-\frac{2\pi}{8}\right) = +\frac{\sqrt{2}}{2} - i\frac{\sqrt{2}}{2},$$

$$Z^2 = \exp\left(-\frac{2\cdot 2i\pi}{8}\right) = -i, \qquad Z^3 = \exp\left(-\frac{2\pi\cdot 3i}{8}\right) = -\frac{\sqrt{2}}{2} - i\frac{\sqrt{2}}{2},$$

$$Z^4 = \exp\left(-\frac{2\pi\cdot 4i}{8}\right) = -Z^0, \qquad Z^5 = \exp\left(-\frac{2\pi\cdot 5i}{8}\right) = -Z^1,$$

$$Z^6 = \exp\left(-\frac{2\cdot 6i\pi}{8}\right) = -Z^2, \qquad Z^7 = \exp\left(-\frac{2\cdot 7i\pi}{8}\right) = -Z^3,$$

$$Z^8 = \exp\left(-\frac{2\pi\cdot 8i}{8}\right) = +Z^0, \qquad Z^9 = \exp\left(-\frac{2\pi\cdot 9i}{8}\right) = +Z^1,$$

$$Z^{10} = \exp\left(-\frac{2\pi\cdot 10i}{8}\right) = +Z^2, \qquad Z^{11} = \exp\left(-\frac{2\pi\cdot 11i}{8}\right) = +Z^3,$$

$$Z^{12} = \exp\left(-\frac{2\pi\cdot 11i}{8}\right) = -Z^0, \dots \qquad\qquad (9.75)$$

When substituted into the definitions of the transforms, we obtain

$$Y_0 = Z^0 y_0 + Z^0 y_1 + Z^0 y_2 + Z^0 y_3 + Z^0 y_4 + Z^0 y_5 + Z^0 y_6 + Z^0 y_7,$$
$$Y_1 = Z^0 y_0 + Z^1 y_1 + Z^2 y_2 + Z^3 y_3 - Z^0 y_4 - Z^1 y_5 - Z^2 y_6 - Z^3 y_7,$$
$$Y_2 = Z^0 y_0 + Z^2 y_1 - Z^0 y_2 - Z^2 y_3 + Z^0 y_4 + Z^2 y_5 - Z^0 y_6 - Z^2 y_7,$$
$$Y_3 = Z^0 y_0 + Z^3 y_1 - Z^2 y_2 + Z^1 y_3 - Z^0 y_4 - Z^3 y_5 + Z^2 y_6 - Z^1 y_7,$$
$$Y_4 = Z^0 y_0 - Z^0 y_1 + Z^0 y_2 - Z^0 y_3 + Z^0 y_4 - Z^0 y_5 + Z^0 y_6 - Z^0 y_7,$$
$$Y_5 = Z^0 y_0 - Z^1 y_1 + Z^2 y_2 - Z^3 y_3 - Z^0 y_4 + Z^1 y_5 - Z^2 y_6 + Z^3 y_7,$$
$$Y_6 = Z^0 y_0 - Z^2 y_1 - Z^0 y_2 + Z^2 y_3 + Z^0 y_4 - Z^2 y_5 - Z^0 y_6 + Z^2 y_7,$$
$$Y_7 = Z^0 y_0 - Z^3 y_1 - Z^2 y_2 - Z^1 y_3 - Z^0 y_4 + Z^3 y_5 + Z^2 y_6 + Z^1 y_7,$$
$$Y_8 = Y_0. \qquad\qquad (9.76)$$

We see that these transforms now require $8 \times 8 = 64$ multiplications of complex numbers, in addition to some less time-consuming additions. We place these equations in an appropriate form for computing by regrouping the terms into sums and differences of the y's:

$$Y_0 = Z^0(y_0 + y_4) + Z^0(y_1 + y_5) + Z^0(y_2 + y_6) + Z^0(y_3 + y_7),$$
$$Y_1 = Z^0(y_0 - y_4) + Z^1(y_1 - y_5) + Z^2(y_2 - y_6) + Z^3(y_3 - y_7),$$
$$Y_2 = Z^0(y_0 + y_4) + Z^2(y_1 + y_5) - Z^0(y_2 + y_6) - Z^2(y_3 + y_7),$$
$$Y_3 = Z^0(y_0 - y_4) + Z^3(y_1 - y_5) - Z^2(y_2 - y_6) + Z^1(y_3 - y_7),$$
$$Y_4 = Z^0(y_0 + y_4) - Z^0(y_1 + y_5) + Z^0(y_2 + y_6) - Z^0(y_3 + y_7),$$
$$Y_5 = Z^0(y_0 - y_4) - Z^1(y_1 - y_5) + Z^2(y_2 - y_6) - Z^3(y_3 - y_7),$$
$$Y_6 = Z^0(y_0 + y_4) - Z^2(y_1 + y_5) - Z^0(y_2 + y_6) + Z^2(y_3 + y_7),$$
$$Y_7 = Z^0(y_0 - y_4) - Z^3(y_1 - y_5) - Z^2(y_2 - y_6) - Z^1(y_3 - y_7),$$
$$Y_8 = Y_0. \qquad\qquad (9.77)$$

Note the repeating factors inside the parentheses, with combinations of the form $y_p \pm y_q$. These symmetries are systematized by introducing the *butterfly operation* (Figure 9.8). This operation takes the y_p and y_q data elements from the left wing and converts them to

Figure 9.8 The basic butterfly operation in which elements y_p and y_q on the left are transformed into $y_p + Z y_q$ and $y_p - Z y_q$ on the right.

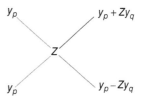

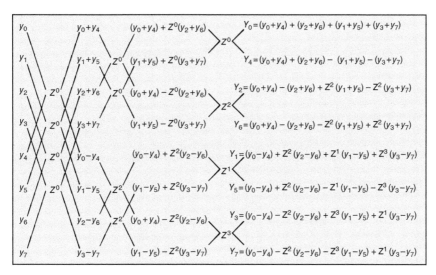

Figure 9.9 The butterfly operations performing an FFT on the eight data on the left leading to eight transforms on the right. The transforms are different linear combinations of the input data.

the $y_p + Z y_q$ elements in the upper- and lower-right wings. In Figure 9.9 we show what happens when we apply the butterfly operations to an entire FFT process, specifically to the pairs (y_0, y_4), (y_1, y_5), (y_2, y_6), and (y_3, y_7). Notice how the number of multiplications of complex numbers has been reduced: For the first butterfly operation there are 8 multiplications by Z^0; for the second butterfly, operation there are 8 multiplications, and so forth, until a total of 24 multiplications are made in four butterflies. In contrast, 64 multiplications are required in the original DFT (9.76).

9.5.1 Bit Reversal

The reader may have observed in Figure 9.9 that we started with 8 data elements in the order 0–7, and that after three butterfly operators we obtained transforms in the order 0, 4, 2, 6, 1, 5, 3, 7. The astute reader may further have observed that these numbers correspond to the bit-reversed order of 0–7. Let us look into this further. We need 3 bits to give the order of each of the 8 input data elements (the numbers 0–7). Explicitly, on the left in Table 10.1, we give the binary representation for decimal numbers 0–7, their bit reversals, and the corresponding decimal numbers. On the right we give the ordering for 16 input data elements, where we need 4 bits to enumerate their order. Notice that the order of the first 8 elements differs in the two cases because the number of bits being reversed differs. Notice too that after the reordering, the first half of the numbers are all even and the second half are all odd.

Dec	Bin	Binary-reversed 0–7		Binary-reversed 0–16	
		Rev	Dec rev	Rev	Dec rev
0	000	000	0	0000	0
1	001	100	4	1000	8
2	010	010	2	0100	4
3	011	110	6	1100	12
4	100	001	1	0010	2
5	101	101	5	1010	10
6	110	011	3	0110	6
7	111	111	7	1110	14
8	1000			0001	1
9	1001			1001	9
10	1010			0101	5
11	1011			1101	13
12	1100			0011	3
13	1101			1011	11
14	1101			0111	7
15	1111			1111	15

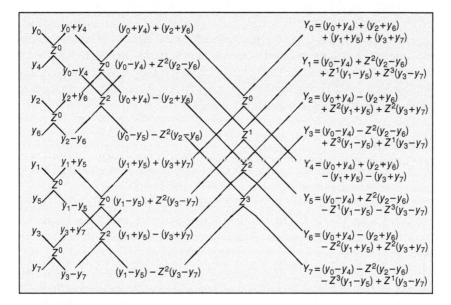

Figure 9.10 A modified FFT in which the eight input data on the left are transformed into eight transforms on the right. The results are the same as in the previous figure, but now the output transforms are in numerical order whereas in the previous figure the input signals were in numerical order.

Table 9.1 Reordering for 16 data complex points.

Order	Input data	New order	Order	Input data	New order
0	$0.0 + 0.0i$	$0.0 + 0.0i$	8	$8.0 + 8.0i$	$1.0 + 1.0i$
1	$1.0 + 1.0i$	$8.0 + 8.0i$	9	$9.0 + 9.0i$	$9.0 + 9.0i$
2	$2.0 + 2.0i$	$4.0 + 4.0i$	10	$10.0 + 10.i$	$5.0 + 5.0i$
3	$3.0 + 3.0i$	$12.0 + 12.0i$	11	$11.0 + 11.0i$	$13.0 + 13.0i$
4	$4.0 + 4.0i$	$2.0 + 2.0i$	12	$12.0 + 12.0i$	$3.0 + 3.0i$
5	$5.0 + 5.0i$	$10.0 + 10.i$	13	$13.0 + 13.0i$	$11.0 + 11.0i$
6	$6.0 + 6.0i$	$6.0 + 6.0i$	14	$14.0 + 14.i$	$7.0 + 7.0i$
7	$7.0 + 7.0i$	$14.0 + 14.0i$	15	$15.0 + 15.0i$	$15.0 + 15.0i$

The fact that the Fourier transforms are produced in an order corresponding to the bit-reversed order of the numbers 0–7 suggests that if we process the data in the bit-reversed order 0, 4, 2, 6, 1, 5, 3, 7, then the output Fourier transforms will be ordered (see Table 10.1). We demonstrate this conjecture in Figure 9.10, where we see that to obtain the Fourier transform for the eight input data, the butterfly operation had to be applied three times. The number 3 occurs here because it is the power of 2 that gives the number of data; that is, $2^3 = 8$. In general, in order for an FFT algorithm to produce transforms in the proper order, it must reshuffle the input data into bit-reversed order. As a case in point, our sample program starts by reordering the 16 (2^4) data elements given in Table 9.1, and then the four butterfly operations produce sequentially ordered output.

9.6 FFT Implementation

The first FFT program we are aware of was written in 1967 in Fortran IV by Norman Brunner at MIT's Lincoln Laboratory [Higgins, 1976], and was hard for us to follow. Our (easier-to-follow) Python version is in Listing 9.3. Its input is $N = 2^n$ data to be transformed (FFTs always require that the number of input data are a power of 2). If the number of your input data is not a power of 2, then you can make it so by concatenating some of the initial data to the end of your input until a power of 2 is obtained; because a DFT is always periodic, this just starts the period a little earlier. Our program assigns complex numbers at the 16 data points

$$y_m = m + mi, \qquad m = 0, \dots, 15, \tag{9.78}$$

reorders the data via bit reversal, and then makes four butterfly operations. The data are stored in the array dt[max,2], with the second subscript denoting real and imaginary parts. We increase speed further by using the 1D array data to make memory access more direct:

$$data[1] = dt[0, 1], \qquad data[2] = dt[1, 1], \qquad data[3] = dt[1, 0], \dots, \tag{9.79}$$

which also provides storage for the output. The FFT transforms data using the butterfly operation and stores the results back in dt[,], where the input data were original.

9.7 FFT Assessment

1) Compile and execute FFT.py. Make sure you understand the output.
2) Take the output from FFT.py, inverse-transform it back to signal space, and compare it to your input. [Checking that the double transform is proportional to itself is adequate, although the normalization factors in (9.35) should make the two equal.]
3) Compare the transforms obtained with an FFT to those obtained with a DFT (you may choose any of the functions studied before). Make sure to compare both precision and execution times.

9.8 Code Listings

Listing 9.1 DFTcomplex.py Uses the built-in complex numbers of Python to compute the discrete Fourier transform for the signal in method f(signal).

```
# DFTcomplex.py:   Discrete Fourier Transform with built in complex

from visual import *;   from visual.graph import *
import cmath                                           # Complex math

N = 100;   twopi = 2.*pi;   h = twopi/N;   sq2pi = 1./sqrt(twopi)
y = zeros(N+1, float);  Ycomplex = zeros(N, complex)    # Declare arrays
SignalGraph = gdisplay(x=0, y=0, width=600, height=250, title ='Signal y(t)',\
      xtitle='x',ytitle='y(t)', xmax=2.*math.pi, xmin=0, ymax=30, ymin=-30)
SignalCurve = gcurve(color=color.yellow, display=SignalGraph)
TransformGraph = gdisplay(x=0,y=250,width=600,height=250,title ='Im Y(omega)',
      xtitle = 'x',ytitle='Im Y(omega)',xmax=10.,xmin=-1,ymax=100,ymin=-250)
TransformCurve = gvbars(delta = 0.05,color=color.red, display = TransformGraph)

def Signal(y):                                         # Signal
    h = twopi/N;          x = 0.
    for i in range(0, N+1):
        y[i] = 30*cos(x) + 60*sin(2*x) + 120*sin(3*x)
        SignalCurve.plot(pos = (x, y[i]))              # Plot
        x += h
def DFT(Ycomplex):                                     # DFT
    for n in range(0, N):
        zsum = complex(0.0,  0.0)
        for  k in range(0, N):
            zexpo = complex(0, twopi*k*n/N)            # Complex exp
            zsum += y[k]*exp(-zexpo)
        Ycomplex[n] = zsum * sq2pi
        if Ycomplex[n].imag != 0:
            TransformCurve.plot(pos=(n,Ycomplex[n].imag))
Signal(y)                                              # Generate signal
DFT(Ycomplex)                                          # Transform signal
```

Listing 9.2 DFTreal.py Computes the discrete Fourier transform for the signal in method f(signal) using real numbers.

```
# DFTreal.py:   Discrete Fourier Transform using real numbers

from visual.graph import *

signgr = gdisplay(x=0,y=0,width=600,height=250, \
      title='Signal y(t)= 3 cos(wt)+2 cos(3wt)+ cos(5wt) ',\
  xtitle='x', ytitle='signal',xmax=2.*math.pi,xmin=0,ymax=7,ymin=-7)
```

```
       sigfig = gcurve(color=color.yellow,display=signgr)
       imagr = gdisplay(x=0,y=250,width=600,height=250,\
            title='Fourier transform imaginary part',xtitle='x',\
11            ytitle='Transf.Imag',xmax=10.0,xmin=-1,ymax=20,ymin=-25)
       impart = gvbars(delta=0.05,color=color.red,display=imagr)
       N = 200
       Np = N
15    signal = zeros((N+1),float)
       twopi = 2.*pi
       sq2pi = 1./sqrt(twopi)
       h = twopi/N
19    dftimag = zeros((Np),float)                          # Im. transform

       def f(signal):
           step = twopi/N
23        t= 0.
           for i in range(0,N+1):
               signal[i] = 3*sin(t*t*t)
               sigfig.plot(pos=(t,signal[i]))
27            t += step
       def fourier(dftimag):                               # DFT
           for n in range(0,Np):
              imag = 0.
31         for k in range(0, N):
               imag += signal[k]*sin((twopi*k*n)/N)
              dftimag[n] = -imag*sq2pi                     # Im transform
              if dftimag[n] !=0:
35                impart.plot(pos=(n,dftimag[n]))
       f(signal)
       fourier(dftimag)
```

Listing 9.3   FFT.py Computes the FFT or inverse transform depending upon the sign of sign.

```
1  # FFT.py:  FFT for complex numbers in Y[][2], returned in Y

   from numpy import *
   max = 2100;   points = 1026; N = 100; Switch = -1  # Switch= -1:Y, 1:y
5  y = zeros(2*(N+4), float);  Y = zeros((N+3,2), float)

   def fft(N,Switch):                     # FFT of Y[n,2]
       n = 2*N
9      for i in range(0,N+1):             # y in Y to y
            j = 2*i+1
            y[j] = Y[i,0]                 # Real Y, odd y[j]
            y[j+1] = Y[i,1]               # Imag Y, even y[j+1]
13     j = 1                              # y in bit reverse order
       for i in range(1,n+2, 2):
           if (i-j) < 0 :                 # Reorder to bit reverse
               tempr = y[j]
17             tempi = y[j+1]
               y[j] = y[i]
               y[j+1] = y[i+1]
               y[i] = tempr
21             y[i+1] = tempi
           m = n/2;
           while (m-2 > 0):
               if (j-m) <= 0 :  break
25             j = j-m
               m = m/2
           j = j+m;
       print("\n Bit-reversed y(t)")
29     for i in range(1,n+1,2): print("%2d y[%2d] %9.5f "%(i,i,y[i]))
       mmax = 2
       while (mmax-n) < 0 :               # Begin transform
           istep = 2*mmax
33         theta = 6.2831853/(1.0*Switch*mmax)
```

```
         sinth = math.sin(theta/2.0)
         wstpr = -2.0*sinth**2
         wstpi = math.sin(theta)
37       wr = 1.0
         wi = 0.0
         for m in range(1,mmax+1,2):
             for i in range(m,n+1,istep):
41               j = i+mmax
                 tempr = wr*y[j]    -wi*y[j+1]
                 tempi = wr*y[j+1] +wi*y[j]
                 y[j]    = y[i]    -tempr
45               y[j+1] = y[i+1] -tempi
                 y[i]    = y[i]    +tempr
                 y[i+1] = y[i+1] +tempi
             tempr = wr
49           wr = wr*wstpr - wi*wstpi + wr
             wi = wi*wstpr + tempr*wstpi + wi;
         mmax = istep
     for i in range(0,N):
53       j = 2*i+1
         Y[i,0] = y[j]
         Y[i,1] = y[j+1]
print('\n Input    \n  i    Re y(t)    Im y(t)')
57 h = 2*pi/N;         x = 0.
   for i in range(0,N+1):                              # Generate signal in Y
       Y[i,0] = 30*cos(x) + 60*sin(2*x) + 120*sin(3*x)    # Real part
       Y[i,1] = 0.                                         # Im part
61     x += h
       print(" %2d %9.5f %9.5f" %(i,Y[i,0],Y[i,1]))
   fft(N, Switch)                              # Call FFT, use global Y[][]
   print '\n Fourier Transform Y(omega)'
65 print("  i      ReY(omega)      ImY(omega)    ")
   for i in range(0,N):
       print(" %2d   %9.5f   %9.5f "%(i,Y[i,0],Y[i,1]))
```

10

Wavelet and Principal Components Analysis

A number of techniques can extend Fourier analysis to signals whose time-dependencies change in time. Part I of this chapter introduces wavelet analysis, a field that has seen extensive development and application in areas as diverse as brain waves, stock-market trends, gravitational waves, and compression of photographic images. Part II of this chapter covers the basics of principal components analysis. This is a powerful tool for situations in which there are very large data sets, and especially those with space-time correlated variables.

10.1 Part I: Wavelet Analysis

Problem   You have sampled the signal in Figure 10.1 that seems to contain an increasing number of frequencies as time increases. Your **problem** is to undertake a spectral analysis of this signal that tells you, in the most compact way possible, the amount of each frequency present at each instant of time. *Hint*: Although we want the method to be general enough to work with numerical data, for pedagogical purposes it is useful to know that the signal is

$$y(t) = \begin{cases} \sin 2\pi t, & \text{for } 0 \le t \le 2, \\ 5\sin 2\pi t + 10\sin 4\pi t, & \text{for } 2 \le t \le 8, \\ 2.5\sin 2\pi t + 6\sin 4\pi t + 10\sin 6\pi t, & \text{for } 8 \le t \le 12. \end{cases} \tag{10.1}$$

The Fourier analysis we used in Chapter 9 reveals the amount of the harmonic functions $\sin(n\omega t)$ and $\cos(n\omega t)$ that are present in a signal. An expansion in periodic functions is fine for *stationary* signals (those whose forms do not change in time), but has shortcomings for the time dependence of our **problem** signal (10.1). One such problem is that the Fourier reconstruction has all frequencies $n\omega$ occurring simultaneously, and so does not contain *time resolution* information indicating when each frequency occurs. Another shortcoming is that all the Fourier components are correlated, and that results in more information being stored than is needed to reconstruct the signal.

There are a number of techniques that extend simple Fourier analysis to nonstationary signals. The idea behind wavelet analysis is to expand a signal in a complete set of functions (wavelets), each of which oscillates for a finite period of time, and each of which

Computational Physics: Problem Solving with Python, Fourth Edition.
Rubin H. Landau, Manuel J. Páez, and Cristian C. Bordeianu.
© 2024 WILEY-VCH GmbH. Published 2024 by WILEY-VCH GmbH.

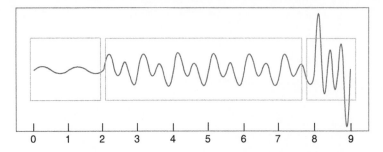

Figure 10.1 The input time signal (10.1) we wish to analyze. The signal is seen to contain additional frequencies as time increases. The boxes are possible placements of windows for short-time Fourier transforms.

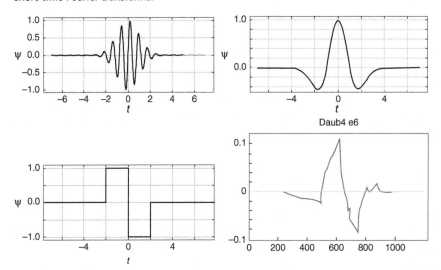

Figure 10.2 Four possible mother wavelets that can be used to generate entire sets of daughter wavelets. *Clockwise from top*: Morlet (real part), Mexican hat, Daub4 e6 (explained later), and Haar. The daughter wavelets are generated by scaling and translating these mother wavelets.

is centered at a different time. To give you a preview before we go into details, we show four sample wavelets in Figure 10.2. Because each wavelet is local in time, it is a wave packet,[1] with its time localization leading to a spectrum with a range of frequencies. These wave packets are called "wavelets" because they exist for only short periods of time [Polikar, 2023].

Although wavelets are required to oscillate in time, they are not restricted to a particular functional form [Addison, 2002; Goswani and Chan, 1999; Graps, 1995]. As a case in point, they may be oscillating Gaussian (Morlet: top left in Figure 10.2),

$$\Psi(t) = e^{2\pi it}e^{-t^2/2\sigma^2} = (\cos 2\pi t + i \sin 2\pi t)e^{-t^2/2\sigma^2} \quad \text{(Morlet)}, \tag{10.2}$$

the second derivative of a Gaussian (Mexican hat, top right),

$$\Psi(t) = -\sigma^2 \frac{d^2}{dt^2} e^{-t^2/2\sigma^2} = \left(1 - \frac{t^2}{\sigma^2}\right) e^{-t^2/2\sigma^2}, \tag{10.3}$$

---

1 We discuss wave packets further in Section 10.2.

an up-and-down step function (lower left), or a fractal shape (bottom right). All of these wavelets are *localized* in both time and frequency, that is, they are large for just a finite time and contain a finite range of frequencies. As we shall see, translating and scaling these *mother wavelet* generates an entire set of *child wavelets* basis functions, with the children covering different frequency ranges at different times.

10.2 Wave Packets and Uncertainty Principle

A *wave packet* or *wave train* is a collection of waves of differing frequencies added together in such a way as to produce a pulse of width Δt. As we shall see, the Fourier transform of a wave packet is a pulse in the frequency domain of width $\Delta\omega$. We'll first study wave packets analytically, and then use them numerically. An example of a simple wave packet is a sine wave that oscillates at frequency ω_0 for N periods (Figure 10.3 left) [Arfken and Weber, 2001]:

$$y(t) = \begin{cases} \sin\omega_0 t, & \text{for} \quad |t| < N\frac{\pi}{\omega_0} \equiv N\frac{T}{2}, \\ 0, & \text{for} \quad |t| > N\frac{\pi}{\omega_0} \equiv N\frac{T}{2}, \end{cases} \tag{10.4}$$

where we relate the frequency to the period via the usual $\omega_0 = 2\pi/T$. In terms of these parameters, the width of the wave packet is

$$\Delta t = NT = N\frac{2\pi}{\omega_0}. \tag{10.5}$$

The Fourier transform of the wave packet (10.4) is a straightforward application of the transform formula (9.17):

$$Y(\omega) = \int_{-\infty}^{+\infty} dt \frac{e^{-i\omega t}}{\sqrt{2\pi}} y(t) = \frac{-i}{\sqrt{2\pi}} \int_0^{N\pi/\omega_0} dt \sin\omega_0 t \sin\omega t \tag{10.6}$$

$$= \frac{(\omega_0 + \omega)\sin\left[(\omega_0 - \omega)\frac{N\pi}{\omega_0}\right] - (\omega_0 - \omega)\sin\left[(\omega_0 + \omega)\frac{N\pi}{\omega_0}\right]}{\sqrt{2\pi}(\omega_0^2 - \omega^2)},$$

where we have dropped a factor of $-i$ that affects only the phase. While at first glance (10.6) appears to be singular at $\omega = \omega_0$, it actually just peaks there (Figure 10.3 right), reflecting the predominance of frequency ω_0. Note that although the signal $y(t)$ appears to have only one frequency, it does drop off sharply in time (Figure 10.3 left), and these corners give $Y(\omega)$ a finite width $\Delta\omega$.

There is a fundamental relation between the widths Δt and $\Delta\omega$ of a wave packet. Although we use a specific example to determine that relation, it is true in general. While there may not be a precise definition of "width" for all functions, one can usually deduce a good measure of the width (say, within 25%). To illustrate, if we look at the right of Figure 10.3, it makes sense to use the distance between the first zeros of the transform $Y(\omega)$ (10.6) as the frequency width $\Delta\omega$. The zeros occur at

$$\frac{\omega - \omega_0}{\omega_0} = \pm\frac{1}{N} \Rightarrow \Delta\omega \simeq \omega - \omega_0 = \frac{\omega_0}{N}, \tag{10.7}$$

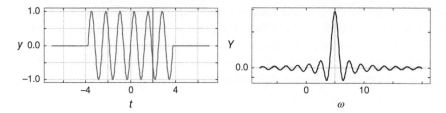

Figure 10.3 *Left*: A wave packet in time corresponding to the functional form (10.4) with $\omega_0 = 5$ and $N = 6$. *Right*: The Fourier transform in frequency of this same wave packet.

where N is the number of cycles in our original wave packet. Because the wave packet in time makes N oscillations each of period T, a reasonable measure of the time width Δt of the signal $y(t)$ is

$$\Delta t = NT = N\frac{2\pi}{\omega_0}. \tag{10.8}$$

When the products of the frequency width (10.7) and the time width (10.8) are combined, we obtain

$$\Delta t\,\Delta\omega \geq 2\pi. \tag{10.9}$$

The greater-than sign is used here to indicate that this is a minimum, that is, that $y(t)$ and $Y(\omega)$ extend beyond Δt and $\Delta\omega$, respectively. Nonetheless, most of the signal and transform should fall within the bounds of (10.9).

A relation of the form (10.9) also occurs in quantum mechanics, where it is known as the *Heisenberg uncertainty principle*, with Δt and $\Delta\omega$ called the uncertainties in t and ω. It is true for transforms in general, and states that as a signal is made more localized in time (smaller Δt), its transform becomes less localized (larger $\Delta\omega$). Conversely, the sine wave $y(t) = \sin\omega_0 t$ is completely localized in frequency, and consequently has an infinite extent in time, $\Delta t \simeq \infty$.

10.2.1 Wave Packet Exercise

Consider the following wave packets:

$$y_1(t) = e^{-t^2/2}, \quad y_2(t) = \sin(8t)e^{-t^2/2}, \quad y_3(t) = (1 - t^2)e^{-t^2/2}. \tag{10.10}$$

For each wave packet:

1) Estimate the width Δt. A good measure might be the *full width at half-maxima* (FWHM) of $|y(t)|$.
2) Use your DFT program to evaluate and plot the Fourier transform $Y(\omega)$ for each wave packet. Make *both* a linear and a semilog plot (small components are often important, yet not evident in linear plots). Make sure that your transform has a good number of closely spaced frequency values over a range that is large enough to show the periodicity of $Y(\omega)$.
3) What are the units for $Y(\omega)$ and ω in your DFT?
4) For each wave packet, estimate the width $\Delta\omega$. A good measure might be the *full width at half-maxima* of $|Y(\omega)|$.

5) For each wave packet determine approximate value for the constant C of the uncertainty principle

$$\Delta t \, \Delta \omega \geq 2\pi C. \qquad (10.11)$$

10.3 Short-Time Fourier Transforms

The constant amplitude of the functions $\sin n\omega t$ and $\cos n\omega t$ for all times can limit the usefulness of Fourier analysis for reproducing signals whose form changes in time. Seeing that these basis functions extend over all times with a constant amplitude, there is considerable overlap among them, and thus the information present in various Fourier components are correlated. This is undesirable for data storage and compression, where you want to store a minimum amount of information, and also want to adjust the amount of information stored dependent on the desired quality of the reconstructed signal.[2] *Lossless compression* exactly reproduces the original signal. You can save space by storing how many times each data element is repeated, and where each element is located. In *lossy compression*, in addition to removing repeated elements, you also eliminate some transform components consistent with the uncertainty relation (10.9) and with the level of resolution required in the reproduction. This leads to even greater compression.

In Section 9.3 we defined the Fourier transform $Y(\omega)$ of signal $y(t)$ as

$$Y(\omega) = \int_{-\infty}^{+\infty} dt \, \frac{e^{-i\omega t}}{\sqrt{2\pi}} y(t) \equiv \langle \omega | y \rangle. \qquad (10.12)$$

As is true for simple vectors, you can think of (10.12) as giving the overlap or scalar product of the basis function $|\omega\rangle = \exp(i\omega t)/\sqrt{2\pi}$ and the signal $y(t)$ [notice that the complex conjugate of the exponential basis function appears in (10.12)]. Another view of (10.12) is the mapping or projection of the signal into ω space. In this latter view, the overlap projects out the amount of the periodic function $\exp(i\omega t)/\sqrt{2\pi}$ in the signal $y(t)$. In other words, the Fourier component $Y(\omega)$ can be thought of as the correlation between the signal $y(t)$ and the basis function $\exp(i\omega t)/\sqrt{2\pi}$. This is the same as what results from filtering the signal $y(t)$ through a frequency filter. If there is no $\exp(i\omega t)$ in the signal, then the integral vanishes and there is no output. If $y(t) = \exp(i\omega t)$, the signal is at only one frequency, and the integral is accordingly singular.

The signal in Figure 10.1 for our problem clearly has different frequencies present at different times, and for different lengths of time. In the past, this signal might have been analyzed with a precursor of wavelet analysis known as the *short-time Fourier transform*. With that technique, the signal $y(t)$ is "chopped up" into different segments along the time axis, with successive segments centered about successive times $\tau_1, \tau_2, \ldots, \tau_N$. For instance, we show three such segments in the boxes of Figure 10.1. Once we have the dissected signal, a Fourier analysis is made for each segment. We are then left with a sequence of transforms $[Y_{\tau_1}^{(ST)}, Y_{\tau_2}^{(ST)}, \ldots, Y_{\tau_N}^{(ST)}]$, one for each short-time interval, where the superscript $^{(ST)}$ indicates short time.

---

2 Wavelets have proven to be a highly effective approach to data compression, with the Joint Photographic Experts Group (JPEG) 2000 standard being based on wavelets.

Rather than chopping up a signal by hand, we can express short-time Fourier trans-forming mathematically by imagining translating a *window function* $w(t - \tau)$, which is zero outside of some chosen interval, over the signal in Figure 10.1:

$$Y^{(\text{ST})}(\omega, \tau) = \int_{-\infty}^{+\infty} dt \, \frac{e^{i\omega t}}{\sqrt{2\pi}} \, w(t - \tau) \, y(t). \tag{10.13}$$

Here the values of the translation time τ correspond to different locations of window w over the signal, and the window function is essentially a transparent box of small size on an opaque background. Any signal within the width of the window is transformed, while the signal lying outside the window is not seen. Note that in (10.13), the extra variable τ in the Fourier transform indicates the location of the time around which the window was placed. Clearly, because the short-time transform is a function of two variables, a surface or 3D plot is needed to view the amplitude as a function of both ω and τ.

10.4 Wavelet Transforms

The wavelet transform of a time signal $y(t)$ is defined as

$$Y(s, \tau) = \int_{-\infty}^{+\infty} dt \, \psi_{s,\tau}^*(t) y(t) \qquad \text{(wavelet transform)}, \tag{10.14}$$

and is similar in concept and notation to a short-time Fourier transform. The difference is rather than using $\exp(i\omega t)$ as the basis functions, here we are using wave packets or wavelets $\psi_{s,\tau}(t)$ localized in time, such as those shown in Figure 10.2. Because each wavelet is localized in time, each acts as its own window function. Because each wavelet is oscillatory, each contains its own limited range of frequencies.

Equation (10.14) says that the wavelet transform $Y(s, \tau)$ is a measure of the amount of basis function $\psi_{s,\tau}(t)$ present in the signal $y(t)$. The τ variable indicates the time portion of the signal being decomposed, while the s variable is equivalent to the frequency present during that time:

$$\omega = \frac{2\pi}{s}, \qquad s = \frac{2\pi}{\omega} \qquad \text{(scale-frequency relation)}. \tag{10.15}$$

Seeing that it is key to much that follows, it is a good idea to think about (10.15) for a moment. If we are interested in the time *details* of a signal, then this is another way of saying that we are interested in what is happening at small values of the *scale s*. Equation (10.15) indicates that small values of s correspond to high-frequency components of the signal. That being the case, the time details of the signal are in the high-frequency, or low-scale, components.

10.4.1 Generating Wavelet Basis Functions

The conceptual discussion of wavelets is over, and it is time to get down to work. We first need a technique for generating wavelet basis functions, and then we need to discretize this technique. As is often the case, the final formulation will turn out to be simple and short, but it will be a while before we get there.

Just as the expansion of an arbitrary function in a complete set of orthogonal functions is not restricted to any particular basis set, so too is the wavelet transform not restricted to any particular wavelet basis set, although some might be better than others for a given signal. The standard way to generate a family of wavelet basis functions starts with $\Psi(t)$, a *mother* or *analyzing* function of the real variable t, and then uses it to generate *daughter* wavelets. As a case in point, we start with the mother wavelet

$$\Psi(t) = \sin(8t)e^{-t^2/2}. \tag{10.16}$$

By scaling, translating, and normalizing this mother wavelet we obtain the set

$$\psi_{s,\tau}(t) \stackrel{\text{def}}{=} \frac{1}{\sqrt{s}}\Psi\left(\frac{t-\tau}{s}\right) = \frac{1}{\sqrt{s}}\sin\left[\frac{8(t-\tau)}{s}\right]e^{-(t-\tau)^2/2s^2}, \tag{10.17}$$

and with it we generate the four wavelet basis functions displayed in Figure 10.4. We see that larger or smaller values of s, respectively, expand or contract the mother wavelet, while different values of τ shift the center of the wavelet. Because the wavelets are inherently oscillatory, the scaling leads to the same number of oscillations occurring in different time spans, which is equivalent to having basis states with differing frequencies. We see that $s < 1$ produces a higher-frequency wavelet, while $s > 1$ produces a lower-frequency one, both of the same shape. As we shall see, we do not need to store much information to outline the large-time-scale s behavior of a signal (its *smooth envelope*), but we do need more information to specify its short-time-scale s behavior (*details*). And if we want to resolve yet finer features in the signal, then we will need to have more information on yet finer details. Here the division by $\sqrt{s}$ is made to ensure that there is equal "power" (or

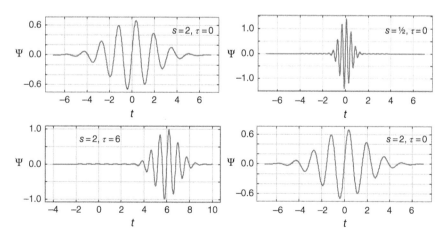

Figure 10.4 Four wavelet basis functions (daughters) generated by scaling (s) and translating (τ) an oscillating Gaussian mother wavelet. *Clockwise from top:* ($s = 1$, $\tau = 0$), ($s = 1/2$, $\tau = 0$), ($s = 1$, $\tau = 6$), and ($s = 2$, $\tau = 60$). Note how $s < 1$ is a wavelet with higher frequency, while $s > 1$ has a lower frequency than the $s = 1$ mother. Likewise, the $\tau = 6$ wavelet is just a translated version of the $\tau = 0$ one directly above it.

energy or intensity) in each region of s, although other normalizations can also be found in the literature. After substituting in the definition of daughters, the wavelet transform (10.14) and its inverse [van den Berg, 1999] are

$$Y(s, \tau) = \frac{1}{\sqrt{s}} \int_{-\infty}^{+\infty} dt \, \Psi^* \left(\frac{t - \tau}{s} \right) y(t) \qquad \textbf{(Wavelet Transform)}, \qquad (10.18)$$

$$y(t) = \frac{1}{C} \int_{-\infty}^{+\infty} d\tau \int_{0}^{+\infty} ds \frac{\psi_{s,\tau}^*(t)}{s^{3/2}} Y(s, \tau) \qquad \textbf{(Inverse Transform)}, \qquad (10.19)$$

where the normalization constant C depends on the wavelet used.

In summary, wavelet bases are functions of the time variable t, as well as of the two parameters s and τ. The t variable is integrated over to yield a transform that is a function of the time scale s (frequency $2\pi/s$) and window location τ. You can think of scale as being like the scale on a map (also discussed in Section 14.4.1 in relation to fractal analysis) or in terms of *resolution*, as might occur in photographic images. Regardless of the words, as we see in Chapter 14, if we have a fractal, then we have a self-similar object that looks the same at all scales or resolutions. Similarly, each wavelet in a set of basis functions is self-similar to the others, but at a different scale or location.

The general requirements for a mother wavelet Ψ are [Addison, 2002; van den Berg, 1999]:

1) $\Psi(t)$ is real.
2) $\Psi(t)$ oscillates around zero such that its average is zero:

$$\int_{-\infty}^{+\infty} \Psi(t) \, dt = 0. \qquad (10.20)$$

3) $\Psi(t)$ is local, that is, a wave packet, and is square-integrable:

$$\Psi(|t| \rightarrow \infty) \rightarrow 0 \quad \text{(rapidly)}, \qquad \int_{-\infty}^{+\infty} |\Psi(t)|^2 \, dt < \infty. \qquad (10.21)$$

4) The transforms of low powers of t vanish, that is, the first p moments:

$$\int_{-\infty}^{+\infty} t^0 \, \Psi(t) \, dt = \int_{-\infty}^{+\infty} t^1 \, \Psi(t) \, dt = \cdots = \int_{-\infty}^{+\infty} t^{p-1} \, \Psi(t) \, dt = 0. \qquad (10.22)$$

This makes the transform more sensitive to details than to general shape.

As an example of how we use the s and τ degrees of freedom in a wavelet transform, consider the analysis of a chirp signal $y(t) = \sin(60t^2)$ in Figure 10.5. We see that a slice at the beginning of the signal is compared to our first basis function. (The comparison is carried out via the *convolution* of the wavelet with the signal.) This first comparison is with a narrow version of the wavelet, that is, at low scale, and yields a single coefficient. The comparison at this scale continues with the next signal slice, and eventually ends when the entire signal has been covered (the top row in the figure). Then the wavelet is expanded to larger s values, and the comparisons are repeated. Eventually, the data are processed at all scales and at all time intervals. The narrow signals correspond to a high-resolution analysis, while the broad signals correspond to low resolution. As the scales get larger (lower frequencies, lower resolution), fewer details of the time signal remain visible, but the overall shape or gross features of the signal become clearer.

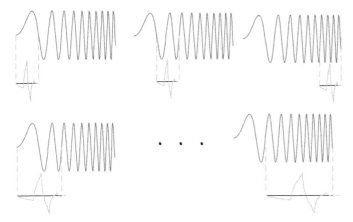

Figure 10.5 A schematic representation of the steps followed in performing a wavelet transformation over all time displacements and scales. The dark grey signal is first analyzed by evaluating its overlap with a narrow wavelet at the signal's beginning. This produces a coefficient that measures the similarity of the signal to the wavelet. The wavelet is successively shifted over the length of the signal and the overlaps are successively evaluated. After the entire signal is covered, the wavelet is expanded and the entire analysis is repeated.

10.4.2 Continuous Wavelet Transforms

We want to develop some intuition as to what wavelet transforms look like before going on to apply them. Accordingly, modify the program you have been using for the Fourier transform so that it now computes the wavelet transform. In contrast to the discrete or digital version, this is a *continuous* wavelet transform.

1) Examine the effect of using different mother wavelets. Accordingly, write a method that calculates the mother wavelet for
   a) a Morlet wavelet (10.2),
   b) a Mexican hat wavelet (10.3),
   c) a Haar wavelet (the square wave in Figure 10.2).
2) Try out your transform for the following input signals and see if the results make sense:
   a) A pure sine wave $y(t) = \sin 2\pi t$,
   b) A sum of sine waves $y(t) = 2.5 \sin 2\pi t + 6 \sin 4\pi t + 10 \sin 6\pi t$,
   c) The nonstationary signal for our problem (10.1)

$$y(t) = \begin{cases} \sin 2\pi t, & \text{for } 0 \le t \le 2, \\ 5 \sin 2\pi t + 10 \sin 4\pi t, & \text{for } 2 \le t \le 8, \\ 2.5 \sin 2\pi t + 6 \sin 4\pi t + 10 \sin 6\pi t, & \text{for } 8 \le t \le 12. \end{cases} \quad (10.23)$$

   d) The half-wave function

$$y(t) = \begin{cases} \sin \omega t, & \text{for } 0 < t < T/2, \\ 0, & \text{for } T/2 < t < T. \end{cases} \quad (10.24)$$

3) ⊙ Use (10.19) to invert your wavelet transform and compare the reconstructed signal to the input signal (you can normalize the two to each other). In Figure 10.6 we show our reconstruction.

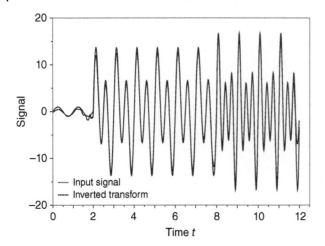

Figure 10.6 Comparison of an input and reconstituted signal (10.23) using Morlet wavelets. The curves overlap nearly perfectly, except at the ends.

In Listing 10.1 we give our *continuous wavelet transformation* `cwt.py` [Lang and Forinash, 1998]. Because wavelets, with their functional dependence on two variables, may be somewhat hard to grasp at first, we suggest that you write your own code and include a portion that does the inverse transform as a check. In Section 10.5, we will describe the *discrete wavelet transformation* that makes optimal discrete choices for the scale and time translation parameters s and τ. Figure 10.7 shows the spectrum produced for the input signal (10.1) in Figure 10.1. And it works! We see predominantly one frequency at short times, two frequencies at intermediate times, and three frequencies at longer times.

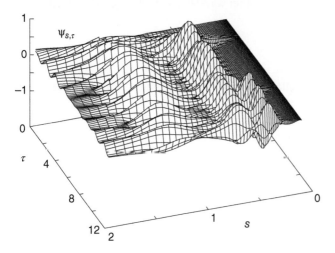

Figure 10.7 The continuous wavelet spectrum obtained by analyzing the input signal with Morlet wavelets. Observe how at small values of time τ there is predominantly one frequency present, how a second, higher-frequency (smaller-scale) component enters at intermediate times, and how at larger times a still higher-frequency components enter. (Figure courtesy of Z. Diabolic.)

10.5 Discrete Wavelet Transforms ⊙

As was true for DFTs, if a time signal is measured at only N discrete times,

$$y(t_m) \equiv y_m, \quad m = 1, \ldots, N, \tag{10.25}$$

then we can determine only N-independent components of the transform Y. The trick is to remain consistent with the uncertainty principle as we compute only the N-independent components required to reproduce the signal. The *discrete wavelet transform* (DWT) evaluates the transforms with discrete values for the scaling parameter s and the time translation parameter τ:

$$\psi_{j,k}(t) = \frac{\Psi\left[(t - k2^j)/2^j\right]}{\sqrt{2^j}} \equiv \frac{\Psi\left(t/2^j - k\right)}{\sqrt{2^j}} \quad \text{(DWT)}, \tag{10.26}$$

$$s = 2^j, \quad \tau = \frac{k}{2^j}, \quad k, j = 0, 1, \ldots. \tag{10.27}$$

Here j and k are integers whose maximum values are yet to be determined, and we measure time in integer values. This choice of s and τ, based on powers of 2, is called a *dyadic grid* arrangement, and will be seen to automatically perform the scalings and translations at the different time scales that are at the heart of wavelet analysis.[3] The DWT now becomes

$$Y_{j,k} = \int_{-\infty}^{+\infty} dt\, \psi_{j,k}(t)\, y(t) \simeq \sum_m \psi_{j,k}(t_m) y(t_m) h \quad \text{(DWT)}, \tag{10.28}$$

where the discreteness here refers to the wavelet basis set and *not* the time variable. For an orthonormal wavelet basis, the inverse discrete transform is then

$$y(t) = \sum_{j,\,k=-\infty}^{+\infty} Y_{j,k}\, \psi_{j,k}(t) \quad \text{(inverse DWT)}. \tag{10.29}$$

This inversion will exactly reproduce the input signal at the N input points, but only if we sum over an infinite number of terms [Addison, 2002]. Practical calculations will be less exact.

Notice in (10.26) and (10.28) that we have kept the time variable t in the wavelet basis functions continuous, despite the fact that s and τ have been made discrete. This is useful in establishing the orthonormality of the basis functions,

$$\int_{-\infty}^{+\infty} dt\, \psi_{j,k}^*(t)\, \psi_{j',k'}(t) = \delta_{jj'}\, \delta_{kk'}, \tag{10.30}$$

where $\delta_{m,n}$ is the Kronecker delta function. Being normalized to 1 means that each wavelet basis has "unit energy"; being orthogonal means that each basis function is independent of the others. And because wavelets are localized in time, the different transform components have low levels of correlations with each other. Altogether, this leads to efficient and flexible data storage.

The use of a discrete wavelet basis makes it clear that we sample the input signal at the discrete values of time determined by the integers j and k. In general, you want time

---

3 Note that some references scale down with increasing j, in contrast to our scaling up.

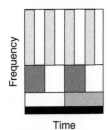

Figure 10.8 A graphical representation of the relation between time and frequency resolutions (the uncertainty relation). Each box represents an equal portion of the time-frequency plane but with different proportions of time and frequency.

steps that sample the signal at enough times in each interval to obtain the desired level of precision. A rule of thumb is to start with 100 steps to cover each major feature. Ideally, the needed times correspond to the times at which the signal was sampled, although this may require some forethought.

Consider Figure 10.8. We measure a signal at a number of discrete times within the intervals (k or τ values) corresponding to the vertical columns of fixed width along the time axis. For each time interval, we want to sample the signal at a number of scales (frequencies or j values). However, as discussed in Section 10.2, the basic mathematics of Fourier transforms indicates that the width Δt of a wave packet $\psi(t)$ and the width $\Delta\omega$ of its Fourier transform $Y(\omega)$ are related by an uncertainty principle

$$\Delta\omega\,\Delta t \geq 2\pi.$$

This relation places a constraint on the time intervals and frequency intervals. Furthermore, while we may want a high-resolution reproduction of our signal, we do not want to store more data than are needed to obtain that reproduction. If we sample the signal for times centered about some τ in an interval of width $\Delta\tau$ (Figure 10.8), and then compute the transform at a number of scales s or frequencies $\omega = 2\pi/s$ covering a range of height $\Delta\omega$, then the relation between the height and width is restricted by the uncertainty relation. All this means that each of the rectangles in Figure 10.8 has the same area $\Delta\omega\,\Delta t = 2\pi$. The increasing heights of the rectangles at higher frequencies means that a larger range of frequencies should be sampled as the frequency increases. The premise here is that the low-frequency components provide the gross or *smooth* outline of the signal which, being smooth, does not require much detail, while the high-frequency components give the details of the signal over a short time interval, and so require many components in order to record these details with high resolution.

Industrial-strength wavelet analyses do not compute explicit integrals, but instead apply a technique known as *multiresolution analysis* (MRA) [Mallat, 1989]. We give an example of this technique in Figure 10.9 and in the code DWT.py in Listing 10.2. It is based on a *pyramid algorithm* that samples the signal at a finite number of times, and then passes it successively through a number of *filters*, with each filter representing a digital version of a wavelet.

Filters were discussed in Chapter 9, where in (9.60) we defined the action of a linear filter as a convolution of the filter response function with the signal. A comparison of the definition of a filter to the definition of a wavelet transform (10.14), shows that the two

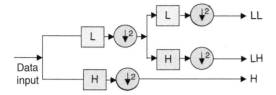

Figure 10.9 A eigenfrequency dyadic (power-of-2) filter tree used for discrete wavelet transformations. The L boxes represent lowpass filters and the H boxes represent highpass filters. Each filter performs a convolution (transform). The circles containing "↓ 2" filter out half of the signal that enters them, which is called *subsampling* or *factor-of-2 decimation*. The signal on the left yields a transform with a single low and two high components (less information is needed about the low components for a faithful reproduction).

are essentially the same. Such being the case, the result of the transform operation is a weighted sum over the input signal values, with each weight the product of the integration weight times the value of the wavelet function at the integration point. Therefore, *rather than tabulate explicit wavelet functions, a set of filter coefficients is all that is needed for DWT.*

In as much as each filter in Figure 10.9 changes the relative strengths of the different frequency components, passing the signal through a series of filters is equivalent, in wavelet language, to analyzing the signal at different scales. This is the origin of the name "multiresolution analysis." Figure 10.9 shows how the pyramid algorithm passes the signal through a series of highpass filters (H) and then through a series of lowpass filters (L). Each filter changes the scale to that of the level below. Notice too, the circles containing ↓ 2 in Figure 10.9. This operation filters out half of the signal and so is called *subsampling* or *factor-of-2 decimation*. It is the way we keep the areas of each box in Figure 10.8 constant as we vary the scale and translation times. We consider subsampling further when we discuss the pyramid algorithm.

In summary, the DWT process decomposes the signal into *smooth* information stored in the low-frequency components and *detailed* information stored in the high-frequency components. Because *high-resolution* reproductions of signals require more information about details than about gross shape, the pyramid algorithm is an effective way to compress data while still maintaining high resolution. In addition, because components of different resolutions are independent of each other, it is possible to lower the number of data stored by systematically eliminating higher-resolution components, if they are not needed. The use of wavelet filters builds in progressive scaling, which is particularly appropriate for fractal-like reproductions.

10.5.1 Pyramid Scheme ⊙

We now implement the pyramid scheme outlined in Figure 10.9. The H and L filters will be represented by matrices, which is an approximate way to perform the integrations or convolutions. Then there is a decimation of the output by one-half, and finally an interleaving of the output for further filtering. This process simultaneously cuts down on the number of points in the data set and changes the scale and the resolution. The decimation

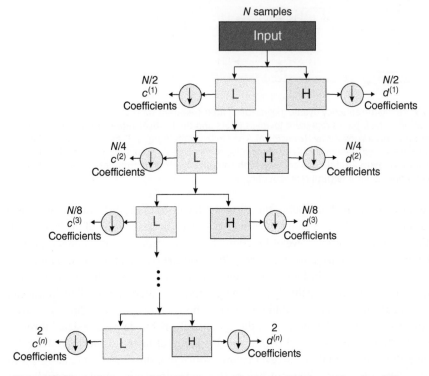

Figure 10.10 An input signal (top) is processed by a tree of high- and low-band filters. The outputs from each filtering are downshifted with half the data kept. The process continues until there are only two data of high-band filtering and two data of low-band filtering.

reduces the number of values of the remaining signal by one-half, with the low-frequency part discarded because the details are in the high-frequency parts.

As indicated in Figure 10.10, the pyramid DWT algorithm follows five steps:

1) Successively applies the (soon-to-be-derived) c matrix (10.41) to the whole N-length vector,

$$\begin{bmatrix} Y_0 \\ Y_1 \\ Y_2 \\ Y_3 \end{bmatrix} = \begin{bmatrix} c_0 & c_1 & c_2 & c_3 \\ c_3 & -c_2 & c_1 & -c_0 \\ c_2 & c_3 & c_0 & c_1 \\ c_1 & -c_0 & c_3 & -c_2 \end{bmatrix} \begin{bmatrix} y_0 \\ y_1 \\ y_2 \\ y_3 \end{bmatrix}. \tag{10.31}$$

2) Applies it to the $N/2$-length smooth vector.
3) Repeats the application until only two smooth components remain.
4) After each filtering, the elements are ordered, with the newest two smooth elements on top, the newest detailed elements below, and the older detailed elements below that.
5) The process continues until there are just two smooth elements left.

To illustrate, here we filter and reorder an initial vector of length $N = 8$:

$$
\begin{bmatrix} y_1 \\ y_2 \\ y_3 \\ y_4 \\ y_5 \\ y_6 \\ y_7 \\ y_8 \end{bmatrix}
\underset{\rightarrow}{\text{filter}}
\begin{bmatrix} s_1^{(1)} \\ d_1^{(1)} \\ s_2^{(1)} \\ d_2^{(1)} \\ s_3^{(1)} \\ d_3^{(1)} \\ s_4^{(1)} \\ d_4^{(1)} \end{bmatrix}
\underset{\rightarrow}{\text{order}}
\begin{bmatrix} s_1^{(1)} \\ s_2^{(1)} \\ s_3^{(1)} \\ s_4^{(1)} \\ d_1^{(1)} \\ d_2^{(1)} \\ d_3^{(1)} \\ d_4^{(1)} \end{bmatrix}
\underset{\rightarrow}{\text{filter}}
\begin{bmatrix} s_1^{(2)} \\ d_1^{(2)} \\ s_2^{(2)} \\ d_2^{(2)} \\ d_1^{(1)} \\ d_2^{(1)} \\ d_3^{(1)} \\ d_4^{(1)} \end{bmatrix}
\underset{\rightarrow}{\text{order}}
\begin{bmatrix} s_1^{(2)} \\ s_2^{(2)} \\ d_1^{(2)} \\ d_2^{(2)} \\ d_1^{(1)} \\ d_2^{(1)} \\ d_3^{(1)} \\ d_4^{(1)} \end{bmatrix}.
\tag{10.32}
$$

The discrete inversion of a transform vector back to a signal vector is made using the transpose (inverse) of the transfer matrix at each stage. For instance,

$$
\begin{bmatrix} y_0 \\ y_1 \\ y_2 \\ y_3 \end{bmatrix}
=
\begin{bmatrix} c_0 & c_3 & c_2 & c_1 \\ c_1 & -c_2 & c_3 & -c_0 \\ c_2 & c_1 & c_0 & c_3 \\ c_3 & -c_0 & c_1 & -c_2 \end{bmatrix}
\begin{bmatrix} Y_0 \\ Y_1 \\ Y_2 \\ Y_3 \end{bmatrix}.
\tag{10.33}
$$

As a more realistic example, imagine that we have sampled the chirp signal $y(t) = \sin(60t^2)$ for 1024 times. The filtering process through which we place this signal is illustrated as a passage from the top to the bottom in Figure 10.10. First the original 1024 samples are passed through a single low band and a single high band (which is mathematically equivalent to performing a series of convolutions). As indicated by the down arrows, the output of the first stage is then downshifted, that is, the number is reduced by a factor of 2. This results in 512 points from the high-band filter as well as 512 points from the low-band filter. This produces the first-level output. The output coefficients from the high-band filters are called $\{d_i^{(1)}\}$ to indicate that they show details, and $\{s_i^{(1)}\}$ to indicate that they show smooth features. The superscript indicates that this is the first level of processing. The detail coefficients $\{d^{(1)}\}$ are stored to become part of the final output.

In the next level down, the 512 smooth data $\{s_i^{(1)}\}$ are passed through new low- and high-band filters using a broader wavelet. The 512 outputs from each are downshifted to form a smooth sequence $\{s_i^{(2)}\}$ of size 256 and a detailed sequence $\{d_i^{(2)}\}$ of size 256. Again the detail coefficients $\{d^{(2)}\}$ are stored to become part of the final output. (Note that this is only half the size of the previously stored details.) The process continues until there are only two numbers left for the detail coefficients and two numbers left for the smooth coefficients. Because this last filtering is carried out with the broadest wavelet, it is of the lowest resolution and therefore requires the least information.

In Figure 10.11, we show the actual effects on the chirp signal of pyramid filtering for various levels in the processing. (The processing is carried out with *Daub4* wavelets, which we will discuss soon.) At the uppermost level, the wavelet is narrow, and so convoluting this wavelet with successive sections of the signal results in smooth components that still

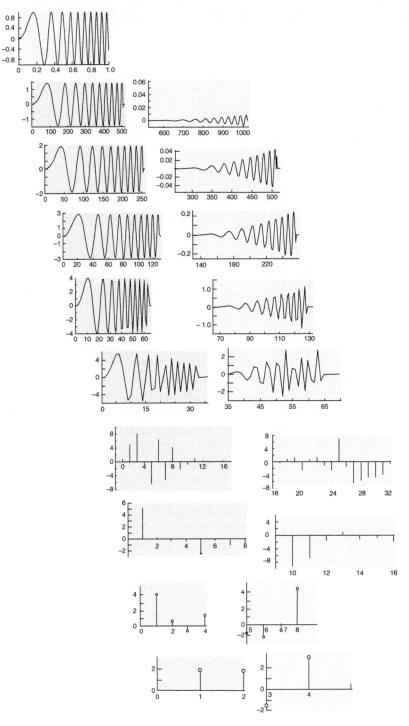

Figure 10.11 In successive passes, the filtering of the original signal at the top goes through the pyramid algorithm and produces the outputs shown. The sampling is reduced by a factor of 2 in each step. Note that in the upper graphs, we have connected the points to emphasize their continuous nature while in the lower graphs, we plot the individual output points as histograms.

contain many large high-frequency parts. The detail components, in contrast, are much smaller in magnitude. In the next stage, the wavelet is dilated to a lower frequency, and the analysis is repeated *on just the smooth (low-band) part*. The resulting output is similar, but with coarser features for the smooth coefficients and larger values for the details. Note that in the upper graphs we have connected the points to make the output look continuous, while in the lower graphs, with fewer points, we have plotted the output as histograms to make the points more evident. Eventually the downshifting leads to just two coefficients output from each filter, at which point the filtering ends.

To reconstruct the original signal (called *synthesis* or *transformation*) a reversed process is followed: Begin with the last sequence of four coefficients, upsample them, pass them through low- and high-band filters to obtain new levels of coefficients, and repeat until all the N values of the original signal are recovered. The inverse scheme is the same as the processing scheme (Figure 10.10), only now the direction of all the arrows is reversed.

10.5.2 Daubechies Wavelets Filters ⊙

We should now be able to understand that digital wavelet analysis has been standardized to the point where classes of wavelet basis functions are specified not by their analytic forms, but rather by their *wavelet filter coefficients*. In 1988, the Belgian mathematician Ingrid Daubechies discovered an important class of such filter coefficients [Daubechies, 1995; Rowe and Abbott, 1995]. We will study just the Daub4 class containing the four coefficients $c_0, c_1, c_2,$ and c_3.

Imagine that our input contains the four elements $\{y_1, y_2, y_3, y_4\}$ corresponding to measurements of a signal at four times. We represent a lowpass filter L and a highpass filter H in terms of the four filter coefficients as

$$L = \begin{bmatrix} c_0 & +c_1 & c_2 & +c_3 \end{bmatrix}, \tag{10.34}$$

$$H = \begin{bmatrix} c_3 & -c_2 & c_1 & -c_0 \end{bmatrix}. \tag{10.35}$$

To see how this works, we form an input vector by placing the four signal elements in a column and then multiply the input by L and H:

$$L \begin{bmatrix} y_0 \\ y_1 \\ y_2 \\ y_3 \end{bmatrix} = \begin{bmatrix} c_0 & c_1 & c_2 & c_3 \end{bmatrix} \begin{bmatrix} y_0 \\ y_1 \\ y_2 \\ y_3 \end{bmatrix} = c_0 y_0 + c_1 y_1 + c_2 y_2 + c_3 y_3,$$

$$H \begin{bmatrix} y_0 \\ y_1 \\ y_2 \\ y_3 \end{bmatrix} = \begin{bmatrix} c_3 & -c_2 & c_1 & -c_0 \end{bmatrix} \begin{bmatrix} y_0 \\ y_1 \\ y_2 \\ y_3 \end{bmatrix} = c_3 y_0 - c_2 y_1 + c_1 y_2 - c_0 y_3.$$

We see that if we choose the values of the c_i's carefully, the result of L acting on the signal vector is a single number that may be viewed as a weighted average of the four input signal elements. Since an averaging process tends to smooth out data, the lowpass filter may be thought of as a *smoothing filter* that outputs the general shape of the signal.

In turn, we see that if we choose the c_i values carefully, the result of H acting on the signal vector is a single number that may be viewed as the weighted differences of the input signal.

Because a differencing process tends to emphasize the variation in the data, the highpass filter may be thought of as a *detail* filter that produces a large output when the signal varies considerably, and a small output when the signal is smooth.

We have just seen how the individual L and H filters, each represented by a single row of the filter matrix, outputs one number when acting upon an input signal containing four elements in a column. If we want the output of the filtering process Y to contain the same number of elements as the input (four y's in this case), we just stack the L and H filters together:

$$\begin{bmatrix} Y_0 \\ Y_1 \\ Y_2 \\ Y_3 \end{bmatrix} = \begin{bmatrix} L \\ H \\ L \\ H \end{bmatrix} \begin{bmatrix} y_0 \\ y_1 \\ y_2 \\ y_3 \end{bmatrix} = \begin{bmatrix} c_0 & c_1 & c_2 & c_3 \\ c_3 & -c_2 & c_1 & -c_0 \\ c_2 & c_3 & c_0 & c_1 \\ c_1 & -c_0 & c_3 & -c_2 \end{bmatrix} \begin{bmatrix} y_0 \\ y_1 \\ y_2 \\ y_3 \end{bmatrix}. \tag{10.36}$$

Of course the first and third rows of the Y vector will be identical, as will the second and fourth, but we will get to that soon.

Now we go about determining the values of the filter coefficients c_i by placing specific demands upon the output of the filter. We start by recalling that in our discussion of discrete Fourier transforms we observed that a transform is equivalent to a rotation from the time domain to the frequency domain. Yet we know from our study of linear algebra that rotations are described by orthogonal matrices, that is, matrices whose inverses are equal to their transposes. In order for the inverse transform to return us to the input signal, the transfer matrix must be orthogonal. For our wavelet transformation to be orthogonal, we must have the 4×4 filter matrix times its transpose equal to the identity matrix:

$$\begin{bmatrix} c_0 & c_1 & c_2 & c_3 \\ c_3 & -c_2 & c_1 & -c_0 \\ c_2 & c_3 & c_0 & c_1 \\ c_1 & -c_0 & c_3 & -c_2 \end{bmatrix} \begin{bmatrix} c_0 & c_3 & c_2 & c_1 \\ c_1 & -c_2 & c_3 & -c_0 \\ c_2 & c_1 & c_0 & c_3 \\ c_3 & -c_0 & c_1 & -c_2 \end{bmatrix} = \begin{bmatrix} 1 & 0 & 0 & 0 \\ 0 & 1 & 0 & 0 \\ 0 & 0 & 1 & 0 \\ 0 & 0 & 0 & 1 \end{bmatrix},$$

$$\Rightarrow \quad c_0^2 + c_1^2 + c_2^2 + c_3^2 = 1, \quad c_2 c_0 + c_3 c_1 = 0. \tag{10.37}$$

Two equations in four unknowns are not enough for a unique solution, so we now include the further requirement that the detail filter $H = (c_3, -c_0, c_1, -c_2)$ must output a zero if the input is smooth. We define "smooth" to mean that the input is constant or linearly increasing:

$$\begin{bmatrix} y_0 & y_1 & y_2 & y_3 \end{bmatrix} = \begin{bmatrix} 1 & 1 & 1 & 1 \end{bmatrix} \quad \text{or} \quad \begin{bmatrix} 0 & 1 & 2 & 3 \end{bmatrix} \tag{10.38}$$

This is equivalent to demanding that the moments up to order p are zero, that is, that we have an "approximation of order p." Explicitly,

$$H \begin{bmatrix} y_0 & y_1 & y_2 & y_3 \end{bmatrix} = H \begin{bmatrix} 1 & 1 & 1 & 1 \end{bmatrix} = H \begin{bmatrix} 0 & 1 & 2 & 3 \end{bmatrix} = 0,$$

$$\Rightarrow \quad c_3 - c_2 + c_1 - c_0 = 0, \quad 0 \times c_3 - 1 \times c_2 + 2 \times c_1 - 3 \times c_0 = 0,$$

$$\Rightarrow \quad c_0 = \frac{1 + \sqrt{3}}{4\sqrt{2}} \simeq 0.483, \quad c_1 = \frac{3 + \sqrt{3}}{4\sqrt{2}} \simeq 0.836, \tag{10.39}$$

$$c_2 = \frac{3-\sqrt{3}}{4\sqrt{2}} \simeq 0.224, \quad c_3 = \frac{1-\sqrt{3}}{4\sqrt{2}} \simeq -0.129. \tag{10.40}$$

These are the basic Daub4 filter coefficients. They are used to create larger filter matrices by placing the row versions of L and H along the diagonal, with successive pairs displaced two columns to the right. For example, for eight elements,

$$\begin{bmatrix} Y_0 \\ Y_1 \\ Y_2 \\ Y_3 \\ Y_4 \\ Y_5 \\ Y_6 \\ Y_7 \end{bmatrix} = \begin{bmatrix} c_0 & c_1 & c_2 & c_3 & 0 & 0 & 0 & 0 \\ c_3 & -c_2 & c_1 & -c_0 & 0 & 0 & 0 & 0 \\ 0 & 0 & c_0 & c_1 & c_2 & c_3 & 0 & 0 \\ 0 & 0 & c_3 & -c_2 & c_1 & -c_0 & 0 & 0 \\ 0 & 0 & 0 & 0 & c_0 & c_1 & c_2 & c_3 \\ 0 & 0 & 0 & 0 & c_3 & -c_2 & c_1 & -c_0 \\ c_2 & c_3 & 0 & 0 & 0 & 0 & c_0 & c_1 \\ c_1 & -c_0 & 0 & 0 & 0 & 0 & c_3 & -c_2 \end{bmatrix} \begin{bmatrix} y_0 \\ y_1 \\ y_2 \\ y_3 \\ y_4 \\ y_5 \\ y_6 \\ y_7 \end{bmatrix}. \tag{10.41}$$

Note that in order not to lose any information, the last pair on the bottom two rows is wrapped over to the left. If you perform the actual multiplications indicated in (10.41), you will note that the output has successive *smooth* and *detailed* information. The output is processed with the pyramid scheme.

The time dependencies of two Daub4 wavelets are displayed in Figure 10.12. To obtain these from our filter coefficients, first imagine that an elementary wavelet $y_{1,1}(t) \equiv \psi_{1,1}(t)$ is input into the filter. This should result in a transform $Y_{1,1} = 1$. Inversely, we obtain $y_{1,1}(t)$ by applying the inverse transform to a Y vector with a 1 in the first position and zeros in all the other positions. Likewise, the *i*th member of the Daubechies class is obtained by applying the inverse transform to a Y vector with a 1 in the *i*th position and zeros in all the other positions.

On the left in Figure 10.12 is the wavelet for coefficient 6 (thus the e6 notation). On the right in Figure 10.12 is the sum of two wavelets corresponding to the coefficients 10 and 58. We see that the two wavelets have different levels of scale as well as different time positions.

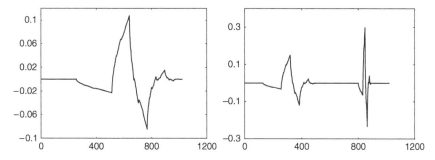

Figure 10.12 *Left*: The Daub4 e6 wavelet constructed by inverse transformation of the wavelet coefficients. This wavelet has been found to be particularly effective in wavelet analyses. *Right*: The sum of Daub4 e10 and Daub4 1e58 wavelets of different scale and time displacements.

So despite the fact that the time dependence of the wavelets is not evident when wavelet (filter) coefficients are used, it is there.

10.5.3 DWT Exercise ⊙

Listing 10.2 gives our program for performing a DWT on the chirp signal $y(t) = \sin(60t^2)$. The method `pyran` calls the `daube4` method to perform the DWT or inverse DWT, depending upon the value of `sign`.

1) Modify the program so that you output to a file the values for the input signal that your code has read in. It is always important to check your input.
2) Try to reproduce the left of Figure 10.11 by using various values for the variable `nend` that controls when the filtering ends. A value `nend=1024` should produce just the first step in the downsampling (top row in Figure 11.10). Selecting `nend=512` should produce the next row, while `nend=4` should output just two smooth and detailed coefficients.
3) Reproduce the scale-time diagram shown on the right in Figure 10.11. This diagram shows the output at different scales and serves to interpret the main components of the signal and the time in which they appear. The timeline at the bottom of the figure corresponds to a signal of length 1 over which 256 samples were recorded. The low-band (smooth) components are shown on the left, and the high-band components on the right.
   a) The bottom most figure results when `nend` = 256.
   b) The figure in the second row up results from `end` = 128, and we have the output from two filterings. The output contains 256 coefficients but divides time into four intervals and shows the frequency components of the original signal in more detail.
   c) Continue with the subdivisions for `end` = 64, 32, 16, 8, and 4.
4) For each of these choices except the topmost, divide the time by 2 and separate the intervals by vertical lines.
5) The topmost spectrum is your final output. Can you see any relation between it and the chirp signal?
6) Change the sign of `sign` and check that the inverse DWT reproduces the original signal.
7) Use the code to visualize the time dependence of the Daubechies mother function at different scales.
   a) Start by performing an inverse transformation on the eight-component signal `[0,0,0,0,1,0,0,0]`. This should yield a function with a width of about 5 units.
   b) Next perform an inverse transformation on a unit vector with $N = 32$ but with all components except the fifth equal to zero. The width should now be about 25 units, a larger scale but still covering the same time interval.
   c) Continue this procedure until you obtain wavelets of 800 units.
   d) Finally, with $N = 1024$, select a portion of the mother wavelet with data in the horizontal interval [590,800]. This should show self-similarity similar to that at the bottom of Figure 10.12.

10.6   Part II: Principal Components Analysis

Problem   Given a dataset describing several properties of irises (flowers), separate the data into groups in order of importance. The properties (in cm) are

```
    ['sepal length', 'sepal width', 'petal length', 'petal width']
    array( [5.1, 3.5, 1.4, 0.2],
3          [4.9, 3.0, 1.4, 0.2],
           [4.7, 3.2, 1.3, 0.2],
           [4.6, 3.1, 1.5, 0.2],
           [5.0, 3.6, 1.4, 0.2],
7          [5.4, 3.9, 1.7, 0.4],
           [4.6, 3.4, 1.4, 0.3],
           [5.0, 3.4, 1.5, 0.2],
           [4.4, 2.9, 1.4, 0.2],
11         [4.9, 3.1, 1.5, 0.1],
           [5.4, 3.7, 1.5, 0.2],
           [4.8, 3.4, 1.6, 0.2],
           [4.8, 3.0, 1.4, 0.1],
15         [4.3, 3.0, 1.4, 0.1],
           [5.8, 4.0, 1.2, 0.2],
           [5.7, 4.4, 1.5, 0.4],
           [5.4, 3.9, 1.3, 0.4],
19         [5.1, 3.5, 1.4, 0.3]   )
```

We have indicated that a shortcoming of Fourier analysis is that it uses an infinite number of components, that all of these components are correlated, and thus are not independent. Consequently, truncation of some of the components leads to difficulties in compression and reconstitution of the input signal. Wavelet analysis, on the other hand, is excellent at data compression, but not appropriate for high-dimensionality data sets, or for non-temporal signals. *Principal Components Analysis (PCA)* is a powerful analysis tool that uses statistics to provide insight into signals that may be contained within a *high-dimensionality*, multivariate dataset. Examples of high dimensionally data include stellar spectra, brain waves, facial patterns, and ocean currents. In these cases there may be hundreds of detectors in space, each of which records several types of signals for weeks on end. Furthermore, these kinds of data are often noisy, and possibly redundant (different detectors recording correlated signals), and so a statistical approach seems appropriate.

Variations of the PCA approach are used in many fields, where it goes by names such as the Kronen-Loèvet transform, the Hostelling transform, the proper orthogonal decomposition, singular value decomposition, factor analysis, empirical orthogonal functions, empirical component analysis, and empirical modal analysis [Wikipedia, 2014]. The approach combines statistics with transformation theory, the latter familiar from linear algebra, to rotate from the basis vectors used to collect the data into new basis vectors known as *principal components* that lie in the direction of maximal signal strength ("power") in the dataspace. This is analogous to the principal axis theorem of mechanics in which the description of solid object rotations is greatly simplified when moments of inertia relative to the principal axes are used. Our references Jackson [1991], Jolliffe [2002], Smith [2002], and Shlens [2003] tend to view PCA as *unsupervised dimensionality reduction,* where "unsupervised" refers to the absence of labels on the data, and "reduction" to the small number of principal components that ultimately result. We prefer to view PCA as the way to extract the dominant dynamics contained in complex datasets.

10.6.1 Multi-dimensional Data Space

It's often helpful when dealing with complex data to imagine an abstract vector space in which the data elements lie. It is in this multi-dimensional *dataspace* that the PCA basis vectors lie. As a simple example, consider the four detectors in Figure 10.13 observing a beam of particles passing by. Each detector records its observations over time, with the measurements at each time considered a separate, individual sample. Furthermore, each detector may record a set of M observables, such as position, angle, intensity, thickness of a track, length of a track, *etc.* This produces an M-dimensional dataspace $\mathcal{R}^M$. Specifically, let's say at each instant of time, detector A records the position (x'_A, y'_A), and so on for detectors B–D. The sample of spatial data at that one instant of time is then represented as an 8-D vector:

$$\mathbf{X}' = [x'_a \ \ y'_a \ \ x'_b \ \ y'_b \ \ x'_c \ \ y'_c \ \ x'_d \ \ y'_d]. \tag{10.42}$$

To get an idea of the sizes of the dataspaces with which we might be dealing, if the detectors make their recordings for 18 minutes (1080 seconds) at 110 Hz, then $1080 \times 110 = 118\,800$ of these vectors are created in $\mathcal{R}^M$. And this is a small problem.

Experimental data usually contain noise in addition to signals of interest. The *variance* $\sigma^2(z)$ in a dataset of N points is a measure of the dispersion of the datum points from their mean $\bar{z}$:

$$\bar{z} = \frac{1}{N} \sum_i^N z_i, \tag{10.43}$$

$$\sigma^2(z) \equiv \text{Var}(z) \overset{\text{def}}{=} \frac{1}{N-1} \sum_i^N (z_i - \bar{z})^2. \tag{10.44}$$

If the data are of high precision, the signal would be much larger than the noise. In practice, measurements contain random and systematic errors, and therefore the signal-to-noise ratio (SNR),

$$SNR = \frac{\sigma^2_{\text{signal}}}{\sigma^2_{\text{noise}}}, \tag{10.45}$$

may not be large. PCA is a good way to deal with small SNR. On the left of Figure 10.14, we present some made-up 2D data (x_A, y_A) from detector A showing the direction of maximum signal variance σ^2_{signal} along PC_1, and the direction of maximum noise (or secondary signal)

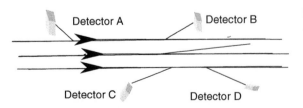

Figure 10.13 A beam of particles being observed by four detectors.

Detector A

Detector B

Detector C

Detector D

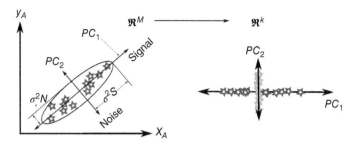

Figure 10.14 *Left*: Samples of 2D data (x_A, y_A) from a detector A showing the direction of maximum signal variance σ_S^2 along the principal component PC_1 basis, and the direction of noise variance σ_N^2 along the secondary PC_2 basis. *Right*: The same data projected onto the principal component axes. (Based on [Shlens, 2003].)

variance σ_{noise}^2 along PC_2. Although a traditional view of statistics may be that a large variance indicates high noise, in the PCA view a large variance indicates that some interesting dynamics may be occurring in that direction in dataspace.

The key PCA assumption is that the direction with the largest variance contains most of the dynamics of interest, and that the deviation of the data from maximum variance may be due to noise, or maybe some less important dynamics. The PC_1 and PC_2 directions in Figure 10.14 are the two PCA basis vectors, and are chosen to maximize the SNR measured along PC_1, relative to that along PC_2 (we'll get to how these are determined shortly). On the right of the figure we show the same data projected along the PC_1 and PC_2 axes. Here $\mathcal{R}^k$ is a lower-dimensional space containing the k orthonormal and uncorrelated principal components vectors.

10.6.2 Wonders of the Covariance Matrix

We have just seen graphically how PCA isolates the signal from the noise for a 2D dataset such as (x'_A, y'_A). However, our sample problem has four detectors, and thus a higher-dimensional space to analyze. We generalize one step at a time by extending the approach to also include detector B. We define two datasets A and B, each centered about their means, $\bar{x}, \bar{y}$, and expressed as the row vectors:

$$\mathbf{A} = [a_1(x_{a,1}, y_{a,1})\ a_2(x_{a,2}, y_{a,2})\ \dots\ a_N(x_{a,N}, y_{a,N})], \tag{10.46}$$

$$\mathbf{B} = [b_1(x_{b,1}, y_{b,1})\ b_2(x_{b,2}, y_{b,2})\ \dots\ b_N(x_{b,N}, y_{b,N})], \tag{10.47}$$

$$x_{a,i} = x'_{a,i} - \bar{x'}, \qquad y_{a,i} = y'_{a,i} - \bar{y'}. \tag{10.48}$$

Next we compute the variances (10.44) for each of the centered ($\bar{a} = \bar{b} = 0$) sets:

$$\sigma_A^2 = \frac{1}{N-1}\sum_i^N a_i^2, \qquad \sigma_B^2 = \frac{1}{N-1}\sum_i^N b_i^2. \tag{10.49}$$

The variance concept is extended to *covariance*, as a measure of the correlation between the centered data in A and B:

$$\text{cov}(A, B) \overset{\text{def}}{=} \sigma_{AB}^2 \overset{\text{def}}{=} \frac{1}{N-1}\sum_i^N a_i b_i. \tag{10.50}$$

A positive covariance indicates that signals within A and B tend to change together and in the same direction. A large covariance indicates a high correlation or redundancy, while a zero value implies no correlation. Note that the variance (10.44) can be viewed as a special case of the covariance, $\mathrm{var}(x) = \mathrm{cov}(x, x)$, and that there is the symmetry $\mathrm{cov}(x, y) = \mathrm{cov}(y, x)$. All of these concepts are combined into the symmetric covariance matrix:

$$C_{AB} = \begin{bmatrix} \mathrm{cov}(A, A) & \mathrm{cov}(A, B) \\ \mathrm{cov}(B, A) & \mathrm{cov}(B, B) \end{bmatrix}. \tag{10.51}$$

These ideas generalize directly to higher dimensions. Start with the sets A (10.46) and B (10.47), where we remind you that the elements may contain a number of measurements. The covariance matrix can be written as the vector direct product (aka dot product, matrix multiplication, or dyadic):

$$C_{AB} \overset{\mathrm{def}}{=} \frac{1}{N-1} \mathbf{AB} \equiv \frac{1}{N-1} \mathbf{A} \otimes \mathbf{B}^T. \tag{10.52}$$

With this notation, we can generalize to higher dimensions by defining new *row* subvectors containing the data from each of the M detectors:

$$\mathbf{x}_1 = \mathbf{A}, \quad \mathbf{x}_2 = \mathbf{B}, \quad \ldots, \quad \mathbf{x}_M = \mathbf{M}. \tag{10.53}$$

We combine these row vectors into an extended $M \times N$ data matrix:

$$X = \begin{bmatrix} \mathbf{x}_1 \\ \vdots \\ \mathbf{x}_M \end{bmatrix} = \begin{bmatrix} \Downarrow \text{All} & \Rightarrow \text{All } A \text{ measurements} \\ \Downarrow \text{one} & \Rightarrow \text{All } B \text{ measurements} \\ \Downarrow \text{time} & \Rightarrow \text{All } C \text{ measurements} \\ \Downarrow \text{measurements} & \Rightarrow \text{All } D \text{ measurements} \end{bmatrix}. \tag{10.54}$$

Each row of this matrix contains all of the measurements from a particular detector, while each column contains all of the measurements for a particular time. With this notation (and $\bar{\mathbf{x}} = 0$), the covariance matrix can be written in the concise form

$$C = \frac{1}{N-1} \mathbf{XX}^T. \tag{10.55}$$

This can be thought of as a generalization of the familiar dot product of two 2D vectors, $\mathbf{x} \cdot \mathbf{x} = \mathbf{x}^T\mathbf{x}$, as a measure of their overlap.

In summary:

- The covariant matrix C_{ij} is the dot product of the centered measurements vector from the ith detector ($i = A, B, \ldots$) with the centered measurements vector from the jth detector.
- For any two variables in the data, C is a square symmetric matrix measuring the relationship between those variables.
- The diagonal elements of C are the variances in the measurements from individual detectors.
- The off-diagonal elements of C are the covariances between the measurements from different detectors, that is, the correlations between detectors.

Steps in a Principal Component Analysis (Easy with NumPy)

1) As indicated in Figure 10.14, there is the assumption that the direction in which the variance is largest indicates the "principal" component in the data, PC_1 or $\mathbf{p}_1$.

Table 10.1 PCA demonstration data.

Data		Adjusted data		In PCA basis	
x	*y*	*x*	*y*	x_1	x_2
2.5	2.4	0.69	0.49	−0.828	−0.175
0.5	0.7	−1.31	−1.21	1.78	0.143
2.2	2.9	0.39	0.99	−0.992	0.484
1.9	2.2	0.09	0.29	−0.274	0.130
3.1	3.0	1.29	1.09	−1.68	−0.209
2.3	2.7	0.49	0.79	0.913	0.175
2	1.6	0.19	−0.31	0.0991	−0.350
1.0	1.1	−0.81	−0.81	1.14	0.464
1.6	1.6	−0.31	−0.31	0.438	0.0178
1.1	0.9	−0.71	−1.01	1.22	−0.163

Consequently, PCA searches for the direction in dataspace for which the variance of **X** is maximized.

2) Once $\mathbf{p}_1$ has been found, the basis vector $\mathbf{p}_2$ is chosen as the orthonormal to $\mathbf{p}_1$.
3) The process is repeated until there are M orthonormal basis vectors. These are the M principal components of the data.
4) The eigenvectors and eigenvalues are ordered according to their corresponding variances.
5) Explicitly, starting with the $M \times N$ data matrix **X**, a matrix **P** is determined such that

$$C_y = \frac{1}{N-1}\mathbf{Y}\mathbf{Y}^{\mathrm{T}} = \text{diagonal}, \qquad \text{where} \quad \mathbf{Y} = \mathbf{PX}. \tag{10.56}$$

6) The rows of **P** are the principal component basis vectors (same as the eigenvectors of $\mathbf{X}\mathbf{X}^{\mathrm{T}}$).
7) The diagonal elements of $\mathbf{C_Y}$ are the variances of **X** along the corresponding $\mathbf{p}_i$'s.

Figure 10.15 Sample data used in our demonstration PCA analysis.

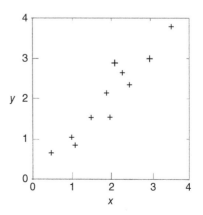

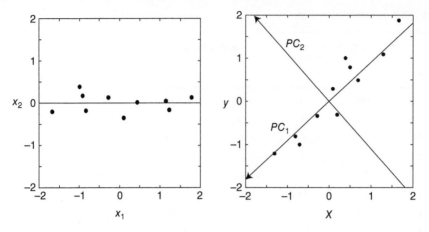

Figure 10.16 *Left*: The PCA basis vectors (eigenvectors of cov(x, y)). *Right*: The normalized data using the PCA eigenvectors as basis.

10.6.3 Demonstration of Principal Component Analysis

We'll leave the analysis of the irises data at the beginning of Part II as *your* **problem**, and, instead, we'll analyze the simpler data in Table 10.1 [Smith, 2002]. These data, are shown in Figure 10.15 as *x versus y*, but they don't have to be spatial. Also shown is the analysis of these data in terms of the first two principal components. As expected, the first eigenvector points in the direction with the largest variance, while the next vector is orthogonal to the first. There clearly is less variance along PC_2 and, consequently, less dynamical importance of that component.

Here are the steps in the analysis:

1) **Enter data as an array**: The first two columns in Table 10.1.
2) **Subtract the mean**: PCA analysis assumes that the data in each dimension has zero mean. Accordingly, as shown in columns two and three in Table 10.1, we calculated the mean for each column, $(\bar{x}, \bar{y})$, and subtracted them from the data. The resulting adjusted data are given in the third and fourth columns of the table.
3) **Calculate the covariance matrix**:

$$\text{var}(x) = \frac{1}{N-1} \sum_{i=1}^{N} (x_i - \bar{x})^2, \tag{10.57}$$

$$\text{cov}(x, y) = \frac{1}{N-1} \sum_{i=1}^{N} (x_i - \bar{x})(y_i - \bar{y}), \tag{10.58}$$

$$C = \begin{bmatrix} \text{cov}(x, x) & \text{cov}(x, y) \\ \text{cov}(y, x) & \text{cov}(y, y) \end{bmatrix} = \begin{bmatrix} 0.6166 & 0.6154 \\ 0.6154 & 0.7166 \end{bmatrix}. \tag{10.59}$$

4) **Compute unit eigenvector and eigenvalues of C (easy with NumPy)**:

$$\lambda_1 = 1.284, \qquad \lambda_2 = 0.4908, \tag{10.60}$$

$$\mathbf{PC}_1 = \begin{bmatrix} -0.6779 \\ -0.7352 \end{bmatrix}, \qquad \mathbf{PC}_2 = \begin{bmatrix} -0.7352 \\ 0.6789 \end{bmatrix}, \tag{10.61}$$

where we have ordered the eigenvalues and eigenvectors. The eigenvector with the largest eigenvalue is the principal component in the data, typically with ~80% of the power in the signal.

In Figure 10.16 we show the translated data and the two PCA eigenvectors of the covariance matrix (scaled to fill the frame). Notice that PC_1, which points in direction of the major variation in the data, is essentially a straight-line fit to the data. The PC_2 eigenvector is clearly orthogonal to PC_1, and contains less signal strength. This is as should be.

5) **Express the data in terms of principal Components**: We next expressed the data in terms of their two principal components by forming two *feature matrices*:

$$F_1 = \begin{bmatrix} -0.6779 \\ -0.7352 \end{bmatrix}, \qquad F_2 = \begin{bmatrix} -0.6779 & -0.7352 \\ -0.7352 & 0.6779 \end{bmatrix}. \tag{10.62}$$

Here F_1 keeps just the major principal component, while F_2 keeps the first two. The matrix gets its name because it focuses on which features of the data are being kept. Next we formed F_2^T, the transpose of the feature matrix, and $\mathbf{X}^T$, the transpose of the translated data matrix $\mathbf{X}$:

$$F_2^T = \begin{bmatrix} -0.6779 & -0.7352 \\ -0.7352 & 0.6779 \end{bmatrix}, \tag{10.63}$$

$$\mathbf{X}^T = \begin{bmatrix} 0.69 & -1.31 & 0.39 & 0.09 & 1.29 & 0.49 & 0.19 & -0.81 & -0.31 & -0.71 \\ 0.49 & -1.21 & 0.99 & 0.29 & 1.09 & 0.79 & -0.31 & -0.81 & -0.31 & -1.01 \end{bmatrix}.$$

We expressed the data in terms of these principal components by multiplying F_2^T and $\mathbf{X}$ together:

$$X^{PCA} = F_2^T \times X^T \tag{10.64}$$

$$= \begin{bmatrix} -0.6779 & -0.7352 \\ -0.7352 & 0.6779 \end{bmatrix}$$

$$\times \begin{bmatrix} 0.69 & -1.31 & 0.39 & 0.09 & 1.29 & 0.49 & 0.19 & -0.81 & -0.31 & -0.71 \\ 0.49 & -1.21 & 0.99 & 0.29 & 1.09 & 0.79 & -0.31 & -0.81 & -0.31 & -1.01 \end{bmatrix}$$

$$= \begin{bmatrix} 0.828 & 1.78 & -0.992 & -0.274 & -1.68 & -0.913 & 0.0991 & 1.15 & 0.438 & 1.22 \\ -0.175 & 0.143 & 0.384 & 0.130 & -0.209 & 0.175 & -0.350 & 0.464 & 0.178 & -0.162 \end{bmatrix}.$$

On the right of Table 10.1, the data are plotted using the PC_1 and PC_2 bases. The plot shows where each datum point sits relative to the trend in the data. If we had plotted only the first principal component, all of the data would fall on a straight line.

10.6.4 PCA Exercises

1) Use just the principal eigenvectors to perform the PCA analysis just completed with two eigenvectors.
2) Store data from 10 cycles of the chaotic pendulum studied in Chapter 8, but do not include transients. Perform a PCA of these data and plot the results using principal component axes.

10.7 Code Listings

Listing 10.1 CWT.py Uses Morlet wavelets to compute the continuous wavelet transform of the sum of sine functions. (Courtesy of Z. Diabolic.)

```
1  # CWT.py  Continuous Wavelet TF.  Based on program by Zlatko Dimcovic

   import matplotlib.pylab as p;
   from mpl_toolkits.mplot3d import Axes3D ;
5  from visual.graph import *;

   originalsignal=gdisplay(x=0, y=0, width=600, height=200, \
         title='Input Signal',xmin=0,xmax=12,ymin=-20,ymax=20)
9  orsigraph=gcurve(color=color.yellow)
   invtrgr = gdisplay(x=0, y=200, width=600, height=200,
         title='Inverted Transform',xmin=0,xmax=12,ymin=-20,ymax=20)
   invtr = gcurve(x=list(range(0,240)), display = invtrgr, color= color.green)
13 iT =   0.0;         fT =  12.0;        W = fT - iT;
   N =  240;           h = W/N        # Need *very* small s for high f
   noPtsSig = N;       noS =  20;         noTau =  90;
   iTau =  0.;         iS =  0.1;         tau = iTau;        s =  iS
17 dTau = W/noTau;     dS = (W/iS)**(1./noS);
   maxY =  0.001;      sig = zeros((noPtsSig), float)                # Signal

   def signal(noPtsSig, y):                             # Signal function
21    t = 0.0;       hs = W/noPtsSig;      t1 = W/6.;      t2 = 4.*W/6.
      for i in range(0, noPtsSig):
          if  t >= iT  and t <=  t1:  y[i] =   sin(2*pi*t)
          elif t >= t1 and t <=  t2: y[i] = 5.*sin(2*pi*t) + 10.*sin(4*pi*t);
25        elif t >= t2 and t <=  fT:
              y[i] = 2.5*sin(2*pi*t) + 6.*sin(4*pi*t) + 10.*sin(6*pi*t)
          else:
              print("In signal(...) : t out of range.")
29            sys.exit(1)
          yy=y[i]
          orsigraph.plot(pos=(t,yy))
          t += hs
33 signal(noPtsSig, sig)                                      # Form signal
   Yn =  zeros( (noS+1, noTau+1), float)                      # Transform
   def morlet(t, s, tau):                                     # Mother
      T =  (t - tau)/s
37    return sin(8*T) * exp( - T*T/2. )
   def transform(s, tau, sig):                          # Find wavelet TF
      integral = 0.
      t = iT;
41    for i in range(0, len(sig) ):
          t += h
          integral += sig[i]*morlet(t, s, tau)*h
      return integral / sqrt(s)
45 def invTransform(t, Yn):                             # Compute inverse
      s - iS                                            # Transform
      tau = iTau
      recSig_t = 0
49    for i in range (0, noS):
          s *= dS                                       # Scale graph
          tau = iTau
          for j in range (0, noTau):
53            tau += dTau
              recSig_t += dTau*dS *(s**(-1.5))* Yn[i,j] * morlet(t,s,tau)
      return recSig_t
   print("working, finding transform, count 20")
57 for i in range( 0, noS):
      s *= dS                                           # Scaling
      tau = iT
      print(i)
61    for j in range(0, noTau):
          tau += dTau                                   # Translate
          Yn[i, j] = transform(s, tau, sig)
```

```
    print("transform found")
65  for i in range( 0, noS):
        for j in range( 0, noTau):
            if Yn[i, j] > maxY or Yn[i, j] < - 1 *maxY :
                maxY = abs( Yn[i, j] )                  # Find max Y
69  tau =   iT
    s =   iS
    print("normalize")
    for i in range( 0, noS):
73      s *= dS
        for j in range( 0, noTau):
            tau += dTau                                 # Transform
            Yn[i, j] = Yn[i, j]/maxY
77      tau = iT
    print("finding inverse transform")                 # Inverse TF
    recSigData =  "recSig.dat"
    recSig =  zeros(len(sig) )
81  t =   0.0;
    print("count to 10")
    kco = 0;              j = 0;           Yinv =  Yn
    for rs in range(0, len(recSig) ):
85      recSig[rs] = invTransform(t, Yinv)              # Invert
        xx=rs/20
        yy=4.6*recSig[rs]
        invtr.plot(pos=(xx,yy))
89      t += h
        if kco %24 == 0:
            j += 1
            print(j)
93      kco += 1
    x = list(range(1, noS + 1))
    y = list(range(1, noTau + 1))
    X,Y = p.meshgrid(x, y)
97
    def functz(Yn):                                     # Transform function
        z = Yn[X, Y]
        return z
101 Z = functz(Yn)
    fig = p.figure()
    ax = Axes3D(fig)
    ax.plot_wireframe(X, Y, Z, color = 'r')
105 ax.set_xlabel('s: scale')
    ax.set_ylabel('Tau')
    ax.set_zlabel('Transform')
    p.show()
109 print("Done")
```

Listing 10.2   DWT.py Uses the Daub4 digital wavelets and the pyramid algorithm to compute the discrete wavelet transform for the chirp signal values stored in f [ ].

```
1  # DWT.py:   Discrete Wavelet Transform, Daubechies, global variables

   from visual import *
   from visual.graph import *
5
   sq3 = sqrt(3);           fsq2 = 4.0*sqrt(2); N = 1024    # N = 2^n
   c0 = (1+sq3)/fsq2;       c1 = (3+sq3)/fsq2            # Daubechies 4
   c2 = (3-sq3)/fsq2;       c3 = (1-sq3)/fsq2
9  transfgr1 = None                                      # Display

   def chirp( xi):                                       # Chirp signal
       y = sin(60.0*xi**2);
13     return y;
   def daube4(f, n, sign):          # DWT if sign >= 0, inverse if sign < 0
       global transfgr1, transfgr2
       tr = zeros( (n + 1), float)                       # Temporary
17     if n < 4 : return
```

```
        mp = n/2
        mp1 = mp + 1                                    # midpoint + 1
        if sign >= 0:                                   # DWT
21          j = 1
            i = 1
            maxx  = n/2
            if n > 128:                                 # Scale
25              maxy = 3.0
                miny = - 3.0
                Maxy = 0.2
                Miny = - 0.2
29              speed = 50                              # Fast rate
            else:
                maxy = 10.0
                miny = - 5.0
33              Maxy = 7.5
                Miny = - 7.5
                speed = 8                               # Lower rate
            if transfgr1:
37              transfgr1.display.visible = False
                transfgr2.display.visible = False
                del transfgr1
                del transfgr2
41          transfgr1 = gdisplay(x=0, y=0, width=600, height=400,\
                         title='Wavelet TF, down sample + low pass', xmax=maxx,\
                         xmin=0, ymax=maxy, ymin=miny)
            transf  = gvbars(delta=2.*n/N,color=color.cyan,display=transfgr1)
45          transfgr2 = gdisplay(x=0, y=400, width=600, height=400,\
                         title='Wavelet TF, down sample + high pass',\
                         xmax=2*maxx, xmin=0, ymax=Maxy, ymin=Miny)
            transf2 = gvbars(delta=2.*n/N,color=color.cyan,display=transfgr2)
49          while j <= n - 3:
                rate(speed)
                tr[i] = c0*f[j] + c1*f[j+1] + c2*f[j+2] + c3*f[j+3]# low-pass
                transf.plot(pos = (i, tr[i]) )          # c coefficients
53              tr[i+mp] = c3*f[j] - c2*f[j+1] + c1*f[j+2] - c0*f[j+3] # high
                transf2.plot(pos = (i + mp, tr[i + mp]) )
                i += 1                                  # d coefficents
                j += 2                                  # Downsampling
57          tr[i] = c0*f[n-1] + c1*f[n] + c2*f[1] + c3*f[2]      # low-pass
            transf.plot(pos = (i, tr[i]) )              # c coefficients
            tr[i+mp] = c3*f[n-1] - c2*f[n] + c1*f[1] - c0*f[2]     # High-pass
            transf2.plot(pos = (i+mp, tr[i+mp]) )
61      else:                                           # Inverse DWT
            tr[1] = c2*f[mp] + c1*f[n] + c0*f[1] + c3*f[mp1]     # Low-pass
            tr[2] = c3*f[mp] - c0*f[n] + c1*f[1] - c2*f[mp1]     # High-pass
            j = 3
65          for i in range (1, mp):
                tr[j] = c2*f[i] + c1*f[i+mp] + c0*f[i+1] + c3*f[i+mp1]   # Low
                j += 1                                  # Upsample
                tr[j] = c3*f[i] - c0*f[i+mp] + c1*f[i+1] - c2*f[i+mp1]  # High
69              j += 1;                                 # Upsampling
        for i in range(1, n+1):
            f[i] = tr[i]                                # Copy TF to array
    def pyram(f, n, sign):                              # DWT, replaces f by TF
73      if (n < 4): return                              # Too few data
        nend = 4                                        # When to stop
        if sign >= 0 :                                  # Transform
            nd = n
77          while nd >= nend:                           # Downsample filtering
                daube4(f, nd, sign)
                nd //= 2
        else:                                           # Inverse TF
81          while nd <= n:                              # Upsampling
                daube4(f, nd, sign)
                nd *= 2
    f = zeros( (N + 1), float)                          # Data vector
85  inxi = 1.0/N                                        # For chirp signal
    xi = 0.0
```

```
for i in range(1, N + 1):
    f[i] = chirp(xi)                       # Function to TF
    xi   += inxi;
n = N                                      # Must be 2^m
pyram(f, n, 1)                                      # TF
# pyram(f, n, - 1)                         # Inverse TF
```

11

Neural Networks and Machine Learning

*Automated systems should provide explanations that are technically valid, meaning-
ful and useful to you and to any operators or others who need to understand the
system, and calibrated to the level of risk based on the context.*
— White House AI Blueprint, 2022

*The human brain, which has evolved (maybe) over six million years, is sometimes effective at
solving problems that traditional computer programming finds hard. This chapter deals with
neural networks, artificial intelligence (AI), and machine learning. Part I deals with simple
models for neurons and neural networks, based on those found in the brain. Part II demon-
strates several state-of-the-art AI software packages, whose innards are based on neural
nets. The applications are from physics, but deliberately simple, in order to demonstrate
what's inside the AI programs.*

Problem   Develop a computer model for a neuron and for a network of these neurons,
and investigate if your network has the capacity to learn.

Artificial intelligence (AI) simulates human cognitive abilities in learning and problem-
solving by capturing the type of tacit knowledge that is very difficult to write into software.
In recent years, AI has made great advances in pattern recognition, decision making, infer-
ence, and generating controversially realistic data.[1] *Machine Learning* (ML) is a subfield
of AI in which neural networks are taught by iteratively and inductively learning from
teaching data. *Deep Learning* is an extension of machine learning that uses layers of neural
networks to pass statistical associations from one layer of the network to the next. Finally,
generative AI uses two neural networks, one to generate data, and a second to evaluate those
data. The output from the evaluation then gets fed back into the first network for further
training and improvement.

---

1 Herbert Simon received the Nobel Memorial Prize in economic sciences in 1978 and the Turing Award
in computer science in 1975, in partial recognition for his developments in AI.

Computational Physics: Problem Solving with Python, Fourth Edition.
Rubin H. Landau, Manuel J. Páez, and Cristian C. Bordeianu.
© 2024 WILEY-VCH GmbH. Published 2024 by WILEY-VCH GmbH.

11.1   Part I: Biological and Artificial Neural Networks

The study of neural networks began with studies of the brain. The human brain weighs about three pounds and contains approximately 10^{11} nerve cells called *neurons*. These neurons are interconnected in complicated networks that somehow provide our mental capacities. A drawing of a neuron is given on the left of Figure 11.1. On the bottom of the neuron's cell body are branch-like dendrites that receive electrical and chemical pulses from the synapses of other neurons. In turn, if properly excited, the cell body sends off a single electrical pulse along the axon to the synaptic terminals on top. The pulse is called an *action potential*, and is typically in the range 0–30 negative millivolts, with a width between 1 and 6 milliseconds.

The pulses entering the cell body may be excitatory or inhibitory, which, respectively, increase or decrease the net voltage that reaches the cell body. The cell body integrates the incoming pulses in a variety of ways, and if some threshold is reached, fires its own pulse along the axon to the synaptic terminals where electrochemical interactions with other dendrites take place. The process is binary in the sense that a pulse is sent, or not sent, each time with essentially the same pulse configuration. After firing, the neuron needs some time to rest.

On the right of Figure 11.1, we see an actual image of the neurons in a mouse brain [Palmer, 2016; Reid *et al.*, 2016]. This is a true biological neural network, and is seen to be a very complicated, net-like structure of neurons connected by axons, dendrites, and synapses. The sense organs pass signals to the outer layer of the brain, where neurons process them there. The output from the outer layer of the brain gets passed on to lower and lower layers in a hierarchical manner, with each layer's processing believed to have its own purpose. The process continues until a final response is reached, for example, the

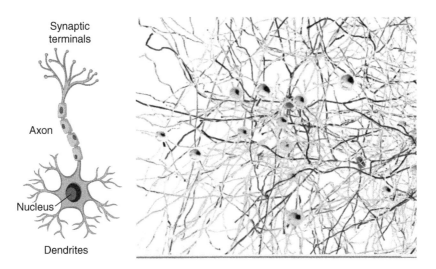

Figure 11.1   *Left*: A sketch of a neuron showing dendrites, a cell body, and synaptic terminals. The dendrites transmit pulses to the cell body, where an electrical action potential originates and is sent along to synaptic terminals. *Right*: An electron microscope image of the network of cortical neurons in a mouse brain (adapted from [Palmer, 2016; Reid *et al.*, 2016]).

recognition of a face. Biological networks appear to be highly parallel, and, possibly for this reason, robust, with no single group of neurons absolutely essential.

11.1.1 Artificial Neural Networks

In 1943, Warren McCullock, a psychiatrist with a deep intellectual interests in how the human brain works, and Walter Pitts, a logician with an interest in biological sciences, proposed a landmark mathematical formulation for a neuron, and for a network composed of such neurons. Based on the functionality of the biological neuron, they went on to develop a symbolic-logical calculus for how their model neurons interact with each other [McCulloch and Pitts, 1943]. McCullock and Pitts thereby proved mathematically that a neural net can be trained and can learn. Their neuron model, the *McCulloch-Pitts neuron*, is still a standard of reference in the field.

The mathematical model of a McCulloch-Pitts neuron is often called a *Perceptron*. However, the term also refers to the electronic version of a neural network based on the actual neurons' biology created by Frank Rosenblatt at Cornell in 1957 [Rosenblatt, 1958].[2] Rosenblatt's Perceptron displayed the ability to learn, and was both sensational and controversial at the time. Specifically, the perceptron was a simulation on an IBM 704 that, after 50 trials, was able to distinguish punched cards marked on the left from cards marked on the right. After each trial, the machine's connections would be tweaked, and it would be noted if there was an improvement in its prediction, and if there was, then the changes would be kept and further trials made in an effort to improve the predictions. However, the computing power of the 1950s and 1960s was orders of magnitude too low to provide the convincing demonstration of machine learning that present-day computers can, and Rosenblatt died in 1971 without seeing the success of his ideas.

In analogy to a biological neuron, an artificial neural network ("net") processes data through multiple layers of neurons or nodes. Each node may accept several inputs, processes them in a computing unit, and, if set criteria or threshold is reached, outputs data down its axon or *edge* to other nodes. The internal algorithm in each neuron used in decision making has changeable parameters, and the network "learns" by iteratively changing the parametric values of each neuron based on the accuracy of overall predictions. In this way, different nodes end up with different parametric values.

11.2 A Simple Neural Network

In Figure 11.2 we see a simple artificial intelligence (AI) neuron, also called a *node*, with two inputs and one output. The *x*'s on the left are the input signals coming from other nodes,

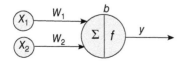

Figure 11.2 An AI neuron with two inputs and one output. The neuron body calculates a weighted sum of the inputs, and then processes the sum, possibly with bias *b* and through a sigmoid function *f*.

---

2 One of us (RHL) recalls fondly that Warren McCullock helped mentor him about graduate school, and that Frank Rosenblatt was one of his undergraduate teachers.

and the Σ in the cell body denotes a weighted summation of the input signals:

$$\Sigma = w_1 x_1 + w_2 x_2. \tag{11.1}$$

Also within the cell body is the *activation* or *sigmoid* (S-shaped) function f that decides, based on the value of Σ, whether or not to fire. Consequently, the output y can be expressed as

$$y = f(x_1 w_1 + x_2 w_2 + b), \tag{11.2}$$

where b is a *bias*. The weights and the biases are the parameters that get changed during learning. As a simple example, let's say our perceptron has weights $w_1 = -1$, $w_2 = 1$, bias $b = 0$, and accepts the two input values,

$$x_1 = 12, \qquad x_2 = 8. \tag{11.3}$$

Then $\Sigma = -1 \times 12 + 1 \times 8 = -4$, and if $f(x) = x$, then this would be the neuron's output y.

The binary nature of the original perceptron neuron, with its output of only 0 or 1, can make building a network out of them challenging to train. A more trainable and robust network would contain *sigmoid neurons* in which the output is no longer restricted to 0 or 1. For example,

$$f(x) = \frac{1}{1 + e^{-x}}, \qquad f(x) = \tanh(x), \qquad f_{\mathrm{ReLU}}(x) = \max(0, x). \tag{11.4}$$

The exponential would produce an output between 0 and 1, the hyperbolic tangent would produce an output between -1 and $+1$, and the Rectified Linear Unit outputs the x, if x is positive, and 0 if the x is negative. For the simple network that we will soon develop, we shall use the exponential sigmoid function.

11.2.1 Coding A Neuron

Exercise Below in Listing 11.1 we give a software model of single neuron coded with NumPy [Zhou, 2022]. Verify that this code reproduces the hand calculation above that gave -4 as output.

11.2.2 Building A Simple Network

In Figure 11.3 we show an AI network containing three layers: an input layer on the left, a *hidden* layer in the middle, and an output layer on the right. Networks with more than three layers produce what's called *deep learning*, and so our network might be called a "shallow learner." The weights for the signals entering each layer are shown, where it should be

Figure 11.3 A simple neural network with two neurons in the input layer, two neurons in the internal hidden layer, and one in the output layer.

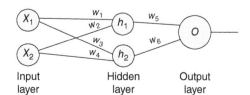

noted that the activation functions within each cell body may be all different. Even a neural network with just a few neurons can have a good number of connections and sigmoid functions. We'll sacrifice computing power for simplicity, and make each neuron identical.

Exercise Calculate by hand the expected output of the network in Figure 11.3 for $x_1 = 2$, $x_2 = 3$, and $w_1 = 0$, $w_2 = 1$. You should get $O = 0.7216$.

The code `NeuralNet.py` in Listing 11.1 is for a neural network and must include the previous neuron class. Check that it produces an output of 0.7216.

11.2.3 Training A Simple Network

Figure 11.4 shows a flowchart for training an AI network. The network is taught by inputting predetermined *training data* for which the "correct" output is known, and comparing the predicted and correct outputs. One then calculates the *Cost* or *Loss* (what a physicist might call *error*):

$$\text{Cost} \equiv \text{Loss} \equiv \mathcal{L} = \text{Correct Output} - \text{Predicted Output.} \tag{11.5}$$

For a perfect network the Cost would be zero. The mathematical relation between the Loss and the network's many parameters is generally unknown, and remains unknown even after training. During training, if the Cost is higher than some set value, the weights are tweaked by an amount based on the numerical evaluation of the derivatives of the Loss function. This process, called *back propagation*, is repeated until a reasonably small Cost is obtained. Then the model is tested on data it has not seen before, and its true accuracy is determined. And if need be, further training might be in order.

Realistic networks may contain several hundred nodes for each feature in the data that we want to understand, with hundreds or more nodes in each hidden layer, but only a small number of nodes for the output. The optimal numbers follow from intuition and trial and error, sometimes with the number increasing until good results are obtained. As you can well imagine, fully trained networks are like black boxes with many parameter values that, in themselves, do not explain the strategy used to obtain the correct answers. So while neural nets are often successful and efficient at recognizing complex patterns, much as the human brain, they do not provide insight into how they made good decisions.

Now that we have a neural net, let's train it. Of course we should not expect high performance from this simple network, but then again, we're not putting much into it! Although

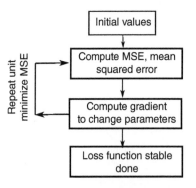

Figure 11.4 A flowchart of the steps in teaching a neural net, with the gradient of the Loss function used to minimize it.

in some sense, "training" the network is programming, it's not the kind of programming we're accustomed to, where we go about modifying statements until we get things right. Rather, we, or some algorithm, will do its own thing and vary the internal parameters' values, with the changes based on how close the network is to producing the right answer. As we have said, once trained, it is essentially impossible to predict what the network does, or how it's doing it, by studying the internal parameters' values.

As a simple example, imagine that you want to identify π mesons from μ mesons based on the length and width of the tracts they have left on a film strip. You start with the characteristics of four tracks that have already been (painstakingly) measured by eye:

Track ID	Length (mm)	Width (mm)	Particle
A	13	6	π
B	16	10	μ
C	15	9	μ
D	12	7	π

In order to make numerical predictions, we label a μ as a 0 and a π as a 1. As is standard in ML and data mining, we convert to *mean-centered data*, which means that we scale the input data to have zero mean (average length 14, average width 8):

Track ID	Length – 14	Width – 8	Particle ID
A	−1	−2	1
B	2	+2	0
C	1	+1	0
D	−2	−1	1

Next we define the *Loss* (11.5) as the mean-squared difference between the correct answers $y_i^{(c)}$ and the predicted ones $y_i^{(p)}$:

$$\mathcal{L} = \frac{1}{N} \sum_i^N \left(y_i^{(c)} - y_i^{(p)} \right)^2, \tag{11.6}$$

where N is the number of input data. Clearly, the smaller the loss, the better the network.

Exercise Run the code below and check that it predicts that all of the input tracks are muons ($y_i^{(p)} \equiv 0$).

```
1 import numpy as np                              # For  NumPy arrays

  def  Loss(y_c, y_p):
       return ((y_c - y_p)**2).mean()            # Auto mean of array
5 y_c = np.array([1, 0, 0, 1])
  y_p = np.array([0, 0, 0, 0])
  print (Loss(y_c, y_p))
```

The code gives $\mathcal{L} = 0.5$, which means we got the right answer half the time (but there were only two choices).

11.2.4 Decreasing the Error

Minimizing the Loss is essentially identical to minimizing a χ^2 fit to data, which we have studied in Chapter 6. Now we want to adjust the weights w_i's and the biases b_i's to minimize $\mathcal{L}$. Even for something as simple as our two-neuron network in Figure 11.3, we have six weights and three biases to adjust. Accordingly, an extremum in $\mathcal{L}$ occurs when

$$\frac{\partial \mathcal{L}}{\partial w_i} = 0, \; i = 1, \dots, 6, \qquad \frac{\partial \mathcal{L}}{\partial b_i} = 0, \; i = 1, 2, 3. \tag{11.7}$$

For a complex network, there might be thousands or more of these equations, with only numerical determinations of the derivatives feasible. This is not a one-time affair; one keeps training the network in hopes that it will move closer to a minimum. Although the parameters w_i and b_i do not appear explicitly in the definition (11.6) of the Loss, the predicted values $y_i^{(p)}$ do, and they are functions of the parameters in some unknown way. Accordingly, we use the chain rule (twice!) to obtain the needed partial derivatives. For example, because the weight w_1 affects only the hidden neuron h_1, and because there is only the single predicted value $y_i^{(p)}$,

$$\frac{\partial \mathcal{L}}{\partial w_1} = \frac{\partial \mathcal{L}}{\partial y_i^{(p)}} \frac{\partial y_i^{(p)}}{\partial h_1} \frac{\partial h_1}{\partial w_1}. \tag{11.8}$$

Focusing on just one weight or bias at a time, in this case w_1, is often used in training, with the network trained sequentially for each of the other parameters.

We now evaluate these partial derivatives. The definition (11.6) of Loss, makes the $\partial y_i^{(p)}$ derivative easy:

$$\frac{\partial \mathcal{L}}{\partial y_i^{(p)}} = \frac{-2}{N} \left(y_i^{(c)} - y_i^{(p)} \right), \tag{11.9}$$

where the correct answers $y_i^{(c)}$'s are known, but not the $y_i^{(p)}$'s. In the present case, the output,

$$y_{\text{out}}^{(p)} = f(w_5 h_1 + w_6 h_2 + b_3), \tag{11.10}$$

makes the derivatives with respect to the weights straightforward:

$$\frac{\partial y_{\text{out}}^{(n)}}{\partial w_5} = h_1 \frac{df(x)}{dx} (x = w_5 h_1 + w_6 h_2 + b_3), \tag{11.11}$$

$$\frac{\partial y_{\text{out}}^{(p)}}{\partial w_6} = h_2 \frac{df(x)}{dx} (x = w_5 h_1 + w_6 h_2 + b_3). \tag{11.12}$$

The derivative of the sigmoid function (11.4) is easy:

$$f(x) = \frac{1}{1 + e^{-x}} \quad \Rightarrow \quad \frac{df(x)}{dx} = \frac{e^{-x}}{(1 + e^{-x})^2}. \tag{11.13}$$

The next derivative we need, $\partial h_1/\partial w_1$, follows from our definition of the h's and their use, as shown in Figure 11.3:

$$h_1 = f(w_1 x_1 + w_2 x_2 + b_1) \tag{11.14}$$

$$\Rightarrow \quad \frac{\partial h_1}{\partial w_1} = x_1 \frac{df}{dx}(x = w_1 x_1 + w_2 x_2 + b_1). \tag{11.15}$$

Now we put all of the pieces together:

$$\frac{\partial \mathcal{L}}{\partial w_i} = \frac{\partial \mathcal{L}}{\partial y_i^{(p)}} \frac{\partial y_i^{(p)}}{\partial w_i} = \frac{-2}{N}\left(y_i^{(c)} - y_i^{(p)}\right)\frac{\partial y_i^{(p)}}{\partial w_i}. \tag{11.16}$$

The derivatives $\partial y_i^{(p)}/\partial w_i$ depends upon the model for the neuron, and in particular, on its sigmoid function f. For our model, its one output is

$$y_{\text{out}}^{(p)} = f(w_5 h_1 + w_6 h_2 + b_3) = \frac{1}{1 + e^{-(w_5 h_1 + w_6 h_2 + b_3)}}. \tag{11.17}$$

For the two-neuron network (Figure 11.3), the h functions are:

$$h_1 = f(w_1 x_1 + w_2 x_2 + b_1), \qquad h_2 = f(w_3 x_1 + w_4 x_2 + b_2). \tag{11.18}$$

Now that we have gone through some of the grubby details, it might be clear that the evaluation of the Loss for a big realistic network is best not done by hand. In order to complete our example with the least amount of pain, let's also limit the network's input to just one of the particle tracks, in this case, track A with length $x_1 = -2$, width $x_2 = -1$, and particle $y^{(c)} = 1$. Furthermore, for simplicity, let's set all of the weights to 1, and all of the biases to 0. We then have

$$h_1 = f(w_1 x_1 + w_2 x_2 + b_1) = f(-2 - 1 + 0) = \frac{1}{1 + e^{-3}} = 0.0474,$$

$$h_2 = f(w_3 x_1 + w_4 x_2 + b_2) = f(-2 - 1 + 0) = \frac{1}{1 + e^{-3}} = 0.0474,$$

$$\Rightarrow \quad y_{\text{out}}^{(p)} = f(w_5 h_1 + w_6 h_2 + b_3) = f(0.0474 + 0.0474)$$

$$= \frac{1}{1 + e^{-0.0948}} = 0.524. \tag{11.19}$$

This prediction says that it is more likely that not that Track A is a π (ID = 1), but not very likely. That being the case, let's adjust the weights and see if it improves the network's prediction. To do that, we need to evaluate the derivative of the loss with respect to the parameter w_1 (11.8):

$$\frac{\partial \mathcal{L}}{\partial w_1} = \frac{\partial \mathcal{L}}{\partial y_{\text{out}}^{(p)}} \frac{\partial y_{\text{out}}^{(p)}}{\partial h_1} \frac{\partial h_1}{\partial w_1},$$

$$\frac{\partial \mathcal{L}}{\partial y_{\text{out}}^{(p)}} = -2(1 - y_{\text{out}}^{(p)}) = -2(1 - 0.524) = -0.952, \tag{11.20}$$

$$\frac{\partial y_{\text{out}}^{(p)}}{\partial h_1} = w_5 \frac{df}{dx}(w_5 h_1 + w_6 h_2 + b_3)$$

$$= 1 \times \frac{df}{dx}(0.0474 + 0.0474 + 0) = \frac{\exp(-0.0948)}{(1 + \exp(0.0948))^2} = 0.249, \tag{11.21}$$

$$\frac{\partial h_1}{\partial w_1} = x_1 \frac{df}{dx}(w_1 x_1 + w_2 x_2 + b_1) = -2\frac{df}{dx}(-2 - 1 + 0) = -0.0904$$

$$\Rightarrow \quad \frac{\partial \mathcal{L}}{\partial w_1} = -0.952 \times 0.249 \times (-0.0904) = 0.0214. \tag{11.22}$$

At last! This tells us that if we decrease w_1, then the Loss $\mathcal{L}$ should get smaller, and thus yield a better prediction.

Exercise Repeat the prediction of $y_{out}^{(p)}$ with an incrementally decreasing w_1, and observe how the prediction changes.

To determine just how much to change the weight, we assume that we are close enough to the correct answer to need only a first-order correction to the weight:

$$w_1^{(new)} \simeq w_1^{(old)} - \eta \frac{\partial \mathcal{L}}{\partial w_1}. \tag{11.23}$$

Here η is called the *learning rate* of the network, and the method is called *stochastic gradient descent* (discussed further in Section 11.6.3). Even for large-scale problems, the process is often much like what we have worked through here, but probably automated: one focuses on a single parameter, as we have done with w_1, and then repeats the process until the Loss stops decreasing, or becomes too slow in its response. Then one goes on and repeats the process with each of the other weights and biases.

11.2.5 Coding and Running A Simple Network

Listing 11.3 at the end of this chapter presents our code for the simple network. It started with Loop n = 0 Loss: 0.164, and ended with Loop n = 960 Loss: 0.002.

1) Run SimpleNet.py and plot the Loss versus the number of learning trials, N.
2) Once you have taught the network enough for it to produce a small Loss, say just a few percent, determine the predictions it makes on some new input data.
3) Extend our simple two-node hidden layer network to a three-node hidden layer.
   a) Repeat the learning exercise used for the two-node hidden layer now for the three-node case.
   b) Compare the effectiveness of learning for the two- and three-node hidden layer networks.
4) Extend our simple two-node hidden layer network to a network with two hidden layers, each containing just two nodes.
   a) Repeat the learning exercise used for a network with a two-node hidden layer, now for one with two two-node hidden layers.
   b) Compare the effectiveness of learning for the single and double two-node hidden layer networks.

11.3 A Graphical Deep Net

We have just built a simple neural network with single hidden layer, and showed that it's capable of learning, if just somewhat. Now, based on Zhou [2018] and Rohrer [2017], we

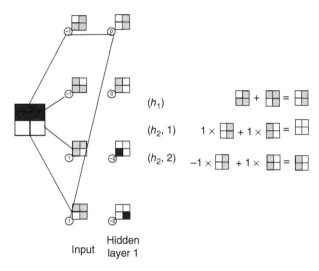

Figure 11.5 A deep neural net that classifies a 4×4 square into different classes.

examine graphically a *deep neural network* with three hidden layers. This network, which is essentially an ML decision tree, has been designed to recognize the orientation of a line, and does it with no error. The point here is to see graphically, without working through the programming, how more complicated tasks can be accomplished by neural networks.

The left of Figure 11.5 shows a three-hidden layer network that has been taught to recognize different patterns within a 4×4 square. This is an example of hierarchical layering in which each successive layer recognizes a more complex pattern. The first hidden layer recognizes a single pixel, the second layer recognizes two-pixel combinations, and so forth. Each cell has two dendrites (edges) entering, and two leaving, with no connections between two neurons in the same layer, or between nonadjacent layers. The numbers next to each cell body are the weights of the edges applied to the signals they have received.

The large square on the left of Figure 11.5 is the four-pixel input, in this case with a horizontal line. The input line can be:

$$\text{horizontal: } [\begin{smallmatrix} x & x \\ \square & \square \end{smallmatrix}], [\begin{smallmatrix} \square & \square \\ x & x \end{smallmatrix}], \text{ vertical: } [\begin{smallmatrix} x & \square \\ x & \square \end{smallmatrix}], [\begin{smallmatrix} \square & x \\ \square & x \end{smallmatrix}], \text{ diagonal: } [\begin{smallmatrix} x & \square \\ \square & x \end{smallmatrix}], [\begin{smallmatrix} \square & x \\ x & \square \end{smallmatrix}], \text{ no line: } [\begin{smallmatrix} x & x \\ x & x \end{smallmatrix}], [\begin{smallmatrix} \square & \square \\ \square & \square \end{smallmatrix}].$$

As indicated by the white squares in Figure 11.5, each of the input layer's four nodes is associated with a different pixel location. Accordingly, the input horizontal line occupying the top two pixels in the figure is not identified with the top two nodes in the input layer (weight -1), but is identified with the bottom two nodes (weight $+1$).

On the left of Figure 11.5, we isolate Hidden Layer 1 and its action in identifying one of the four pixels. Based on the input to it from the edges, this layer combines single pixels into two-pixel combinations, and determines appropriate weights. For example, in Figure 11.5 left we see how edges from the top and the bottom nodes of the input layer are fed into the top node of Hidden Layer 1. On the (h_1) line in Figure 11.5b, we see how these two edges are combined into the recognition of a vertical line. There being no vertical line in the input picture, a weight of 0 is recorded. This is expressed analytically as

$$x_1 w_1 + x_4 w_4 = (-1)(1) + (1)(1) = 0. \tag{11.24}$$

On the right of Figure 11.5 we demonstrate how the top two nodes in Hidden Layer 2 are activated. The top node on line $(h_2,1)$ performs the combination

$$1 \times \begin{bmatrix} \square & x \\ \square & x \end{bmatrix} + 1 \times \begin{bmatrix} x & \square \\ x & \square \end{bmatrix} = \begin{bmatrix} \square & \square \\ \square & \square \end{bmatrix}. \tag{11.25}$$

The second down node $(h_2,2)$ performs the combination

$$-1 \times \begin{bmatrix} \square & x \\ \square & x \end{bmatrix} + 1 \times \begin{bmatrix} x & \square \\ x & \square \end{bmatrix} = \begin{bmatrix} x & \square \\ x & \square \end{bmatrix}, \tag{11.26}$$

where the negation of white is defined as black.

The cell bodies in the hidden layers contain the *activation functions* that determine the neurons' actions based on weighted values of the inputs. As long as the signal transmitted to the node is nonzero, it remains active and a signal gets transmitted onward. A zero-input signal places the node in an inactive state with no transmission.

Hidden Layer 3 uses the *ReLU* activation function (rectified linear unit) that transmits positive signals, but turns the neuron off if the input is negative. We leave it as an exercise to work through the actions in Hidden Layer 3.

Problem Here are four combinations: $[X][\square]$, $[X][X]$, $[\square][X]$, $[\square][\square]$. Build a neural network that can distinguish these combinations.

11.4 Part II: Machine Learning Software

In Part II of this chapter, we give examples of using Python with several industrial-strength ML software packages [Campesato, 2020], [Yalcin, 2021]. Preparing data for ML is often a time-consuming, and, accordingly, we also discuss a number of tools for preprocessing data.

TensorFlow is a free and powerful package of software for machine learning via deep neural networks. It was developed by *Google Brain* for their own AI research and development, but in 2015 was made available as open-source software. The more user-friendly TensorFlow 2 was released in 2019. When Google uses TensorFlow they employ their own TensorFlow CPU (TPU), which is also available in the cloud. (Microsoft has also invested heavily into the computing power needed for AI, but it's not free.) Google's TPUs are capable of some four trillion operations per second, much more than anything we will need. However, that amount of computing power is valuable for tasks like exoplanet recognition, at which AI excels, and, indeed, the absence of that power was the reason AI did not prosper in the 1950s. In general, neural networks are probably best suited to big problems, not for the small pedagogical problems we'll look at.

TensorFlow uses *dataflow* graphs as its basic computational element, with each graph composed of nodes and edges (cell bodies and axons like those in Figure 11.2). The "tensor" aspect of TensorFlow refers to its use of arrays with multiple indices, like tensors in physics, to represent the edges. The arrays are compatible with Python's NumPy, with which you are familiar from Chapter 7. The nodes in TensorFlow perform the mathematical operations, and the edges transfer the data.

11.4.1 TensorFlow Installation and Execution

Even though some of these directions are a repeat of those in Chapter 1, some are new, and so for the sake of completeness, we repeat them here.

In order to run TensorFlow interactively in a notebook environment, you'll need a few things:

- Set up a notebook environment such as Jupyter [2022]. This gives you a web-based, free, interactive computing platform that combines live code, equations, text, visualizations, and much else.
- Install an up-to-date version of Python, as available from AnaConda [2022].
- Install a *package manager*, which helps in the installation of all the associated bits and pieces of packages. We recommend [Conda, 2023] within a Jupyter Notebook. To do this, use a shell (the `Command` shell or `PowerShell` on Windows, or the `Terminal` on Macs) to create the Conda environment:

```
conda create -name MyEnv
```

  where you may use a name other than `MyEnv` for your environment.
- Finally, you'll need TensorFlow. Follow the instructions in [Tensor, 2022].
- Activate your environment by entering:

```
conda activate tensorflow
```

- Next tell Conda to use the GitHub repository `conda-forge` for needed packages:

```
conda install -c conda-forge tensorflow
```

- Once TensorFlow is installed, call the *Anaconda Navigator* and select `MyEnv` from `Applications on/base(root)`. At the top of the navigator, there are three boxes. Select the middle one, which by default is `base(root)`, and change it to `MyEnv`, or whatever you have defined as your tensorflow environment in Jupyter. Also, when writing a new `.ipynb` notebook in the box `New`, select your TensorFlow environment (`MyEnv`).
- Once Jupyter is launched, select *New/Python3(ipykernel)/MyEnv*.
- In a notebook cell enter:

```
import tensorflow as tf
```

  and then run `tf`.
- In the next cell enter:

```
print(tf. __version__).
```

  Ensure that you have `TensorFlow 2` or `T2.x` and not `T1.x`.

11.5 TensorFlow and SkLearn Examples

Before we start using TensorFlow for some AI work, we'll run through several simple calculations with it as a check that it's installed and working properly.

Problem   Use TensorFlow to compute the mass number A, given the atomic number Z and the neutron number N.

As we all know, the atomic number Z is the number of protons in a nucleus, and the mass number $A = Z + N$ is the sum of the number of protons and neutrons in a nucleus. This program calculates A, given Z and N:

```
# TensorTest.py:   Test TensorFlow
[1] import tensorflow as tf
[2]   Z = tf.constant(1)                                  #  Hydrogen
[3]   N = tf.constant(2)                    # Two neutrons => tritium
[4]   A = tf.add(Z,N)
[5]   print("A:", A)
      A: tf.tensor(3, shape=(), dtype=int32)
```

To understand this output, here's some TensorFlow speak:

- **Size:** Total number of elements in a tensor.
- **Axis or Dimension:** A particular dimension of a tensor.
- **Rank 0 (scalar):** A tensor with one value, no axis.
- **Rank 1 (vector):** A tensor with a list of values on one axes.
- **Rank 2 (matrix):** A tensor with a list of values on two axes.
- **Rank N:** A tensor with N indices (aka order, degree, or ndims).
- **Shape**: The length (number of elements) on each axes of a tensor. A constant has shape (), a tensor with dimensions [2,3] has shape (2,3).
- **Data types:** Line 6 (L6) of the listing shown as `int32`. Other data types are:

  `tf.float32`   `tf.float64`   `tf.int8`   `tf.int16`

  `tf.int64`   `tf.uint8`   `tf.string`   `tf.bool`

Problem Use Tensorflow to compute the mass excess of the hydrogen isotopes.

Problem Find the binding energy for each of the seven H isotopes, and plot the binding energies versus A.

Figure 11.6 shows the TensorFlow calculation of the hydrogen isotope mass excess, ran within a Jupyter notebook. The Z protons and N neutrons within a nucleus are bound by

```
In [1]: >|  import tensorflow as tf

In [2]: >|  # atomic masses in u, dtype float32
            am = tf.constant([1.007827032, 2.01401778, 3.016049278,
3.016049278,
                       4.026, 5.035,6.045, 7.05])
            # Atomic number: Z = 1 + N
            A = tf.constant([1,2.,3.,4.,5.,6.,7.])  # as dtype float32

in [3]: >|  for i in range(7):
                mas = am[i]
                nA = A[i]
                di = tf.subtract(mas,na)
                fi - tf.multiply(di,932.494028)
                print("dif :", fi)

                dif : tf.Tensor(7.290844,   shape=(), dtype=float32)
                dif : tf.Tensor(13.057516,  shape=(), dtype=float32)
                dif : tf.Tensor(14.949906,  shape=(), dtype=float32)
                dif : tf.Tensor(24.21.8866, shape=(), dtype=float32)
                dif : tf.Tensor(32.60215,   shape=(), dtype=float32)
                dif : tf.Tensor(41.9173,    shape=(), dtype=float32)
                dif : tf.Tensor(46.57488,   shape=(), dtype=float32)
```

Figure 11.6 A screenshot of a TensorFlow calculation of the mass excess of the hydrogen isotopes within a notebook environment.

the nuclear force. As a consequence, the rest energy (mc^2) of the nucleus is less than the sum of its constituent masses by the binding energy B:

$$B = \left[Zm(^1H) + Nm_n - M_{nuc} \right] c^2. \tag{11.27}$$

Atomic masses are usually stated in Daltons (Da, or u), with u defined as 1/12 the mass of a ^{12}C atom $= 1.660538782 \times 10^{-27}$ kg ≈ 931.5MeV/c^2. The mass excess is the difference between the atomic mass M and u times the atomic number:

$$\text{Mass excess} \stackrel{\text{def}}{=} M - A\text{u}. \tag{11.28}$$

Hydrogen exists as seven isotopes, three of which occur naturally [Haynes, 2017]:

N	Atomic mass (u)	N	Atomic mass (u)
0	1.007827032	4	5.035
1	2.014101778	5	6.045
2	3.016049278	6	7.05
3	4.026		

Here's our program `TensorBE.py` that calculates the mass excess and produces the left part of Figure 11.7:

```
1  # TensorBE.py:  TensorFlow calc H isotope binding E's

   import tensorflow as tf
   import matplotlib.pyplot as mpl
5  import numpy as np
   B = np.zeros(7)#[0,0,0,0,0,0,0]
   mP = tf.constant(938.2592)                        # Proton mass
   mN = tf.constant(939.5527)                        # Neutron mass
9  mH = tf.multiply(1.00784, 931.494028)         # H mass in MeV/c2
   am = tf.constant([1.007825032, 2.01401778,\
                     3.016049278, 4.026, 5.035, 6.045, 7.05])   # Masses
   A = tf.constant([1, 2., 3., 4., 5., 6., 7.])         # Atomic numbers
13 for i in range(7):
       C = mH + (i) * mN - am[i]*931.494028
       AN = A[i]
       B[i]= C/AN
17     print("BN :",B[i] )
   mpl.ylabel('Binding energy per nucleon (MeV)')
   mpl.xlabel('Atomic mass number')
   mpl.plot(A,B)
21 mpl.show()
```

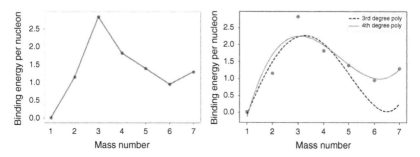

Figure 11.7 *Left*: TensorFlow's linear regression fit to the binding energies of seven hydrogen isotopes. *Right*: Third- and fourth-degree polynomial fits to the same binding energies.

11.5.1 Preprocessing with Scikit-learn

The Python package *scikit-learn*, aka *sklearn*, is a library of algorithms used in ML for the classification, regression (fitting), and clustering of data. It is often used in the processing of large datasets. The package is installed from a shell with the command:

`pip install scikit-learn.`

Here `pip` is Python's package installer, and it's also useful for ensuring that your software is up to date:

`pip upgrade scikit-learn.`

Problem Make a best fit (regression) of a polynomial to the hydrogen isotopes' binding energies as a function of atomic mass number.

Here's our program `SkPolyFit.py` that uses sklearn's polynomial and linear regression methods, and which we'll discuss below:

```
# SkPolyFit.py: Polynomial regression with sklearn

import numpy as np
from sklearn.preprocessing import PolynomialFeatures
from sklearn.linear_model import LinearRegression
import matplotlib.pyplot as plt
import numpy as np

poly = PolynomialFeatures(degree=6, include_bias=False)    # Degree 6 poly
mA = np.array([1, 2, 3, 4, 5, 6, 7])                # Atomic mass number
poly_features = poly.fit_transform(mA.reshape(-1, 1))          # Data
B = [0.0140, 1.1520, 2.8235, 1.8150, 1.3871, 0.9465, 1.2971]   # BE/N
poly_reg_model = LinearRegression()
poly_reg_model.fit(poly_features, B)
b_predicted = poly_reg_model.predict(poly_features)
intcp = poly_reg_model.intercept_,
print(intcp)
coefs = poly_reg_model.coef_                    # Polynomial coefficients
print(coefs)

def predict_y_value(x):
    y = -1.91 + 1.72*x + 0.288*(x**2) - 0.182*(x**3) + 0.016* (x**4)
    return y
def pred_y_val(x):
    y = intcp + coefs[0]*x + coefs[1]*x*x + coefs[2]*x**3
    return y
xx = np.linspace(1,7,50)                        # Plot polynomial
yy = predict_y_value(xx)
y4 = pred_y_val(xx)
fig, ax = plt.subplots()
ax.scatter(mA,B)                                # Plot points
plt.xlabel('Mass Number')
plt.ylabel('Binding Energy per nucleon')
plt.plot(xx, yy, c = "red", label="3rd degree poly")     # Solid line
plt.legend()
plt.plot(xx, y4, label ="4th degree poly")
plt.legend()
plt.show()
```

Although life would be simpler if there were a standard way to store matrices and arrays, it is not our choice to make. Specifically, scikit-learn works with 2D *vertical arrays*, which are different from the familiar NumPy arrays. These vertical arrays are more efficient when dealing with *sparse matrices* containing many zeros. Here's a NumPy array and its SciPy sparse matrix version using compressed row storage (CSR) format:

```
     NumPy Array:             SciPy Sparse CSR Matrix:
2  [ [1.  0.  0.  0.]              (0, 0)   1.0
     [0.  1.  0.  0.]              (1, 1)   1.0
     [0.  0.  1.  0.]              (2, 2)   1.0
     [0.  0.  0.  1.] ]            (3, 3)   1.0
```

Returning to our code `skPolyFit.py` in the listing above, notice the column of `B` values, with the mass array `mA` reshaped into a 2D column (the -1 on L11). On L16, `intcp = poly_reg_model.intercept_` and `coefs = poly_reg_model.coef_` give the intercept and the coefficients of the fitted polynomial. A third- and fourth-degree polynomial fit are shown on the right of Figure 11.7. Neither fit is very good, which is to be expected since nuclear binding is not a simple process.

11.5.1.1 Gradient Tape

We have already seen in our work with a simple neural network that ML involves so-called *backward pass* operations that repeat a calculation using the output from a prior execution. To avoid having to redefine "new" and "old" versions of variables and functions, Tensor-Flow has a `GradientTape` command that acts like a tape recorder that stores intermediate results for future use. Here, and in the programs to follow, are examples of its use:

```
   # GradTape.py:  Use of Tensor Flow's GradientTape
   import tensorflow as tf
3  m = tf.Variable(1.5)
   b = tf.Variable(2.2)
   x = tf.Variable(0.5)
   y = tf.Variable(1.8)
7  with tf.GradientTape() as tape:
       z = tf.add(tf.multiply(m, x), b)          # z = mx +b
       loss = tf.reduce_sum(tf.square(y - z))    # (y-(mx+ b))**2
   dloss_dx = tape.gradient(loss, x)                 # Gradient
11 tf.print('dL/dx:', dloss_dx)        # Output dL/dx: 3.45000029
   tf.print(2*(-m)*(y-(m*x+b)))        # Check: output 3.45000029
```

We see that `GradientTape.py` has recorded the loss function $\mathcal{L} = [y - (mx + b)]^2$, and evaluated the gradient of $\mathcal{L}$.

11.5.2 Linear Fit to Hubble's Data

In 1924, Hubble fitted a straight line of slope ~ 500 (km/s)/Mpc to his measurements of the recessional velocity of nebulae versus their distance from Earth [Hubble, 1929]. We repeat that fitting using TensorFlow's minimization of the Loss function. Listing 11.6 gives our program `Hubble.py` that does the fitting, and in Figure 11.8 we show the initial and final fits. Note in the program:

- On L6-11 the data are entered explicitly via the `tf.Variable` command.
- On L17 `x_train` is assigned to r, and on L19 `y_train` is set equal to the equation of a straight line, $y = mx + b$.
- On L25 `tf.reduce_mean(tf.square(y_pred - y_true))` is used to predict the mean-square error.

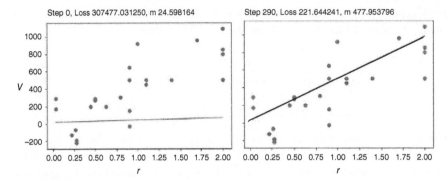

Figure 11.8 TensorFlow's first (left) and final (right) fits to Hubble's data of velocities v of nebulae versus their distance r.

- The fitting is done by predicting values for `y`, computing the resulting `loss`, and then using the computed gradient of the loss function to guess new values for `m` and `b`. The process is repeated some 300 times.

11.6 ML Clustering

A key element in ML's interpretation of data, and in its training of a neural network, is grouping data into *clusters* based upon some common features. This can reveal similarities, or differences, in the data elements, and can be used to highlight unusual elements within the data. If the data are given with labels, as in Table 11.2, then this is *supervised learning*; otherwise, it's *unsupervised learning*. Not to think that this is just bookkeeping; the *clustering* problem is classified as computationally difficult (*NP-hard*), which means it cannot be solved in polynomial time, which means it takes a lot of computing time to solve it.

The *Scikit-learn* package, which we have introduced in Section 11.5.1, clusters data via unsupervised ML. You tell Scikit-learn the number k of clusters you want to form, and the program searches through the elements to find the most meaningful clusters. An integral

Table 11.1 Data for 18 elementary particles tabulated by Index, Name, and Mass [PDG, 2023].

Index	Name	Mass (MeV/c^2)	Index	Name	Mass (MeV/c^2)
1	ν	0.8×10^{-6}	10	$K^\pm$	493.677
2	e	0.5110041	11	p	938.2721
3	μ^-	105.65	12	n	939.5654
4	μ	105.6583	13	Λ	1115.683
5	π^0	134.98	14	Σ^+	1180.37
6	π^+	139.57	15	Σ^-	1197.449
7	π^-	139.57	16	Ξ^0	1314.86
8	η	547.862	17	Ξ^-	1321.71
9	K^0	497.611	18	Ω^-	1672.45

Table 11.2 Fourteen elementary particles and their masses.

Number	Name	Masse	Number	Name	Mass	Number	Name	Mass
0	neutrino	0	7	eta	548	14	Sigma−	1197
1	electron	0.5	8	K0	498	15	Xi0	1315
2	mu−	106	9	K+−	494	16	Xi−	1322
3	mu	106	10	p	938	17	Omega−	1672
4	pi0	1345	11	n	940	12	Lambda	1115
5	pi+	140	6	pi−	140	13	Sigma+	1180

part of that search is the calculation of the centroid of each cluster, that is, the location within each cluster that minimizes the squared-distance to that cluster's elements. The learning process has the program repeatedly trying out different clusters, with the process ending when the new centroids do not move, or after a fixed number of cycles.

Problem You are given Table 11.1 containing the masses of 18 elementary particles [PDG, 2023]. Form three clusters of these particles based on their masses.

Our program `KmeansCluster.py` is given in Listing 11.4, and Figure 11.9 shows the three clusters, and their centroids, that it found. There clearly is a low-mass cluster, a high-mass cluster, and a small medium-mass cluster. Notice in the program:

- On L7-12 the data are entered directly into the program as a NumPy array.
- L13 informs `KMeans` that we want three centroids, with initial random elements.
- L14 determines the initial clusters by `kmeans.fit`.
- The program tries out different clusters, and predicts new locations for the centroids, as it looks for a minimum in the Loss function.
- The program creates a scatterplot with the centroid locations shown as diamonds, and the data elements as separate colors for each cluster.

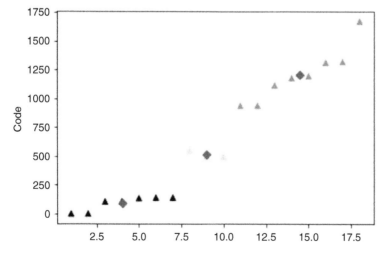

Figure 11.9 Clustered elementary particles from KmeansCluster.py showing three clusters and their centroids as diamonds.

11.6.1   Reading Files with Panda

In the previous exercise, we entered data directly into the program KmeansCluster.py. This really wouldn't do for large datasets, or for analysing a number of datasets. The Python package *Pandas* is what we need, as it provides a number of tools for manipulating and analyzing data. It is particularly useful for inputting data in tabular (column) form, in contrast to NumPy, which works with data in array form.

Problem   Repeat the last problem using Pandas to read in the data from a file.

Our program PandaRead.py is given in Listing 11.7, where you may notice that L7 reads in the entire file C:ElemnPart.dat from Table 11.1 using "whitespace" as column separators. L8 eliminates the superfluous "Name" column, while on L10 the x variable is assigned to "Number," and the y variable to "Mass." Then, as before, Kmeans uses these variables to find three clusters based on Mass.

11.6.2   Clustering with Perceptrons

In Section 11.1.1 we introduced Perceptrons as the historical, artificial neural net in which a neuron fires or not, depending on some threshold value. Although perceptrons are not state-of-the-art AI, they are useful for smaller dataframes (data structures with the data arranged in a 2D table of rows and columns).

Problem   You are given Table 11.2 containing 14 elementary particles and several of their properties. Use a Perceptron to cluster the particles into four groups, labeled by the type index T, based on their properties.

The program Perceptron.py in Listing 11.5 uses Python's sklearn package to create a perception, and then proceeds to find clusters of the particles. What's new here is the use of a classifier algorithm that assumes an approximate linear behavior of the Loss function:

$$\mathcal{L} \simeq w^T x + b, \tag{11.29}$$

and determines w^T and b from the training data. The learning routine iterates over the data, updating the weights w via:

$$w \rightarrow w - \eta \frac{\partial \mathcal{L}(w^T x_i + b, y_i)}{\partial w}, \tag{11.30}$$

where η is the *learning rate* parameter. To provide stability and precision, the learning rate for cycle t is made to decrease gradually through the training data:

$$\eta(t) = \frac{1}{\alpha(t_0 + t)}. \tag{11.31}$$

Some explicit steps of Perceptron.py program are:

- L8 uses pandas to read in the columnar data, and L9-10 assigns x to "Mass" and y ("Name") to the type index T.

- L13-16 splits the data into training and test groups, and places them into a dataframe `d` with columns labeled Type and mass. L15 specifies a seed for the random numbers, and a learning rate `eta` between 0 and 1.
- The data are scaled into standard form (zero means, variance 1) on L21-22, and reassigned to new training and testing variables on L24-25.
- L26-29 imports the `perceptron`, and uses it to make an ML fit to the data. As is typical for AI, we are not given details about just how that is done. The `linear_model` specification on L26 tells the `perceptron` to combine the weights of the input signals linearly, and compare the result to a threshold θ in order to decide whether a neuron should fire or not:

$$\text{if} \qquad \phi(w_1 x_1 + w_2 x_2 + \cdots + w_m x_m) > \theta, \qquad \text{fire.} \qquad (11.32)$$

The fit is made with the `ppn.fit(X_train_std,y_train)` command, and obtains an accuracy 0.909 (“`missclasified`“) on group 1 of the testing data.
- The rest of the program determines the accuracy of the fit, and then outputs the results using Matplotlib and colors to designate the clusters.
- Figure 11.10 left shows the four classes predicted by the `perceptron`. The data are contained well within their clusters, though not perfectly so.

11.6.3 Clustering with Stochastic Gradient Descent

In Section 11.2.4 we incorporated *Stochastic Gradient Descent* (SGD) in our simple network as an optimization technique to minimize the Loss. "Stochastic" refers to the presence of randomness in the iterative search for the minimum, and "gradient descent" to the use of the direction of the gradient of the Loss function as the direction in which to move in the search.

As given in Listing 11.8, we again analyze the dataset of the 14 elementary particles in Table 11.2, but now with *supervised learning*. The Perceptron's clustering of the particles is now based on Mass and Type (Number) as the label. The training data are input and placed in random order, and then shuffled after each training period to avoid cycles. The final output is shown in Figure 11.10 right, where the dashed lines, called *hyperplanes*, are the dividing lines between subspaces. It is seen that the clustering is similar to that found with the previous perceptron (`Perceptron.py`) on the left, but not identical.

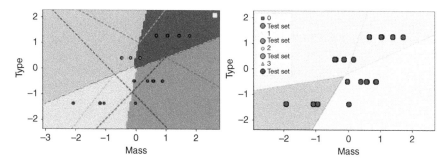

Figure 11.10 *Left*: Clustering of the data from Table 11.2. The Perceptron's clustering of particles based on their Type (Number) and Mass is shown in shades of grey. *Right*: The clusters found with the SGD algorithm.

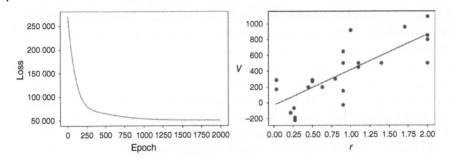

Figure 11.11 *Left*: The decrease in Loss with increasing epochs. *Right*: The linear regression fit to Hubble's data.

11.7 Keras: Python's Deep Learning API

[Keras, 2023] is Python's Application Program Interface (API) for building and training *deep learning* neural nets. As we indicated in Section 11.3, deep learning refers to neural nets with multiple layers of neurons through which data are transferred successively down through the layers. A *layer* is the basic element in a deep neural network; it receives input information, processes it with various activation functions, biases, and weights, and then passes its output on to a lower layer. A *dense* layer is one in which each neuron in the layer receives input from *all* of the neurons in a previous layer. Computationally, the input to a layer is fed the number of neurons (units) in the previous layer, the weights for the neurons' inputs, their activation functions, constraints on the weights, and regularizers to optimize the output. Here's the Keras command to do all that:

```
tf.keras.layers.Dense(
   units, activation = None, use_bias = True, kernel_initializer = "glorot_uniform",
      bias_initializer = "zeros", kernel_regularizer = None, bias_regularizer =
      None, activity_regularizer = None,
   kernel_constraint = None, bias_constraint = None,**kwargs        )
```

Our program `keras.py` in Listing 11.9 again fits a straight line to Hubble's data, now with one dense layer [TechBrij, 2020]. Figure 11.11 left shows its output, where you will notice a rapid decrease in Loss as the training goes through thousands of epochs. Figure 11.11 right shows the final fit to the data. Here's what the variables mean:

- **units**: The dimension of the output vector.
- **activation**: The neurons' activation functions: `sigmoid`, `relu` (rectified linear), `tanh` (hyperbolic tangent), `selu` (scaled exponential linear unit).
- **bias**: the number added in to the neurons' responses.
- **kernel_initializer**: initializes the weights matrix.

11.8 Image Processing with OpenCV

Problem   Separate ripe strawberries from green ones using images of them.

Figure 11.12 *Left*: Ripe strawberries. *Right*: Not so ripe strawberries.

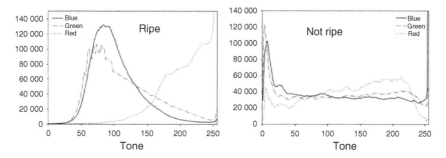

Figure 11.13 *Left*: The 256 tones in each of three colors for ripe strawberries. *Right*: The 256 tones in each of three colors for not quite ripe strawberries.

ML is popular for image recognition and processing, and *OpenCV* is a library of computer vision (CV) programs including modules for machine learning and neural networks. OpenCV is installed from a shell with whichever of these commands work best:

```
pip install opencv-python
pip install -user opencv-contrib-python
```

Computer images are composites of pixels using combinations of Red, Green, and Blue (RGB) colors. Typically, the amount of each color present is represented by a single byte, which permits $2^8 = 256$ levels (leaving off 0). That being the case, there can be $256 \times 256 \times 256 = 16,777,216$ RGB tones. OpenCV analyzes an image and determines how many pixels are present in each of the RGB tones.

To solve our fruit problem, consider Figure 11.12 showing some ripe, and some not so ripe, strawberries. We want to use visual processing and ML to separate strawberries into their different states of ripeness (or coffee beans into different levels of roastings, if you are not a fruit person.) We do it by forming histograms showing the amount of each of the 255 tones present, for each of the three colors. We do this in Figure 11.13, and imagine using the differences in the histograms to separate the fruit. Here is our program that reads in the image `ripe2.jpg` and produces the histograms in Figure 11.13:

```
1  import numpy as np
   import cv2 as cv
   import matplotlib.pyplot as plt
   image = cv.imread("c:/ripe2.jpg")                        # Read image
5  fig, ax =plt.subplots()
   hist = cv.calcHist([image], [0], None, [256], [0,256])   # 3 colors
   ax.plot(hist, color ='b', linestyle='-')
   hist = cv.calcHist([image],[1], None,[256], [0,256])
9  ax.plot(hist, color ='g',linestyle='-.')
   hist = cv.calcHist([image], [2], None, [256], [0,256])
   ax.plot(hist, color ='r', linestyle=':')
   plt.legend(["blue", "green", "red"])
13 plt.title("ripe2")
   plt.xlim([0,256])
   plt.ylim([0,150000])
   plt.show()
```

11.8.1 Background Subtraction

Another type of image processing removes a static background from a video recording, as might be needed in the search for exoplanets or supernovas. This is accomplished by looking at the difference between successive image frames, and removing the parts that do not change. Our program, below, processed an avi file, and produced Figure 11.14, where you will see that, other than the rising smoke and the moving piston, the background has been removed. The program was written by one of the authors (MJP) for a virtual physics course in the Program@udea at La Universidad de Antioquia.

```
   import cv2 as cv
   sub_backg = cv.createBackgroundSubtractorMOG2()
   cap = cv.VideoCapture('c:/vapor.avi')
4  while(1):
       ret, frame = cap.read()
       imgNoBg =  sub_backg.apply(frame)
       #fgmask = fgbg.apply(frame)
8      cv.imshow('frame', frame)
       cv.imshow("no bkgr", imgNoBg)
       k = cv.waitKey(30) & 0xff
       if k == 27:  break
```

Figure 11.14 *Left*: One frame from an animation in which the piston is moving back and forth in front of a background image. *Right*: An image in which the stationary part (background) of the video has been removed.

11.9 Explore ML Data Repositories

Try using some of the tools presented here on real datasets. Find one that interests you, or look here (some of which are used in competitions):

Deep learning physics open data: www.deeplearnphysics.org/DataChallenge/
MLPhysics portal: mlphysics.ics.uci.edu/
Particle tracking challenge: www.kaggle.com/c/trackml-particle-identification
17 datasets for physics: paperswithcode.com/datasets?mod=physics
Carbon nanotubes: www.kaggle.com/inancigdem/carbon-nanotubes
Public datasets – IML – CERN: iml.web.cern.ch/public-datasets
Molecular properties: www.kaggle.com/c/champs-scalar-coupling
Steel defect detection: www.kaggle.com/c/severstal-steel-defect-detection

11.10 Code Listings

Listing 11.1   Neuron.py, An AI neuron.

```
# Neuron.py:  An AI neuron

import numpy as np

def f(x) : return 1./ (1. + np.exp(-x))      # Activation function
class Neuron :
    def __init__(self, weights, bias) :
        self.weights = weights
        self.bias = bias

    def feedforward(self, inputs) :              # Process input
        Sum = np.dot (self.weights, inputs) + self.bias
        return f(Sum)

weights = np.array([-1., 1.])                    # w1 = -1, w2 = 1
bias = 0
n = Neuron(weights,bias)
x = np.array([12,8])                             # x1 = 12, x2 = 8
print(n.feedforward(x))
# output: 0.01798620996209156
```

Listing 11.2   NeuralNet.py A simple AI neural network.

```
# NeuralNet.py:  A simple AI neural network

import numpy as np

def f(x) : return 1./ (1. + np.exp(-x))      # Activation function
class Neuron :
    def __init__(self, weights, bias) :
        self.weights = weights
        self.bias = bias
    def feedforward(self, inputs) :              # Process input
        Sum = np.dot (self.weights, inputs) + self.bias
        return f(Sum)

weights = np.array([-1., 1.])                    # w1 = -1, w2 = 1
bias = 0
n = Neuron(weights,bias)
x = np.array([12,8])                             # x1 = 12, x2=8
```

```
18  print(n.feedforward(x))                  # Output: 0.01798620996209156

    class NeuralNetwork:    # 2-neuron network, 2 hidden layers, 1 output
      def __init__(self):
22        weights = np.array([0,1])
          bias = 0
          self.h1 = Neuron(weights, bias)          # Neuron class as before
          self.h2 = Neuron(weights, bias)
26        self.O  = Neuron(weights, bias)

      def feedforward(self, x):
          out_h1 = self.h1.feedforward(x)
30        out_h2 = self.h2.feedforward(x)
          out_out  = self.O.feedforward(np.array([out_h1, out_h2]))
          return out_out
    network = NeuralNetwork()
34  x = np.array([2, 3])
    print(network.feedforward(x))          #output: 0.7216325609518421
```

Listing 11.3   SimpleNet.py A python code for our simple neural network.

```
1  # SimpleNet.py:   A simple neuron network

   import numpy as np
   def f(x):   return 1/(1 + np.exp(-x))          # Sigmoid activation function
5  def fprime(x):    return np.exp(-x)/(1 + np.exp(-x))**2 # Sigmoid derivs
   def Loss(y_true, y_out):
       los = ((y_true - y_out)**2).mean()
       #print(los)
9      return los

   class SimpleNet:                       # x_1, x_2 in, hidden h1, h2, y_out
     def __init__(self):                                  # Random inits
13       self.w1=np.random.normal()                        # Weights
         self.w2=np.random.normal()
         self.w3=np.random.normal()
         self.w4=np.random.normal()
17       self.w5=np.random.normal()
         self.w6=np.random.normal()
         self.b1=np.random.normal()                        # Biases
         self.b2=np.random.normal()
21       self.b3=np.random.normal()

     def feedfwd(self, x):
         h1  = f(self.w1*x[0]  + self.w2*x[1] + self.b1)
25       h2  = f(self.w3*x[0]  + self.w4*x[1] + self.b2)
         out = f(self.w5*h1    + self.w6*h2   + self.b3)
         return out

29   def train (self, data, all_y_trues):
         learn_rate = 0.1
         N = 1000                                # Number of learning loops
         for n in range(N):
33         for x, y_true in zip(data, all_y_trues):
               sum_h1   = self.w1*x[0] + self.w2*x[1] + self.b1
               h1       = f(sum_h1)
               sum_h2   = self.w3*x[0] + self.w4*x[1] + self.b2
37             h2       = f(sum_h2)
               sum_out  = self.w5*h1 + self.w6*h2 + self.b3
               out      = f(sum_out)
               y_out    = out
41             d_L_d_yout =-2*(y_true-y_out)            # Partial deriv
               d_yout_d_w5 = h1*fprime(sum_out)         # Output neuron
               d_yout_d_w6 = h2*fprime(sum_out)
               d_yout_d_b3 = fprime(sum_out)
45             d_yout_d_h1 = self.w5*fprime(sum_out)
               d_yout_d_h2 = self.w6*fprime(sum_out )
               d_h1_d_w1 = x[0]*fprime(sum_h1)         # Hidden Neuron h1
               d_h1_d_w2 = x[1]*fprime(sum_h1)
```

```
49        d_h1_d_b1 = fprime(sum_h1)
          d_h2_d_w3 = x[0]*fprime(sum_h2)              # Hidden Neuron h2
          d_h2_d_w4 = x[1]*fprime(sum_h2)
          d_h2_d_b2 = fprime(sum_h2)
53        # Update weights and biases
          self.w1 -= learn_rate*d_L_d_yout*d_yout_d_h1*d_h1_d_w1      # h1
          self.w2 -= learn_rate*d_L_d_yout*d_yout_d_h1*d_h1_d_w2
          self.b1 -= learn_rate*d_L_d_yout*d_yout_d_h1*d_h1_d_b1
57        self.w3 -= learn_rate*d_L_d_yout*d_yout_d_h2*d_h2_d_w3      # h2
          self.w4 -= learn_rate*d_L_d_yout*d_yout_d_h2*d_h2_d_w4
          self.b2 -= learn_rate*d_L_d_yout*d_yout_d_h2*d_h2_d_b2
          self.w5 -= learn_rate * d_L_d_yout * d_yout_d_w5           # Out n's
61        self.w6 -= learn_rate * d_L_d_yout * d_yout_d_w6
          self.b3 -= learn_rate * d_L_d_yout * d_yout_d_b3
        if (n%10) == 0:                                # Loss at loop ends
          y_outs = np.apply_along_axis(self.feedfwd,1,data)
65        TotLoss =Loss(all_y_trues,y_outs)
          #print("resta",((all_y_trues- y_outs)**2).mean())
          #print("y_trues",all_y_trues)
          print(" Loop n = %d Loss: %.3f" % (n, TotLoss))
69 data = np.array ([[-2, -1], [25, 6], [17, 4],[-15, -6] ])    # Input Data
   all_y_trues = np.array ([ 1, 0, 0, 1 ])
   network = SimpleNet()                                        # Train net
   network.train(data, all_y_trues)
```

Listing 11.4 KmeansCluster.py Clustering of data with sklearn's Kmeans.

```
# KmeansCluster.py:  Clustering with sklearn's KMeans
2
   from sklearn.cluster import KMeans
   import matplotlib.pyplot as plt
   import numpy as np
6  %matplotlib inline
   X = np.array([
   [1, 0], [2, 0.511], [3, 105.65], [4, 105.6583], [5, 134.98],
   [6, 139.57],[7,139.57],[8,547.86],[9,497.68],[10,493.677],
10 [11,938.2721],[12,939.5654],[13,1115.68],[14,1180.37],
   [15,1197.5],[16,1314.86], [17,1321.71],[18,1672.45]
   ])
   kmeans = KMeans(n_clusters=3, random_state=42)      # 3 random centroids
14 kmeans.fit(X)                                        # Compute clustering
   kmeans.predict(X)                                    # Predict closest cluster
   kmeans.labels_
   cc = kmeans.cluster_centers_                         # Cluster centers
18 print("cc:",cc)
   fig, ax = plt.subplots()
   plt.xlabel("N")
   plt.ylabel("Code")
22 plt.scatter(X[:,0],X[:,1],c=kmeans.labels_ , marker="^")
   plt.scatter(cc[:,0],cc[:,1],c='red', marker="D") #with diamonds
   plt.show()
```

Listing 11.5 Perceptron.py Sklearn's `Perceptron` command creates a perceptron.

```
# Perceptron.py: Creat perceptron with sklearn
2
   import pandas as pd                                  # To read dataset
   import matplotlib.pyplot as plt
   import numpy as np
6  %matplotlib inline

   parts = pd.read_table("C:particle.dat",delim_whitespace=True)
   X = parts["Mass"]                                    # X: masses
10 y = parts['T']                                       # y: Type
   print('Class labels:', np.unique(y))                # The 4 classes
   d = {'col1':X, 'col2':y}                             # d: 2-D array of X & y
```

```
     dfrom sklearn.model_selection import train_test_split        # Split array
14   X_train, X_test, y_train, y_test = train_test_split(
     df, y, test_size=0.3, random_state=1, stratify=y)   # Form 2-D dataframe
     from sklearn.model_selection import train_test_split        # Split array

18   # Shuffle data dataf=pd.DataFrame(d)
     X_train, X_test, y_train, y_test = train_test_split(
     df, y, test_size=0.3, random_state=1, stratify=y)
     from sklearn.preprocessing import StandardScaler
22   sc = StandardScaler()
     sc.fit(X_train)
     X_train_std = sc.transform(X_train)
     X_test_std = sc.transform(X_test)
26   from sklearn.linear_model import Perceptron
     ppn = Perceptron(eta0=0.1, random_state=1)
     ppn.fit(X_train_std,y_train)                # Fit data
     y_pred = ppn.predict(X_test_std)
30   print('Misclassified examples: %d' % (y_test != y_pred).sum())
     from sklearn.metrics import accuracy_score
     print('Accuracy: %.3f' % accuracy_score(y_test, y_pred))
     print('Accuracy: %.3f' % ppn.score(X_test_std, y_test))
34   from matplotlib.colors import ListedColormap
     fig, ax = plt.subplots()
     plt.xlabel("mass")
     plt.ylabel("Type")
38
     for i in range(36): # Plot spin (0, 1, 3/2, 1/2) vs mass
         if y[i] == 0: plt.scatter(X[i],y[i], c='red',marker='x',s=150)
         if y[i] == 1: plt.scatter(X[i],y[i], c='blue',marker="^",s=150)
42       if y[i] == 3: plt.scatter(X[i],y[i], c='brown', marker=">",s=150)
         if y[i] == 2: plt.scatter(X[i],y[i], c='magenta', marker="<",s=150)
     from matplotlib.colors import ListedColormap

46   def plot_decision_regions(X, y, classifier, test_idx=None, resolution=0.01):
         markers = ('s', 'x', 'o', '^', 'v')          # Markers for and color map
         colors = ('brown', 'peachPuff', 'lightgreen', 'gold', 'cyan')
         cmap = ListedColormap(colors[:len(np.unique(y))])
50       x1_min, x1_max = X[:, 0].min() - 1, X[:, 0].max() + 1
         x2_min, x2_max = X[:, 1].min() - 1, X[:, 1].max() + 1
         xx1, xx2 = np.meshgrid(np.arange(x1_min, x1_max, resolution),
                      np.arange(x2_min, x2_max, resolution)) # Decision surface
54       Z = classifier.predict(np.array([xx1.ravel(), xx2.ravel()]).T)
         Z = Z.reshape(xx1.shape)
         plt.contourf(xx1, xx2, Z, alpha=0.3, cmap=cmap) # Alpha: Transp
         plt.xlim(xx1.min(), xx1.max())
58       plt.ylim(xx2.min(), xx2.max())
         for idx, cl in enumerate(np.unique(y)):
             plt.scatter(x=X[y == cl, 0], y=X[y == cl, 1],alpha=0.8,\
                c=colors[idx], marker=markers[idx], label=cl,edgecolor='black')
62           if test_idx:                             # Highlight test examples
                 X_test, y_test = X[test_idx, :], y[test_idx]
                 plt.scatter(X_test[:, 0], X_test[:, 1],edgecolor='black',\
                    alpha=1.0,linewidth=1, marker='o',s=100, label='test set')
66   plt.show()
```

Listing 11.6   Hubble.py A linear fit to Hubble's data using TensorFlow.

```
     # Hubble.py:  Fit to Hubble dat, adapted from Campesato tensorflow 2 primer
     import tensorflow as tf
     import numpy as np
4    import matplotlib.pyplot as plt

     r = tf.Variable([0.032,0.034,0.214,0.263, 0.275, 0.275, 0.45, 0.5, 0.5,\
            0.63,0.8,0.9,0.9,0.9,0.9, 1.0,1.1,1.1,1.4,1.7,2.0,2.0,2.0,2.0]) # R
8    v = tf.Variable([170.,290.,-130.,-70.,-185.,-220.,200.,290.,270.,200.,300.,
            -30.,650.,150.,500.,920.,450.,500.,500.,960. ,500.,850.,800.,1090.])
     m = tf.Variable(0.)     # Init m, b;   y = mx + b
     b = tf.Variable(0. )
12   slope = 500.
```

```
     bias = 0.0
     step = 10
     learning_rate = 0.02
16   steps = 300
     x_train = r
     print(x_train)
     y_train = slope*x_train + bias
20
     def predict_y_value(x):                                    # y(x)
         y = m*x + b
         return y
24   def squared_error(y_pred, y_true):              # Sum squared errors
         return tf.reduce_mean(tf.square(y_pred - y_true))

     loss = squared_error(predict_y_value(x_train), y_train)
28   for i in range(steps):
       with tf.GradientTape() as tape:
         predictions = predict_y_value(x_train)
         loss = squared_error(predictions, y_train)
32     gradients = tape.gradient(loss, [m, b])
       m.assign_sub(gradients[0] * learning_rate)
       b.assign_sub(gradients[1] * learning_rate)
       if(i % step) == 0:
36         print("Step %d, Loss %f, m %f " % (i, loss.numpy(),m))
         y = m*x_train + b
         plt.xlabel("r Mpc")
         plt.ylabel("v km/s")
40         plt.scatter(r, v)
         plt.plot(x_train, y)
       plt.show()
```

Listing 11.7 PandaRead.py A table read with pandas, and cluster ID with kmeans.

```
# PandaRead.py:  Read table with pandas and use kmeans to find clusters

import pandas as pd
4 from sklearn.cluster import KMeans
import matplotlib.pyplot as plt
import numpy as np
parts = pd.read_table("C:\ElemnPart.dat", delim_whitespace = True)
8 data = parts.drop("Name", axis=1)                          # Drop this column
data.head()
X = np.array(data["Number"])                            # 1st column
y = np.array(data['Mass'])                              # 2nd column
12 kmeans = KMeans(n_clusters = 3, random_state = 42)    # Random init clusters
kmeans.fit(data)                                         # Computes clusters
kmeans.predict(data)                               # Predict closest cluster
kmeans.labels_
16 cc = kmeans.cluster_centers_                              # Centroids
print(cc)                                               # Show centroids
fig, ax = plt.subplots()
plt.xlabel("N")
20 plt.ylabel("Code")
plt.scatter(X[:],y[:],c=kmeans.labels_, marker="^")              # Arrows
plt.scatter(cc[:,0],cc[:,1],c='red', marker="D")            # Diamonds
plt.show()
```

Listing 11.8 SGDclass.py Supervised ML classification via a stochastic gradient descent algorithm fit to the Loss function.

```
# SGDclass.py:  ML via Stochastic Gradient Descent

3 from sklearn.linear_model import SGDClassifier              # StochGradDescent
from sklearn.inspection import DecisionBoundaryDisplay        # Def region
import pandas as pd                                      # To read dataset
```

```
   import matplotlib.pyplot as plt
 7 import numpy as np
   %matplotlib inline                              # Set matplot for notebook
   parts = pd.read_table("part.dat",delim_whitespace=True)     # Read data
   X = parts["Mass"]                                           # X: masses
11 y = parts['Type']                               #  Types (integers)
   print('Class labels:', np.unique(y))                        # 4 classes
   d = {'col1':x, 'col2':y}                        # 2 column X,y array
   df = pd.DataFrame(d)                            # Form 2d DataFrame
15 X = np.array(df)                                # DataFrame to numpy array
   idx = np.arange(X.shape[0])                                 # Index 0-35
   np.random.seed(13)
   np.random.shuffle(idx)                          # Random index shuffle
19 X = X[idx]                                          # Random X order
   y = y[idx]                                          # Random Y order
   colors = "bryg"                                     # 4 class colors
   mean = X.mean(axis=0)                                    # Calc mean
23 std  = X.std(axis=0)
   X = (X - mean)/std                                  # Now mean = 0
   print("mean std", mean,std)
   lrgd = SGDClassifier(alpha=0.001, max_iter=100).fit(X,y)
27 print(lrgd)                                     # Alpha: regularization strength
   ax = plt.gca()
   disp = DecisionBoundaryDisplay.from_estimator(lrgd ,X, cmap = plt.cm.Paired, ax =
       ax, response_method = "predict", xlabel = "massMeV/c2",ylabel="Type")
   plt.axis("tight")
31 print("1classes", lrgd.classes_)                            # 4 classes
   for i, color in zip(lrgd.classes_, colors):     # Plot training points
       idx = np.where(y == i)
       print("scatter", X[idx,0], X[idx,1])
35     plt.scatter(X[idx,0],X[idx,1], c=color, cmap=plt.cm.Paired,
                   edgecolor="black",s=20)
   plt.axis("tight")
   xmin, xmax = plt.xlim()
39 ymin, ymax = plt.ylim()
   coef = lrgd.coef_                               # Average weights for all steps
   intercept = lrgd.intercept_

43 def plot_hyperplane(c, color):
       def line(x0):
           return (-(x0 * coef[c,0]) - intercept[c]) / coef[c,1]
       plt.plot([xmin, xmax], [line(xmin), line(xmax)],
47                         ls="--", color=color)
   print(lrgd.classes_)
   for i, color in zip(lrgd.classes_, colors):                 # Plot lines
       print(i,color)
51     plot_hyperplane(i, color)
   plt.legend()
   plt.show()
```

Listing 11.9 Keras.py Linear fir to Hubble data using Keras.

```
 1 # Keras.py: Linear regression fit to Hubble data with Keras

   import matplotlib.pyplot as plt
   import tensorflow as tf
 5 from tensorflow import keras
   from keras import layers
   from keras import Sequential
   from keras.layers import Dense
 9 import numpy as np
                       # Data
   r = [0.032,0.034,0.214,0.263,.275,.275,.45,.5,.5,.63,.8,.9,.9,.9,.9,
       1.0,1.1,1.1,1.4,1.7,2.0,2.0,2.0,2.0]              # Distance Mparse
13 v = [170.,290.,-130.,-70.,-185.,-220.,200.,290.,270.,200.,300.,
       -30.,650.,150.,500.,920.,450.,500.,500.,960.
       ,500.,850.,800.,1090.]  # Recession velocity km/s
   # Create the model: Sequential() only 1 dense layer
17 layer0 = tf.keras.layers.Dense(units=1,input_shape=[1])
```

```
    model = tf.keras.Sequential([layer0])
    model.compile(loss='mean_squared_error',
                  optimizer=tf.keras.optimizers.Adam(1))
21  history = model.fit(r,v,epochs=2000,verbose=0)
    plt.plot(history.history['loss'])
    plt.xlabel("Epochs number")
    plt.ylabel("Loss")
25  plt.show()
    weights = layer0.get_weights()
    weight = weights[0][0]
    bias = weights[1]
29  print('weight: {} bias: {}'.format(weight, bias))
    y_learned = r * weight + bias
    plt.scatter(r, v, c='blue')
    plt.plot(r, y_learned,color='r')
33  plt.show()
    weights = layer0.get_weights()
    weight = weights[0][0]
    bias = weights[1]
37  print('weight: {} bias: {}'.format(weight, bias))
    y_learned = r * weight + bias
    # Output: weight: [448.52048] bias: [-34.726036]
    plt.scatter(r, v, c='blue')
41  plt.plot(r, y_learned,color='r')
    plt.show()
```

12

Quantum Computing (G. He, Coauthor)

> *Although this is our most-recently added chapter, it is by no means the last word on Quantum Computing (QC). Seeing that QC employs its own version of Dirac notation, we start the chapter with the quantum-mechanical version of Dirac notation.[1] We then discuss the foundation of QC, namely, qubits, entanglement, and quantum gates. We introduce quantum programming and execution using the Google Cirq framework, and conclude by using the physical IBM Quantum Computer to solve some realistic problems. You may find that, much like in the early days of traditional computing, the "programs" for QC, using gates to process bits, are at a (painfully) low level. Further material on QC can be found in our sources [Hidary, 2021; Stolze and Suter, 2004; Nielsen and Chuang, 2010; IBMqc, 2023; Cirq, 2023].*

Problem   Develop several computer programs that use quantum mechanical states for storage of information and computation.

12.1   Dirac Notation in Quantum Mechanics

Quantum computing (QC) uses a language based on Dirac's quantum mechanical notation. In Dirac's formalism, a quantum state is represented by a *ket* $|\psi\rangle$, which is a ray in an abstract, infinite, or finite, dimensional, complex Hilbert space. (In QC with n qubits, the dimension would be 2^n.) The familiar wave function $\psi(x)$ is the concrete, coordinate-space representation of that abstract state, and is obtained from it by the inner product:

$$\psi(x) = \langle x|\psi\rangle. \tag{12.1}$$

Here we have formed the product of the *ket* $|\psi\rangle$ with the *bra* $\langle x|$, to form a "bra-ket" (bracket). In this view, the wave function $\psi(x)$, which is the probability amplitude of finding the state $|\psi\rangle$ at x, can be thought of as the projection of $|\psi\rangle$ onto the x basis vectors.

---

1 For more of a review, Section 7.6 uses spin states and matrix operators to calculate the hyperfine structure of hydrogen.

Computational Physics: Problem Solving with Python, Fourth Edition.
Rubin H. Landau, Manuel J. Páez, and Cristian C. Bordeianu.
© 2024 WILEY-VCH GmbH. Published 2024 by WILEY-VCH GmbH.

The Dirac formalism also includes a *dual adjoint* or *covector* space in which the state $|\psi\rangle$ is represented by the bra $\langle\psi|$. The $1:1$ correspondence between kets and bras is expressed with the adjoint operation

$$\langle\psi| = |\psi\rangle^\dagger. \tag{12.2}$$

The *scalar* or *inner product* of the two states $|\phi\rangle$ and $|\psi\rangle$ is given by the bracket

$$\langle\phi|\psi\rangle \equiv (\phi,\psi) = \langle\psi|\phi\rangle^*. \tag{12.3}$$

In contrast, the juxtaposition $O = |\phi\rangle\langle\psi|$ is an *operator*, and not a simple scalar product, since it changes one state into another:

$$O|\psi\rangle = |\phi\rangle = |O\psi\rangle. \tag{12.4}$$

Here we use O to denote an operator, assume $\psi\rangle$ is normalized, and note that operators are often represented as matrices, and that, in general, $O|\psi\rangle$ is *not* proportional to $|\psi\rangle$.

If we want to be consistent with Dirac notation, then a description of states as vectors in a spin $1/2$ space S would express the states as $\langle S|\psi\rangle$. In common practice the bra $\langle S|$ is left off, and the up and down spin $1/2$ states are written as

$$\psi_+ = \left|+\tfrac{1}{2}\right\rangle = \begin{bmatrix} 1 \\ 0 \end{bmatrix} \equiv |0\rangle, \tag{12.5}$$

$$\psi_- = \left|-\tfrac{1}{2}\right\rangle = \begin{bmatrix} 0 \\ 1 \end{bmatrix} \equiv |1\rangle, \tag{12.6}$$

where $|0\rangle$ and $|1\rangle$ are the symbols used in QC (presumably as an analogy to the traditional bits 0 and 1 being represented as spin up and spin down). Likewise, operators like O are represented by 2×2 matrices, for example by the *direct product*:

$$\left|\tfrac{1}{2}\right\rangle\left\langle\tfrac{1}{2}\right| = \begin{bmatrix} 1 \\ 0 \end{bmatrix} \begin{bmatrix} 1 & 0 \end{bmatrix} = \begin{bmatrix} 1 & 0 \\ 0 & 0 \end{bmatrix}. \tag{12.7}$$

12.2 From Bits to Qubits

Quantum computing (QC) is based on storing information in quantum mechanical states, and then manipulating these states to perform numerical operations. This is fundamentally different from, and potentially more powerful than, the traditional approach to computing, and may be especially applicable in areas such as cryptography and simulation of quantum systems.

In the traditional approach to a computer's memory, information is stored using a number system based on the binary integers (*bits*) 0 and 1. [Originally, a 0 was stored in a magnetic core pointing up, and a 1 in a core pointing down, like the qubits in (12.5).] All the rest of what gets stored consists of arrays of these bits. In the (simplest) quantum computer memory system, information is stored in states that are combinations of elementary spin-like states, called *quantum bits* or *qubits*. Although it would be a major advance in miniaturization if the energy levels of neutral atoms were used for qubits, in practice the storage uses electronic devices, such as superconducting AC Josephson junctions.

The smallest unit of information in QC is the quantum bit or *qubit*. A single qubit storage unit is expressed in terms of the same bases vectors used for "spin-up" and "spin-down"

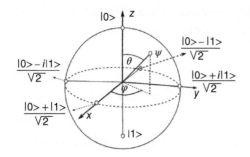

Figure 12.1 The Bloch sphere, a geometric representation of a two-level quantum system (modified www.pngwing.com).

quantum states, but with the names changed to 0 and 1:

$$|0\rangle \overset{\text{def}}{=} \left|+\tfrac{1}{2}\right\rangle = \begin{bmatrix} 1 \\ 0 \end{bmatrix}, \qquad |1\rangle \overset{\text{def}}{=} \left|-\tfrac{1}{2}\right\rangle = \begin{bmatrix} 0 \\ 1 \end{bmatrix}. \tag{12.8}$$

A qubit is defined as a linear combination of these two basis states:

$$|\psi\rangle = u\,|0\rangle + v\,|1\rangle \equiv \begin{bmatrix} u \\ v \end{bmatrix}. \tag{12.9}$$

Here u and v complex numbers satisfying the normalization condition:

$$|u|^2 + |v|^2 = 1. \tag{12.10}$$

Although probability conservation is important in quantum mechanics, the normalization of states is often just an arbitrary overall constant applied to the wave function, when needed. In QC, however, states are uniformly assumed to be normalized.

As illustrated in Figure 12.1, because the state $|\psi\rangle$ is a ray in an abstract vector space, it can have a concrete representation as the direction of a ray on a *Bloch sphere* with the polar angle representation:

$$|\psi\rangle = \cos\tfrac{\theta}{2}\,|0\rangle + e^{i\phi}\sin\tfrac{\theta}{2}\,|1\rangle, \quad \theta \in [0, \pi], \quad \phi \in [0, 2\pi). \tag{12.11}$$

Accordingly, a pure $|0\rangle$ state ($\theta = 0$) lies on the $+z$ axis, and a pure $|1\rangle$ state lies along the $-z$ axis ($\theta = \pi$), with complex combinations of the two lying someplace on the surface of the sphere.

12.2.1 Multiple Qubit States

Consider a state $|\psi_A\rangle$ within the Hilbert space H_A, and a separate state $|\psi_B\rangle$ within the Hilbert space H_B. If we wish to combine these two kets into a single state, then the composite would exist within an expanded Hilbert space created by the tensor product of H_A and H_B, with the state vector also a direct product:

$$H_{AB} = H_A \otimes H_B, \tag{12.12}$$

$$\Rightarrow |\psi_{AB}\rangle = |\psi_A\rangle \otimes |\psi_B\rangle, \tag{12.13}$$

$$\text{where} \quad \begin{bmatrix} a \\ b \end{bmatrix} \otimes \begin{bmatrix} c \\ d \end{bmatrix} \overset{\text{def}}{=} \begin{bmatrix} ac \\ ad \\ bc \\ bd \end{bmatrix}. \tag{12.14}$$

As an explicit example, if we start with the states

$$|\psi_A\rangle = u_1 |0\rangle + v_1 |1\rangle, \qquad |\psi_B\rangle = u_2 |0\rangle + v_2 |1\rangle, \quad \text{then} \tag{12.15}$$

$$|\psi_A\rangle \otimes |\psi_B\rangle \equiv |\psi_A\rangle |\psi_B\rangle = \left(u_1 |0\rangle + v_1 |1\rangle\right)\left(u_2 |0\rangle + v_2 |1\rangle\right) \tag{12.16}$$

$$= u_1 u_2 |00\rangle + u_1 v_2 |01\rangle + v_1 u_2 |10\rangle + v_1 v_2 |11\rangle, \tag{12.17}$$

$$\text{where} \quad |00\rangle = \begin{bmatrix} 1 \\ 0 \\ 0 \\ 0 \end{bmatrix}, \quad |01\rangle = \begin{bmatrix} 0 \\ 1 \\ 0 \\ 0 \end{bmatrix}, \quad |10\rangle = \begin{bmatrix} 0 \\ 0 \\ 1 \\ 0 \end{bmatrix}, \quad |11\rangle = \begin{bmatrix} 0 \\ 0 \\ 0 \\ 1 \end{bmatrix}. \tag{12.18}$$

The 4-D vectors in (12.18) are the appropriate basis vectors for a two-qubit system.

12.3 Entangled and Separable States

States formed with a direct product, such as in (12.16), are called *separable*. For example, a qubit in a $|0\rangle$ state and a different qubit also in a $|0\rangle$ state form the separable state $|0_A\rangle \otimes |0_B\rangle$, which is usually written as just $|00\rangle$. Yet qubits do not live on isolated qubit islands, so they can interact with each other. If two interacting systems are otherwise isolated, but cannot be expressed as the direct product of the two states, these qubits are *entangled*. If the two systems are not entangled, then they are *separable*.

Entanglement may lead to some profound consequences. For example, the spin-up state $|0\rangle$ and the spin-down state $|1\rangle$ can be physically far from each other, yet still be entangled. This means that the up state cannot be described as just a single particle state, but must be described as part of the full state vector including the down state, wherever it may be. So if the total state has spin zero, and one particle is spin up, then the other particle, even if far away, must be correlated and must have spin down. Just how the two particles communicate with each other quantum mechanically at a macroscopic distances is a subject of current discussion and debate (and the 2022 Nobel Prize).

Let's be more explicit about this entanglement concept. Here are two states in the 2D Hilbert space of complex numbers $\mathbb{C}^2$,

$$|\psi_A\rangle = \begin{bmatrix} a \\ b \end{bmatrix}, \qquad |\psi_B\rangle = \begin{bmatrix} c \\ d \end{bmatrix}. \tag{12.19}$$

The *tensor* or *direct product* of these states is

$$|\Psi\rangle \stackrel{\text{def}}{=} |\psi_A\rangle \otimes |\psi_B\rangle = \begin{bmatrix} a \\ b \end{bmatrix} \otimes \begin{bmatrix} c \\ d \end{bmatrix} = \begin{bmatrix} ac \\ ad \\ bc \\ bd \end{bmatrix}. \tag{12.20}$$

This product state is in the 4-D Hilbert space of complex numbers $\mathbb{C}^4$:

$$|\psi\rangle = \begin{bmatrix} w \\ x \\ y \\ z \end{bmatrix}. \tag{12.21}$$

The state is separable, if, and only if, wz = xy. For the product state in (12.20), separability thus requires $acbd = adbc$; which is in fact the case, and so (12.20) is separable. A famous example of entanglement is the two-qubit *Bell States*:

$$|\beta_{00}\rangle = \frac{1}{\sqrt{2}}(|00\rangle + |11\rangle), \qquad |\beta_{01}\rangle = \frac{1}{\sqrt{2}}(|01\rangle + |10\rangle), \tag{12.22}$$

$$|\beta_{10}\rangle = \frac{1}{\sqrt{2}}(|00\rangle - |11\rangle), \qquad |\beta_{11}\rangle = \frac{1}{\sqrt{2}}(|01\rangle - |10\rangle). \tag{12.23}$$

Use the definition of separability and the basis vectors (12.18) to prove that the Bell states are entangled.

A powerful way to describe the quantum state of a system is in terms of the *density matrix* ρ. It can be used to calculate observables without resorting to wave functions, and is particularly useful when dealing with an ensemble of pure states. The density matrix or operator is defined as

$$\rho = \sum_i p_i |\psi_i\rangle \langle \psi_i|. \tag{12.24}$$

Here p_i is the probability of the pure state $|\psi_i\rangle$ being present in the ensemble, and, just to remind you, the product of a ket times a bra is an operator. A system consisting of just a pure state would have $p_i = 1$.

12.3.1 Physics Exercise: Two Entangled Dipoles

Two interacting magnetic dipoles σ_A and σ_B, separated by a distance r, have the interaction Hamiltonian:

$$H = \frac{\mu^2}{r^3}\left(\sigma_A \cdot \sigma_B - 3\sigma_A \cdot \hat{r}\,\sigma_B \cdot \hat{r}\right), \tag{12.25}$$

$$\sigma_A = X_A\hat{\mathbf{i}} + Y_A\hat{\mathbf{j}} + Z_A\hat{\mathbf{k}}, \qquad \sigma_B = X_B\hat{\mathbf{i}} + Y_B\hat{\mathbf{j}} + Z_B\hat{\mathbf{k}}. \tag{12.26}$$

Here we employ **QC notation** that labels the Pauli matrices as X, Y, and Z:

$$X \stackrel{\text{def}}{=} \sigma_x = \begin{bmatrix} 0 & 1 \\ 1 & 0 \end{bmatrix}, \quad Y \stackrel{\text{def}}{=} \sigma_y = \begin{bmatrix} 0 & -i \\ i & 0 \end{bmatrix}, \quad Z \stackrel{\text{def}}{=} \sigma_z = \begin{bmatrix} 1 & 0 \\ 0 & -1 \end{bmatrix}. \tag{12.27}$$

And in yet more QC notation, the two dipoles can be in the four, direct product states:

$$|0_A0_B\rangle = |0_A\rangle|0_B\rangle, \qquad |0_A1_B\rangle = |0_A\rangle|1_B\rangle, \tag{12.28}$$

$$|1_A0_B\rangle = |1_A\rangle|0_B\rangle, \qquad |1_A1_B\rangle = |1_A\rangle|1_B\rangle. \tag{12.29}$$

As discussed in Section 12.1, the Pauli 4×4 matrices (12.27) are operators that transform states. As we shall see, in QC they represent Boolean *logic gates* with the properties:

$$X|0\rangle = |1\rangle, \quad X|1\rangle = +|0\rangle, \quad Y|0\rangle = i|1\rangle, \quad Y|1\rangle = -i|0\rangle, \tag{12.30}$$

$$Z|0\rangle = |0\rangle, \quad Z|1\rangle = -|1\rangle. \tag{12.31}$$

Exercises

1) Show that these direct product states form a basis for $\mathbb{C}^4$:

$$|00\rangle = \begin{bmatrix} 1 \\ 0 \end{bmatrix} \otimes \begin{bmatrix} 1 \\ 0 \end{bmatrix}, \quad |01\rangle = \begin{bmatrix} 1 \\ 0 \end{bmatrix} \otimes \begin{bmatrix} 0 \\ 1 \end{bmatrix}, \tag{12.32}$$

$$|10\rangle = \begin{bmatrix} 0 \\ 1 \end{bmatrix} \otimes \begin{bmatrix} 1 \\ 0 \end{bmatrix}, \quad |11\rangle = \begin{bmatrix} 0 \\ 1 \end{bmatrix} \otimes \begin{bmatrix} 0 \\ 1 \end{bmatrix}. \tag{12.33}$$

Hint: $|11\rangle = \begin{bmatrix} 0 \\ 0 \\ 0 \\ 1 \end{bmatrix}.$ \hfill (12.34)

2) Consider P and Q as the operators in separate Hilbert spaces,

$$P = \begin{bmatrix} p_{11} & p_{12} \\ p_{21} & p_{22} \end{bmatrix}, \quad Q = \begin{bmatrix} q_{11} & q_{12} \\ q_{21} & q_{22} \end{bmatrix}. \tag{12.35}$$

Show that their direct product is

$$P \otimes Q = \begin{bmatrix} p_{11}q_{11} & p_{11}q_{12} & p_{12}q_{11} & p_{12}q_{12} \\ p_{11}q_{21} & p_{11}q_{22} & p_{12}q_{21} & p_{12}q_{22} \\ p_{21}q_{11} & p_{21}q_{12} & p_{22}q_{11} & p_{22}q_{12} \\ p_{21}q_{21} & p_{21}q_{22} & p_{22}q_{21} & p_{22}q_{22} \end{bmatrix}. \tag{12.36}$$

3) Show that for $\hat{r} = \hat{k}$, the Hamiltonian (12.25) in direct product space is

$$H = \frac{\mu^2}{r^3} \left(X_A \otimes X_B + Y_A \otimes Y_B + Z_A \otimes Z_B - 3Z_A \otimes Z_B \right). \tag{12.37}$$

4) Show that the direct product:

$$X_A \otimes X_B = \begin{bmatrix} 0 & 0 & 0 & 1 \\ 0 & 0 & 1 & 0 \\ 0 & 1 & 0 & 0 \\ 1 & 0 & 0 & 0 \end{bmatrix}. \tag{12.38}$$

5) Evaluate the direct products $Y_A \otimes Y_B$ and $Z_A \otimes Z_B$ as 4×4 matrices, and thereby show that the Hamiltonian in the direct product space is:

$$H = \frac{\mu^2}{r^3} \begin{bmatrix} -2 & 0 & 0 & 0 \\ 0 & 2 & 2 & 0 \\ 0 & 2 & 2 & 0 \\ 0 & 0 & 0 & -2 \end{bmatrix}. \tag{12.39}$$

6) Use a linear algebra package to show that the eigenvalues of $H/(\mu^2/r^3)$ are 4, 0, -2, and -2, and that the corresponding eigenvectors are:

$$\phi_1 = \frac{1}{\sqrt{2}} \begin{bmatrix} 0 \\ +1 \\ +1 \\ 0 \end{bmatrix} = \frac{|01\rangle + |10\rangle}{\sqrt{2}}, \quad \phi_2 = \begin{bmatrix} 0 \\ 0 \\ 0 \\ 1 \end{bmatrix} = |11\rangle, \tag{12.40}$$

$$\phi_4 = \frac{1}{\sqrt{2}} \begin{bmatrix} 0 \\ 1 \\ -1 \\ 0 \end{bmatrix} = \frac{|01\rangle - |10\rangle}{\sqrt{2}}, \qquad \phi_3 = \begin{bmatrix} 1 \\ 0 \\ 0 \\ 0 \end{bmatrix} = |00\rangle. \tag{12.41}$$

7) Recall the discussion of entanglement. Of the four eigenstates just obtained, determine which ones are separable and which ones are entangled.

8) Use these eigenvectors states as basis states to evaluate the Hamiltonian matrix:

$$H = \begin{bmatrix} \langle\phi_1|H|\phi_1\rangle & \langle\phi_1|H|\phi_2\rangle & \langle\phi_1|H|\phi_3\rangle & \langle\phi_1|H|\phi_4\rangle \\ \langle\phi_2|H|\phi_1\rangle & \langle\phi_2|H|\phi_2\rangle & \langle\phi_2|H|\phi_3\rangle & \langle\phi_2|H|\phi_4\rangle \\ \langle\phi_3|H|\phi_1\rangle & \langle\phi_3|H|\phi_2\rangle & \langle\phi_3|H|\phi_3\rangle & \langle\phi_3|H|\phi_4\rangle \\ \langle\phi_4|H|\phi_1\rangle & \langle\phi_4|H|\phi_2\rangle & \langle\phi_4|H|\phi_3\rangle & \langle\phi_4|H|\phi_4\rangle \end{bmatrix}. \tag{12.42}$$

If you have done this correctly, the Hamiltonian should now be diagonal with the eigenvalues as the diagonal elements.

In Listing 12.1, we present the program `Entangle.py` that performs the necessary linear algebra using the `numpy` package. It produces the results:

```
    Hamiltonian without mu^2/r^3 factor
    [[-2   0   0   0]
     [ 0   2   2   0]
4    [ 0   2   2   0]
     [ 0   0   0  -2]]
    Eigenvalues
     [  4.0000000e+00   4.4408921e-16  -2.0000000e+00  -2.0000000e+00]
8   Eigenvectors(incolumns)
     [[ 0.          0.          1.          0.         ]
     [  0.70710678  0.70710678  0.          0.         ]
     [  0.70710678 -0.70710678  0.          0.         ]
12   [  0.          0.          0.          1.         ]]
    Hamiltonian in Eigenvector Basis
    [[  4.00000000e+00   0.00000000e+00   0.00000000e+00   6.66133815e-16]
     [  0.00000000e+00  -2.00000000e+00   0.00000000e+00   0.00000000e+00]
16   [  0.00000000e+00   0.00000000e+00  -2.00000000e+00   0.00000000e+00]
     [  6.28036983e-16   0.00000000e+00   0.00000000e+00   9.86076132e-32]]
```

12.4 Logic Gates

Recall that traditional computers use electronic circuits called *logic gates* to perform basic, logical operations on bits. More complex operations are created by combining multiple gates. There are six basic logic gates:

AND, NAND, NOT, OR, NOR, XOR (exclusive OR).

For example, here are the symbols that represent the AND and XOR gates, as well as the *truth tables* that define their outputs according to their inputs:

As an instance of how these gates can be combined, here we construct a *half-adder* from xor and and gates:

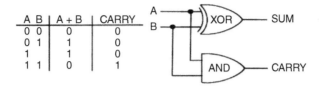

A B	A + B	CARRY
0 0	0	0
0 1	1	0
1 0	1	0
1 1	0	1

The half-adder adds two bits, and if the answer is greater than 1, it carries over a bit to a higher memory position.

12.4.1 1-Qubit Gates

In similarity with classical computers, quantum computers employ *quantum logic gates* to perform elementary operations on qubits. These gates are represented in Hilbert space as *unitary operators*, $U |\psi_{in}\rangle = |\psi_{out}\rangle$, which means they preserve probability. These are the gates:

State gate U: $|\psi_{in}\rangle$—$\boxed{U}$—$|\psi_{out}\rangle$ Determines the state of a ket.

Pauli matrix gates: Our old friends the *Pauli spin matrices* given in (12.27) are used as QC gates, where they are renamed as $X = \sigma_x$, $Y = \sigma_y$, and $Z = \sigma_z$.

NOT gate X: The quantum NOT gate flips $|0\rangle$ (formerly spin up) to $|1\rangle$ (formerly spin down), and vise versa. Used as a gate, the Pauli matrix X acts as the NOT operator $\overset{\text{NOT}}{—\oplus—}$, changing one state into another:

$$X \overset{\text{def}}{=} \sigma_x = |0\rangle\langle 1| + |1\rangle\langle 0| = \begin{bmatrix} 0 & 1 \\ 1 & 0 \end{bmatrix}, \tag{12.43}$$

$$\Rightarrow X |0\rangle = |1\rangle, \quad X |1\rangle = |0\rangle. \tag{12.44}$$

Y gate: The Pauli matrix σ_y acts as the Y gate:

$$Y \overset{\text{def}}{=} \sigma_y = i(|1\rangle\langle 0| - |0\rangle\langle 1|) = \begin{bmatrix} 0 & -i \\ i & 0 \end{bmatrix}. \tag{12.45}$$

Z gate: The Pauli matrix σ_z acts as the Z gate. It flips the sign of the $|1\rangle$ state, but leaves the $|0\rangle$ state unchanged:

$$Z \overset{\text{def}}{=} \sigma_z = |0\rangle\langle 0| - |1\rangle\langle 1| = \begin{bmatrix} 1 & 0 \\ 0 & -1 \end{bmatrix}, \tag{12.46}$$

$$Z |z\rangle = (-1)^z |z\rangle. \tag{12.47}$$

As with the quantum spin, the states $|0\rangle$ and $|1\rangle$ are the eigenstates of Z.

Hadamard gate H: converts qubits that are eigenstates of Z to ones that are eigenstates of X:

$$H |0\rangle = \frac{1}{\sqrt{2}}(|0\rangle + |1\rangle) \equiv |+\rangle, \quad H |1\rangle = \frac{1}{\sqrt{2}}(|0\rangle - |1\rangle) \equiv |-\rangle, \tag{12.48}$$

$$H = (|+\rangle\langle 0| + |-\rangle\langle 1|) = \frac{1}{\sqrt{2}} \begin{bmatrix} 1 & 1 \\ 1 & -1 \end{bmatrix}. \tag{12.49}$$

The H gate also creates equal mixtures of the $|0\rangle$ and $|1\rangle$ basis states, and thus is useful in transforming clustered qubits into states with uniform superpositions:

$$H|0\rangle = \frac{1}{\sqrt{2}}(|0\rangle + |1\rangle), \quad H|1\rangle = \frac{1}{\sqrt{2}}(|0\rangle - |1\rangle). \tag{12.50}$$

$\mathbf{R_\varphi}$: ——$\boxed{R_\varphi}$——: also called the P or phase gate, rotates $|1\rangle$ by an angle φ about the z-axis, while leaving $|0\rangle$ untouched:

$$R_\varphi|0\rangle = |0\rangle, \quad R_\varphi|1\rangle = e^{i\varphi}|1\rangle, \quad R_\varphi = \begin{bmatrix} 1 & 0 \\ 0 & e^{i\varphi} \end{bmatrix}. \tag{12.51}$$

S and T gates: Are special cases of the R_φ gate. S rotates a ket by $\varphi = \pi/2$ ($e^{i\varphi} = i$), and T by $\phi = \pi/4$:

$$\boxed{S} \qquad S|0\rangle = |0\rangle, \quad S|1\rangle = i|1\rangle, \tag{12.52}$$

$$\boxed{T} \qquad T|0\rangle = |0\rangle, \quad T|1\rangle = e^{i\pi/4}|1\rangle, \tag{12.53}$$

$$T = \begin{bmatrix} 1 & 0 \\ 0 & \exp(i\pi/4) \end{bmatrix}, \quad S = T^2 = \begin{bmatrix} 1 & 0 \\ 0 & i \end{bmatrix}. \tag{12.54}$$

$\mathbf{R_x, R_y, R_z}$: These gates perform a general rotation of qubits on the Bloch sphere by the angle α around the x, y, or z axes, respectively:

$$R_x(\alpha) = e^{-i\alpha\sigma_x/2}, \quad R_y(\alpha) = e^{-i\alpha\sigma_y/2}, \quad R_z(\alpha) = e^{-i\alpha\sigma_z/2}. \tag{12.55}$$

Measurement operation: $|\psi\rangle$——$\boxed{\measuredangle}$== While not a gate because it's not unitary, the classical measurement of a quantum state is an oft-used operation.

Exercise Use the Bloch sphere of Figure 12.1 and equation (12.11) to determine:

- What gate transforms $|0\rangle$ to $(|0\rangle + |1\rangle)/\sqrt{2}$?
- What gate rotates $(|0\rangle + |1\rangle)/\sqrt{2}$ by $\pi/2$ into $(|0\rangle + i|1\rangle)/\sqrt{2}$?
- What gate rotates $|0\rangle$ by π into $|1\rangle$?

12.4.2 2-Qubit Gates

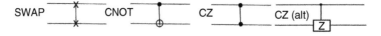

There is a whole set of two-qubit gates with the property that their combined actions can approximate a 4×4 unitary matrix to arbitrary precision. In addition to the Hadamard gate, already defined for one-qubit use, two-qubit gates include SWAP, CNOT, and CZ:

SWAP: Transforms $|01\rangle$ to $|10\rangle$, that is, it swaps the two kets:

$$\text{SWAP}\,|01\rangle = \begin{bmatrix} 1 & 0 & 0 & 0 \\ 0 & 0 & 1 & 0 \\ 0 & 1 & 0 & 0 \\ 0 & 0 & 0 & 1 \end{bmatrix} \begin{bmatrix} 0 \\ 1 \\ 0 \\ 0 \end{bmatrix} = \begin{bmatrix} 0 \\ 0 \\ 1 \\ 0 \end{bmatrix} = |10\rangle. \tag{12.56}$$

In turn, SWAP transforms $|10\rangle$ to $|01\rangle$.

Controlled gates: Controlled gates act on 2-qubit states $|c\rangle\,|t\rangle$, where $|c\rangle$ is the *control* bit and $|t\rangle$ is the *target*. If a 1 qubit gate U is used in a 2 qubit controlled gate combination CU, it has the following effect:

$$CU\,|c\rangle\,|t\rangle = |c\rangle\,U^c\,|t\rangle, \tag{12.57}$$

where U^c is a version of U modified by the control.

Controlled NOT, CNOT: ⊥ uses one qubit to control its action on the target qubit. If the *control* qubit is 0, then the target is unchanged; if the control qubit is 1, then the *target* qubit is flipped. With the left qubit as control:

$$\mathrm{CNOT}(x,y) = \begin{cases} (x,y) \text{ if } x = 0 \\ (x, 1-y) \text{ if } x = 1, \end{cases} \tag{12.58}$$

$$\mathrm{CNOT}\,|10\rangle = \begin{bmatrix} 1&0&0&0 \\ 0&1&0&0 \\ 0&0&0&1 \\ 0&0&1&0 \end{bmatrix}\begin{bmatrix}0\\0\\1\\0\end{bmatrix} = \begin{bmatrix}0\\0\\0\\1\end{bmatrix} = |11\rangle. \tag{12.59}$$

Controlled Z, CZ: The controlled Z gate reverses the sign of the $|11\rangle$ qubit:

$$CZ\,|00\rangle = |00\rangle,\ CZ\,|01\rangle = |01\rangle,\ CZ\,|10\rangle = |10\rangle,\ CZ\,|11\rangle = -|11\rangle, \tag{12.60}$$

$$CZ = \begin{bmatrix} 1&0&0&0 \\ 0&1&0&0 \\ 0&0&1&0 \\ 0&0&0&-1 \end{bmatrix}. \tag{12.61}$$

Exercise Determine the effect of the CNOT gate on: $|10\rangle$, $|01\rangle$, $|00\rangle$, and $|11\rangle$.

Exercise Verify the above effects of the CZ gate.

12.4.3 Entanglement via Gates

The entangled Bell states (12.22) can be created with the aforementioned gates. For example, Figure 12.2 shows a quantum circuit creating the Bell state $|\beta_{00}\rangle$ by employing an H (Hadamard) gate followed by a CNOT gate. Here we start with the two qubit state $|00\rangle$, and use H on the first qubit, leaving the second qubit unchanged. Then the CNOT gate, which uses the first qubit for control, and the second as the target:

$$(1)\ H\,|0\rangle\,|0\rangle = \frac{1}{\sqrt{2}}\,(|0\rangle + |1\rangle)\,|0\rangle, \tag{12.62}$$

$$(2)\ CX\,\frac{1}{\sqrt{2}}\,(|0\rangle + |1\rangle)\,|0\rangle = \frac{1}{\sqrt{2}}\,(|0\rangle\,|0\rangle + |1\rangle\,|1\rangle). \tag{12.63}$$

Figure 12.2 A quantum circuit for creating an entangled state $|\beta_{00}\rangle$.

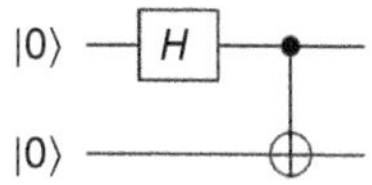

12.4.4 3-Qubit Gates

Three-qubit states use basis vectors created by the direct products of three kets:

$$|ijk\rangle = |i\rangle\,|j\rangle\,|k\rangle \equiv |i\rangle \otimes |j\rangle \otimes |k\rangle. \tag{12.64}$$

There are, accordingly, eight such states, $|000\rangle$, $|001\rangle$, $|010\rangle$, $|011\rangle$, $|100\rangle$, $|101\rangle$, $|110\rangle$, $|111\rangle$. For example,

$$|111\rangle = \begin{bmatrix} 0 \\ 0 \\ 0 \\ 0 \\ 0 \\ 0 \\ 0 \\ 1 \end{bmatrix}, \quad |110\rangle = \begin{bmatrix} 0 \\ 0 \\ 0 \\ 0 \\ 0 \\ 0 \\ 1 \\ 0 \end{bmatrix}. \tag{12.65}$$

This means that the three-qubit gates that operate on these 8-D basis vectors are represented by 8×8 matrices.

TOFFOLI, CCNOT GATE: is an extension of the CNOT gate that flips the third qubit, if and only if the two first two qubits are in the $|011\rangle$ state:

$$T\,|110\rangle = \begin{bmatrix} 1 & 0 & 0 & 0 & 0 & 0 & 0 & 0 \\ 0 & 1 & 0 & 0 & 0 & 0 & 0 & 0 \\ 0 & 0 & 1 & 0 & 0 & 0 & 0 & 0 \\ 0 & 0 & 0 & 1 & 0 & 0 & 0 & 0 \\ 0 & 0 & 0 & 0 & 1 & 0 & 0 & 0 \\ 0 & 0 & 0 & 0 & 0 & 1 & 0 & 0 \\ 0 & 0 & 0 & 0 & 0 & 0 & 0 & 1 \\ 0 & 0 & 0 & 0 & 0 & 0 & 1 & 0 \end{bmatrix} \begin{bmatrix} 0 \\ 0 \\ 0 \\ 0 \\ 0 \\ 0 \\ 1 \\ 0 \end{bmatrix} = \begin{bmatrix} 0 \\ 0 \\ 0 \\ 0 \\ 0 \\ 0 \\ 0 \\ 1 \end{bmatrix} = |111\rangle \tag{12.66}$$

12.5 An Intro to QC Programming

Now that we have the building blocks for QC, namely, qubits and gates, it's time to put them together into programs, aka *circuits*. We start by using Google's 2018 release of *Cirq*, a "Python software library for writing, manipulating and optimizing quantum circuits, then running them on quantum computers and quantum simulators" [Cirq, 2023]. After the simple examples in this section implementing the various gates, we will use the powerful *IBM Quantum Computer* for some more advanced programming and running on a physical quantum computer.

To install Cirq using Anaconda, we recommend using the *package manager conda*, which helps to install all the associated bits and pieces of packages. To do that, use a shell (the Command shell or PowerShell on Windows, or the Terminal on Macs) and issue the command:

```
conda install -c psi4 cirq
```

Alternatively, you can use `pip` to install cirq:

```
python -m pip install -upgrade, pip python -m pip install cirq
```

- **Hadamard gate:**

  Recall, an H gate converts eigenstates of Z to eigenstates of X. Enter and run your first Cirq program to create an H gate that operates on $|0\rangle$:

```
# Hadamard.py: Cirq program to create H gate

 3  import cirq                               # Import Cirq

    circuit = cirq.Circuit()                  # Build circuit
    qubit = cirq.GridQubit(0,0)               # Create qubit at (0,0)
 7  circuit.append(cirq.H(qubit))             # Append Hadamard gate
    s = cirq.Simulator()                      # Initialize simulator
    print('Simulate the circuit:')
    print(circuit)                            # Output circuit
11  results = s.simulate(circuit)             # Run simulator
    print(results)                            # Output resulting kets
    ───────────Cirque's Output ------------------------------------
    Simulate the circuit:
15  (0,0):              H
    output vector:              0.707 |0> + 0.707|1>
```

Not only was that pretty easy, but it also gave the correct answer:

$$H\,|0\rangle = \frac{1}{\sqrt{2}}\begin{bmatrix}1 & 1\\ 1 & -1\end{bmatrix}\begin{bmatrix}1\\ 0\end{bmatrix} = \frac{1}{\sqrt{2}}\begin{bmatrix}1\\ 1\end{bmatrix} \equiv 0.707\begin{bmatrix}1\\ 0\end{bmatrix} + 0.707\begin{bmatrix}0\\ 1\end{bmatrix}. \qquad (12.67)$$

- **Two Hadamard gates:**

  The application of two H gates to a state should act as the identity operator. Extend the previous program to append a second Hadamard gate, now using the command `a = cirq.QubitNamed("a")` to define a qubit:

```
# TwoHgates.py: Cirq program to create 2 H gates on one line

    import cirq
 4
    circuit = cirq.Circuit()                  # Build circuit
    a = cirq.NamedQubit('a')                  # Define named qubit
    circuit.append(cirq.H(a))                 # Append H gate to a
 8  circuit.append(cirq.H(a))                 # Append another H gate to a
    s = cirq.Simulator()                      # Initialize simulator
    print('Simulate the circuit:')
    print(circuit)                            # Output circuit
12  results = s.simulate(circuit)             # Run simulator
    print(results)                            # Output resulting kets
    ───────────Cirque's Output-------------------
    Simulate the circuit:
16                      a:          H          H
    output vector:                  |0>
```

Exercise Use the output of `Hadamard.py` to check that `TwoHgates.py` acts as the identity operator.

- **X and H gates:**

  Write a program that uses an X gate to convert $|0\rangle$ to $|1\rangle$, and then an H gate to create an eigenstate of X:

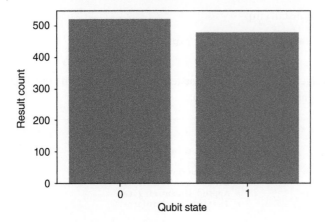

Figure 12.3 Histogram of the state formed by application of X, Z, and H gates. (If there was no noise, the heights would be equal.)

```
# XplusH.py: Cirque program, setup X & H gates and |1>

3 import cirq

  circuit = cirq.Circuit()             # Build circuit
  a = cirq.NamedQubit('a')             # Define qubit
7 circuit.append(cirq.X(a))            # Append X (NOT) gate
  circuit.append(cirq.H(a))            # Append H gate
  s = cirq.Simulator()                 # Initialize simulator
  print('Simulate the circuit:')
11 print(circuit)
  results = s.simulate(circuit)        # Run simulator
  print(results)
  --------Cirque Output----------------------------------
15 Simulate the circuit:use
  a:_____X_____H_____
  output vector:               0.707|0> - 0.707|1>
```

Exercise   Verify that the output is an eigenstate of X.

- **X, Z, and H Gates and M Op:**
  A *measurement* operator acting on a quantum state provides a classical output of the probability distribution of a particular measurement. (Recall, "gates" are unitary operators, while measurement operators are not unitary, and thus not gates.) Extend your circuit to include H, X, and Z gates, as well as a measurement operator on the result. Run the simulation 1000 times in order to accumulate some statistics. Because Cirq runs within Python, it's easy to use Matplotlib to produce the visualization of probability distribution, as we have done in Figure 12.3.

```
# XZHM.py: Cirq Simulation with H, X, Z Gates + Measurement Op

3 import cirq

  import matplotlib.pyplot as plt
  circuit = cirq.Circuit()             # Build circuit
7 a = cirq.NamedQubit('a')             # Create qubit
  circuit.append(cirq.X(a))            # Append X gate
  circuit.append(cirq.Z(a))            # Append Z gate
  circuit.append(cirq.H(a))            # Append H gate
11 s = cirq.Simulator()                # Initialize simulator
```

```
    print('Simulate the circuit:')
    results = s.simulate(circuit)          # Run simulator
    circuit.append(cirq.measure(a, key ='result'))
15  samples = s.run(circuit,repetitions =1000)
    print(circuit)
    print(resultsC
    print(samples)
19  cirq.plot_state_histogram(samples)
    plt.show()
    ──────────Cirq Output─────────────
    Simulate the circuit:
23  a:                        _____X_____Z_____H_____M_____
    output vector:                          −0.707|0>   +   0.707|1>
    result=10011100010011010011101110001000001....
```

Exercise   Deduce what should be the effect of successive X, Z, and H gates, and compare
with the above output.

Exercise   Make sense of the M operator's output:
```
result=10011100010011010011101110001000001 . . . ..
```

- **A 2-qubit Cirq circuit:**
  Cirq contains the command `q0, q1 = cirq.LineQubit.range(2)` that creates a two-
  qubit circuit.

  Exercise
  1) Use Cirq to create the circuit:

     q_0 ──⊕──

     q_1 ─┤ Z ├─

  2) Run the circuit and verify that its output is |10⟩.
  3) Explain why this is the expected output.
  4) Recall the SWAP gate (a vertical line connecting qubits) that swaps one qubit with
     another. Extend your program to include a *SWAP* gate between q0 and q1:

     q_0 ──⊕──✕──

     q_1 ─┤ Z ├─✕──

  Our program `cirqSwap.py` created this circuit:

```
    # CirqSwap.py: Cirq program to create & swap 2 qubits

3   import cirq

    circuit = cirq.Circuit()
    q0, q1 = cirq.LineQubit.range(2)              # Create two qubits
7   circuit.append(cirq.X(q0))                    # Append X to q0
    circuit.append(cirq.Z(q1))                    # Append Z to q1
    circuit.append(cirq.SWAP(q0,q1))              # Swap qubits
    print(circuit)
11  s = cirq.Simulator()                          # Initialize simulator
    print('Simulate the circuit:')
    results = s.simulate(circuit)                 # Run simulator
    print(results)
15  ──────────Cirque Output ──────────────────────
    Simulate the circuit:
    output vector:                    |01>
```

- CNOT **on 2 qubits:**

$$CNOT(q0,\ q1) = \begin{cases} (q0,\ q1) & \text{if } q0 = 0, \\ (q0,\ 1 - q1) & \text{if } q0 = 1. \end{cases} \tag{12.68}$$

```
# CirqCNOT.py:   Cirq program with CNOT gate

3  import cirq

   circuit = cirq.Circuit()
   q0, q1 = cirq.LineQubit.range(2)        # Create two qubits
7  circuit.append(cirq.X(q0))              # Append X to q0
   circuit.append(cirq.Z(q1))              # Append Z to q1
   circuit.append(cirq.CNOT(q0, q1))       # Append CNOT, q0 = control
   print(circuit)
11 s = cirq.Simulator()                    # Initialize Simulator
   print('Simulate the circuit:')
   results = s.simulate(circuit)           # Run simulator
   print(results)
15 ------------Output------------
   Simulate the circuit:
   measurements: (no measurements)
   output vector:                |11>
```

Exercise Deduce what should be the effect of CNOT on the qubits, and compare with the output.

Exercise Add a SWAP gate before CNOT, and see if the output agrees with what you would expect.

- **3-Qubit** TOFFOLI **gate:**
Create a circuit that implements the TOFFOLI/CCNOT (controlled-controlled-not) gate. It should take three bits as input, and invert the third bit iff the first two bits are 1's:

```
1  #   CirToffoli.py: Cirq program with 3 qubit CCNOT gate

   import cirq

5  q0, q1, q2 = cirq.LineQubit.range(3)    # Create 3 qubits
   circuit = cirq.Circuit()                # Build circuit
   circuit.append(cirq.X(q0))              # Append X to q0
   circuit.append(cirq.Z(q2))              # Append Z to q2
9  circuit.append(cirq.Toffoli(q0,q1,q2))  # Connect all 3 wi Toffoli
        print(circuit)                     # Output circuit
   s = cirq.Simulator()                    # Initialize Simulator
   print('Simulate the circuit:')
   results = s.simulate(circuit)           # Run simulator
13 print(results)
   ------------Output------------
   Simulate the circuit:
   output vector:                |101>
```

Exercise Try all possible values for q0 and q2, and compare the output with the expected CCNOT effect.

12.5.1 Half and Full Adders

- **Half adder:**

  A half-adder adds the qubits, q0 and q1, and outputs the sum. It also outputs a carry bit q2 = 1, if q0 = q1 = 1, else q2 = 0. Create a half-adder circuit using three qubits and a TOFFOLI gate followed by a CNOT gate:

```
 # CirqHalfAdder.py:     Cirq circuit for half adder

   import cirq
4
   q0, q1, q2 = cirq.LineQubit.range(3)      # Create 3 qubits
   circuit = cirq.Circuit()                  # Build circuit
   circuit.append(cirq.X(q0))                # Append X to q0
8  circuit.append(cirq.X(q1))                # Append X to q1
   circuit.append(cirq.Toffoli(q0, q1,q2))   # Append Toffoli to 3 qs
   circuit.append(cirq.CNOT(q0, q1))         # Append CNOT to q0 & q1
   print(circuit)                            # Output circuit
12 s = cirq.Simulator()                      # Initialize Simulator
   print('Simulate the circuit:')
   results = s.simulate(circuit)             # Run simulator
   print(results)
16 ──────────Output──────────
   Simulate the circuit
   output vector:                  |101>
```

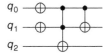

Exercise Verify the additions: $1 + 1, 1 + 0, 0 + 1$.

- **Full adder:**

  Design a full adder that adds q0 + q1, with q2 as the sum and q3 as the carry. It can be implemented with the program in FullAdder.py for the circuit:

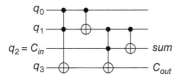

```
 # FullAdder.py: Cirq q0+q1 full adder program
2
   import cirq

   circuit = cirq.Circuit()# Build circuit
6  q0, q1, q2,q3 = cirq.LineQubit.range(4)     # Create 4 qubits
   circuit = cirq.Circuit()                    # Build circuit with qubits
   circuit.append(cirq.X(q0))                  # Append X to q0
   circuit.append(cirq.X(q1))                  # Append X to q1
10 circuit.append(cirq.Toffoli(q0, q1,q2))     # Append Toffoli
   circuit.append(cirq.CNOT(q0, q1))           # Append CNOT to q0,q1
   circuit.append(cirq.Toffoli(q1, q2,q3))     # Append Toffoli
```

```
   circuit.append(cirq.CNOT(q1, q2))          # Append CNOT to q1,q2
14 circuit.append(cirq.CNOT(q0, q1))          # Append CNOT to q0,q1
   print(circuit)
   s = cirq.Simulator()                       # Initialize Simulator
   print('Simulate the circuit:')
18 results = s.simulate(circuit)              # Run simulator
   print(results)
   ——————————Output——————————
   Simulate the circuit:
22 output vector:              |1110>
```

Use this program to fill this table:

q0	q1	Cin	Sum	Cout
$|0\rangle$	$|0\rangle$			
$|0\rangle$	$|1\rangle$			
$|1\rangle$	$|0\rangle$			
$|1\rangle$	$|1\rangle$			

12.6   Accessing the *IBM Quantum Computer*

We have just now checked, and since Amazon has yet to start selling personal quantum computers, we, instead, will go online and use the *IBM Quantum* [IBMqc, 2023]. Before you can use it, however, there are a number of steps to follow:

1) Go to QUANTUM-COMPUTING.IBM.COM/LOGIN
2) You will need to create an *IBMid*. To do that, have your cell phone and its QR reader in hand, click on

   Create an IBMid account.[2]

   This will let you log in to the *IBM Quantum*, as well as give you access to tutorials and programming tools. Follow the instructions on that page for creating and authenticating your account. Alternatively, you may be able to use your Google or GitHub account to log into IBM Quantum.
3) You can find instructions on how to run IBM Quantum codes at

   QUANTUM-COMPUTING.IBM.COM/LAB/DOCS/IQL/RUNTIME/START.

   You will be given a token to use to run a program.

12.6.1   IBM Quantum Composer

After logging in to the IBM Quantum, you will be presented with a *dashboard* containing several access routes to the computer. We will demonstrate one of them. Pushing the button on the dashboard labeled Launch Composer, brings up the *IBM Quantum Composer*, which is shown in Figure 12.4. The Composer is a graphical tool for creating quantum circuits (programs) by dragging and dropping operators, and then running them on the IBM Quantum, or a simulator. As seen in the figure, the Composer is divided into several

———
2 This might not work for all countries.

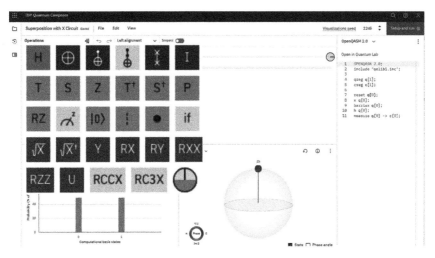

Figure 12.4 A screenshot of the dashboard for the *IBM Quantum Composer, with the gates section magnified.*

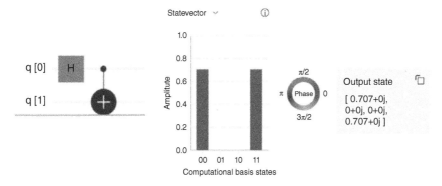

Figure 12.5 *Left*: A Quantum Composer circuit for generating the Bell state $|\beta_{00}\rangle$. *Right*: The generated $|\beta_{00}\rangle$ state vector in histogram and numerical forms.

panels. The top left panel, which we have magnified, presents you with a set of colorful little icons, each representing a different quantum gate, most of which we have just discussed.

As an example of how this all works, in Figure 12.5, we show the Composer's output from graphical program that we used to create the β_{00} Bell state (12.22),

$$|\beta_{00}\rangle = \frac{1}{\sqrt{2}}(|00\rangle + |11\rangle). \tag{12.69}$$

On the left is the circuit that generated the state, and on the right are the two generated computational basis states, shown as both a histogram and as numbers. Here's how our dragging and dropping created the circuit:

- We started by selecting File/New, which gave us the four qubits, q[0], q[1], q[2], q[3], and a single classical, 4-bit register c4 (c for "classical").
- Seeing that we need only two qubits to construct the β_{00} state, we eliminated q[2], q[3], and c4.

- Next, we dragged the *Hadamard* H gate (defined in Section 12.4) to the q[0] line.
- Then we dragged the controlled-NOT gate ⊕ to the q[0] line after the H gate, with the target symbol placed on the q[1] line.
- Lastly, we saved the circuit as the file beta_00.
- As was to be hoped for, the output in Figure 12.5 agrees with (12.69).

A Small Caveat The IBM Quantum employs a reversed Dirac notation in which qubits are ordered from right to left, that is, as $|q_{n-1}, \dots q_1 q_0\rangle$. So in IBM speak:

$$|01\rangle = |q_1 = 0\rangle \otimes |q_0 = 1\rangle = |01\rangle = \begin{bmatrix} 1 \\ 0 \end{bmatrix} \otimes \begin{bmatrix} 0 \\ 1 \end{bmatrix} = \begin{bmatrix} 0 \\ 1 \\ 0 \\ 0 \end{bmatrix}. \tag{12.70}$$

12.7 Qiskit Plus IBM Quantum

Another avenue to QC is [Qiskit, 2023], an open-source software development kit (SDK) that may be used with the IBM Quantum, or with its built-in quantum simulator. Running on both a simulator and a physical QC is revealing, as the simulator, being software, may well give more accurate results than an imperfect physical machine.

There are different ways to use Qiskit. Here we used an Anaconda window in a *Jupyter Notebook*, all in MS Windows, to set up the virtual environment qiskit. We activated the environment and installed the Qiskit package, including visualization and the qiskit_ibm_provider package:

```
  conda create --name qiskit python jupyter notebook
2 conda activate qiskit
  pip install qiskit[visualization] qiskit_ibm_provider
```

If you want to access IBM Quantum from your local computer, then you will need an *API token* that authenticates your external use. You can copy the token from the IBM Quantum dashboard, or you can run the following code to save the token, api_token, on your local disk for future use:

```
1   from qiskit_ibm_provider import IBMProvider
    IBMProvider.save_account(api_token)
```

You can now use your local computer to build a quantum circuit and to run it on the IBM Quantum. Place this code in a Jupyter Notebook named beta_00:

```
    from qiskit import QuantumCircuit      # Load needed package
2
    circuit = QuantumCircuit(2)            # Create a 2-q circuit
    # Apply H to q0,   then CNOT,  q0 = control, q1 = target
    circuit.h(0)
6   circuit.cx(0, 1)
    circuit.draw('mpl')                     # Draw circuit
```

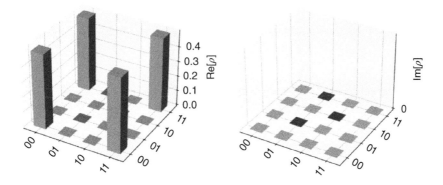

Figure 12.6 The real and imaginary part of the density matrix for the $|\beta_{00}\rangle$ state computed with Qisket and the IBM Quantum.

This should produce the same results as shown in Figure 12.5. Next, we'll use this same circuit to compute the state vector on Qiskit's quantum *simulator* **Aer**, and visualize it with Qisket's **plot_state_city** visualization tool:

```
from qiskit import Aer
from qiskit.visualization import plot_state_city

backend = Aer.get_backend("statevector_simulator")
job = backend.run(circuit)
result = job.result()
statevector = result.get_statevector(circuit, decimals=3)
statevector.draw(output="latex")
plot_state_city(statevector)
```

As seen on the left of Figure 12.6, this yields the real and imaginary parts of the state vector's *density matrix* [(12.24) in Section 12.3].

We now will run the same circuit on the IBM Quantum Computer. We start by finding the least busy device:

```
from qiskit_ibm_provider import IBMProvider

# Select hub/group/project
provider = IBMProvider(instance="ibm-q/open/main")
# Get the least busy backend
from qiskit.providers.ibmq import least_busy
device = least_busy(provider.backends(
    filters=lambda x: int(x.configuration().n_qubits) >= 3
                      and not x.configuration().simulator
                      and x.status().operational is True))
print("Running on current least busy device: ", device)
```

Next we *transpile* (translate from one compiler to another) the quantum circuit and run it on our defined **device**:

```
from qiskit import transpile
from qiskit.tools.monitor import job_monitor

circuit.measure_all()  # Measure the two qubits
transpiled_circuit = transpile(circuit, device)
```

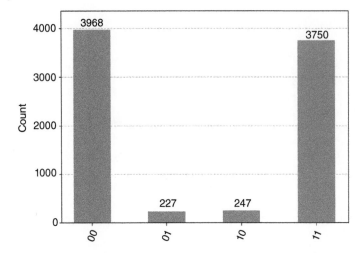

Figure 12.7 Histogram of the $|\beta_{00}\rangle$ Bell state found using Qiskit and the IBM Quantum.

```
    job = device.run(tranpiled_circuit, shots=8192)
    job_monitor(job, interval=2)
    result = job.result()
9   counts = result.get_counts(circuit)
    from qiskit.visualization import plot_histogram
    plot_histogram(counts)
```

The output is shown in Figure 12.7. You may note that the physical Quantum Computer is not a perfect quantum device, as witnessed by the experimental error, namely, the small, but nonzero, counts for the $|01\rangle$ and $|10\rangle$ states.

12.7.1 A Full Adder

In Section 12.5.1 we used Google's Cirq simulator to build a circuit that adds two bits, using two TOFFOLI gates, two CNOT gates, and four qubits. Now we'll build an IBM Quantum circuit that does the same thing, namely adds x to y. The simple three-qubit circuit that does that is shown on the left of Figure 12.8. On the right of the figure we show a more advanced IBM Quantum implementation for adding 01 + 10 using three TOFFOLI (CCX) gates, three CNOT (CX) gates, three measurement ops, three measurement ops, and five qubits. The program, given below, starts by initializing the q_0, q_1, and q_2 qubits that are used to represent $|x\rangle$, $|y\rangle$, and $|0\rangle$ respectively. It then uses the first TOFFOLI gate, taking q_0 and q_1 as the control and q_2 as the target for the carry bit. Then a controlled-NOT gate is used with q_0 as the control and q_1 as the target. The readouts of the states of q_2 and q_1 provide the result $x + y$:

```
1   def adder_circuit(x_in: int, y_in: int) -> QuantumCircuit:

        # q[0, 1, 2, 3, 4] --> x[0,1], y[0,1], c
        s = f"0{y_in:02b}{x_in:02b}"
5       qc = QuantumCircuit(5, 3)
        qc.initialize(s)
        qc.ccx(0, 2, 4)
        qc.cx(0, 2)
9       qc.reset(0)
        qc.ccx(1, 3, 0)
```

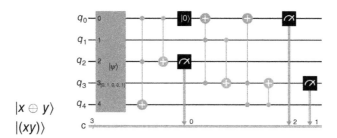

Figure 12.8 *Left*: A quantum circuit for adding two bits. *Right*: The IBM Quantum version of an adder for 01 + 10.

```
13   qc.cx(1, 3)
     qc.ccx(3, 4, 0)
     qc.cx(4, 3)
     qc.measure([2, 3, 0], [0, 1, 2])
     return qc
```

12.7.2 IBM Quantum Exercises

Use the IBM Quantum Composer to perform these exercises.

1) Prove that

$$CZ\,|00\rangle = |00\rangle, \quad CZ\,|01\rangle = |01\rangle, \quad CZ\,|10\rangle = |10\rangle, \quad CZ\,|11\rangle = -\,|11\rangle.$$

2) Determine the effect of the CNOT gate on: $|10\rangle$, $|01\rangle$, $|00\rangle$, and $|11\rangle$.
3) Create a quantum circuit for creating the entangled Bell state

$$|\beta_{11}\rangle = \frac{1}{\sqrt{2}}(|01\rangle - |10\rangle). \tag{12.71}$$

4) Create a circuit that demonstrates the effect of an H gate on $|0\rangle$:

$$H\,|0\rangle = \frac{1}{\sqrt{2}} \begin{bmatrix} 1 \\ 1 \end{bmatrix}. \tag{12.72}$$

5) Proved that a circuit with two Hadamard gates acts as the identity operator.
6) Verify the effect of the SWAP and CNOT gates acting on two qubits.
7) Create a circuit that shows how the Toffoli/ccnot gate acting on three qubits inverts the third qubit, if the first two qubits are 1's.
8) Create a circuit for a *half adder* that adds the qubits, q0 and q1, outputs the sum, as well as a carry bit q2 = 1 if q0 = q1 = 1, else q2 = 0. Verify the additions for $1 + 1, 1 + 0, 0 + 1$.

12.8 The Quantum Fourier Transform

As studied in Chapter 9, the discrete Fourier transform (DFT) transforms N values of a "signal" y_k, $k = 0, 1, \ldots, N$, measured at N equally-spaced times $t_k = kh$, into N complex, transform components Y_n. The transform and its inverse can be written in a concise and insightful way, and evaluated efficiently, by introducing a complex variable Z raised to various powers:

$$Y = DFT(y), \quad y = DFT^{-1}Y \tag{12.73}$$

$$Y_n = \frac{1}{\sqrt{2\pi}}\sum_{k=1}^{N} Z^{nk}y_k, \quad y_k = \frac{\sqrt{2\pi}}{N}\sum_{n=1}^{N} Z^{-nk}Y_n, \tag{12.74}$$

$$Z \overset{\text{def}}{=} e^{-2\pi i/N} \qquad Z^{nk} \equiv [Z^n]^k. \tag{12.75}$$

With this formulation, only one explicit computation of the exponential is required.

We now generalize the DFT to a *Quantum Fourier Transform* (QFT) where n qubits are used to computer 2^n components. The QFT transforms a signal space state $|y\rangle$ into a transform space state $|Y\rangle$:

$$|Y\rangle = QFT_N|y\rangle, \quad |y\rangle = QFT_N^{-1}|Y\rangle \tag{12.76}$$

$$|Y\rangle = \frac{1}{\sqrt{N}}\sum_{k=0}^{N-1}\sum_{l=0}^{N-1} y_l Z_N^{-kl}|k\rangle, \quad |y\rangle = \frac{1}{\sqrt{N}}\sum_{k=0}^{N-1}\sum_{l=0}^{N-1} Y_l Z_N^{kl}|k\rangle. \tag{12.77}$$

Here the subscript N on Z indicates the number of basis vectors being used and l the components of y and Y. *Note, we have switched to the computer science convention of starting the sums at 0, Qiskit's convention for the sign of the power of Z, and the absence of the $\sqrt{2\pi}$ normalization factors used in Chapter 9.*

12.8.1 1-Qubit QFT

Our 1-qubit QFT computes $N = 2^1 = 2$ components via (12.77):

$$QFT_2|0\rangle = \frac{1}{\sqrt{2}}(|0\rangle + |1\rangle), \quad QFT_2|1\rangle = \frac{1}{\sqrt{2}}(|0\rangle - |1\rangle), \tag{12.78}$$

$$\Rightarrow QFT_2|q\rangle = \frac{1}{\sqrt{2}}\sum_{p=0}^{2} Z_2^{-pq}|p\rangle. \tag{12.79}$$

Note, since $Z_2 = -1$, the QFT_2 is the same as the Hadamard gate H, (12.48).

12.8.2 2-Qubit QFT

Our 2-qubit QFT computes $N = 2^2 = 4$ components via (12.77):

$$|Y\rangle = QFT_4|y\rangle = \frac{1}{\sqrt{4}}\sum_{k=0}^{3}\sum_{l=0}^{3} y_l Z^{-kl}|k\rangle \tag{12.80}$$

$$= \frac{1}{2}\Big[(y_0 Z^0 + y_1 Z^0 + y_2 Z^0 + y_3 Z^0)|0\rangle$$

$$+ (y_0 Z^0 + y_1 Z^{-l} + y_2 Z^{-2} + y_3 Z^{-3})|1\rangle$$

$$+ (y_0 Z^0 + y_1 Z^{-2} + y_2 Z^{-4} + y_3 Z^{-6})|2\rangle$$

$$+ (y_0 Z^0 + y_1 Z^{-3} + y_2 Z^{-6} + y_3 Z^{-9})|3\rangle\Big]$$

$$\Rightarrow QFT_4 = \begin{bmatrix} \langle 0|\, QFT_4\, |0\rangle & \langle 0|\, QFT_4\, |1\rangle & \langle 0|\, QFT_4\, |2\rangle & \langle 0|\, QFT_4\, |3\rangle \\ \langle 1|\, QFT_4\, |0\rangle & \langle 1|\, QFT_4\, |1\rangle & \langle 1|\, QFT_4\, |2\rangle & \langle 1|\, QFT_4\, |3\rangle \\ \langle 2|\, QFT_4\, |0\rangle & \langle 2|\, QFT_4\, |1\rangle & \langle 2|\, QFT_4\, |2\rangle & \langle 2|\, QFT_4\, |3\rangle \\ \langle 3|\, QFT_4\, |0\rangle & \langle 3|\, QFT_4\, |1\rangle & \langle 3|\, QFT_4\, |2\rangle & \langle 3|\, QFT_4\, |3\rangle \end{bmatrix}$$

$$= \frac{1}{2}\begin{bmatrix} 1 & 1 & 1 & 1 \\ 1 & Z^{-1} & Z^{-2} & Z^{-3} \\ 1 & Z^{-2} & 1 & Z^{-2} \\ 1 & Z^{-3} & Z^{-2} & Z^{-1} \end{bmatrix}. \tag{12.81}$$

Note, for clarity we have left off the subscript 4 on Z used to indicate the number of Fourier components, that is, here $Z \equiv Z_4 = e^{-i\pi/2} = -i$. We have also multiplied out some powers of Z in (12.81). In the Qiskit convention of Section 12.7, the states are

$$|0\rangle = |00\rangle, \qquad |1\rangle = |01\rangle, \qquad |2\rangle = |10\rangle, \qquad |3\rangle = |11\rangle. \tag{12.82}$$

The rules for the QFT are thus:

$$QFT_4\, |00\rangle = H\,|0\rangle \otimes H\,|0\rangle, \qquad QFT_4\, |01\rangle = H\,|1\rangle \otimes P(\pi/2)H\,|0\rangle, \tag{12.83}$$

$$QFT_4\, |10\rangle = H\,|0\rangle \otimes H\,|1\rangle, \qquad QFT_4\, |11\rangle = H\,|1\rangle \otimes P(\pi/2)H\,|1\rangle, \tag{12.84}$$

where H is the Hadamard gate, and $P(\theta)$ is the phase gate:

$$P(\theta) = \begin{bmatrix} 1 & 0 \\ 0 & e^{i\theta} \end{bmatrix}. \tag{12.85}$$

Putting all of the pieces together, we have QFT_4 being accomplished with just two qubits plus four operations (three gates and a direct product):

$$QFT_4\, |q_1 q_0\rangle = H\,|q_0\rangle \otimes P^{q_0}(\pi/2)\, H\,|q_1\rangle. \tag{12.86}$$

Figure 12.9 shows the circuit based on (12.86), where you may note that the swapping of the results of two qubits, $H\,|q_0\rangle$ and $P^{q_0}(\pi/2)H\,|q_1\rangle$, is accomplished by the SWAP gate at the end. The same circuit can be implemented in Qiskit, as we do in Listing 12.2. The correctness of the circuit is verified by its output matching the matrix in (12.81).

Exercise Take four sample values, y_k's, of the function

$$y(t) = 3\cos(\omega t) + 2\cos(3\omega t). \tag{12.87}$$

1) Use QFT4 to determine $y(t)$'s Fourier components.
2) What are the frequencies of the deduced components?
3) Check that the summation of the components reproduces the input signal.

12.8.3 *n*-Qubit QFT ⊙

Given n qubits to work with, we can compute $N = 2^n$ transform components. The basis vectors for n-qubits are the direct products in an N-dimensional Hilbert space:

Figure 12.9 Quantum Fourier transform circuit for 2-qubits.

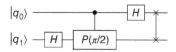

$$|y_{n-1} \cdots y_0\rangle = |y_{n-1}\rangle \otimes \cdots \otimes |y_0\rangle, \qquad y_k \in \{0,1\}, \tag{12.88}$$

where we are following the Qiskit convention. This product can also be labeled by its integer representation y:

$$y = \sum_{k=0}^{n-1} y_k 2^k, \qquad \in \{0, 1, \ldots, N-1\}, \tag{12.89}$$

$$|0\rangle = |0 \ldots 0\rangle, \qquad |2\rangle = |0 \ldots 010\rangle, \quad \ldots . \tag{12.90}$$

The QFT of the basis vector $|y\rangle$ is then

$$QFT_{2^n} |y\rangle = \frac{1}{\sqrt{N}} \sum_{k=0}^{N-1} e^{2\pi i y k/N} |k\rangle \tag{12.91}$$

$$= \frac{1}{\sqrt{N}} \sum_{k_0=0}^{1} \cdots \sum_{k_{n-1}=0}^{1} e^{2\pi i y k/N} |k_{n-1} \cdots k_0\rangle \tag{12.92}$$

$$= \frac{1}{\sqrt{N}} \sum_{k_0=0}^{1} \cdots \sum_{k_{n-1}=0}^{1} \otimes_{l=0}^{n-1} e^{2\pi i y k_l 2^{l-n}} |k_l\rangle \tag{12.93}$$

$$= \frac{1}{\sqrt{N}} \otimes_{l=0}^{n-1} \left(|k_l = 0\rangle + e^{2\pi i 0.y_{n-l-1}\cdots y_0} |k_l = 1\rangle \right)$$

$$= \frac{1}{\sqrt{N}} \left(|0\rangle + e^{2\pi i 0.y_0} |1\rangle \right) \otimes \left(|0\rangle + e^{2\pi i 0.y_1 y_0} |1\rangle \right)$$

$$\otimes \cdots \otimes \left(|0\rangle + e^{2\pi i 0.y_{n-1}\cdots y_0} |1\rangle \right) \tag{12.94}$$

$$= H |y_0\rangle \otimes P^{y_0}(\pi/2) H |y_1\rangle \otimes P^{y_1}(\pi/2) P^{y_0}(\pi/2^2) H |y_2\rangle$$

$$\otimes \cdots \otimes P^{y_{n-2}}(\pi/2) \cdots P^{y_0}(\pi/2^{n-1}) H |y_{n-1}\rangle . \tag{12.95}$$

Here we have employed the binary fraction notation:

$$0.x_m \cdots x_0 \stackrel{\text{def}}{=} \sum_{l=0}^{m} x_l 2^{l-m-1}. \tag{12.96}$$

The Python–Qiskit implementation of this n-qubit QFT is amazingly short and efficient, and given in Listing 12.3. Of course, Qiskit already has a QFT method, and we could have just used It. In fact, we did use it to verify this implementation.

12.9   Oracle + Diffuser = Grover's Search Algorithm

Problem   You are given a database containing $N = 2^n$ elements, each referenced by a specific index j, and the function f for which

$$f(j) = \delta_{ij}, \tag{12.97}$$

where we have employed the Kronecker delta function. Use n qubits to find item i in the database.

Of course, you could just start out at the first element and keep testing if $f(j) = i$ until you find $j = i$; but we want something faster than that. As our example, let's say we have a database with 16 elements:

j	V_j	j	V_j	j	V_j	j	V_j
2	30	6	70	10	110	14	150
0	10	4	50	8	90	12	130
3	40	7	80	11	120	15	160
1	20	5	60	9	100	13	140

If we want to search through these $16 = 2^4$ elements, then we need $n = 4$ qubits. And if we want the element with $j = 8$, then we want to come up with the value 90.

Now we must translate this search into a quantum computation. First, we initialize the 4-qubit system into $|0\rangle^{\otimes n}$, and then we use a direct product of Hadamard operators to transform the system into a uniform superposition of states:

$$|\psi\rangle = H^{\otimes n}|0\rangle^{\otimes n} = \frac{1}{\sqrt{N}}\sum_{k=0}^{N-1}|k\rangle. \tag{12.98}$$

Now we have the stage set for a little magic. In literature, an oracle is a divine communication or revelation that provides advice or prophecy. In QC, an *oracle* $\mathcal{O}$ is a unitary operator or circuit that provides a way to distinguish between different states. On the left of Figure 12.10, we show a schematic circuit that constructs an oracle $\mathcal{O}$ for this particular database, and for which

$$\mathcal{O}\,|k\rangle = (-1)^{f(k)}\,|k\rangle, \tag{12.99}$$

where $f(k)$ is the function in (12.97). In the middle ($i = 15$) and right ($i = 9$) of Figure 12.10 we show what's inside the black (actually white) box shown on the left of Figure 12.10. In Listing 12.4 we give a Qiskit simulator code for generating an *oracle* circuit. In Listing 12.5, we have another version of that code including the calls needed to run it on the IBM Quantum.

Applying the oracle to the expansion (12.98) yields:

$$\mathcal{O}\,|\psi\rangle = \frac{1}{\sqrt{N}}\sum_{k\neq i}|k\rangle - \frac{1}{\sqrt{N}}\,|i\rangle. \tag{12.100}$$

We see that the oracle distinguishes the desired state $|i\rangle$ in the expansion by flipping its sign, while leaving the other states untouched. However, this hardly isolates the desired state from its brethren. Not to worry, we will now use the oracle to construct the *Grover*

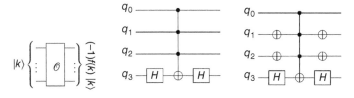

Figure 12.10 *Left*: A schematic quantum circuit for an *oracle*. *Right*: An *oracle* circuit for $i = 15$. (c) An *oracle* circuit for $i = 9$.

algorithm, whose repeated application will amplify the amplitude of $|i\rangle$ so that it stands out to a desired level of precision. The algorithm incorporates the *diffuser operator* U_ψ:

$$U_\psi \overset{\text{def}}{=} 2|\psi\rangle\langle\psi| - I, \qquad \text{where} \qquad |\psi\rangle = \frac{1}{\sqrt{N}}\sum_{k=0}^{N-1}|k\rangle. \tag{12.101}$$

The diffuser has the unusual property:

$$U_\psi \sum_k \alpha_k |k\rangle = \sum_k \left[\bar{\alpha} + (\bar{\alpha} - \alpha_k)\right]|k\rangle, \qquad \text{where} \qquad \bar{\alpha} \overset{\text{def}}{=} \frac{1}{N}\sum_k \alpha_k. \tag{12.102}$$

For those amplitude α_k greater (less) than the average $\bar{\alpha}$, the diffuser decreases (increases) the amplitude below (above) the average $\bar{\alpha}$, to $\bar{\alpha} - (\alpha_k - \bar{\alpha})$. Geometrically, the diffuser "reflects" each amplitude α_k with respect to the average $\bar{\alpha}$. And so a single application of an oracle + diffuser combo amplifies the amplitude of $|i\rangle$ relative to that of the others:

$$U_\psi \mathcal{O} |\psi\rangle = \frac{3N - 4}{N^{3/2}}|i\rangle + \frac{N - 4}{N^{3/2}}\sum_{k \neq \omega}|k\rangle, \tag{12.103}$$

(need we point out that $3N > N$?). The combination $U_\psi \mathcal{O}$, of the diffuser and oracle is called the *Grover operator*. Repeated applications of the operator will keep increasing $|i\rangle$'s relative amplitude, at least until the number of applications reaches $\pi\sqrt{N}/4$ [Nielsen and Chuang, 2010].

12.9.1 Grover's Implementation

We now use Qiskit to construct a 4-qubit implementation of Grover's algorithm that will single out the $i = 15$ elements in the dataset. First, we create an oracle that will flip the sign for the state $|15\rangle \equiv |1111\rangle$. That is accomplished by placing a triple-controlled-Z gate, HXH = Z, between two Hadamard gates, as shown on the left of Figure 12.10. The oracle for other values of i can be built by adding a pair of X gates before and after the triple-controlled-Z gate on the corresponding qubits. For example, on the right of Figure 12.10, we form an oracle for $|9\rangle \equiv |1001\rangle$ by adding a pair of X gates onto qubit-0.

In Figure 12.11 we give a circuit for the Grover's algorithm. In Listing 12.4 we give a program that generates a circuit for the algorithm, and then runs it on a simulator. (The diffuser code within is based on Shor's Algorithm [2023].) In Listing 12.5 we give a driver version of this same code, now including the calls needed to run it on the IBM Quantum.

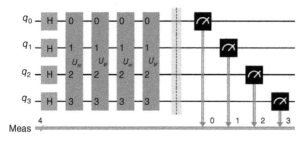

Figure 12.11 A circuit for Grover's Algorithm that combines an oracle (O) and a diffuser algorithm (U) that singles out the $i = 15$ elements in a dataset.

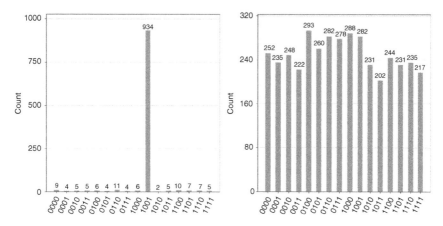

Figure 12.12 *Left*: Output of Grover's search for $i = 9$ as run on a simulator. *Right*: Output of running the same program on the IBM Quantum Computer *ibmq_lima*.

Figure 12.12 left shows the histogram resulting from running the Grover algorithm for $i = 9$ on the simulator. We see that 926 out of 1000 trials resulted in the state $|1001\rangle$ (the desired $i = 9$); this is a clear indication of the power of the algorithm. In contrast, Figure 12.12-right shows the histogram resulting from running the code on ibmq_lima, a physical IBM Quantum computer. You will notice that the physical computer does *not* yield a clearly outstanding peak for the $|1001\rangle$ state. This is a consequence of noise in the electronics. The discussion of quantum noise is an active research field that is beyond the scope of this book. Interested readers are referred to Wang and Krstic [2020] and reference therein.

Exercise Use Grover's algorithm and four qubits to search the table given at the beginning of this section. See if you come up with 90 as the answer.

12.10 Shor's Factoring ⊙

What with its reliance on the computational difficulty of factoring large integers, prime numbers are an essential element in cryptology. The *prime factors* of a number are the prime numbers that when multiplied together result in the original number. For example, the prime factors of 30 are 2, 3, and 5, but not 6 since it's not prime. *Shor's algorithm* is a computational technique for finding the prime factors of an integer using quantum gates. Due to the efficiency of the QC algorithm, the quantum computation is nearly exponentially faster than the classical algorithm. Here are the steps in Shor's algorithm [Wikipedia, 2023]:

1) Pick a random integer $1 < r < N$.
2) Compute the greatest common divisor (the Python command, $K = \gcd(r, N)$).
3) If $K \neq 1$, then K is a factor of N, and you have found a factor.
4) Use a period-finding method, to find the smallest period T of the function

$$f(x) = r^x(\bmod N), \quad f(x + T) = f(x), \tag{12.104}$$

where the (mod) operator makes f a periodic function of x.

5) If T is odd, or if $r^{T/2} = -1 (\text{mod } N)$, return to step 1.

6) Else, $\gcd(r^{T/2} + 1, N)$ or $\gcd(r^{T/2} - 1, N)$, or both, are nontrivial factors of N, and you have found at least one factor.

To illustrate the algorithm, we have chosen $N = 15$ and written the method `Amod15`. It is given in Listing 12.6, where it creates a unitary operator U such that

$$U |y\rangle = |Cy (\text{mod } N)\rangle, \quad \text{for all } C \text{ coprime with 15.} \tag{12.105}$$

Phase Estimation Before we can go on and implement Shor's algorithm, we need QC methods for *phase estimation* and *period finding*. We start with phase estimation. Let U be a unitary operator with eigenvectors $|u\rangle$ and eigenvalues $e^{2\pi i\phi}$:

$$U |u\rangle = e^{2\pi i\phi} |u\rangle . \tag{12.106}$$

Without actually solving this eigenvalue problem, the quantum phase algorithm provides an estimate of the phase ϕ. Figure 12.13 shows a circuit that implements the phase algorithm using two quantum registers (computing units). The number of qubits needed for the t register, the unit on top, is determined by the required accuracy. The number of qubits in the second register below is the same as the number of qubits that U operates on. As we see in the figure, the t register starts with the state $|0^t\rangle$, and then the H-gates transform each qubit in the state into $(|0\rangle + |1\rangle)/\sqrt{2}$. The control register U^{2^j} leaves the second qubit unchanged, but changes the jth qubit in the first register into the state $(|0\rangle + e^{2\pi i 2^j \phi} |1\rangle)/\sqrt{2}$ (a "phase kick back"). After performing all of the control operations, the t register will be left in the state

$$\frac{|0\rangle + e^{2\pi i 2^{t-1}\phi} |1\rangle}{2} \otimes \cdots \otimes \frac{|0\rangle + e^{2\pi i 2^0 \phi} |1\rangle}{2} = \frac{1}{2^{t/2}} \sum_{k=0}^{2^t - 1} e^{2\pi i k\phi} |k\rangle . \tag{12.107}$$

Yet we also know that for an integer $0 \le s < 2^t$, the QFT acting on the basis state $|s\rangle$ gives us

$$QFT_{2^t} |s\rangle = \frac{1}{2^{t/2}} \sum_{k=0}^{2^t - 1} e^{2\pi i k s/2^t} |k\rangle . \tag{12.108}$$

If $2^t \phi = s$ is an integer, then we see that the RHS of (12.107) is exactly the transform for 2^t components, $QFT_{2^t} |s\rangle$. The inverse QFT on $QFT_{2^t} |s\rangle$ reveals the value s:

$$|s\rangle = QFT_{2^t}^{-1} \left[\frac{|0\rangle + e^{2\pi i 2^{t-1}\phi} |1\rangle}{2} \otimes \cdots \otimes \frac{|0\rangle + e^{2\pi i 2^0 \phi} |1\rangle}{2} \right], \quad \phi = \frac{s}{2^t} .$$

Even if $2^t \phi$ is not an integer, the analysis of Nielsen and Chuang [2010] shows that the circuit estimates ϕ to an accuracy of $t - \left\lceil \log \left(2 + \frac{1}{2\epsilon} \right) \right\rceil$ bits, where ϵ is the probability that the algorithm fails.

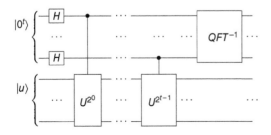

Figure 12.13 A quantum circuit for phase estimation.

Period Finding The second algorithm we need to implement Shor's algorithm is one that determines the period of the function

$$f(x) = r^x (\text{mod } N). \tag{12.109}$$

Here $r < N$ is a positive integer, r and N are coprime (have 1 as the only common factor), and the (mod N) operator makes f a periodic function of x. We want to find the smallest value of T such that

$$f(x + T) = f(x), \quad \text{i.e.,} \quad r^T (\text{mod } N) = 1. \tag{12.110}$$

Let U be a unitary operator such that

$$U |y\rangle = |ry (\text{mod } N)\rangle. \tag{12.111}$$

We define an eigenstate of U:

$$|u_S\rangle = \frac{1}{\sqrt{T}} \sum_{k=0}^{r-1} e^{-2\pi i Sk/T} \left| r^k (\text{mod } N) \right\rangle, \quad 0 \le S \le T - 1, \tag{12.112}$$

$$U |u_S\rangle = e^{2\pi i S/T} |u_S\rangle. \tag{12.113}$$

Even though we do not know the value of T, or all of the details of the $|u_S\rangle$ state, we do know a neat identity, namely:

$$\frac{1}{\sqrt{T}} \sum_{S=0}^{T-1} |u_S\rangle = |1\rangle. \tag{12.114}$$

So if we set the initial state of the second quantum register to $|1\rangle$, which is a superposition of $|u_S\rangle$'s, the quantum phase estimation will measure the phase

$$\phi = \frac{S}{T}, \tag{12.115}$$

where s is a random integer between 0 and $T - 1$. We can now use the *continued fractions algorithm* to determine the value of T.

In Listing 12.6, we finally give our code for a quantum circuit that computes Shor's algorithm [Shor's Algorithm, 2023]. It combines quantum circuits for phase estimation and period-finding `qpe`, with the unitary operator `amod15`, and uses them to factor the number 15. Here's a typical output from it:

```
 1  Attempt # 0
    Random a = 7
    Register reading: 00000000
    Corresponding phase: 0.000000
 5  Phase: 0.0
    r = 1
    Attempt # 1
    Random a = 8
 9  Register reading: 01000000
    Corresponding phase: 0.250000
    Phase: 0.25
    r = 4
13  Found factor: 5
    Found factor: 3
```

12.11 Code Listings

Listing 12.1 Entangle.py computes Hamiltonian, eigenvalues, and eigenvectors for entangled quantum states using numpy.

```
# Entangle.py: Calculate entangled quantum states
2
   from numpy import *;  from numpy.linalg import*

   nmax = 4
6  H = zeros((nmax,nmax), float)
   XAXB = array([[0,0,0,1],[0,0,1,0],[0,1,0,0],[1,0,0,0]])       # sigxA.sigxB
   YAYB = array([[0,0,0,-1],[0,0,1,0],[0,1,0,0],[-1,0,0,0]])     # sigyA.sigyA
   ZAZB = array([[1,0,0,0],[0,-1,0,0],[0,0,-1,0],[0,0,0,1]])     # sigzA.sigzA
10 SASB = XAXB + YAYB + ZAZB -3*ZAZB                  # Hamiltonian/factor
   print('Hamiltonian without mu^2/r^3 factor \n',SASB)
   es,ev = eig(SASB)                                  # Eigenvalues & vectors
   print('Eigenvalues \n',es)
14 print('Eigenvectors(incolumns)\n',ev)
   phi1 = (ev[0,0], ev[1,0], ev[2,0], ev[3,0])            # Extract vectors
   phi4 = (ev[0,1], ev[1,1], ev[2,1], ev[3,1])
   phi3 = (ev[0,2], ev[1,2], ev[2,2], ev[3,2])
18 phi2 = (ev[0,3], ev[1,3], ev[2,3], ev[3,3])
   basis = [phi1,phi2,phi3,phi4]                       # List eigenvectors
   for i in range ( 0 , nmax ) :              # Hamiltonian in new basis
      for j in range ( 0 ,nmax ) :
22        term = dot (SASB, basis [i ] )
          H[i,j] = dot (basis [j],term )
   print('Hamiltonian in Eigenvector Basis \n' , H)
```

Listing 12.2 QFT4.py A 2-qubit circuit that computes four Fourier components using Qiskit.

```
# QFT4.py: A Qiskit program to compute a 2-qubit QFT for 4 components
2
   import math
   from qiskit import QuantumCircuit
   import qiskit.quantum_info as qi
6  import numpy as np

   def qft2(inverse=False) -> QuantumCircuit:
      angle = math.pi/2
10    if inverse is True:  angle = -angle
      qc = QuantumCircuit(2)                  # Create a 2-qubit circuit
      qc.h(1)                                 # H gate on qubit-1
      qc.cp(angle, 0, 1)                      # Controlled phase gate
14    qc.h(0)
      qc.swap(0, 1)                           # Swap the qubits
      return qc                               # Return circuit as gate

18 if __name__ == "__main__":
      print(np.around(qi.Operator(qft2()).data, 3)) # Circuit matrix
```

Listing 12.3 QFTn.py An n-qubit circuit that computes 2^n Fourier components using Qiskit.

```
# QFT.py: A Qiskit program using n-qubits for 2^n component QFT
3  import math
   from qiskit import QuantumCircuit
   import qiskit.quantum_info as qi
   from qiskit.circuit.library import QFT
7  import numpy as np
```

```
def qft(n: int, inverse: bool = False, skip_swap: bool = False) -> QuantumCircuit:
        # build a n-qubit qft circuit
11      angle = np.pi/2
        if inverse is True:
            angle = -angle
        qc = QuantumCircuit(n)
15      for i in reversed(range(n)):
            qc.h(i)
            for j in range(i):
                qc.cp(angle/2**(i-j-1), j, i)
19      if skip_swap is False:
            for i in range(math.floor(n/2)):
                qc.swap(i, n-i-1)
        return qc
23
    if __name__ == "__main__":
        for k in range(10):
            print(np.max(np.abs(qi.Operator(qft(k)).data-qi.Operator(QFT(k)).data)))
```

Listing 12.4 OracleSim.py A quantum circuit for Grover's algorithm for i=0-15 on a simulator.

```
# OracleSim.py: Simulator version of Qiskit code for Oracle circuit, i=0-15

from qiskit import QuantumCircuit, Aer, transpile, assemble
4 from qiskit.visualization import plot_histogram
from numpy import math

def oracle(omega: int):                            # With removed->Gate
8   if omega < 0 or omega >= 16:                    # Flip sign if |omega>
        raise ValueError("Input should be betwn"+"0 & 15, got", omega)
    bit_string =f"{omega: 04b}"                     # Convert omega to bit pattern
    quantum_circuit = QuantumCircuit(4)             # 4 bit quantum circuit
12  [quantum_circuit.x(3-idx) for idx in range(4) if bit_string[idx]=='0']
    quantum_circuit.h( 3 )
    quantum_circuit.mcx([0,1,2],3)
    quantum_circuit.h(3)
16  [quantum_circuit.x( 3-idx) for idx in range(4) if bit_string[idx]=='0']
    cap_u_omega = quantum_circuit.to_gate()
    cap_u_omega.name="omega"                        # Differs
    return cap_u_omega
20
# diffuser.py: circuit for a diffuser with n qubits
def diffuser(n_qubits: int): # remove Gate, I - 22|psi><psi|
    quantum_circuit = QuantumCircuit(n_qubits)      # n qubit circuit
24  quantum_circuit.h( range( n_qubits))            # Map|psi> to |0...0 >
    quantum_circuit.x(range(n_qubits))              # Map|0...0> to |1...1 >
    quantum_circuit.h( n_qubits -1) # Multi cntrl-z, flips sign |1...1>
    quantum_circuit.mcx(list(range(n_qubits -1)),n_qubits -1)
28  quantum_circuit.h(n_qubits -1)                  # Map back to |0...0 >
    quantum_circuit.x(range( n_qubits))
    quantum_circuit.h( range( n_qubits))            # Map back to |psi>
    cap_u_psi = quantum_circuit.to_gate()
32  cap_u_psi.name="$U_{\\psi}$"
    return cap_u_psi

# Grover.py: Driver code for QC Grover algorithm on simulator
36 if __name__ == "__main__" :
    cap_n=4
    qc = QuantumCircuit(cap_n)
    qc.h(range(cap_n))                              # put into|\psi>
40  # Run
    cap_r = math.ceil(math.pi*math.sqrt(cap_n)/4) # Iterate Grover R times
    for i in range(cap_r):
        qc.append(oracle(9), range(cap_n))
44      qc.append(diffuser(cap_n), range(cap_n))
    qc.measure_all()
```

```
      qc.draw(output="mpl",filename="grover4_circuit.png")
      backend=Aer.get_backend("aer_simulator")              # Run on simulator
48    transpiled_circuit = transpile(qc,backend=backend)
      job = backend.run(transpiled_circuit)
      result = job.result()
      histogram = result.get_counts()
52    plot_histogram(histogram,filename="grover4_sim_histogram.png")
```

Listing 12.5 OracleIBM.py An IBM Quantum circuit for Grover's algorithm for i=0-15.

```
   # OracleIBM.py: IBM QC Qiskit code for Oracle circuit, i=0-15.
2
   from qiskit import QuantumCircuit, Aer, transpile
   from qiskit.visualization import plot_histogram
   from qiskit.tools import job_monitor
6  from qiskit_ibm_provider import IBMProvider, least_busy
   from numpy import math

   def oracle(omega: int):   # remove ->Gate
10     # Flip the sign if state is |omega>
       if omega < 0 or omega >= 16:
           raise ValueError("Need input" + "0 - 15, got ", omega)
       # Convert omega into bit pattern
14     # bit_string = f"{omega:    04b}"   ########################
       bit_string = f"{omega:04b}"
       # print("bit-string",bit_string)
       # Start a  quantum circuit of 4 qubits
18     quantum_circuit = QuantumCircuit(4)
       [quantum_circuit.x(3 - idx) for idx in range(4) if bit_string[idx] == '0']
       quantum_circuit.h(3)
       quantum_circuit.mcx([0, 1, 2], 3)
22     quantum_circuit.h(3)
       [quantum_circuit.x(3 - idx) for idx in range(4) if bit_string[idx] == '0']
       cap_u_omega = quantum_circuit.to_gate()
       cap_u_omega.name = "$U_\\omega$"
26     return cap_u_omega

   # diffuser.py: a quantum circuit for a general diffuser with n qubits
   def diffuser(n_qubits: int):   # remove Gate
30     # Where|psi>is the uniform superposition state
       # Create a circuit with n_qubits
       quantum_circuit = QuantumCircuit(n_qubits)
       # Map|psi> to |0...0>
34     quantum_circuit.h(range(n_qubits))
       # Map|0...0> to |1...1>
       quantum_circuit.x(range(n_qubits))
       # Multiply controlled-z
38     # To flip sign for |1...1>
       quantum_circuit.h(n_qubits - 1)
       quantum_circuit.mcx(list(range(n_qubits - 1)), n_qubits - 1)
       quantum_circuit.h(n_qubits - 1)
42     # Map back from |1...1> to |0...0>
       quantum_circuit.x(range(n_qubits))
       # Map back |0...0> to |psi>
       quantum_circuit.h(range(n_qubits))
46     cap_u_psi = quantum_circuit.to_gate()
       cap_u_psi.name = "$U_{\\psi}$"
       return cap_u_psi

50 # Grover.py:Driver code for QCGrover algorithm on simulator&IBMQuantum
   if __name__ == "__main__":
       # these 2 commented lines only need to run once at the very beginning
       # token ='***********'
54     # QiskitRuntimeService.save_account(channel="ibm_quantum", token=token,
            overwrite=True)
       cap_n = 4  # number of qubits
       qc = QuantumCircuit(cap_n)
       qc.h(range(cap_n))  # put into|\psi>
58     # Run Grover iteration for R times
```

```
        cap_r = math.ceil(math.pi * math.sqrt(cap_n) / 4)
        for i in range(cap_r):
            qc.append(oracle(9), range(cap_n))
62          qc.append(diffuser(cap_n), range(cap_n))
        qc.measure_all()
        qc.draw(output="mpl", filename="grover4_circuit.png")
        # Run on simulator
66      backend = Aer.get_backend("aer_simulator")
        transpiled_circuit = transpile(qc, backend=backend)
        job = backend.run(transpiled_circuit)
        result = job.result()
70      histogram = result.get_counts()
        plot_histogram(histogram, figsize=(7, 7), filename="grover4_sim_histogram.png")
        print(max(histogram, key=histogram.get))
        # Load account and get provider
74      provider = IBMProvider(instance="ibm-q/open/main")
        device = least_busy(provider.backends(
                filters=lambda x: int(x.configuration().n_qubits) >= cap_n
                and not x.configuration().simulator
78              and x.status().operational is True))
        print("Running on least busy device:", device)
        # Transpile and run
        transpiled_circuit = transpile(qc, device)
82      job = device.run(transpiled_circuit)
        job_monitor(job, interval=2)
        # Get result
        result = job.result()
86      histogram = result.get_counts(qc)
        plot_histogram(histogram, figsize=(7, 7), filename="grover4_histogram.png")
```

Listing 12.6   Shor.py A Quantum circuit for Shor's Algorithm.

```
    # Shor.py: Shor's algorithm
    # https://qiskit.org/textbook/ch-algorithms/shor.html
3
    import random
    from fractions import Fraction
    from math import gcd
7   from typing import List
    from qiskit import QuantumCircuit, Aer, transpile, assemble
    from qiskit.circuit.library import QFT

11  def amod15(a_in: int, p_in: int) -> QuantumCircuit: # Mult x a_in mod 15
        if a_in not in [2, 4, 7, 8, 11, 13, 14]:
            raise ValueError("'a_in' must be 2,4,7,8,11,13 or 14")
        quantum_circuit = QuantumCircuit(4)
15      for iteration in range(p_in):
            if a_in in [2, 13]:
                quantum_circuit.swap(2, 3)
                quantum_circuit.swap(1, 2)
19              quantum_circuit.swap(0, 1)
            if a_in in [7, 8]:
                quantum_circuit.swap(0, 1)
                quantum_circuit.swap(1, 2)
23              quantum_circuit.swap(2, 3)
            if a_in in [4, 11]:
                quantum_circuit.swap(1, 3)
                quantum_circuit.swap(0, 2)
27          if a_in in [7, 11, 13, 14]:  # I added 14 here
                for i in range(4):
                    quantum_circuit.x(i)
        quantum_circuit.name = "%i^%i mod 15" % (a_in, p_in)
31      return quantum_circuit  # return the circuit

    def qpe(u_list: List[QuantumCircuit]) -> float:  # Build phase circuit
        # u_list: a list of QuantumCircuit
35      # [U^(2^0), U^(2^1), ... U^(2^(t-1))]
        t = len(u_list)
        num_qubits_u = u_list[0].num_qubits         # N qubits for cap_u gate
```

```
         qc = QuantumCircuit(t + num_qubits_u, t)
39       # put the first t_count qubits into superposition
         for i in range(t):
             qc.h(i)
         # put the last n_u qubit into |1> state
43       qc.x(t)                                          # qiskit convention
         for i in range(t):                        # Add contr-U^{2^j} gate
             qc.append(u_list[i].to_gate().control(), [i]
                       + [j+t for j in range(num_qubits_u)])
47       qc.append(QFT(t, inverse=True).to_gate(), range(t))  # Inverse QFT
         qc.measure(range(t), range(t))                      # Finally, measure
         simulator = Aer.get_backend("aer_simulator")     # Run on simulator
         q_obj = assemble(transpile(qc, simulator), shots=1)
51       result = simulator.run(q_obj, memory=True).result()
         readings = result.get_memory()
         print("Register reading: " + readings[0])
         phase = int(readings[0], 2)/(2**t)
55       print("Corresponding phase: %f" % phase)
         return phase

     if __name__ == "__main__":
59       cap_n = 15
         factor_found = False
         attempt = 0
         while not factor_found:
63           print("Attempt #", attempt)
             attempt += 1
             a = random.randint(2, cap_n-1)
             print("Random a = ", a)
67           k = gcd(a, cap_n)
             if k != 1:
                 factor_found = 1
                 print("Found factor: ", k)
71           else:
                 p = qpe([amod15(a, 2**j) for j in range(8)])
                 print("Phase: ", p)
                 fraction = Fraction(p).limit_denominator(cap_n)
75               s, r = fraction.numerator, fraction.denominator
                 print("r = ", r)
                 if r % 2 == 0:  # r is even
                     guesses = [gcd(a**(r//2)+1, cap_n),
79                              gcd(a**(r//2)-1, cap_n)]
                     for g in guesses:
                         if g not in [1, cap_n] and (cap_n % g) == 0:
                             print("Found factor: %i" % g)
83                           factor_found = True
```

Part III

Applications

13

ODE Applications; Eigenvalues, Scattering, Trajectories

Now that we have developed reliable methods to solve ODEs, we apply them to some challenging problems. First, we combine our ODE solver with a search algorithm to solve the quantum eigenvalue problem for an arbitrary potential. Then we study classical scattering in a system that becomes chaotic. Finally, we look upward to balls falling out of the sky and planets that do not.

13.1 Quantum Eigenvalues for Arbitrary Potentials

Problem   What is the energy of a particle bound by a potential that confines it to an atomic distance?

Quantum mechanics describes phenomena that occur at atomic and subatomic (particle) scales. It is a statistical theory in which the prime observable is the probability that a particle is located in a region dx around point x. Yet, usually, we solve for a *wave function* $\psi(x)$, and then calculate the probability as $\mathcal{P} = |\psi(x)|^2 \, dx$. If a particle of energy E is moving in one dimension and experiences a potential $V(x)$, that wave function is determined by an ordinary differential equation, the time-independent Schrödinger equation:[1]

$$\frac{-\hbar^2}{2m} \frac{d^2 \psi(x)}{dx^2} + V(x)\psi(x) = E\psi(x). \tag{13.1}$$

In practice, we solve for the *wave vector* κ, where it is related to bound states ($E < 0$) by:

$$\kappa^2 = -\frac{2m}{\hbar^2} E. \tag{13.2}$$

The Schrödinger equation now takes the form

$$\frac{d^2 \psi(x)}{dx^2} - \frac{2m}{\hbar^2} V(x)\psi(x) = \kappa^2 \psi(x). \tag{13.3}$$

The **problem** states that the particle is bound, which means that it is confined to some finite region of space, which, in turn, implies that $\psi(x)$ is normalizable. (Unbound particles do

---

1 The equation for 2D or 3D motion, or with time-dependence, requires the solution of a partial differential equation, as discussed in Chapter 24.

Computational Physics: Problem Solving with Python, Fourth Edition.
Rubin H. Landau, Manuel J. Páez, and Cristian C. Bordeianu.
© 2024 WILEY-VCH GmbH. Published 2024 by WILEY-VCH GmbH.

not have normalizable wave functions.) The only way to have a normalizable wave function, is if $\psi(x)$ decays exponentially as $x \to \pm\infty$, where $V = 0$:[2]

$$\psi(x) \to \begin{cases} e^{-\kappa x}, & \text{for } x \to +\infty, \\ e^{+\kappa x}, & \text{for } x \to -\infty. \end{cases} \quad (13.4)$$

In summary, although we know how to solve the ODE (13.1) with our numerical tools, we must also figure out a technique to do so within the constraints of the boundary conditions (13.4). This extra condition turns the ODE problem into an *eigenvalue problem*, which has solutions (*eigenvalues*) for only certain values of the energy E or κ. The ground-state energy corresponds to the smallest (most negative) eigenvalue. Because the greater the number of oscillations in a wave function, the greater is the kinetic energy of the bound particle, the more nodes in a wave function, the higher the energy. Accordingly, we expect that the bound state with the least energy (ground-state) will have a nodeless wave function.

13.1.1 Model: Nucleon in a Box

The numerical methods we describe are capable of handling the most realistic potential shapes. Yet to make a connection with the standard textbook case, and to permit some analytic checking, we will use a simple model in which the potential $V(x)$ in (13.1) is a finite square well (Figure 13.1):

$$V(x) = \begin{cases} -V_0 = -83 \, \text{MeV}, & \text{for } |x| \le a = 2 \, \text{fm}, \\ 0, & \text{for } |x| > a = 2 \, \text{fm}, \end{cases} \quad (13.5)$$

where values of 83 MeV for the depth, and 2 fm for the radius, are typical for nuclear-bound states. With this potential the Schrödinger equation (13.3) becomes

$$\frac{d^2\psi(x)}{dx^2} + \left(\frac{2m}{\hbar^2}V_0 - \kappa^2\right)\psi(x) = 0, \quad \text{for } |x| \le a, \quad (13.6)$$

$$\frac{d^2\psi(x)}{dx^2} - \kappa^2\psi(x) = 0, \quad \text{for } |x| > a. \quad (13.7)$$

To evaluate the ratio of constants here, we insert c^2, the speed of light squared, into both the numerator and the denominator and then some familiar values:

$$\frac{2m}{\hbar^2} = \frac{2mc^2}{(\hbar c)^2} \simeq \frac{2 \times 940 \, \text{MeV}}{(197.32 \, \text{MeV fm})^2} = 0.0483 \, \text{MeV}^{-1} \, \text{fm}^{-2}. \quad (13.8)$$

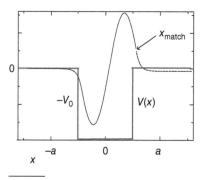

Figure 13.1 A square well in bold, and the wave function within it. At the location x_m near the right edge of the well, the wave function computed by integration in from the left is matched to the one computed by integration in from the right (dashed curve).

---

2 If we were working with a Coulomb potential, its very slow falloff would require using Coulomb functions at $\pm\infty$, not exponentials [Landau, 1996].

13.2   Algorithm: ODE Solver + Search

A key element in solving for bound states is the realization that the second-order ODE will have two solutions: one that decays as $x \to \pm\infty$, and a second one that grows as $x \to \pm\infty$. Consequently, a numerical solution, being an approximation, will always contain some of each. Yet if we integrate step-by-step on a solution that is increasing in value, then the bit of the decreasing function that is mixed in will get smaller and smaller, while the increasing function keeps increasing, and thus the solution becomes more accurate. Likewise, if we integrate step-by-step on a solution that is decreasing in value, then the bit of the increasing function that is mixed in will get greater and greater, while the decreasing function keeps decreasing, and thus the solution becomes less accurate. *Always try to integrate an ODE as the function is increasing!*

The solution to our eigenvalue problem combines the numerical solution of the ordinary differential equation (13.3) with a trial-and-error search for a wave function that also satisfies the boundary conditions (13.4). This is carried out in several steps:

1) Start at the far *left* at $x = -x_\infty \simeq -\infty$, where $x_\infty \gg a$. Since the potential $V = 0$ in this region, the analytic solution here is $e^{\pm\kappa x}$. Accordingly, assume that the wave function there satisfies the left-hand boundary condition:

$$\psi_L(x = -x_\infty) = e^{+\kappa x} = e^{-\kappa x_\infty}. \tag{13.9}$$

2) Use your `rk4` ODE solver to integrate $\psi_L(x)$ toward the origin (to the right), from $x = -x_\infty$, until you reach the *matching radius x_m*. In this way, we are integrating, step-by-step, over an increasing function. The exact value of this matching radius is not important, and our final solution should be independent of it. In Figure 13.1, we show a sample solution with $x_m \simeq a$, that is, we match just beyond the right edge of the potential. In Figure 13.2, we see some guesses that do not match.

3) Start at the extreme *right*, that is, at $x = +x_\infty \simeq +\infty$, with a wave function that satisfies the right-hand boundary condition:

$$\psi_R(x = \kappa x_\infty) = e^{-\kappa x} = e^{-\kappa x_\infty}. \tag{13.10}$$

4) Use your `rk4` solver to step $\psi_R(x)$ in toward the origin (to the left), from $x = +x_\infty$, until you reach the matching radius x_m. In this way, we are integrating, step-by-step, over an increasing function. This means that we have integrated up to the potential well (Figure 13.1).

Figure 13.2   Two guesses for the energy that are either too low or too high to be an eigenvalue. We see that the low-E guess does not oscillate fast enough to match a dying exponential, while the high-E guess oscillates too fast.

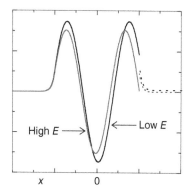

5) In order for probability and current to be continuous at $x = x_m$, $\psi(x)$, and $\psi'(x)$ must be continuous there. Requiring the ratio $\psi'(x)/\psi(x)$, called the *logarithmic derivative*, to be continuous there encapsulates both continuity conditions into a single condition, and is independent of ψ's normalization.

6) Although we do not know ahead of time what value for κ will be an eigenvalue, we still need a starting value for it in order to use our ODE solver. Such being the case, we start the solution with a guess. A good guess for ground-state energy would be a value somewhat up from that at the bottom of the well, $E > -V_0$.

7) Because it is unlikely that a guess will be correct, the left- and right-wave functions will not quite match at $x = x_m$ (Figure 13.2). This is fine because we can use the amount of mismatch to improve the next guess. We measure how well the right and left wave functions match by calculating the difference in logarithmic derivatives:

$$\Delta(E, x) = \frac{\psi_L'(x)/\psi_L(x) - \psi_R'(x)/\psi_R(x)}{\psi_L'(x)/\psi_L(x) + \psi_R'(x)/\psi_R(x)}\Bigg|_{x=x_m}, \tag{13.11}$$

where the denominator is there to avoid overly large or small numbers. Next, we try a different energy, note how much $\Delta(E)$ has changed, and use the change to deduce a better guess for the energy. The search continues until the left and right ψ'/ψ match within some set tolerance that depends on the precision in energy desired.

13.2.1 Not Recommended: Matchless Searching

A simpler approach to finding bound states is to try out a bunch of energies, and for each just integrate out from 0 to infinity. Then, if the wave function has not blown up, you have found a bound state. The problem with this approach is that it integrates step-by-step on a decreasing function. This means that the small amount of increasing function that is mixed in from the start keeps getting larger, while the desired solution keeps getting smaller, and so you will end up with a wave function containing a large percentage of error. *Always try to integrate on increasing functions!*

13.2.2 Numerov Algorithm for Schrödinger ODE ⊙

We generally recommend the fourth-order Runge–Kutta method for solving ODEs. However, if the ODE being solved does not contain any first derivatives (such as our Schrödinger equation), then it is possible to use the Numerov algorithm, which is specialized for just this situation. While this algorithm is not as general as rk4, it is of $\mathcal{O}(h^6)$, and thus speeds up the calculation by providing additional precision.

We start by rewriting the Schrödinger equation (13.3) in the generic form,

$$\frac{d^2\psi}{dx^2} + k^2(x)\psi = 0, \quad k^2(x) = \frac{2m}{\hbar^2}\begin{cases} E + V_0, & \text{for } |x| < a, \\ E, & \text{for } |x| > a, \end{cases} \tag{13.12}$$

where $k^2 = -\kappa^2$ for bound states. Observe that although (13.12) is specialized to a square well, other potentials would have $V(x)$ in place of $-V_0$. The trick in the Numerov method is to get extra precision in the second derivative by taking advantage of there being no first derivative $d\psi/dx$ in (13.12). We start with the Taylor expansions of the wave function,

$$\psi(x+h) \simeq \psi(x) + h\psi^{(1)}(x) + \frac{h^2}{2}\psi^{(2)}(x) + \frac{h^3}{3!}\psi^{(3)}(x) + \frac{h^4}{4!}\psi^{(4)}(x) + \cdots$$

$$\psi(x-h) \simeq \psi(x) - h\psi^{(1)}(x) + \frac{h^2}{2}\psi^{(2)}(x) - \frac{h^3}{3!}\psi^{(3)}(x) + \frac{h^4}{4!}\psi^{(4)}(x) + \cdots,$$

where $\psi^{(n)}$ signifies the nth derivative $d^n\psi/dx^n$. Because the expansion of $\psi(x-h)$ has odd powers of h appearing with negative signs, all odd powers cancel when we add $\psi(x+h)$ and $\psi(x-h)$ together:

$$\psi(x+h) + \psi(x-h) \simeq 2\psi(x) + h^2\psi^{(2)}(x) + \frac{h^4}{12}\psi^{(4)}(x) + \mathcal{O}(h^6),$$

$$\Rightarrow \quad \psi^{(2)}(x) \simeq \frac{\psi(x+h) + \psi(x-h) - 2\psi(x)}{h^2} - \frac{h^2}{12}\psi^{(4)}(x) + \mathcal{O}(h^4).$$

To obtain an algorithm for the second derivative, we eliminate the fourth-derivative term by applying the operator $1 + \frac{h^2}{12}\frac{d^2}{dx^2}$ to the Schrödinger equation (13.12):

$$\psi^{(2)}(x) + \frac{h^2}{12}\psi^{(4)}(x) + k^2(x)\psi + \frac{h^2}{12}\frac{d^2}{dx^2}[k^2(x)\psi^{(4)}(x)] = 0. \tag{13.13}$$

We eliminate the $\psi^{(4)}$ terms by substituting the derived expression for $\psi^{(2)}$:

$$\frac{\psi(x+h) + \psi(x-h) - 2\psi(x)}{h^2} + k^2(x)\psi(x) + \frac{h^2}{12}\frac{d^2}{dx^2}[k^2(x)\psi(x)] \simeq 0. \tag{13.14}$$

Now we use a central-difference approximation for the second derivative:

$$h^2\frac{d^2[k^2(x)\psi(x)]}{dx^2} \simeq [(k^2\psi)_{x+h} - (k^2\psi)_x] + [(k^2\psi)_{x-h} - (k^2\psi)_x]. \tag{13.15}$$

After this substitution, we obtain the Numerov algorithm:

$$\psi(x+h) \simeq \frac{2\left[1 - \frac{5}{12}h^2k^2(x)\right]\psi(x) - \left[1 + \frac{h^2}{12}k^2(x-h)\right]\psi(x-h)}{1 + h^2k^2(x+h)/12}. \tag{13.16}$$

We see that the Numerov algorithm uses the values of ψ at the two previous steps x and $x-h$ to move ψ forward to $x+h$. To step backward in x, we need only to reverse the sign of h. Our implementation of this algorithm, Numerov.py, is given in Listing 13.1.

13.2.3 Implementation: Eigenvalues via ODE Solver + Bisection Algorithm

1) Combine your bisection algorithm search program with your rk4 or Numerov ODE solver program to create an eigenvalue solver. Start with a step size $h = 0.04$.
2) Write a method that calculates the matching function $\Delta(E,x)$ as a function of energy and matching radius. This method will be called by the bisection algorithm program to search for the energy at which $\Delta(E,x=2)$ vanishes.
3) As a first guess, take $E \simeq -65\,\text{MeV}$.
4) Search until $\Delta(E,x)$ changes in only the fourth decimal place. We do this in the code QuantumEigen.py given in Listing 13.2.
5) Print out the value of the energy for each iteration. This will give you a feel as to how well the procedure converges, as well as a measure of the precision obtained. Try different values for the tolerance of the logarithmic derivative until you are confident that you are obtaining three good decimal places in the energy.

6) Build in a limit to the number of iterations you permit, with a warning if the iteration scheme fails.

7) Plot the wave function and potential on the same graph (you will have to scale one ordinate).

8) Deduce, by counting the number of nodes in the wave function, whether the solution found is a ground state (no nodes) or an excited state (with nodes) and whether the solution is even or odd about the origin (the ground state must be even).

9) Include in your version of Figure 13.1 a horizontal line within the potential indicating the energy of the ground state relative to the potential's depth.

10) Increase the value of the initial energy guess and search for excited states. Make sure to examine the wave function for each state found to establish that it is continuous, and to count the number of nodes to see if you have missed any states.

11) Add each new state found as another horizontal bar within the potential.

13.2.4 Explorations

1) Check to see how well your search procedure works by using arbitrary values for the starting energy. For example, because no bound-state energies can lie below the bottom of the well, try $E \geq -V_0$, as well as some arbitrary fractions of V_0. In every case examine the resulting ground state wave function and check that it is both symmetric and continuous.

2) Increase the depth of your potential progressively until you find several bound states. Look at the wave function in each case, and correlate the number of nodes in the wave function with the position of the bound state in the well.

3) Explore how a bound-state energy changes as you change the depth V_0 of the well. In particular, as you keep decreasing the depth, watch the eigenenergy move closer to $E = 0$ and see if you can find the potential depth at which the bound state has $E \simeq 0$.

4) For a fixed well depth V_0, explore how the energy of a bound state changes as the well radius a is varied. Larger radius should give increased binding.

5) Solve for the wave function of a linear potential:

$$V(x) = -V_0 \begin{cases} |x|, & \text{for } |x| < a, \\ 0, & \text{for } |x| > a. \end{cases} \tag{13.17}$$

There is less potential here than for a square well, so you may expect smaller binding energies and a less confined wave function. (For this potential, there are no analytic results with which to compare.)

6) Compare the results obtained, and the time the computer took to get them, using both the Numerov and rk4 methods.

7) **Newton–Raphson extension:** Extend the eigenvalue search by using the Newton–Raphson method in place of the bisection algorithm. Determine the speedup.

13.3 Classical Chaotic Scattering

One might expect that the classical scattering of a projectile from a passive target will vary smoothly. Yet experiments (Figure 13.3 left) have found that when a projectile undergoes multiple internal scatterings, its final trajectory appears unrelated to its initial one.

 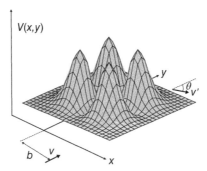

Figure 13.3 *Left*: A classic pinball machine in which the bumpers lead to multiple scatterings. A potential model that may support multiple internal scatterings. The incident velocity v is in the y direction at an impact parameter b. After scattering, the particle moves off at an angle θ.

Problem Determine if multiple internal scatterings may lead to such a chaotic situation. (Chaos is discussed more fully in Chapter 15.)

13.3.1 Model and Theory

Our model for scattering from the bumpers in pinball machines is a point particle scattering from the stationary 2D potential [Bleher *et al.*, 1990]

$$V(x,y) = \pm x^2 y^2 e^{-(x^2+y^2)}. \tag{13.18}$$

As seen on the right of Figure 13.3, this potential has four circularly symmetric peaks in the xy plane. The minus sign in (13.18) (which we will drop from now on) would produce attractive potential wells. Due to there being four peaks, it seems possible to have multiple scatterings in which the projectile bounces back and forth between the peaks, like in a pinball machine.

The **theory** for this problem is classical dynamics. Visualize a scattering experiment in which a projectile starts out at $(x = b, y = -\infty)$ with velocity $\mathbf{v}$ (Figure 13.3). The distance b is called the *impact parameter*. After scattering and moving out to $y = +\infty$, the projectile is observed at the scattering angle θ. Because a fixed potential does not recoil and carry off energy, the speed of the projectile does not change, only its direction.

An experiment would measure the number of particles scattered at each scattering angle θ. The analysis would convert the measurements into the differential cross section $\sigma(\theta)$:

$$\sigma(\theta) = \lim_{\Delta\Omega,\ \Delta A \to 0} \frac{N_{\text{scatt}}(\theta)/\Delta\Omega}{N_{\text{in}}/\Delta A_{\text{in}}}. \tag{13.19}$$

Here $N_{\text{scatt}}(\theta)$ is the number of particles per unit of time scattered into the detector at angle θ that subtends a solid angle $\Delta\Omega$, and N_{in} is the number of particles per unit of time incident on the target of cross-sectional area ΔA_{in}.

The definition (13.19) for the cross section is the one used by experimentalists. We need not worry about applying it. Instead, we need to solve for the trajectory $[x(t), y(t)]$ of the projectile scattering from the potential (13.18), and from that deduce the dependence of scattering angle $\theta(b)$ on the impact parameter b. Once we have that we can calculate the differential cross section [Marion and Thornton, 2019]:

$$\sigma(\theta) = \left| \frac{d\theta}{db} \right| \frac{b}{\sin \theta(b)}. \tag{13.20}$$

As your computation should show, there are parameter values for which $d\theta/db$ gets very large, or even discontinuous, and this leads to chaotic cross sections.

We need to solve Newton's law for $[x(t), y(t)]$ in the potential (13.18):

$$\mathbf{F} = m\mathbf{a}$$

$$-\frac{\partial V}{\partial x}\hat{i} - \frac{\partial V}{\partial y}\hat{j} = m\frac{d^2\mathbf{x}}{dt^2}, \tag{13.21}$$

$$\Rightarrow \quad -2y^2x(1-x^2)e^{-(x^2+y^2)} = m\frac{d^2x}{dt^2}, \tag{13.22}$$

$$-2x^2y(1-y^2)e^{-(x^2+y^2)} = m\frac{d^2y}{dt^2}. \tag{13.23}$$

The peaks of the potential in Figure 13.3 right are at $x = \pm 1$ and $y = \pm 1$, where $V_{max} = e^{-2}$. This sets the energy scale for the problem.

13.3.2 Implementation

Although (13.22) and (13.23) are simultaneous second-order ODEs, we can still use our standard `rk4` ODE solver and formalism by extending them from two to four dimensions:

$$\frac{d\mathbf{y}(t)}{dt} = \mathbf{f}(t, \mathbf{y}), \tag{13.24}$$

$$y^{(0)} \stackrel{\text{def}}{=} x(t), \quad y^{(1)} \stackrel{\text{def}}{=} y(t), \quad y^{(2)} \stackrel{\text{def}}{=} \frac{dx}{dt}, \quad y^{(3)} \stackrel{\text{def}}{=} \frac{dy}{dt}. \tag{13.25}$$

(The order in which the $y^{(i)}$s are assigned is arbitrary.) When applied to (13.22)–(13.23), and expressed in terms of the $y^{(i)}$'s, we have:

$$f^{(0)} = y^{(2)}, \qquad\qquad\qquad f^{(1)} = y^{(3)}, \tag{13.26}$$

$$f^{(2)} = \frac{-1}{m}2y^2x(1-x^2)e^{-[x^2+y^2]} \tag{13.27}$$

$$= \frac{-1}{m}2y^{(1)^2}y^{(0)}(1-y^{(0)^2})e^{-[y^{(0)^2}+y^{(1)^2}]}, \tag{13.28}$$

$$f^{(3)} = \frac{-1}{m}2x^2y(1-y^2)e^{-[x^2+y^2]} \tag{13.29}$$

$$= \frac{-1}{m}2y^{(0)^2}y^{(1)}(1-y^{(1)^2})e^{-[y^{(0)^2}+y^{(1)^2}]}. \tag{13.30}$$

To deduce the scattering angle from our calculation, we examine the trajectory of the projectile at $y \simeq \infty$, which we take to be the y value for which the potential has essentially vanished, $|PE|/E \le 10^{-10}$. The scattering angle is deduced from the components of velocity,

$$\theta = \tan^{-1}\left(\frac{v_y}{v_x}\right) = \texttt{math.atan2(y, x)}. \tag{13.31}$$

Here `atan2` is a function that computes the arctangent in the correct quadrant, without a division by v_x which can cause an overflow.

13.3.3 Assessment

1) Apply the `rk4` method to solve the simultaneous ODEs (13.22) and (13.23).
2) The **initial conditions** are $(x = b, y = y_\infty)$, where $|\text{PE}(y_\infty)|/E \leq 10^{-10}$.
3) Good starting parameters are $m = 0.5$, $v_y(0) = 0.5$, $v_x(0) = 0.0$, $\Delta b = 0.05$, $-1 \leq b \leq 1$. You may want to lower the energy and use a finer step size once you have found regions of rapid variation in the cross section.
4) Plot a number of trajectories $[x(t), y(t)]$ that show usual and unusual behaviors. In particular, plot those for which back angle scattering occurs, and, consequently, for which there must have been significant multiple scatterings.
5) Plot a number of phase space trajectories $[x(t), \dot{x}(t)]$ and $[y(t), \dot{y}(t)]$. How do these differ from those of bound states?
6) Determine the scattering angle $\theta =$ `atan2(Vx,Vy)` by determining the velocity components of the scattered particle after it has left the interaction region, that is, when $\text{PE}/E \leq 10^{-10}$.
7) Identify which characteristics of a trajectory lead to discontinuities in $d\theta/db$ and thus $\sigma(\theta)$.
8) Run the simulations for both attractive and repulsive potentials, and for a range of energies less than and greater than $V_{\text{max}} = \exp(-2)$.
9) **Time delay:** Another way to find unusual behavior in scattering is to compute the *time delay* $T(b)$ as a function of the impact parameter b. The time delay is the increase in the time it takes a particle to travel through the interaction region due to its interactions. Look for highly oscillatory regions in the semilog plot of $T(b)$, and once you find some, repeat the simulation at a finer scale by setting $b \simeq b/10$ (the structures are fractals, as discussed in Chapter 14).
10) OK, now go back and do this all again, but with an attractive potential that may not want to let the projectile go free!

13.4 Projectile Motion with Drag

Golf and baseball players claim that balls appear to fall out of the sky at the end of their trajectories (sort of like the solid curve in Figure 13.4, which was computed with the program `ProjectileAir.py` in Listing 13.3).

Your **problem** is to determine whether there is a physics explanation for this effect, or whether it is "all in the mind's eye."

Figure 13.4 shows the initial velocity V_0 and inclination θ for a projectile launched from the origin. If we ignore air resistance, the projectile has only the force of gravity acting on it, a

Figure 13.4 The trajectories of a projectile fired with initial velocity V_0 in the θ direction. The lower curve includes air resistance.

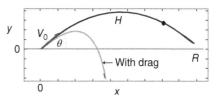

constant acceleration $a_y = -g = -9.8\,\text{m/s}^2$, and the familiar analytic solutions:

$$x(t) = V_0 \cos \theta t, \qquad y(t) = V_0 \sin \theta t - \tfrac{1}{2}gt^2, \tag{13.32}$$

$$v_x(t) = V_{0x}, \qquad v_y(t) = V_{0y} - gt, \tag{13.33}$$

$$y(x) = \frac{V_{0y}}{V_{0x}}x - \frac{g}{2V_{0x}^2}. \tag{13.34}$$

Likewise, it is easy to show that the range $R = 2V_0^2 \sin \theta \cos \theta / g$ and the maximum height $H = \tfrac{1}{2}V_0^2\sin^2\theta/g$.

The parabola (13.34) for frictionless motion is symmetric about its midpoint, which does not resemble a ball falling out of the sky. To see if air resistance will change that, we include a frictional force $\mathbf{F}^{(f)}$ in Newton's second law:

$$\mathbf{F}^{(f)} - mg\hat{\mathbf{e}}_y = m\frac{d^2\mathbf{x}(t)}{dt^2}, \tag{13.35}$$

$$\Rightarrow \quad F_x^{(f)} = m\frac{d^2x}{dt^2}, \qquad F_y^{(f)} - mg = m\frac{d^2y}{dt^2}. \tag{13.36}$$

As a model for what is really more complicated, we assume that the frictional force is proportional to some power n of the projectile's speed [Marion and Thornton, 2019]:

$$\mathbf{F}^{(f)} = -km\,|v|^n\,\frac{\mathbf{v}}{|v|}, \tag{13.37}$$

where the $-\mathbf{v}/|v|$ factor ensures that the frictional force is always in a direction opposite that of the velocity. Experiments indicate that the power n is noninteger and varies with velocity. The equations of motion are thus

$$\frac{d^2x}{dt^2} = -k\,v_x^n\,\frac{v_x}{|v|}, \qquad \frac{d^2y}{dt^2} = -g - k\,v_y^n\,\frac{v_y}{|v|}, \qquad |v| = \sqrt{v_x^2 + v_y^2}. \tag{13.38}$$

Or in dynamical form:

$$\frac{dy^{(0)}}{dt} = y^{(1)}, \qquad \frac{dy^{(1)}}{dt} = \frac{1}{m}F_x^{(f)}(\mathbf{y}) \tag{13.39}$$

$$\frac{dy^{(2)}}{dt} = y^{(3)}, \qquad \frac{dy^{(3)}}{dt} = \frac{1}{m}F_y^{(f)}(\mathbf{y}) - g, \tag{13.40}$$

$$f^{(0)} = y^{(1)}, \qquad f^{(1)} = \tfrac{1}{m}F_x^{(f)}, \qquad f^{(2)} = y^{(3)}, \qquad f^{(3)} = \tfrac{1}{m}F_y^{(f)} - g.$$

Consider three values for n, each of which represents a different model for the air resistance: (i) $n = 1$ for low velocities; (ii) $n = 3/2$, for medium velocities; and (iii) $n = 2$ for high velocities.

13.4.1 Assessment

1) Modify your rk4 program so that it solves the simultaneous ODEs for projectile motion (13.38) with friction ($n = 1$).
2) Check that you obtain graphs similar to those in Figure 13.4.
3) Use (13.37) with $n = 1$ for low velocities, $n = 3/2$, for medium-velocities, and $n = 2$ for high-velocities. Adjust the value of k for the latter two cases such that the initial force of friction $k\,v_0^n$ is the same for all three cases.
4) What is your conclusion about balls falling out of the sky?

13.5   2- and 3-Body Planetary Orbits

13.5.1   Planets via Two of Newton's Laws

Newton's explanation of the motion of the planets in terms of a universal law of gravitation is one of the greatest achievements of science. He was able to prove that planets traveled in elliptical orbits with the sun at one vertex, and then go on to predict the periods of the motions. All Newton needed to do was invent calculus and postulate that the force between a planet of mass m and the sun of mass M is

$$F_g = -\frac{GmM}{r^2}.$$ (13.41)

Here r is the planet-sun CM distance, G is the universal gravitational constant, and the attractive force lies along the line connecting the planet and the sun (Figure 13.5 left). The hard part for Newton was solving the resulting differential equations. In contrast, the numerical solution is straightforward. Even for planets, the equation of motion is still

$$\mathbf{F} = m\mathbf{a} = m\frac{d^2\mathbf{x}}{dt^2}.$$ (13.42)

In Cartesian components (Figure 13.5):

$$F_x = F_g \cos\theta = F_g\frac{x}{r} = F_g\frac{x}{\sqrt{x^2+y^2}},$$ (13.43)

$$F_y = F_g \sin\theta = F_g\frac{y}{r} = F_g\frac{y}{\sqrt{x^2+y^2}}.$$ (13.44)

The equation of motion (13.42) is thus two simultaneous second-order ODEs:

$$\frac{d^2x}{dt^2} = -GM\frac{x}{(x^2+y^2)^{3/2}}, \qquad \frac{d^2y}{dt^2} = -GM\frac{y}{(x^2+y^2)^{3/2}}.$$ (13.45)

1) Assume units such that $GM = 1$ and the initial conditions

$$x(0) = 0.5, \ y(0) = 0, \quad v_x(0) = 0.0, \quad v_y(0) = 1.63.$$ (13.46)

2) Modify your ODE solver program to solve (13.45).
3) Make sure to use small enough time steps to achieve high precision. Then you should find that the orbits are closed and fall upon themselves.
4) Experiment with the initial conditions until you find the ones that produce a circular orbit (a special case of an ellipse).
5) Note the effect of progressively increasing the initial velocity until the orbits open up and the planets become unbound.

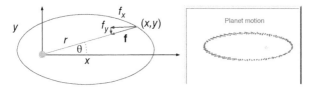

Figure 13.5   *Left:* The components of the gravitational force on a planet at a distance r from the sun. *Right:* The precession of a planet's orbit for a gravitational force $\propto 1/r^4$.

6) For the same initial conditions that produced the ellipse, investigate the effect of the power in (13.41) being $1/r^{2+\alpha}$ with $\alpha \neq 0$. Even for small values for α, you should find that the ellipses now rotate or precess (Figure 13.5). (A small value for α is predicted by general relativity (Chapter 19).)

13.5.2 The Discovery of Neptune

The planet Uranus was discovered in 1781 by William Herschel and found to have an orbital period of approximately 84 years. And yet, by 1846, when Uranus had not even completed a full orbit around the sun, something seemed wrong. This could be explained if Uranus was being perturbed by a yet-to-be-discovered planet lying about 50% further away from the sun than Uranus. The planet Neptune was thus discovered theoretically and confirmed experimentally. (If Pluto is discarded as just a dwarf planet, then Neptune is the most distant planet in the solar system.)

Assume that the orbits of Neptune and Uranus are circular and coplanar, and that the initial angular positions with respect to the x-axis are as given in this table:

	Mass	Distance	Orbital period	Angular position
	($\times 10^{-5}$ Solar masses)	(AU)	(Years)	(in 1690)
Uranus	4.366244	19.1914	84.0110	$\sim 205.64^0$
Neptune	5.151389	30.0611	164.7901	$\sim 288.38^0$

Use these data and rk4 to find the variation in angular position of Uranus with respect to the Sun as a result of the influence of Neptune during one complete orbit of Neptune. Consider

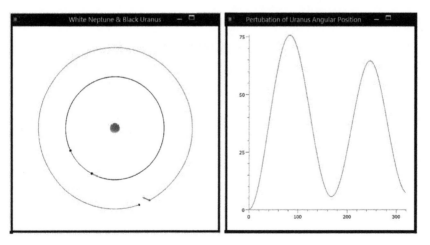

Figure 13.6 A snapshot from the animated output of the code `UranusNeptune.py` showing: *Left*: The orbits of Uranus (inner circle) and of Neptune (outer circle) with the sun in the center. The arrows indicate the Uranus–Neptune force that causes a perturbation in the orbits. *Right*: The perturbation in the angular position of Uranus as a function of time resulting from the presence of Neptune.

only the forces of the Sun and Neptune on Uranus. In the astronomical units, $M_s = 1$ and $G = 4\pi^2$. Figure 13.6 shows the output of our program that used these constants:

```
   G = 4*pi*pi              # AU, Msun=1
   mu = 4.366244e-5         # Uranus mass
 3 M = 1.0                  # Sun mass
   mn = 5.151389e-5         # Neptune mass
   du = 19.1914             # Uranus Sun distance
   dn = 30.0611             # Neptune sun distance
 7 Tur = 84.0110            # Uranus Period
   Tnp = 164.7901           # Neptune Period
   omeur = 2*pi/Tur         # Uranus angular velocity
   omennp = 2*pi/Tnp        # Neptune angular velocity
11 omreal = omeur
   urvel = 2*pi*du/Tur      # Uranus orbital velocity UA/yr
   npvel = 2*pi*dn/Tnp      # Neptune orbital velocity UA/yr
   radur = (205.64)*pi/180. # in radians
15 urx = du*cos(radur)      # init x Uranus in 1690
   ury = du*sin(radur)      # init y Uranus in 1690
   urvelx = urvel*sin(radur)
   urvely = -urvel*cos(radur)
19 radnp = (288.38)*pi/180. # Neptune angular pos.
```

13.6 Code Listings

Listing 13.1 QuantumNumerov.py Solves the time-independent Schrödinger equation for bound-state energies using a Numerov method.

```
 1 # QuantumNumerov.py: Solve quantum bound state via Numerov algorithm
   # hbarc* omega=hbarc*sqrt(k/m)=19.733,  r mc**2=940 MeV, k=9.4
   # E =(N+1/2)hbarc*omega = (N+1/2)19.733, N=0,2,4, change if N odd

 5 from numpy import *
   import numpy as np, matplotlib.pyplot as plt

   n = 1000; m = 2;  imax = 100;  Xleft0 = -10; Xright0 = 10; h  = 0.02
 9 amin= 81.; amax = 92.;  e = amin;  de  = 0.01;   eps= 1e-4; im = 500
   nl = im + 2;  nr = n - im + 2; xmax = 5.0
   print("nl, nr",nl, nr)
   print(h)
13 xLeft = arange(-10,0.02,0.02); xRight = arange(10,0.02,-0.02)
   xp = arange(-10,10,0.02)              # Bisection interval
   uL =  zeros((503),float);   uR =  zeros([503],float)
   k2L = zeros([1000],float);  k2R = zeros([1000],float)
17 uL[0] = 0; uL[1] =0.00001;  uR[0] = 0; uR[1] = 0.00001

   def V(x):                       # Potential harmonic oscillator
       v = 4.7*x*x
21     return v
   def setk2(e):                    # Set k2L=(sqrt(e-V))^2, k2R
       for i in range(0,n):
           xLeft = Xleft0 + i*h
25         xr = Xright0 - i*h
           fact=0.04829              # 2m*c**2/hbarc**2
           k2L[i] = fact*(e-V(xLeft))
           k2R[i] = fact*(e-V(xr))
29 def Numerov (n,h,k2,u,e):
       setk2(e)
       b=(h**2)/12.0                 # L & R wave functions
       for i in range(1,n):
33         u[i+1]=(2*u[i]*(1-5.*b*k2[i])-(1+b*k2[i-1])*u[i-1])/(1+b*k2[i+1])
   def diff(e):
       Numerov (nl,h,k2L,uL,e)       # Left wf
       Numerov (nr,h,k2R,uR,e)       # Right wf
```

```
37      f0 = (uR[nr-1] + uL[nl-1] - uR[nr-3] - uL[nl-3])/(h*uR[nr-2])
        return f0

    istep = 0
41  x1 = arange(-10,.02,0.02);    x2 = arange(10,-0.02,-0.02)
    fig = plt.figure()
    ax = fig.add_subplot(111)
    ax.grid()
45  while abs(diff(e)) > eps :          # Bisection algorithm
        e =(amin + amax)/2
        print(e,istep)
        if diff(e)*diff(amax) > 0: amax = e
49      else: amin = e
        ax.clear()
        plt.text(3,-200,'Energy= %10.4f'%(e),fontsize=14)
        plt.plot(x1,uL[:-2])
53      plt.plot(x2,uR[:-2])
        plt.xlabel('x')
        plt.ylabel('ψ(x)' ,fontsize=18)
        plt.title('R & L Wavefunctions Matched at x = 0')
57      istep = istep+1
        plt.pause(0.8)  # Pause to delay figures
    plt.show()
```

Listing 13.2   QuantumEigen.py Solves the time-independent Schrödinger equation for bound-state energies using the rk4 algorithm.

```
    # QuantumEigen.py:   Finds E and psi via rk4 + bisection

3   # m/(hbar*c)**2= 940MeV/(197.33MeV-fm)**2 =0.4829, well width=20 fm
    # well depth 10 MeV, Wave function not normalized

    from visual import *
7
    psigr = display(x=0,y=0,width=600,height=300, title='R & L Wavefunc')
    Lwf = curve(x=list(range(502)),color=color.red)
    Rwf = curve(x=list(range(997)),color=color.yellow)
11  eps        = 1E-3                                      # Precision
    n_steps    = 501
    E          = -17.0                                     # E guess
    h          = 0.04
15  count_max  = 100
    Emax       = 1.1*E                                     # E limits
    Emin       = E/1.1

19  def f(x, y, F,E):
        F[0] = y[1]
        F[1] = -(0.4829)*(E-V(x))*y[0]
    def V(x):
23      if (abs(x) < 10.):  return (-16.0)                 # Well depth
        else:               return (0.)
    def rk4(t, y,h,Neqs,E):
        F   = zeros((Neqs),float)
27      ydumb      = zeros((Neqs),float)
        k1 = zeros((Neqs),float)
        k2 = zeros((Neqs),float)
        k3 = zeros((Neqs),float)
31      k4 = zeros((Neqs),float)
        f(t, y, F,E)
        for i in range(0,Neqs):
            k1[i] = h*F[i]
35          ydumb[i] = y[i] + k1[i]/2.
        f(t + h/2., ydumb, F,E)
        for i in range(0,Neqs):
            k2[i] = h*F[i]
39          ydumb[i] = y[i] + k2[i]/2.
        f(t + h/2., ydumb, F,E)
        for i in range(0,Neqs):
```

```
                k3[i]=   h*F[i]
43              ydumb[i] = y[i] + k3[i]
            f(t + h, ydumb, F,E);
            for i in range(0,Neqs):
                k4[i]=h*F[i]
47              y[i]=y[i]+(k1[i]+2*(k2[i]+k3[i])+k4[i])/6.0
    def diff(E, h):
            y = zeros((2),float)
            i_match = n_steps//3                        # Matching radius
51          nL = i_match + 1
            y[0] = 1.E-15;                              # Initial left wf
            y[1] = y[0]*sqrt(-E*0.4829)
            for ix in range(0,nL + 1):
55              x = h * (ix  -n_steps/2)
                rk4(x, y, h, 2, E)
            left = y[1]/y[0]                            # Log  derivative
            y[0] = 1.E-15;             # slope for even;  reverse for odd
59          y[1] = -y[0]*sqrt(-E*0.4829)               # Initialize R wf
            for ix in range(n_steps,nL+1,-1):
                x = h*(ix+1-n_steps/2)
                rk4(x, y, -h, 2, E)
63          right = y[1]/y[0]                           # Log derivative
            return( (left - right)/(left + right) )
    def plot(E, h):                             # Repeat integrations for plot
            x = 0.
67          n_steps = 1501                             # # integration steps
            y = zeros((2),float)
            yL = zeros((2,505),float)
            i_match = 500                              # Matching point
71          nL = i_match + 1;
            y[0] = 1.E-40                              # Initial left wf
            y[1] = -sqrt(-E*0.4829) *y[0]
            for ix in range(0,nL+1):
75              yL[0][ix] = y[0]
                yL[1][ix] = y[1]
                x = h * (ix -n_steps/2)
                rk4(x, y, h, 2, E)
79          y[0] = -1.E-15                      # - slope: even;  reverse for odd
            y[1] = -sqrt(-E*0.4829)*y[0]
            j=0
            for ix in range(n_steps -1,nL + 2,-1):      # right wave function
83              x = h * (ix + 1 -n_steps/2)              # Integrate in
                rk4(x, y, -h, 2, E)
                Rwf.x[j] = 2.*(ix + 1 -n_steps/2)-500.0
                Rwf.y[j] = y[0]*35e-9 +200
87              j +=1
            x = x-h
            normL = y[0]/yL[0][nL]
            j=0
91          # Renormalize L wf & derivative
            for ix in range(0,nL+1):
                x = h * (ix-n_steps/2 + 1)
                y[0] = yL[0][ix]*normL
95              y[1] = yL[1][ix]*normL
                Lwf.x[j] = 2.*(ix  -n_steps/2+1)-500.0
                Lwf.y[j] = y[0]*35e-9+200              # Factor for scale
                j +=1
99  for count in range(0,count_max+1):
            rate(1)                                   # Slow rate to show changes
            # Iteration loop
            E = (Emax + Emin)/2.                      # Divide E range
103         Diff = diff(E, h)
            if (diff(Emax, h)*Diff > 0): Emax = E      # Bisection algorithm
            else:                        Emin = E
            if ( abs(Diff) < eps ):      break
107         if count >3:                              # First iterates too irregular
                rate(4)
                plot(E, h)
            elabel      = label(pos=(700, 400), text='E=', box=0)
111         elabel.text = 'E=%13.10f' %E
            ilabel      = label(pos=(700, 600), text='istep=', box=0)
```

```
        ilabel.text = 'istep=%4s' %count
     elabel      = label(pos=(700, 400), text='E=', box=0)      # Last iteration
115  elabel.text = 'E=%13.10f' %E
     ilabel      = label(pos=(700, 600), text='istep=', box=0)
     ilabel.text = 'istep=%4s' %count
     print("Final eigenvalue E = ",E)
119  print("iterations, max = ",count)
```

Listing 13.3 ProjectileAir.py Solves for projectile motion with air resistance as well as analytically for the frictionless case.

```
# ProjectileAir.py: Order dt^2 projectile trajectory + drag

3  from visual import *
   from visual.graph import *

   v0 = 22.;   angle = 34.;   g = 9.8;   kf = 0.8;   N = 5
7  v0x = v0*cos(angle*pi/180);   v0y = v0*sin(angle*pi/180)
   T = 2*v0y/g;   H = v0y*v0y/2/g;   R = 2*v0x*v0y/g
   graph1 = gdisplay(title='Projectile with & without Drag',
    xtitle='x', ytitle='y', xmax=R, xmin=-R/20.,ymax=8,ymin=-6.0)
11 funct = gcurve(color=color.red)
   funct1 = gcurve(color=color.yellow)
   print('No Drag T =',T,', H =',H,', R =',R)

15 def plotNumeric(k):
     vx = v0*cos(angle*pi/180.)
     vy = v0*sin(angle*pi/180.)
     x = 0.0
19   y = 0.0
     dt = vy/g/N/2.
     print("\n With Friction  ")
     print("        x              y")
23   for i in range(N):
         rate(30)
         vx = vx - k*vx*dt
         vy = vy - g*dt - k*vy*dt
27       x = x + vx*dt
         y = y + vy*dt
         funct.plot(pos=(x,y))
         print(" %13.10f  %13.10f "%(x,y))
31 def plotAnalytic():
     v0x = v0*cos(angle*pi/180.)
     v0y = v0*sin(angle*pi/180.)
     dt = 2.*v0y/g/N
35   print("\n  No Friction  ")
     print("        x              y")
     for i in range(N):
         rate(30)
39       t = i*dt
         x = v0x*t
         y = v0y*t -g*t*t/2.
         funct1.plot(pos=(x,y))
43       print(" %13.10f  %13.10f"%(x ,y))

   plotNumeric(kf)
   plotAnalytic()
```

14

Fractals and Statistical Growth Models

In this chapter we implement models that create fractals. We emphasize the simple underlying rules, the statistical aspects of the rules, and the meaning of self-similarity. To the extent that these models generate structures that look like those in nature, it is reasonable to assume that the natural processes may be following similar rules arising from some basic physics or biology.

It is common to notice regular and eye-pleasing natural objects, such as plants and sea shells, that do not have well-defined geometric patterns. When analyzed mathematically, some of these patterns have a dimension that is a fractional number. Benoit Mandelbrot, who first studied fractional-dimension figures with supercomputers at IBM Research, gave them the name *fractals* [Mandelbrot, 1982]. Some geometric objects, such as Koch curves, are exact fractals with the same dimension for all their parts. Other objects, such as bifurcation curves of Chapter 15, are statistical fractals in which elements of randomness are mixed in, in which case there may be different dimensions for each part of the object.

Consider an abstract object such as the density of charge within an atom. There are an infinite number of ways to define the "size" of this object. For example, each moment $\langle r^n \rangle$ is a measure of the size, and there is an infinite number of moments. Likewise, when we deal with complicated objects, there are different definitions of dimension, and each may give a somewhat different value.

The *Hausdorff–Besicovitch dimension* d_f, is based on our knowledge that a line has dimension 1, a triangle has dimension 2, and a cube has dimension 3. It seems perfectly reasonable, then, to take a mathematical formula that agrees with our experience with regular objects, and apply it to irregular objects. For simplicity, let us consider objects that have the same length L on each side, as do equilateral triangles and squares, and that have uniform density. We postulate that the dimension of an object is determined by the dependence of its total mass upon its length:

$$M(L) \propto L^{d_f}, \tag{14.1}$$

where the power d_f is the *fractal dimension*. As you may verify, this rule works for the 1D, 2D, and 3D figures we are familiar with, so it is a reasonable to try it elsewhere. When (14.1) is applied to irregular objects, we end up with fractional values for d_f. Actually, we will find it easier to determine the fractal dimension, not from an object's mass, which is *extensive*

Computational Physics: Problem Solving with Python, Fourth Edition.
Rubin H. Landau, Manuel J. Páez, and Cristian C. Bordeianu.
© 2024 WILEY-VCH GmbH. Published 2024 by WILEY-VCH GmbH.

(depends on size), but rather from its density, which is *intensive*. The density is defined as mass/length for a linear object, mass/area for a planar object, and mass/volume for a solid object. That being the case, for a planar object we hypothesize that

$$\rho = \frac{M(L)}{\text{area}} \propto \frac{L^{d_f}}{L^2} \propto L^{d_f-2}. \tag{14.2}$$

14.1 The Sierpiński Gasket

To generate our first fractal, shown in Figure 14.1, we play a game of chance in which we place dots at points placed randomly within a triangle [Bunde and Havlin, 1991]. Here are the rules (which you should try out in the margins right now).

1) Draw an equilateral triangle with vertices and coordinates:

   vertex 1: (a_1, b_1);   vertex 2: (a_2, b_2);   vertex 3: (a_3, b_3).

2) Place a dot at a random point $P = (x_0, y_0)$ within this triangle.
3) Find the next point by selecting randomly the integer 1, 2, or 3:
   a) If 1, place a dot halfway between P and vertex 1.
   b) If 2, place a dot halfway between P and vertex 2.
   c) If 3, place a dot halfway between P and vertex 3.
4) Repeat the process using the last dot as the new P.

Mathematically, the coordinates of successive points are given by the formulas

$$(x_{k+1}, y_{k+1}) = \frac{(x_k, y_k) + (a_n, b_n)}{2}, \quad n = \text{integer}\,(1 + 3r_i), \tag{14.3}$$

where r_i is a random number between 0 and 1, and where the *integer* function outputs the closest integer smaller than or equal to the argument. After 15,000 points, you should obtain a collection of dots like those on the left in Figure 14.1.

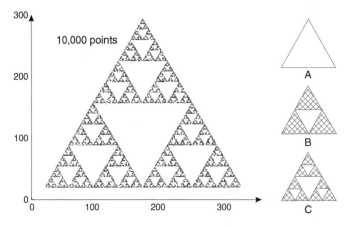

Figure 14.1 *Left*: The Sierpiński gasket, a statistical fractal containing 10,000 points. Note the self-similarity at different scales. *Right*: A geometric Sierpiński gasket constructed by successively connecting the midpoints of the sides of each equilateral triangle. The first three steps in the process are labeled as A, B, and C.

Exercise Write a program to produce a Sierpiński gasket. Determine empirically the fractal dimension of your figure. Assume that each dot has mass 1 and that $\rho = CL^\alpha$. (You can have the computer do the counting by defining an array box of all 0 values, and then by changing a 0 to a 1 when a dot is placed there.)

14.1.1 Measuring Fractal Dimension

The topology in Figure 14.1 was first analyzed by the Polish mathematician Sierpiński. Observe the same structure occurs in a small region as occurs in the entire figure. In other words, if the figure had infinite resolution, any part of the figure could be scaled up in size, and would be similar to the whole. This property is called *self-similarity*.

We construct a non-statistical form of the Sierpiński gasket by removing an inverted equilateral triangle from the center of all filled equilateral triangles (Figure 14.1 right). This creates the next figure to work on. We repeat the process ad infinitum, scaling up the triangles so each one has side $r = 1$ after each step. To see what is unusual about this type of object, we look at how its density (mass/area) changes with size, and then apply (14.2) to determine its fractal dimension. Assume that each triangle has mass m and assign unit density to the single triangle:

$$\rho(L = r) \propto \frac{M}{r^2} = \frac{m}{r^2} \stackrel{\text{def}}{=} \rho_0 \qquad \text{(Figure 14.1A).} \tag{14.4}$$

Now, for the equilateral triangle with side $L = 2$, the density is

$$\rho(L = 2r) \propto \frac{(M = 3m)}{(2r)^2} = \frac{3}{4}mr^2 = \frac{3}{4}\rho_0 \qquad \text{(Figure 14.1B).} \tag{14.5}$$

We see that the extra white space in Figure 14.1B leads to a density that is $\frac{3}{4}$ that of the previous stage. For the structure in Figure 14.1C, we obtain

$$\rho(L = 4r) \propto \frac{(M = 9m)}{(4r)^2} = \frac{9}{16}\frac{m}{r^2} = \left(\frac{3}{4}\right)^2 \rho_0. \qquad \text{(Figure 14.1C).} \tag{14.6}$$

We see that as we continue the construction process, the density of each new structure is $\frac{3}{4}$ that of the previous one. Interesting. Yet in (14.2) we derived that

$$\rho \propto CL^{d_f - 2}. \tag{14.7}$$

Equation (14.7) implies that a plot of the logarithm of the density ρ *versus* the logarithm of the length L for successive structures yields a straight line of slope

$$d_f - 2 = \frac{\Delta \log \rho}{\Delta \log L}. \tag{14.8}$$

As applied to our problem,

$$d_f = 2 + \frac{\Delta \log \rho(L)}{\Delta \log L} = 2 + \frac{\log 1 - \log \frac{3}{4}}{\log 1 - \log 2} \simeq 1.58496. \tag{14.9}$$

As is evident in Figure 14.1, as the gasket grows larger (and consequently more massive), it contains more open space. So despite the fact that its mass approaches infinity as $L \to \infty$, its density approaches zero! Because the Sierpiński gasket has a slope $d_f - 2 \simeq -0.41504$, it fills space to a lesser extent than a 2D object, but more than a 1D object; it is a fractal with dimension of ~1.6.

14.2   Growing Plants

It seems paradoxical that natural processes subject to chance can produce objects of such high regularity, symmetry, and beauty. For example, it is hard to believe that something as graceful as a fern (Figure 14.2 left) has random elements in it. Nonetheless, there is a clue here that much of the fern's beauty arises from the similarity of each part to the whole (self-similarity), with different ferns similar, but not identical to each other. These are all characteristics of fractals. Your **problem** is to discover if a simple algorithm including some randomness can draw regular ferns. If the algorithm produces objects that resemble ferns, then, presumably, you have uncovered some underlying mathematics similar to those responsible for the shapes of ferns.

14.2.1   Self-Affine Connection

In (14.3), which defines mathematically how a Sierpiński gasket is constructed, a *scaling factor* of $\frac{1}{2}$ is part of the relation of one point to the next. A more general transformation of a point $P = (x, y)$ into another point $P' = (x', y')$ via *scaling* is

$$(x', y') = s(x, y) = (sx, sy) \qquad \text{(scaling)}. \tag{14.10}$$

If the scale factor $s > 0$, an amplification occurs, whereas if $s < 0$, a reduction occurs. In our definition (14.3) of the Sierpiński gasket, we also added in a constant a_n. This is a *translation operation* of the general form

$$(x', y') = (x, y) + (a_x, a_y) \qquad \text{(translation)}. \tag{14.11}$$

Another operation, not used in the Sierpiński gasket, is a *rotation* by angle θ:

$$x' = x \cos\theta - y \sin\theta, \quad y' = x \sin\theta + y \cos\theta \qquad \text{(rotation)}. \tag{14.12}$$

This entire set of transformations, scalings, rotations, and translations, defines an *affine transformation* (affine denotes a close relation between successive points). The

Figure 14.2   *Left*: A fractal fern generated by 30,000 iterations of the algorithm (14.13). Enlarging this fern shows that each frond has a similar structure. *Right*: A fractal tree created with the algorithm (14.16).

transformation is still considered affine even if there are contractions and reflections. What is important is that the object created with these rules turns out to be self-similar; each step leads to new parts of the object that bear the same relation to the ancestor parts as the ancestors did to theirs. This is what makes the object look similar at all scales.

14.2.2 Barnsley's Fern

We obtain a Barnsley's fern [Barnsley and Hurd, 1992] by extending the dots game to one in which new points are selected using an affine connection, but with some elements of chance mixed in:

$$(x,y)_{n+1} = \begin{cases} (0.5, 0.27y_n), & \text{with 2\% probability,} \\ (-0.139x_n + 0.263y_n + 0.57 \\ 0.246x_n + 0.224y_n - 0.036), & \text{with 15\% probability,} \\ (0.17x_n - 0.215y_n + 0.408 \\ 0.222x_n + 0.176y_n + 0.0893), & \text{with 13\% probability,} \\ (0.781x_n + 0.034y_n + 0.1075 \\ -0.032x_n + 0.739y_n + 0.27), & \text{with 70\% probability.} \end{cases} \tag{14.13}$$

To select a transformation with probability $\mathcal{P}$, we select a uniform random number $0 \leq r \leq 1$, and then perform the transformation if r is in a range proportional to $\mathcal{P}$:

$$\mathcal{P} = \begin{cases} 2\%, & r < 0.02, \\ 15\%, & 0.02 \leq r \leq 0.17, \\ 13\%, & 0.17 < r \leq 0.3, \\ 70\%, & 0.3 < r < 1. \end{cases} \tag{14.14}$$

The rules (14.13) and (14.14) can be combined into one:

$$(x,y)_{n+1} = \begin{cases} (0.5, 0.27y_n), & r < 0.02, \\ (-0.139x_n + 0.263y_n + 0.57 \\ 0.246x_n + 0.224y_n - 0.036), & 0.02 \leq r \leq 0.17, \\ (0.17x_n - 0.215y_n + 0.408 \\ 0.222x_n + 0.176y_n + 0.0893), & 0.17 < r \leq 0.3, \\ (0.781x_n + 0.034y_n + 0.1075, \\ -0.032x_n + 0.739y_n + 0.27), & 0.3 < r < 1. \end{cases} \tag{14.15}$$

Although (14.13) makes the basic idea clearer (14.15), is easier to program.

The starting point in Barnsley's fern (Figure 14.2) is $(x_1, y_1) = (0.5, 0.0)$, and the points are generated by repeated iterations. An important property of this fern is that it is not completely self-similar, as you can see by noting how different are the stems and the fronds. Nevertheless, the stem can be viewed as a compressed copy of a frond, and the fractal obtained with (14.13) is still *self-affine*, yet with a dimension that varies from part to part of the fern. Our code `Fern3D.py` is given in Listing 14.1.

14.2.3 Self-Affine Trees

Now that you know how to grow ferns, look around and notice the regularity in trees (such as in Figure 14.2 right). Can it be that this also arises from a self-affine structure? Write a program, similar to the one for the fern, starting at $(x_1, y_1) = (0.5, 0.0)$, and iterate the following self-affine transformation:

$$(x_{n+1}, y_{n+1}) = \begin{cases} (0.05x_n, 0.6y_n), & 10\% \text{ probability,} \\ (0.05x_n, -0.5y_n + 1.0), & 10\% \text{ probability,} \\ (0.46x_n - 0.15y_n, 0.39x_n + 0.38y_n + 0.6), & 20\% \text{ probability,} \\ (0.47x_n - 0.15y_n, 0.17x_n + 0.42y_n + 1.1), & 20\% \text{ probability,} \\ (0.43x_n + 0.28y_n, -0.25x_n + 0.45y_n + 1.0), & 20\% \text{ probability,} \\ (0.42x_n + 0.26y_n, -0.35x_n + 0.31y_n + 0.7), & 20\% \text{ probability.} \end{cases} \tag{14.16}$$

14.3 Ballistic Deposition

There are a number of natural and manufacturing processes in which particles are deposited on a surface and form a film. If the particles are evaporated from a hot filament, there would be a randomness in the emission process, even though the produced films seem quite regular. Again we suspect fractals. Your **problem** is to develop a model that simulates this growth process, and compare your produced structures to those observed.

The idea of simulating random depositions was first reported in Vold [1959] in their simulation of the sedimentation of moist spheres in hydrocarbons. We shall examine Family and Vicsek [1985]'s method of simulation that results in the deposition shown in Figure 14.3.

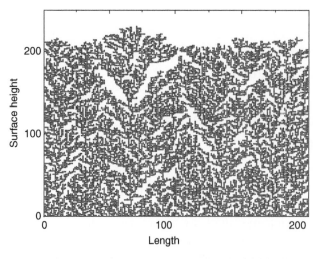

Figure 14.3 A simulation of the ballistic deposition of 20,000 particles onto a substrate of length 200. The vertical height increases in proportion to the length of deposition time, with the top being the final surface.

Consider particles falling onto, and sticking to, a horizontal line of length L composed of 200 deposition sites. All particles start from the same height, but to simulate their different emission velocities, we assume they start at random distances from the left side of the line. The simulation consists of generating uniform random sites between 0 and L, and having a particle stick to the site on which it lands. Seeing that the physical situation would have columns of aggregates of different heights, the particle may be stopped before it gets to a line, or it may bounce around and fall into a hole. We therefore assume that if the column height at which the particle lands is greater than that of both its neighbors, it will add to that height. If the particle lands in a hole, or if there is an adjacent hole, it will fill up the hole. We speed up the simulation by setting the height of the hole equal to the maximum of its neighbors. Here are the steps:

1) Choose a random site r.
2) Let the array h_r be the height of the column at site r.
3) Make the decision:

$$
h_r = \begin{cases} h_r + 1, & \text{if } h_r \geq h_{r-1}, \ h_r > h_{r+1}, \\ \max\left[h_{r-1}, h_{r+1}\right], & \text{if } h_r < h_{r-1}, \ h_r < h_{r+1}. \end{cases} \tag{14.17}
$$

The essential loop in the simulation is:

```
     spot = int(random)
     if (spot == 0)
         if ( coast[spot] < coast[spot+1] )
4            coast[spot] = coast[spot+1];
         else coast[spot]++;
     else if ( spot == coast.length - 1 )
         if ( coast[spot] < coast[spot-1] ) coast[spot] = coast[spot-1];
8        else coast[spot]++;
     else if ( coast[spot]<coast[spot-1] && coast[spot]<coast[spot+1] )
     if ( coast[spot-1] > coast[spot+1] ) coast[spot] = coast[spot-1];
         else coast[spot] = coast[spot+1];
12   else coast[spot]++;
```

The results of this simulation show several empty regions scattered throughout the line (Figure 14.3), which is an indication of the statistical nature of growing films. Simulations by Fereydoon produced fractal surfaces that reproduced the experimental observation that the average height increases linearly with time. (You will be asked to determine the fractal dimension of a similar surface as an exercise.)

Exercise Extend the simulation of random deposition to two dimensions, so rather than making a line of particles you now deposit upon an entire surface.

14.4 Length of British Coastline

In 1967 Benoit Mandelbrot asked a classic question, "How long is the coast of Britain?" [Mandelbrot, 1967]. If Britain had the shape of Colorado or Wyoming, both of which have straight-line boundaries, its perimeter would be a curve of dimension 1 with finite length. However, coastlines are geographic, and not geometric curves, with each portion of the coast appearing somewhat self-similar to the entire coast. If the perimeter of the coast is, in fact, a fractal, then its length is either infinite or meaningless. Mandelbrot deduced the dimension of the west coast of Britain to be $d_f = 1.25$, which implies infinite length. In your **problem**, we ask you to determine the dimension of the perimeter of one of your fractal simulations.

The length of the coastline of an island is the perimeter of that island. While the concept of perimeter is clear for geometric figures, some thought is required to give it meaning for an object that may be infinitely self-similar. Let us assume that a map maker has a ruler of length r. If she walks along the coastline and counts the number of times N that she must place the ruler down in order to *cover* the coastline, she will obtain a value for the length L of the coast as Nr. Imagine now that the map maker keeps repeating her walk with smaller and smaller rulers. If the coast were a geometric figure, or a *rectifiable curve*, at some point the length L would become essentially independent of r and would approach a constant. Nonetheless, as discovered empirically by Richardson [1961] for natural coastlines, such as those of South Africa and Britain, the perimeter appears to be an unusual function of r:

$$L(r) \simeq Mr^{1-d_f}, \tag{14.18}$$

where M and d_f are empirical constants. For a geometric figure, or for Colorado, $d_f = 1$, and the length approaches a constant as $r \to 0$. Yet for a fractal with $d_f > 1$, the perimeter $L \to \infty$ as $r \to 0$. This means that as a consequence of self-similarity, fractals may be of finite size, but have infinite perimeters. Physically, at some point, there may be no more details to discern as $r \to 0$ (say, at the quantum or Compton size limit), and so the limit may not be physically meaningful.

14.4.1 Box Counting Algorithm

Consider a line of length L broken up into segments of length r (Figure 14.4 left). The number of segments or "boxes" needed to cover the line is related to the size r of the box by

$$N(r) = \frac{L}{r} = \frac{C}{r}, \tag{14.19}$$

where C is a constant. One definition of fractional dimension is the power of r in this expression as $r \to 0$. In our example, it tells us that the line has dimension $d_f = 1$. If we now ask how many little circles of radius r it would take to *cover* or fill a circle of area A (Figure 14.4 middle), we will find that

$$N(r) - \lim_{r \to 0} \frac{A}{\pi r^2} \quad \Rightarrow \quad d_f = 2, \tag{14.20}$$

as expected. Likewise, counting the number of little spheres or cubes that can be packed within a large sphere tells us that a sphere has dimension $d_f = 3$. In general, if it takes N little spheres or cubes of side $r \to 0$ to cover some object, then the fractal dimension d_f can be deduced as

$$N(r) = C\left(\frac{1}{r}\right)^{d_f} = C' s^{d_f} \qquad (\text{as } r \to 0), \tag{14.21}$$

$$\log N(r) = \log C - d_f \log(r) \qquad (\text{as } r \to 0), \tag{14.22}$$

$$\Rightarrow \quad d_f = -\lim_{r \to 0} \frac{\Delta \log N(r)}{\Delta \log r}. \tag{14.23}$$

Here $s \propto 1/r$ is called the *scale* in geography, so $r \to 0$ corresponds to an infinite scale. To illustrate, you may be familiar with the low scale on a map being 10,000 m to a centimeter, while the high scale is 100 m to a centimeter. If we want the map to show small details (sizes), we need a map of high scale.

For the coastline problem, we'll use box counting to determine the dimension of a perimeter, and not of an entire figure. Once we have a value for the dimension, we will go on and determine the length of the perimeter via (14.18).

14.4.2 Coastline Exercise

Rather than ruin your eyes focusing on a geographic map, we suggest using something at hand that looks like a natural coastline, namely, the top portion of Figure 14.3. Determine d_f by covering this figure, or one you have generated, with a semitransparent piece of graph paper,[1] and counting the number of boxes containing any part of the coastline (Figures 14.4 and 14.5).

1) Print your coastline graph with the same physical scale (*aspect ratio*) for the vertical and horizontal axes. This is required because the graph paper you will use for box counting has square boxes and so you want your graph to also have the same vertical and horizontal scales. Place a piece of graph paper over your printout and look through the graph paper at your coastline. If you do not have a piece of graph paper available, or if you are unable to obtain a printout with the same aspect ratio for the horizontal and vertical axes, add a series of closely spaced horizontal and vertical lines to your coastline printout

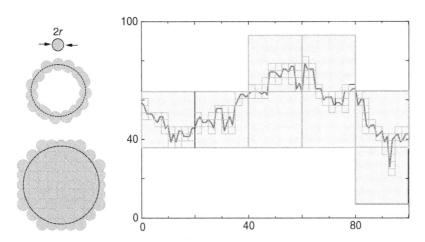

Figure 14.4 Examples of the use of box counting to determine fractal dimension. In the top left the "boxes" are circles and the perimeter is being covered. In the bottom left an entire figure is being covered, and on the right a "coastline" is being covered by boxes of two different sizes (scales). The fractal dimension can be deduced by recording the number of boxes of different scales needed to cover the figures.

---

1 Yes, we are suggesting a painfully analog technique based on the theory that trauma leaves a lasting impression. If you prefer, you can store your output as a matrix of 1 and 0 values, and let the computer do the counting, but this will take more of your time than being analog!

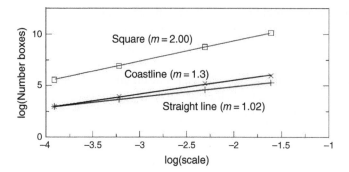

Figure 14.5 Fractal dimensions of a line, box, and coastline determined by box counting. The slope at vanishingly small scale determines the dimension.

and use these lines as your graph paper. (Box counting should still be accurate if both your coastline and your graph paper have the same aspect ratios.)

2) The vertical height in our printout was 17 cm, and the largest division on our graph paper was 1 cm. This sets the scale of the graph as 1:17, or $s = 17$ for the largest divisions (lowest scale). Measure the vertical height of your fractal, compare it to the size of the biggest boxes on whatever you are using as your piece of graph paper, and thus determine your lowest scale.

3) With our largest boxes of 1×1 cm, we found that the coastline passed through $N = 24$ boxes, that is, 24 large boxes covered the coastline at $s = 17$. Determine how many of the largest boxes (lowest scale) are needed to cover your coastline.

4) With our next smaller boxes of 0.5×0.5 cm, we found that 51 boxes covered the coastline at a scale of $s = 34$. Determine how many of the midsize boxes (midrange scale) are needed to cover your coastline.

5) With our smallest boxes of 1×1 mm, we found that 406 boxes covered the coastline at a scale of $s = 170$. Determine how many of the smallest boxes (highest scale) are needed to cover your coastline.

6) Equation (14.23) tells us that as the box sizes get progressively smaller, we have

$$\log N \simeq \log A + d_f \log s, \tag{14.24}$$

$$\Rightarrow \quad d_f \simeq \frac{\Delta \log N}{\Delta \log s} = \frac{\log N_2 - \log N_1}{\log s_2 - \log s_1} = \frac{\log(N_2/N_1)}{\log(s_2/s_1)}. \tag{14.25}$$

Clearly, only the relative scales matter because the proportionality constants cancel out in the ratio. A plot of $\log N$ *versus* $\log s$ should yield a straight line with a slope of d_f (1.23 for us). Determine the fractal dimension for your coastline. Although only two points are needed to determine the slope, use your lowest scale point as an important check. (Because the fractal dimension is defined as a limit for infinitesimal box sizes, the highest scale points are most significant.)

7) Using (14.18), we find that the length of our coastline for our s value is

$$L \propto s^{1.23-1} = s^{0.23}. \tag{14.26}$$

If we keep making the boxes smaller and smaller, so that we are looking at the coastline at higher and higher scale, *and* if the coastline is self-similar at all levels, then the scale

s will keep getting larger and larger with no limits (or at least until we get down to some quantum limit on small sizes), and thus

$$L \propto \lim_{s \to \infty} s^{0.23} = \infty. \tag{14.27}$$

Does your fractal imply an infinite coastline? Does it make sense that a small island like Britain, which you can walk around, has an infinite perimeter?

14.5 Correlated Growth

It is an empirical fact that there is increased likelihood that a plant will grow if there is another one nearby (Figure 14.6 left). This type of *correlation* also seems to occur in the deposition of surface films. Your **problem** is to include correlations in your surface simulation and observe the change it makes.

A variation of the ballistic deposition, known as the *correlated ballistic deposition*, simulates mineral deposition onto substrates on which dendrites form [Tait *et al.*, 1990; Sander *et al.*, 1994]. We extend the ballistic deposition algorithm to include the likelihood that a freshly deposited particle will attract another particle. The extension is to assume that the probability of sticking P depends inversely on the distance d that the added particle is from the last one (Figure 14.6 right):

$$P = c\,d^{-\eta}. \tag{14.28}$$

Here η is a parameter and c is a constant that sets the probability scale.[2] For our implementation we choose $\eta = 2$, which means that there is an inverse square attraction between the particles (decreased probability as they get farther apart).

As in our study of uncorrelated deposition, a uniform random number in the interval $[0, L]$ determines the column in which the particle will be deposited. We use the same rules

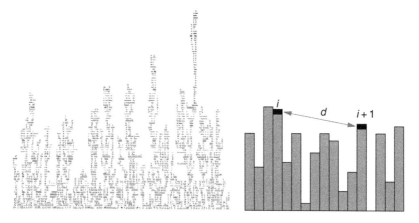

Figure 14.6 *Left*: A view of what might be the undergrowth of a forest or dendrites formed during surface deposition. *Right*: The probability of particle $i + 1$ sticking in one column depends upon the distance d from the previously deposited particle i.

---

2 The absolute probability, of course, must be less than one, but it is nice to choose c so that the relative probabilities produce a graph with easily seen variations.

about the heights as before, but now a second random number is used in conjunction with
(14.28) to decide if the particle will stick. For instance, if the computed probability is 0.6
and if $r < 0.6$, the particle will be accepted (sticks), if $r > 0.6$, the particle will be rejected.
Our code `column.py` is given in Listing 14.2.

14.6   Diffusion-Limited Aggregation

Consider a bunch of grapes on an overhead vine. Your **problem** is to create a model of how
its tantalizing shape might arise. In a flash of divine insight, you realize that these shapes, as
well as others, such as those of colloids and thin-film structures, may result from an aggregation process with particles diffusing around each other. In fact, a model of diffusion-limited
aggregation (DLA) has successfully explained the relation between a cluster's perimeter
and mass [Witten and Sander, 1981].

Exercise   Follow these steps to construct your model:

1) Define a 2D lattice of points represented by the array `grid[400,400]`, with all elements
   initially zero.
2) Place a seed particle at the center of the lattice by setting `grid[199,199] = 1`.
3) Imagine a circle of radius 180 lattice spacings centered at `grid[199,199]`. This is the
   circle from which you release particles.
4) Determine the angular location on the circle's circumference from which to release a
   particle by generating a uniform random angle between 0 and 2π.
5) You are about to release a new particle, and have it execute a random walk, much like
   the one we studied in Chapter 4, but restricted to vertical or horizontal jumps between
   lattice sites:
   a) Generate a uniform random number $0 < r_{xy} < 1$.
   b) if $r_{xy} < 0.5$, the motion will be vertical.
   c) if $r_{xy} \geq 0.5$, the motion will be horizontal.
6) Make the model more realistic by letting the length of each step vary according to a
   random Gaussian distribution. Generate a Gaussian-weighted random number in the
   interval $[-\infty, \infty]$. This is the size of the step, with the sign indicating direction. (*Hint*:
   The sum of a uniform random distribution provides a Gaussian distribution.)
7) We now know the total distance and direction the particle will travel. Have it jump one
   lattice spacing at a time until this total distance is covered.
8) Before a jump, check whether a nearest-neighbor site is occupied:
   a) If occupied, the particles stick together and stay in that position. The walk for that
      particle is over.
   b) If the site is unoccupied, the particle jumps one lattice spacing.
9) Continue the checking and jumping until the calculated distance is covered, until the
   particle sticks, or until it leaves the circle and is lost from our grip.
10) Once one random walk is over, release another particle, and repeat the process as often
    as desired. Because many particles are lost, you may need to generate hundreds of
    thousands of particles to form a cluster of several hundred particles. Your results should
    look like Figure 14.7.

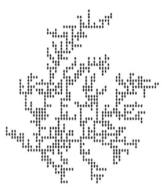

Figure 14.7 A globular cluster of particles of the type that might occur in a colloid.

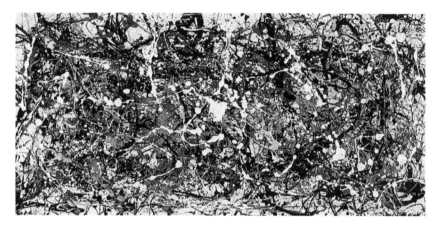

Figure 14.8 Number 8 by the American painter Jackson Pollock. (Used with permission State University of New York.) Some researchers claim that Pollock's paintings exhibit a characteristic fractal structure, while some others question this [Kennedy, 2006].

14.6.1 Fractal of DLA or Pollock

A cluster generated with the DLA technique is shown in Figure 14.7. We wish to analyze it to see if the structure is a fractal, and, if so, to determine its dimension. (As an alternative, you may analyze the fractal nature of the Pollock painting in Figure 14.8, a technique used to determine the authenticity of this sort of art.) As a control, *simultaneously* analyze a geometric figure, such as a square or circle, whose dimension is known. The analysis is a variation of the one used to determine the length of the coastline of Britain.

1) If you have not already done so, use the box-counting method to determine the fractal dimension of a simple square.
2) Draw a square of length L, small relative to the size of the cluster, around the seed particle. (Small might be seven lattice spacings to a side.)
3) Count the number of particles within the square.
4) Compute the particle density ρ by dividing the number of particles by the number of sites available in the box (49 in our example).
5) Repeat the procedure using larger and larger squares.

6) Stop when the cluster is covered.
7) The fractal dimension d_f is estimated from a log-log plot of the density ρ *versus* L. If the cluster is a fractal, then (14.2) tells us that $\rho \propto L^{d_f-2}$, and the graph should be a straight line of slope $d_f - 2$.

The graph we generated had a slope of -0.36, which corresponds to a fractal dimension of 1.66. Seeing that random numbers are involved, the graph you generate will be different, but the fractal dimension should be similar. (Actually, the structure is multifractal, and so the dimension also varies with location in the cluster.)

14.7 Fractals in Bifurcations

In the next chapter there is a project involving the logistics map where we plot the values of the number of bugs *versus* the growth parameter μ. Take one of the bifurcation graphs produced there and determine the fractal dimension of different parts of the graph by using the same technique that was applied to the coastline of Britain.

14.8 Cellular Automata Fractals

There is a class of statistical models known as *cellular automata* that produce complex behaviors from very simple rules. Cellular automata were developed by von Neumann and Ulam in the early 1940s (von Neumann was also working on the theory behind modern computers then). Though very simple, cellular automata have found applications in many branches of science [Peitgen *et al.*, 1994; Sipper, 1997]. Their definition [Barnsley and Hurd, 1992]:

> *A cellular automaton is a discrete dynamical system in which space, time, and the states of the system are discrete. Each point in a regular spatial lattice, called a cell, can have any one of a finite number of states, and the states of the cells in the lattice are updated according to a local rule. That is, the state of a cell at a given time depends only on its own state one time step previously, and the states of its nearby neighbors at the previous time step. All cells on the lattice are updated synchronously, and so the state of the entice lattice advances in discrete time steps.*

A cellular automaton in two dimensions consists of a number of square cells that grow upon each other. A famous one is *Conway's Game of Life*. In this, cells with value 1 are alive, while cells with value 0 are dead. Cells grow according to the rules:

1) If a cell is alive, and if two or three of its eight neighbors are alive, then the cell remains alive.
2) If a cell is alive, and if more than three of its eight neighbors are alive, then the cell dies due to overcrowding.
3) If a cell is alive, and only one of its eight neighbors is alive, then the cell dies of loneliness.
4) If a cell is dead, and more than three of its neighbors are alive, then the cell revives.

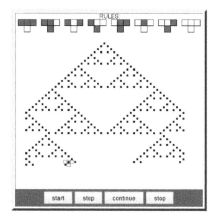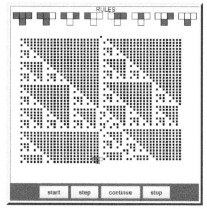

Figure 14.9 The rules for two versions of the Game of Life. The rules, given graphically on the top row, create the gaskets below. Our code `Gameoflife.py` is given in Listing 14.3.

Early studies of the statistical mechanics of cellular automata were made by Wolfram [1983], who indicated how one can be used to generate a Sierpiński gasket. Because we have already seen that a Sierpiński gasket exhibits fractal geometry (Section 14.1), this represents a microscopic model of how fractals may occur in nature. This model uses eight rules, given graphically at the top of Figure 14.9, to generate new cells from old. We see all possible configurations for three cells in the top row, and the begetted next generation in the row below. At the bottom of Figure 14.9, a Sierpiński gasket is so generated.

14.9 Perlin Noise Adds Realism ⊙

We have seen how statistical fractals are able to generate objects with a striking resemblance to those in nature. This appearance of realism may be further enhanced by including a type of coherent randomness known as *Perlin noise*. This technique was developed by Ken Perlin of New York University, who won an Academy Award (an Oscar) in 1997 for it and has continued to improve it Perlin [2023]. This type of coherent noise has found use in important physics simulations of stochastic media [Tickner, 2004], as well as in video games, and motion pictures like *Tron*.

The inclusion of Perlin noise in a simulation adds both randomness and a type of coherence among points in space that tends to make dense regions denser and sparse regions sparser. This is similar to our correlated ballistic deposition simulations (Section 14.3), and is related to chaos in its long-range randomness with short-range correlations. We start with some known functions of x and y, and add noise to them. For this purpose, Perlin used the mapping or *ease* function (Figure 14.12 right)

$$f(p) = 3p^2 - 2p^3. \tag{14.29}$$

As a consequence of its S shape, this mapping makes regions close to 0, even closer to 0, while making regions close to 1, even closer to 1 (in other words, it increases the tendency to clump, which shows up as higher contrast). We then break space up into a uniform rectangular grid of points (Figure 14.10), and consider a point (x, y) within a square with vertices

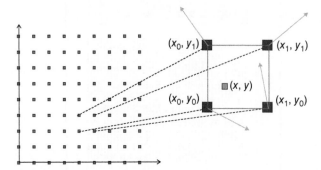

Figure 14.10 The coordinates used in adding Perlin noise. The rectangular grid is used to locate a square in space and a corresponding point within the square. As shown with the arrows, unit vectors $\mathbf{g}_i$ with random orientation are assigned at each grid point.

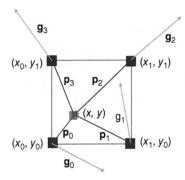

Figure 14.11 The coordinates used in adding Perlin noise. A point within each square is located by drawing the four $\mathbf{p}_i$. The $\mathbf{g}_i$ vectors are the same as on the left.

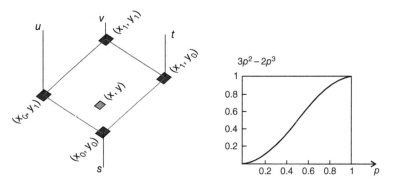

Figure 14.12 The mapping used in adding Perlin noise. *Left*: The numbers s, t, u, and v are represented by perpendiculars to the four vertices, with lengths proportional to their values. *Right*: The function $3p^2 - 2p^3$ is used as a map of the noise at a point like (x,y) to others close by.

(x_0, y_0), (x_1, y_0), (x_0, y_1), and (x_1, y_1). We next assign unit gradients vectors $\mathbf{g}_0$ to $\mathbf{g}_3$ with random orientation at each grid point. A point within each square is located by drawing the four $\mathbf{p}_i$ vectors (Figure 14.11):

$$\mathbf{p}_0 = (x - x_0)\mathbf{i} + (y - y_0)\mathbf{j}, \quad \mathbf{p}_1 = (x - x_1)\mathbf{i} + (y - y_0)\mathbf{j}, \tag{14.30}$$

$$\mathbf{p}_2 = (x - x_1)\mathbf{i} + (y - y_1)\mathbf{j}, \quad \mathbf{p}_3 = (x - x_0)\mathbf{i} + (y - y_1)\mathbf{j}. \tag{14.31}$$

Next, the scalar products of the $\mathbf{p}'$s and the $\mathbf{g}'$s are formed:

$$s = \mathbf{p}_0 \cdot \mathbf{g}_0, \; t = \mathbf{p}_1 \cdot \mathbf{g}_1, \; v = \mathbf{p}_2 \cdot \mathbf{g}_2, \; u = \mathbf{p}_3 \cdot \mathbf{g}_3. \tag{14.32}$$

As shown on the left in Figure 14.12, the numbers s, t, u, and v are assigned to the four vertices of the square and represented there by lines perpendicular to the square with lengths proportional to the values of s, t, u, and v (which can be positive or negative).

The actual mapping proceeds via a number of steps (Figure 14.13):

1) Transform the point (x, y) to (s_x, s_y),

$$s_x = 3x^2 - 2x^3, \quad s_y = 3y^2 - 2y^3. \tag{14.33}$$

2) Assign the lengths s, t, u, and v to the vertices in the mapped square.

Perlin noise

3) Obtain the height a (Figure 14.13) via linear interpolation between s and t.
4) Obtain the height b via linear interpolation between u and v.
5) Obtain s_y as a linear interpolation between a and b.
6) The vector c so obtained is now the two components of the noise at (x, y).

14.9.1 Ray Tracing Algorithms

Ray tracing is a technique that renders an image of a scene by simulating the way rays of light travel [Pov-Ray, 2023]. To avoid tracing rays that do not contribute to the final image, ray-tracing programs start at the viewer, trace rays backward onto the scene, and then back again onto the light sources. You can vary the location of the viewer and light sources and the properties of the objects being viewed, as well as atmospheric conditions such as fog, haze, and fire.

As an example of what this can do, on the right in Figure 14.14, we show the output from the ray-tracing program Pov-Ray [2023], using as input the coherent random noise

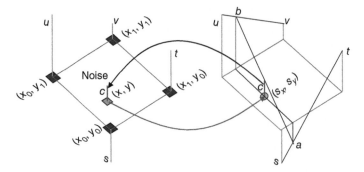

Figure 14.13 Perlin noise mapping. *Left*: The point (x, y) is mapped to point (s_x, x_y). *Right*: Using (14.33). Then three linear interpolations are performed to find c, the noise at (x, y).

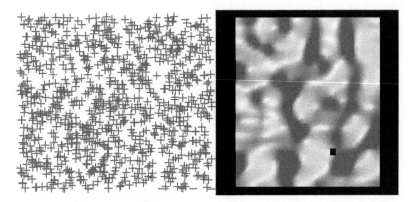

Figure 14.14 After the addition of Perlin noise, the random scatterplot on the left becomes the clusters on the right.

on the left in Figure 14.14. The program options we used are given in Listing 14.4, and are seen to include commands to color the islands, to include waves, and to give textures to the sky and the sea. Pov-Ray also allows the possibility of using Perlin noise to give textures to the objects to be created. For example, the stone cup on the right of the inset above has a marble-like texture produced by Perlin noise.

14.10 Code Listings

Listing 14.1 Fern3D.py Simulates the growth of ferns in 3D.

```
# Fern3D.py:  Fern in 3D, see Barnsley, "Fractals Everywhere"

from visual import *
from visual.graph import *
import random

imax = 20000
x = 0.5;    y = 0.0;   z = -0.2;   xn = 0.0;   yn = 0.0
graph1 = display(width=500, height=500, forward=(-3,0,-1),\
        title='3D Fractal Fern (rotate via right mouse button)', range=10)
graph1.show_rendertime = True # Pts/sphs: cycle=27/750 ms, render=6/30
```

```
12  pts = points(color=color.green, size=0.01)
    for i in range(1,imax):
       r = random.random();
       if ( r <= 0.1):                              # 10% probability
16        xn = 0.0
          yn = 0.18*y
          zn = 0.0
       elif ( r > 0.1 and r <= 0.7):                # 60% probability
20        xn =  0.85 * x
          yn =  0.85 * y + 0.1 * z + 1.6
          zn = -0.1  * y + 0.85 * z
       elif ( r > 0.7 and r <= 0.85):               # 15 % probability
24        xn =  0.2 * x - 0.2 * y
          yn =  0.2 * x + 0.2 * y + 0.8
          zn=   0.3 * z
       else:
28          xn = -0.2 * x +0.2 * y                  # 15% probability
            yn =  0.2 * x +0.2 * y + 0.8
            zn =  0.3 * z
       x = xn
32     y = yn
       z = zn
       xc = 4.0*x                                   # linear TF for plot
       yc = 2.0*y-7
36     zc = z
       pts.append(pos=(xc,yc,zc))
```

Listing 14.2   Column.py Simulates correlated ballistic deposition of minerals onto substrates on which dendrites form.

```
1  # Column.py: Fractal growth of columns

   from visual import *;
   import random
5
   maxi = 100000;   npoints = 200                # Number iterations, spaces
   i = 0;    dist = 0;    r = 0;    x = 0;    y = 0
   oldx = 0;    oldy = 0;    pp = 0.0;    prob = 0.0
9  hit = zeros( (200), int)
   graph1 = display(width = 500, height = 500, range=250,
                    title = 'Correlated Ballistic Deposition')
   pts = points(color=color.green, size=2)
13 for i in range(0, npoints): hit[i] = 0               # Clear array
   oldx = 100;            oldy = 0
   for i in range(1, maxi + 1):
      r = int(npoints*random.random() )
17    x = r - oldx
      y = hit[r] - oldy
      dist = x*x  + y*y
      if (dist == 0): prob = 1.0          # Sticking prob depends on last x
21    else: prob = 9.0/dist
      pp = random.random()
      if (pp < prob):
         if(r>0 and r<(npoints - 1) ):
25          if( (hit[r] >=  hit[r - 1]) and (hit[r] >=  hit[r + 1]) ):
               hit[r] = hit[r] + 1
            else:
               if (hit[r - 1] > hit[r + 1]):
29               hit[r] = hit[r - 1]
               else: hit[r] = hit[r + 1]
         oldx = r
         oldy = hit[r]
33       olxc = oldx*2 - 200                          # TF for plot
         olyc = oldy*4 - 200
         pts.append(pos=(olxc,olyc))
```

Listing 14.3   Gameoflife.py Is an extension of Conway's Game of Life in which cells always revive if one out of eight neighbors is alive.

```
# Gameoflife.py:          Cellular automata in 2 dimensions

'''* Rules: a cell can be either dead (0) or alive (1)
   * If a cell is alive:
   * on next step will remain alive if
   * 2 or 3 of its closer 8 neighbors are alive.
   * If > 3 of 8 neighbors are alive, cell dies of overcrowdedness
   * If less than  2 neighbors are alive the cell dies of loneliness
   * A dead cell will be alive if  3 of its 8 neighbors are alive'''

from visual import *
from visual.graph import * ; import random

scene = display(width= 500,height= 500, title= 'Game of Life')
cell  = zeros((50,50));          cellu = zeros((50,50))
curve(pos= [(-49,-49),(-49,49),(49,49),(49,-49),(-49,-49)],color=color.white)
boxes = points(shape='square', size=8, color=color.cyan)

def drawcells(ce):
    boxes.pos = []                              # Erase previous cells
    for j in range(0,50):
        for i in range(0,50):
            if ce[i,j] == 1:
                xx = 2*i-50
                yy = 2*j-50
                boxes.append(pos=(xx,yy))
def initial():
    for j in range (20,28):
        for  i in range(20, 28):
            r= int(random.random()*2)
            cell[j,i] = r
    return cell
def gameoflife(cell):
    for i in range(1,49):
        for j in range(1,49):
            sum1 = cell[i-1,j-1] + cell[i,j-1] + cell[i+1,j-1] # neighb
            sum2 = cell[i-1,j] + cell[i+1,j] + cell[i-1,j+1] \
                 + cell[i,j+1] + cell[i+1,j+1]
            alive = sum1+sum2
            if cell[i,j] == 1:
                if  alive == 2 or alive == 3:                   # Alive
                    cellu[i,j] = 1                              # Lives
                if  alive > 3 or alive < 2:     # Overcrowded or solitude
                    cellu[i,j] = 0                               # dies
            if  cell[i,j] == 0:
                if  alive == 3:
                    cellu[i,j] = 1                            # Revives
                else:
                    cellu[i,j] = 0                       # Remains dead
    alive = 0
    return cellu
temp = initial()
drawcells(temp)
while True:
    rate(6)
    cell = temp
    temp = gameoflife(cell)
    drawcells(cell)
```

Listing 14.4   Islands.pov The Pov-Ray ray-tracing commands needed to convert the coherent noise random plot of Figure 14.14 into the mountain-like image in Figure 14.14.

```
   // Islands.pov   Pov-Ray program to create Islands , by Manuel J Paez

   plane {
4    <0, 1, 0>, 0                                              // Sky
     pigment { color rgb <0, 0, 1> }
     scale 1
     rotate <0, 0, 0>
8    translate y*0.2
   }
   global_settings {
     adc_bailout 0.00392157
12   assumed_gamma 1.5
     noise_generator 2
   }
   #declare Island_texture = texture {
16   pigment {
       gradient <0, 1, 0>                                      // Vertical direction
       color_map {                                            // Color the islands
         [ 0.15 color rgb <1, 0.968627, 0> ]
20       [ 0.2   color rgb <0.886275, 0.733333, 0.180392>  ]
         [ 0.3   color rgb <0.372549, 0.643137, 0.0823529> ]
         [ 0.4   color rgb <0.101961, 0.588235, 0.184314>  ]
         [ 0.5   color rgb <0.223529, 0.666667, 0.301961>  ]
24       [ 0.6   color rgb <0.611765, 0.886275, 0.0196078> ]
         [ 0.69  color rgb <0.678431, 0.921569, 0.0117647> ]
         [ 0.74  color rgb <0.886275, 0.886275, 0.317647>  ]
         [ 0.86  color rgb <0.823529, 0.796078, 0.0196078> ]
28       [ 0.93  color rgb <0.905882, 0.545098, 0.00392157> ]
       }
     }
     finish {
32     ambient rgbft <0.2, 0.2, 0.2, 0.2, 0.2>
       diffuse 0.8
     }
   }
36 camera {                                    // Camera characteristics and location
     perspective
     location <-15, 6, -20>                                    // Located here
     sky <0, 1, 0>
40   direction <0, 0, 1>
     right <1.3333, 0, 0>
     up <0, 1, 0>
     look_at <-0.5, 0, 4>                                   //looking at that point
44   angle 36
   }
   light_source {<-10, 20, -25>, rgb <1, 0.733333, 0.00392157>}       // Light

48 #declare Islands = height_field {          // Takes  gif and finds heights
     gif "d:\pov\montania.gif"                      // Windows directory naming
     scale <50, 2, 50>
     translate <-25, 0, -25>
52 }
   object {                                                          // Islands
     Islands
     texture {
56     Island_texture
       scale 2
     }
   }
60 box {                                    // Upper face of the box is the sea
     <-50, 0, -50>, <50, 0.3, 50>            // Location of 2 opposite vertices
     translate <-25, 0, -25>
     texture {                                              // Simulate waves
64     normal {
         spotted
         0.4
         scale <0.1, 1, 0.1>
```

```
68        }
        pigment { color rgb <0.164706, 0.556863, 0.901961> }
      }
    }
72  fog {                                      // A constant fog is defined
      fog_type 1
      distance 30
      rgb <0.984314, 1, 0.964706>
76  }
```

15

Nonlinear Population Dynamics

We view nonlinear dynamics as one of the success stories of computational physics. It has been explored by scientists and engineers with computers as an essential tool, often then followed by mathematicians [Motter and Campbell, 2013]. The computations have led to the discovery of new phenomena such as chaos, solitons, and fractals; the first of which we cover in this chapter, and the last, later on. Here we look at discrete and continuous models of population dynamics that are simple, yet which yield surprising complex behavior. In Chapter 16, we explore nonlinear behavior in classical oscillations.

15.1 The Logistic Map, A Bug Population Model

Populations of bugs and patterns of weather do not appear to follow any simple laws.[1] At times, the population patterns appear stable, at other times they vary periodically, and at other times they appear chaotic, with no discernable regularity, only to settle back down to something simple again.

Problem Deduce if a simple law can produce such complicated behaviors.

Imagine a bunch of bugs reproducing generation after generation. We start with a N_0 bugs, then in the next generation we have to live with N_1 of them, and after n generations, there are N_n of them around to bug us. We want to develop a model of how N_n varies with the generation number n. Clearly, if the rates of breeding and dying are the same, then a stable population occurs. Yet bugs cannot live on love alone, they must also eat, and bugs, not being farmers, must compete for the available food supply. This tends to restrict their number to lie below some maximum population N_*. We want to build all of these observations into our model.

For guidance, we look to the radioactive decay simulation in Chapter 4 where the discrete decay law,

$$\Delta N / \Delta t = -\lambda N, \tag{15.1}$$

---

1 Except maybe in Oregon, where storm clouds come to spend their weekends.

Computational Physics: Problem Solving with Python, Fourth Edition.
Rubin H. Landau, Manuel J. Páez, and Cristian C. Bordeianu.
© 2024 WILEY-VCH GmbH. Published 2024 by WILEY-VCH GmbH.

led to exponential-like decay. Being clever, we start our modeling by reversing the sign of λ, which should give us *growth*:

$$\frac{\Delta N_i}{\Delta t} = \lambda N_i. \tag{15.2}$$

Yet we know that exponential growth eventually tapers off, with the population reaching a maximum N_* (the *carrying capacity*). Consequently, we modify the growth model (15.2) by changing the growth rate parameter λ to one that decreases as the population approaches N_*:

$$\lambda = \lambda'(N_* - N_i), \tag{15.3}$$

$$\Rightarrow \frac{\Delta N_i}{\Delta t} = \lambda'(N_* - N_i)N_i. \tag{15.4}$$

We expect that when N_i is small compared to N_*, the population will grow nearly exponentially. Yet we also expect that as N_i approaches N_*, the growth rate will decrease, eventually becoming negative if N_i exceeds the carrying capacity N_*. We can imagine the two possibilities leading to oscillations.

Equation (15.4) is a version of the *logistic map*. It is usually written in a form that relates the number of bugs in the future to the number in the present generation:

$$N_{i+1} = N_i + \lambda'\Delta t(N_* - N_i)N_i, \tag{15.5}$$

$$= N_i \left(1 + \lambda'\Delta t N_*\right) \left[1 - \frac{\lambda'\Delta t}{1 + \lambda'\Delta t N_*}N_i\right]. \tag{15.6}$$

This relation looks simpler when expressed in terms of dimensionless variables:

$$x_{i+1} = \mu x_i(1 - x_i), \tag{15.7}$$

$$\mu \stackrel{\text{def}}{=} 1 + \lambda'\Delta t N_*, \tag{15.8}$$

$$x_i \stackrel{\text{def}}{=} \frac{\lambda'\Delta t}{1 + \lambda'\Delta t N_*}N_i \simeq \frac{N_i}{N_*}. \tag{15.9}$$

Here x_i is a dimensionless population variable and μ is a (yet another) dimensionless growth parameter. Observe from (15.8), that if the number of bugs born per generation $\lambda'\Delta t$ is large, then $x_i \simeq N_i/N_*$, that is, x_i is essentially the fraction of the maximum population N_*. Consequently, realistic x values generally lie in the range $0 \leq x_i \leq 1$, with $x = 0$ corresponding to no bugs, and $x = 1$ corresponding to the carrying capacity. Also note that the *growth rate* μ equals 1, only if the breeding rate λ' equals 0, and is otherwise expected to be larger than 1.

The map (15.7) is seen to be the sum of linear and quadratic dependencies on x_i. It is called a map because it converts one number in a sequence to the next,

$$x_{i+1} = f(x_i). \tag{15.10}$$

For the logistic map, $f(x) = \mu x(1 - x)$, with the quadratic dependence on x making this a nonlinear map, and the dependence on only the one variable x making it a *one-dimensional* map.

Just by looking at (15.7), we really would not expect that anything as simple as this might be realistic description of bug population dynamics. However, if it exhibits some features similar to those found in nature, then it may well form the foundation for a more complete description (as we will develop in Section 15.5).

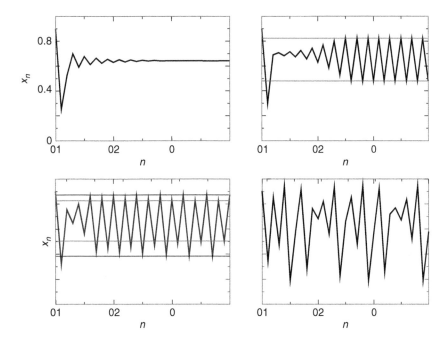

Figure 15.1   The bug population x_n *versus* the generation number n for the four growth rates: (a) $\mu = 2.8$, a single attractor; (b) $\mu = 3.3$, a double attractor; (c) $\mu = 3.5$, a quadruple attractor; (d) $\mu = 3.8$, a chaotic regime.

15.1.1 Exploring Map Properties

Rather than reading about how fancy mathematical analyses deduce the properties of the logistic map [Rasband, 1990], we here ask you to explore it yourself by generating and plotting sequences of x_i values. You should get results similar to those shown in Figure 15.1.

15.1.1.1 Stable Populations
A stable population is one that remains the same from generation to generation.

1) Start with an initial population $x_0 = 0.75$, called the *seed*. You should find that the dynamical effects are not sensitive to it. Also, as a check on the model, start with some negative and zero values for μ, which should produce decaying populations. Make plots of x_i versus i.
2) Now look for stable populations with $\mu = 0, 0.5, 1, 1.5, 2$.
3) Take note of the *transient* behaviors that occur for early generations before more regular behaviors set in.
4) For a fixed value of μ, try different values for the seed x_0, and thereby verify that while the transients may differ, the regular behaviors do not.

You should have found that this model yields stable populations for positive growth rates μ, with the maximum population reached more rapidly as μ gets larger. This is a good validation of the model. Some typical behaviors are shown in Figure 15.1. In Figure 15.1a, we see equilibration into a single population; in Figure 15.1b, we see oscillation between two population levels; in Figure 15.1c, we see oscillation among four levels; and in Figure 15.1d, we see a chaotic system.

15.1.2 Fixed Points

An important property of the map (15.7) is the possibility of the sequence x_i reaching one, or more, *fixed* points x_*, that is, populations at which the system remains, or returns to regularly. At a *one-cycle* fixed-point, there would be no change in the population from generation i to generation $i + 1$, that is,

$$x_{i+1} = x_i = x_*. \tag{15.11}$$

Substituting this relation into the logistic map (15.7) yields a quadratic equation we can easily solve:

$$\mu x_*(1 - x_*) = x_*, \tag{15.12}$$

$$\Rightarrow \quad x_* = 0, \quad \text{or} \quad x_* = \frac{\mu-1}{\mu}. \tag{15.13}$$

The nonzero fixed-point, $x_* = (\mu - 1)/\mu$, corresponds to a stable population in which there is a balance between birth and death, as in Figure 15.1a. In contrast, the $x_* = 0$ point is unstable since the population remains static only as long as no bugs exist; if even a few bugs are introduced, exponential growth occurs. Further analysis, which we are about to explore computationally, tells us that the stability of a population is determined by the magnitude of the derivative of the mapping function $f(x_i)$ at the fixed-point [Rasband, 1990]:

$$\left|\frac{df}{dx}\right|_{x_*} < 1 \quad \text{(stable)}. \tag{15.14}$$

For the one cycle of the logistic map (15.7), the derivative is

$$\frac{df}{dx}\bigg|_{x_*} = \mu - 2\mu x_* = \begin{cases} \mu, & \text{stable at } x_* = 0 \text{ if } \mu < 1, \\ 2 - \mu, & \text{stable at } x_* = \frac{\mu-1}{\mu} \text{ if } \mu < 3. \end{cases} \tag{15.15}$$

15.1.3 Period Doubling, Bifurcations

Equation (15.15) tells us that there will not be any stable populations for $\mu > 3$. In this case, the system undergoes *bifurcations* into two populations, a so-called *two-cycle*. The effect is known as *period doubling*, and is evident in Figure 15.1b. Because the system now moves between these two populations, the populations are called *attractors* or *cycle points*. We can easily predict the x values for two-cycle attractors by demanding that generation $i + 2$ has the same population as generation i:

$$x_i = x_{i+2} = \mu x_{i+1}(1 - x_{i+1}), \tag{15.16}$$

$$\Rightarrow \quad x_* = \frac{1 + \mu \pm \sqrt{\mu^2 - 2\mu - 3}}{2\mu}. \tag{15.17}$$

We see that as long as $\mu > 3$, the square root produces a real number and thus that physical solutions. We leave it to your explorations to discover how the system continues to bifurcate as μ is increased further. In all cases, the behavior repeats, with a single population bifurcating into two.

15.1.4 Mapping Implementation

It is now time to carry out a more careful investigation of the logistic map, following the original path of Feigenbaum [1979] and his hand calculator:

1) Confirm that you obtain the different patterns shown in Figure 15.1 for $\mu = (0.4, 2.4, 3.2, 3.6, 3.8304)$ and seed $x_0 = 0.75$.
2) Identify the following in your graphs:
   a) **Transients**: Irregular behaviors before reaching a regular behavior, and that the transients differ for different seeds.
   b) **Asymptotes**: In some cases, the steady state is reached after only 20 generations, while for larger μ values, hundreds of generations may be needed. These steady-state populations are independent of the seed.
   c) **Extinction**: If the growth rate is too low, $\mu \leq 1$, the population dies off.
   d) **Stable states**: The stable single-population states attained for $\mu < 3$ should agree with the prediction (15.13).
   e) **Multiple cycles**: Examine populations for a growth parameter μ increasing continuously through 3. Observe how the system continues to bifurcate. For example, Figure 15.1c with $\mu = 3.5$ contains four attractors (a *four-cycle*).
   f) **Intermittency**: Observe simulations for $3.8264 < \mu < 3.8304$. Here the system appears stable for a finite number of generations and then jumps all around, only to become stable again. (Old radios tended to do this.)

15.2 Chaos

"Chaos" has different meanings to different people. For present purposes, we define chaos as *the deterministic behavior of a system displaying no discernible regularity*. This may seem contradictory; if a system is deterministic, it must have step-to-step correlations, which, when added up, means long-range correlations. But when the behavior is chaotic, the complexities of the behavior may hide the determinism within. In an operational sense, *a chaotic system is one with an extremely high sensitivity to parameter values or initial conditions*. This sensitivity to even minuscule changes is so high that, in a practical sense, it is impossible to predict the long-range behavior without knowing the parameters to infinite precision, a physical impossibility. Yet because the system is mathematically deterministic, it is not random. As you may recall from Chapter 4, a random sequence has no correlation from one step to the next, whereas a chaotic one does.

1) Explore the long-term behaviors of the logistic map in the chaotic region starting with the two, essentially identical, seeds $x_0 = 0.75$ and $x_0' = 0.75(1 + \epsilon)$, where $\epsilon \simeq 2 \times 10^{-14}$.
2) Repeat the simulation with $x_0 = 0.75$ and two essentially identical survival parameters, $\mu = 4$ and $\mu = 4(1 - \epsilon)$, where $\epsilon \simeq 2 \times 10^{-14}$. Both simulations should start off the same, but eventually diverge.

15.3 Bifurcation Diagrams

Watching the population change as a function of generation number provides a good picture of the basic dynamics at work, at least until things get too complicated to discern patterns.

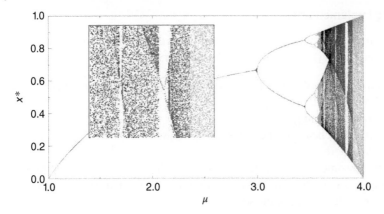

Figure 15.2 A bifurcation plot of attractor population x^* *versus* growth rate μ for the logistic map. The inset shows some details of a three-cycle window. (Gray scales indicate the regimes over which Hans Kowallik distributed the work on different CPUs when run in parallel.)

In particular, as the number of bifurcations keeps increasing, the output may seem too complicated for you to discern any pattern. One way to visualize what is going on, is to concentrate on the attractors, that is, those populations that appear to attract the solutions, and to which the solutions continuously return (long-term iterates). A plot of the x values of these attractors as a function of the growth parameter μ, turns out to be an illuminating window into the dynamics.

One such *bifurcation diagram* for the logistic map is shown in Figure 15.2. A corresponding, and similar, diagram for a very different map, the Gaussian map, is given in Figure 15.3. (More maps are given in Section 15.3.3.) To generate such a diagram, you have your calculation proceed through all values of μ in small steps. For each μ value, you wait while the system goes through hundreds (or more) of iterations, so that the transients die out; at this, you are at a fixed point x_*. Next, you write the pair (x_*, μ) to a file, and continue the iteration for hundreds of cycles, without changing μ. If the system falls into an n-cycle for this μ value, then there should predominantly be n different x_* values written to the file. Next, the value of the initial population x_0 is changed slightly, and the entire procedure is repeated to ensure that no fixed-points are missed. When finished, your program will have stepped through all the values of μ and x_0.

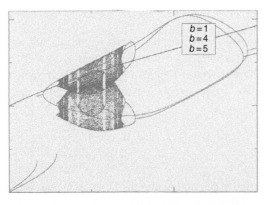

Figure 15.3 A bifurcation plot, x_* *versus* λ, for the Gaussian map. (W. Hager.)

15.3.1 Bifurcation Diagram Implementation

Our sample program `Bugs.py` is given in Listing 15.1. We ask you to reproduce Figure 15.2 at various levels of detail. You create a visualization of this sort by plotting individual points, with the density in each region of the screen determined by the number of points plotted there. When thinking about plotting many points, it is important to keep in mind that your monitor and printer can display only a finite number of pixels (picture elements). At present, an HD monitor has $1920 \times 1080 = 2{,}073{,}600$ pixels, but you do not need to use every pixel; in any case, printing at a finer resolution is a waste of time. Here's how to do it:

1) Break up the range $1 \leq \mu \leq 4$ into 1000 steps. These are the "bins" into which you will place the x_* values.
2) In order not to miss any structures in your bifurcation diagram, loop through a range of initial x_0 values.
3) Wait at least 200 generations for transients to die out, and then output the next several hundred (μ, x_*) values to a file.
4) Output your x_* values to no more than three or four decimal places. You will not be able to resolve more places than this on your plot, and this restriction will reduce the number of duplicate entries. You can use formatted output to control the number of decimal places, or you can do it via a simple conversion: multiply the x_i values by 1000, and then throw away the part to the right of the decimal point: `Ix[i]= int(1000*x[i])`. Then divide by 1000 if you want floating-point numbers.
5) Plot x_* *versus* μ using small symbols for the points, with no connections between points.
6) Enlarge (zoom in on) sections of your plot, and notice how a similar bifurcation diagram tends to be contained within each magnified portion (this is *self-similarity*).
7) Look over the series of bifurcations occurring

$$\mu_k \simeq 3,\ 3.449,\ 3.544,\ 3.5644,\ 3.5688,\ 3.569692,\ 3.56989,\ \dots. \tag{15.18}$$

8) Note how the end of this series is in a region of chaotic behavior, and that the system sometimes enters into the chaotic regions quickly. Accordingly, you may have to make plots over a very small range of μ values to see all of the structures there. A close examination of Figure 15.2 shows regions where, for a slight increase in μ, a very large number of populations suddenly change to very few populations. Whereas these may appear to be artifacts of the video display, this is a real effect, and these regions are called *windows*. Check that around $\mu = 3.828427$ chaos moves into a three-cycle window.

15.3.2 Feigenbaum Constants

Feigenbaum [1979] discovered that the sequence of μ_k values (15.18) at which bifurcations occur follows a regular pattern. Specifically, the μ values converge geometrically when expressed in terms of the distance between bifurcations δ:

$$\mu_k \to \mu_\infty - \frac{c}{\delta^k}, \qquad \delta = \lim_{k \to \infty} \frac{\mu_k - \mu_{k-1}}{\mu_{k+1} - \mu_k}. \tag{15.19}$$

Use your sequence of μ_k values to determine the three constants in (15.19), and compare them to those found by Feigenbaum:

$$\mu_\infty \simeq 3.56995, \quad c \simeq 2.637, \quad \delta \simeq 4.6692. \tag{15.20}$$

Amazingly, the value of δ is universal for all second-order maps.

15.3.3 Other Maps

Bifurcations and chaos are typical characteristics of nonlinear systems. Yet systems can be nonlinear in a number of ways. The table below lists four maps that generate x_i sequences containing bifurcations.

Name	$f(x)$	Name	$f(x)$
Logistic	$\mu x(1-x)$	Tent	$\mu(1-2\,|x-1/2|)$
Ecology	$xe^{\mu(1-x)}$	Quartic	$\mu[1-(2x-1)^4]$
Gaussian	$e^{-bx^2}+\mu$		

The tent map derives its nonlinear dependence from the absolute value operator, while the logistic map is seen to be a subclass of the ecology map. Explore the properties of these other maps and note the similarities and differences.

15.4 Measures of Chaos

Our definition of chaos in terms of unpredictability seems rather subjective, or maybe hard to apply analytically. Accordingly, several analytic measures of chaos have been developed, and in this section, we examine two, the Lyapunov coefficients and Shannon entropy.

15.4.1 Lyapunov Coefficients ⊙

The Lyapunov coefficient λ provides an analytic signal of chaos [Wolf *et al.*, 1985; Ramasubramanian and Sriram, 2000; Williams, 1997; Manneville, 1990]. Specifically, the coefficient is the rate parameter in the exponent describing the exponential growth of x_* versus μ. For 1D problems, there is only one such coefficient, whereas, in general, there is a coefficient for each degree of freedom. The essential assumption is that neighboring paths x_n near an attractor x_* have a time dependence $L \propto \exp(\lambda t)$. Consequently, if $\lambda > 0$, the number of fixed points grows exponentially, which is chaotic; if $\lambda = 0$, we have a marginally stable population; while $\lambda < 0$ implies a stable and periodic population. Mathematically, the Lyapunov coefficient is defined as

$$\lambda = \lim_{t \to \infty} \frac{1}{t} \log \frac{L(t)}{L(t_0)}, \tag{15.21}$$

where $L(t)$ is the distance between neighboring phase space trajectories at time t.

As an example, we'll calculate the Lyapunov exponent for a general 1D map,

$$x_{n+1} = f(x_n), \tag{15.22}$$

where the generation number n now replaces the time t. To determine stability, we examine perturbations about a reference trajectory x_0 by adding a small perturbation, and iterating once:

$$\hat{x}_0 = x_0 + \delta x_0, \qquad \hat{x}_1 = x_1 + \delta x_1. \tag{15.23}$$

We substitute this into (15.22) and expand f in a Taylor series around x_0:

$$x_1 + \delta x_1 = f(x_0 + \delta x_0) \simeq f(x_0) + \left.\frac{\delta f}{\delta x}\right|_{x_0} \delta x_0 = x_1 + \left.\frac{\delta f}{\delta x}\right|_{x_0} \delta x_0,$$

$$\Rightarrow \delta x_1 \simeq \left(\frac{\delta f}{\delta x}\right)_{x_0} \delta x_0. \tag{15.24}$$

This is the proof of our earlier statement that a negative df/dx indicates stability. To deduce the general result, we examine one iteration:

$$\delta x_2 \simeq \left(\frac{\delta f}{\delta x}\right)_{x_1} \delta x_1 = \left(\frac{\delta f}{\delta x}\right)_{x_0} \left(\frac{\delta f}{\delta x}\right)_{x_1} \delta x_0, \tag{15.25}$$

$$\Rightarrow \delta x_n = \prod_{i=0}^{n-1} \left(\frac{\delta f}{\delta x}\right)_{x_i} \delta x_0. \tag{15.26}$$

This last relation tells us how trajectories differ on the average after n steps:

$$|\delta x_n| = L^n |\delta x_0|, \qquad L^n = \prod_{i=0}^{n-1} \left|\left(\frac{\delta f}{\delta x}\right)_{x_i}\right|. \tag{15.27}$$

We now solve for the L and take its logarithm to obtain the Lyapunov coefficient:

$$\lambda = \ln(L) = \lim_{n \to \infty} \frac{1}{n} \sum_{i=0}^{n-1} \ln \left|\left(\frac{\delta f}{\delta x}\right)_{x_i}\right|. \tag{15.28}$$

For the logistic map, $f(x) = \mu x(1 - x)$, we obtain

$$\lambda = \frac{1}{n} \sum_{i=0}^{n-1} \ln |\mu - 2\mu x_i|, \tag{15.29}$$

where the sum is over iterations.

The code `LyapLog.py` in Listing 15.2 computes the Lyapunov exponents for the logistic map. In Figures 15.4 and 15.5 we show its output. Note the sign changes in λ where the

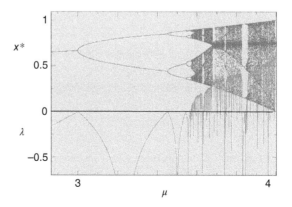

Figure 15.4 Fixed point bifurcations (top) and Lyapunov coefficient (bottom) for the logistic map as functions of the growth rate μ. Notice how the Lyapunov coefficient, a measure of chaos, changes abruptly at the bifurcations with positive values indicating instabilities.

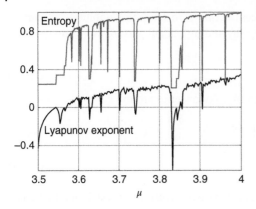

Figure 15.5 Shannon entropy (top) and Lyapunov coefficient (bottom) for the logistic map. Notice the close relation between the thermodynamic measure of disorder (entropy) and the nonlinear dynamics measure of chaos (Lyapunov).

system becomes chaotic, and the abrupt changes in slope at the bifurcations. (A similar curve is obtained for the fractal dimension of the logistic map, and, indeed the two are proportional.)

15.4.2 Shannon Entropy

Another measure that can indicate chaotic behavior is the Shannon entropy. Entropy is a measure of uncertainty (garbled signal) that has proven useful in communication theory [Shannon, 1948; Ott, 2002; Gould *et al.*, 2006]. Imagine that an experiment has N possible outcomes. If the probability of each is $p_1, p_2, \ldots, p_N$, with normalization such that $\sum_{i=1}^{N} p_i = 1$, then the Shannon entropy is defined as

$$S_{\text{Sh}} = -\sum_{i=1}^{N} p_i \ln p_i. \tag{15.30}$$

If $p_i \equiv 0$, there is no uncertainty, and $S_{\text{Sh}} = 0$, as you might expect. If all N outcomes have equal probability, $p_i \equiv 1/N$, we obtain the expression familiar from statistical mechanics, $S_{\text{Sh}} = \ln N$.

The code `Entropy.py` in Listing 15.3 computes the Shannon entropy for the logistic map as a function of the growth parameter μ. The results (Figure 15.5, top) are seen to be quite similar to the Lyapunov exponent, again with discontinuities occurring at the bifurcations.

15.5 Coupled Predator–Prey Models ⊙

We have seen complicated behavior arising from a population model in which we imposed a population limit. Now we extend that model to describe coexisting predator and prey populations [Lotka, 1925; Volterra, 1926].

Problem Calculate what it may take to control a population of pests (prey) by introducing predators. Include in your considerations the interaction between the populations, as well as the competition for food and the time needed for predation.

15.5.1 Lotka–Volterra Model

We extend the logistic map to the Lotka–Volterra model (LVM) that describes coexisting predator and a prey population by introducing an additional population:

$$p(t) = \text{prey (bug) density}, \quad P(t) = \text{Predator density}. \tag{15.31}$$

In the absence of interactions between the species, we assume that the prey population p breeds at a per-capita rate of a:

$$\frac{\Delta p}{\Delta t} = ap \quad \text{(Discrete)}, \tag{15.32}$$

$$\frac{dp}{dt} = ap, \quad \text{(Continuous)} \Rightarrow p(t) = p(0)e^{at}. \tag{15.33}$$

Here we give both the discrete and continuous versions of the model, and will work with the continuous model, where we see the exponential growth explicitly. However, if there are predators that "interact with" (gobble up) an abundance of prey, then this may affect the prey growth rate. We assume that the interaction (gobble) rate is proportional to their joint probability:

$$\text{Interaction rate} = bpP, \tag{15.34}$$

where b is a constant. This leads to a prey growth rate including both predation and breeding:

$$\frac{dp}{dt} = ap - bpP, \quad \text{(LVM-I for prey)}. \tag{15.35}$$

If left to themselves, predators P will breed and increase their population exponentially. Yet we all need to eat, and if there is no prey around, predators will eat each other (or their young) at a per-capita mortality rate m:

$$\left.\frac{dP}{dt}\right|_{\text{mort}} = -mP, \quad \Rightarrow P(t) = P(0)e^{-mt}. \tag{15.36}$$

However, if we also include the possibility that there are prey to "interact with" at a rate bpP, the predator population will grow at the rate

$$\frac{dP}{dt} = \epsilon bpP - mP \quad \text{(LVM-I for predators)}. \tag{15.37}$$

Here ϵ is a constant that measures the efficiency with which predators convert prey interactions into food.

Equations (15.35) and (15.37) are two simultaneous ODEs that define the first LVM model. They can be solved after placing them in the standard dynamic form:

$$\begin{aligned} &d\mathbf{y}/dt = \mathbf{f}(\mathbf{y}, t), \\ &y_0 = p, \qquad\qquad f_0 = ay_0 - by_0y_1, \\ &y_1 = P, \qquad\qquad f_1 = \epsilon by_0y_1 - my_1. \end{aligned} \tag{15.38}$$

Our code to solve these equations is `PredatorPrey.py` in Listing 15.4, with results shown in Figure 15.6. On the left, we see two populations that oscillate out of phase with each other in time. When there are many prey, the predator population eats them and grows, yet then the predators face a decreased food supply, and so their population decreases; that, in turn, permits the prey population to grow, and so forth. On the right in

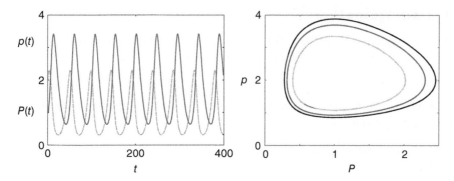

Figure 15.6 *Left*: The time dependencies of the prey population of *p(t)* (solid curve) and of the predator population *P(t)* (dashed curve) for the Lotka–Volterra model. *Right*: A "phase space" plot of *p(t) versus P(t)*. The different orbits correspond to different initial populations.

Figure 15.6, we plot a "phase space" plot of $P(t)$ *versus* $p(t)$.[2] A closed orbit in the phase space plot indicates a limit cycle that repeats indefinitely. Although increasing the initial number of predators does decrease the maximum number of pests and keeps their behavior in check, it is not a satisfactory control since their number shows a large variation.

15.5.2 Predator–Prey Chaos

It seems that introducing prey has kept the predator populations from becoming chaotic. Mathematical analyses tell us that, in addition to nonlinearity, a system must contain a number of degrees of freedom before chaos will occur. For a predator–prey model, introducing another species or two may do the trick. And so we extend our previous to include four species, each with population p_i competing for the same finite set of resources [Vano *et al.*, 2006]. This extends (15.37) to:

$$\frac{dp_i}{dt} = a_i p_i \left(1 - \sum_{j=1}^{4} b_{ij} p_j \right), \qquad i = 1, 4. \tag{15.39}$$

Here a_i is a measure of the growth rate of species i, and b_{ij} is a measure of the rate at which species j consumes the resources needed by species i. Since four species covers a very large parameter space, we suggest that you start your exploration using the same parameters that Vano *et al.* [2006] found produce chaos:

$$a_i = \begin{bmatrix} 1 \\ 0.72 \\ 1.53 \\ 1.27 \end{bmatrix}, \qquad b_{ij} = \begin{bmatrix} 1 & 1.09 & 1.52 & 0 \\ 0 & 1 & 0.44 & 1.36 \\ 2.33 & 0 & 1 & 0.47 \\ 1.21 & 0.51 & 0.35 & 1 \end{bmatrix}. \tag{15.40}$$

With chaotic systems being hypersensitive to exact parameter values, you may want to modify these somewhat. Note that the self-interaction terms $b_{ii} = 1$, which is a consequence of measuring the population of each species in units of its individual carrying capacity.

---

2 We discuss phase space plots at length in Chapter 16.

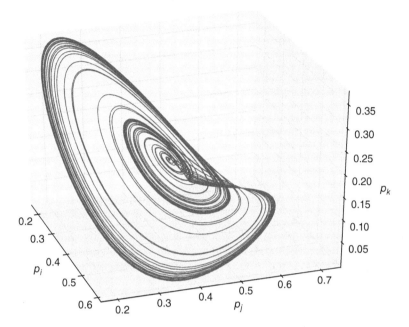

Figure 15.7 A chaotic attractor for the 4D Lotka–Volterra model projected onto three axes.

We solve (15.39) with initial conditions corresponding to an equilibrium point at which all species coexist:

$$p_i(t = 0) = \begin{bmatrix} 0.3013 \\ 0.4586 \\ 0.1307 \\ 0.3557 \end{bmatrix}. \tag{15.41}$$

An illuminating way to visualize the behavior of this system would be to create a 4D phase-space plot, $[p_1(t_i), p_2(t_i), p_3(t_i), p_4(t_i)]$ for $i = 1, N$, where N is the number of time steps in the numerical solution. The geometric structures so created may have a smooth and well-defined shape. Unfortunately, we have no way to show such a plot, and so in its stead, as seen in Figure 15.7, we project the 4D structure onto 2D and 3D axes. We see a classic type of chaotic attractor, with the 3D structure folded over into a nearly 2D structure.

Exercise

1) Visualize the solution to (15.39) by plotting $p_1(t)$, $p_2(t)$, $p_3(t)$, and $p_4(t)$ *versus* time.
2) Construct the 4D chaotic attractor formed by the solutions of (15.39). Output the values $[p_1(t_i), p_2(t_i), p_3(t_i), p_4(t_i)]$, for each time step i. In order to avoid needlessly long files, you may want to skip a number of time steps.
   a) Plot all possible 2D phase space plots, that is, plots of p_i *versus* p_j, $i \neq j = 1 - 3$.
   b) Plot all possible 3D phase space plots, that is, plots of p_i *versus* p_j *versus* p_k.
      Note: you have to adjust the parameters or initial conditions slightly to obtain truly chaotic behavior.

15.5.3 LVM with Prey Limit

The initial assumption in the LVM that prey grow without limit in the absence of predators is unrealistic. As with the logistic map, we include a limit on prey numbers that accounts for depletion of the food supply as the prey population grows. Accordingly, we modify the constant growth rate from a to $a(1 - p/K)$, so that growth vanishes when the population reaches the *carrying capacity K*:

$$\frac{dp}{dt} = ap\left(1 - \frac{p}{K}\right) - bpP, \qquad \text{(LVM-II)}. \tag{15.42}$$

$$\frac{dP}{dt} = \epsilon bpP - mP. \tag{15.43}$$

The behavior of this model with prey limitations is shown in Figure 15.8. We see that both populations exhibit damped oscillations as they approach their equilibrium values, and that, as hoped for, the equilibrium populations are independent of the initial conditions. Note how the phase-space plot spirals inward to a single close limit cycle on which it remains, with little variation in prey number. At last, the desired biological "control."

15.5.4 LVM with Predation Efficiency

Another unrealistic assumption in the original LVM is that the predators immediately eat all the prey with which they interact. As anyone who has watched a cat hunt a mouse knows, predators also spend their time finding, chasing, killing, eating, and digesting prey. This *handling time* decreases the rate of bpP at which prey are eliminated. We define the *functional response* p_a as the probability of one predator finding one prey. If a single predator spends time t_{search} searching for prey, then

$$p_a = bt_{\text{search}}P \Rightarrow t_{\text{search}} = \frac{p_a}{bp}. \tag{15.44}$$

If we call t_h the time a predator spends handling a single prey, then the effective time a predator spends handling a prey is $p_a t_h$. Such being the case, the total time T that a predator spends finding and handling a single prey is

$$T = t_{\text{search}} + t_{\text{handling}} = \frac{p_a}{bp} + p_a t_h, \tag{15.45}$$

$$\Rightarrow \frac{p_a}{T} = \frac{bp}{1 + bpt_h}, \tag{15.46}$$

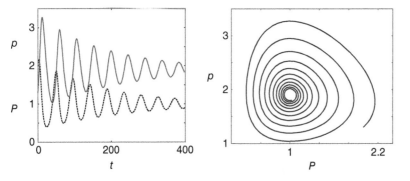

Figure 15.8 The Lotka–Volterra model including a limit on prey population. *Left*: Solid curve: of prey population *p(t)*; dashed curve: predator population *P(t)*. *Right*: A phase space plot of prey population *p* as a function of predator population *P*.

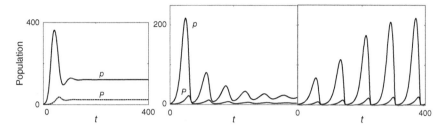

Figure 15.9 Lotka–Volterra model with predation efficiency and prey limitations. From left to right: overdamping, $b = 0.01$; damped oscillations, $b = 0.1$, and limit cycle, $b = 0.3$.

where p_a/T is the effective *rate* of eating prey. We see that as the number of prey $p \to \infty$, the efficiency in eating them $\to 1$. We include this efficiency in (15.42) by modifying the rate b at which a predator eliminates prey to $b/(1 + bpt_h)$:

$$\frac{dp}{dt} = ap\left(1 - \frac{p}{K}\right) - \frac{bpP}{1 + bpt_h}, \qquad \text{(LVM-III)}. \tag{15.47}$$

To be still more realistic about the predator growth, we also place a limit on the predator carrying capacity, but make it proportional to the number of prey:

$$\frac{dP}{dt} = mP\left(1 - \frac{P}{kp}\right), \qquad \text{(LVM-III)}. \tag{15.48}$$

Solutions for the extended model (15.47) and (15.48) are shown in Figure 15.9. Observe the existence of three dynamic regimes as a function of b:

- **Small** b: No oscillations, no overdamping,
- **Medium** b: Damped oscillations that converge to a stable equilibrium,
- **Large** b: Limit cycle.

The transition from equilibrium to a limit cycle is called a *phase transition*.

We finally have a satisfactory solution to our **problem**. Although the prey population is not eliminated, it can be kept from getting too large and from fluctuating widely. Nonetheless, changes in the parameters can lead to large fluctuations or to nearly vanishing predators.

15.5.5 LVM Implementation and Assessment

1) Solve all three LVM models using the following parameter values:

Model	a	b	ϵ	m	K	k
LVM-I	0.2	0.1	1	0.1	0	
LVM-II	0.2	0.1	1	0.1	20	
LVM-III	0.2	0.1		0.1	500	0.2

2) For each of the three models, construct
    a) a time series for prey and predator populations,
    b) phase space plots of predator *versus* prey populations.

3) **LVM-I**: Compute the equilibrium values for the prey and predator populations. Do you think that a model in which the cycle amplitude depends on the initial conditions can be realistic? Explain.

4) **LVM-II**: Calculate numerical values for the equilibrium values of the prey and predator populations. Make a series of runs for different values of prey carrying capacity K. Can you deduce how the equilibrium populations vary with prey carrying capacity?

5) Make a series of runs for different initial conditions for predator and prey populations. Do the cycle amplitudes depend on the initial conditions?

6) **LVM-III**: Make a series of runs for different values of b and reproduce the three regimes present in Figure 15.9.

7) Calculate the critical value for b corresponding to a phase transition between the stable equilibrium and the limit cycle.

15.5.6 Two Predators, One Prey

1) Another version of the LVM includes the possibility that two populations of predators P_1 and P_2 may "share" the same prey population p. Investigate the behavior of a system in which the prey population grows logistically in the absence of predators:

$$\frac{dp}{dt} = ap\left(1 - \frac{p}{K}\right) - \left(b_1 P_1 + b_2 P_2\right)p,$$ (15.49)

$$\frac{dP}{dt} = \epsilon_1 b_1 pP_1 - m_1 P_1, \qquad \frac{dP_2}{dt} = \epsilon_2 b_2 pP_2 - m_2 P_2.$$ (15.50)

a) Use the following values for the model parameters and initial conditions: $a = 0.2$, $K = 1.7$, $b_1 = 0.1$, $b_2 = 0.2$, $m_1 = m_2 = 0.1$, $\epsilon_1 = 1.0$, $\epsilon_2 = 2.0$, $p(0) = P_2(0) = 1.7$, and $P_1(0) = 1.0$.

b) Determine the time dependence for each population.

c) Vary the characteristics of the second predator and calculate the equilibrium population for the three components.

d) What is your answer to the question, "Can two predators that share the same prey coexist?"

15.6 Code Listings

Listing 15.1 Bugs.py Produces the bifurcation diagram of the logistic map. A full program requires finer grids, a scan over initial values, and removal of duplicates.

```
# Bugs.py The Logistic map

from visual.graph import *
m_min = 1.0;       m_max = 4.0;        step = 0.01
graph1 = gdisplay(width=600, height=400, title='Logistic Map',\
        xtitle='m', ytitle='x', xmax=4.0, xmin=1., ymax=1., ymin=0.)
pts = gdots(shape = 'round', size = 1.5, color = color.green)
lasty = int(1000 * 0.5)                 # Eliminates some points
count = 0                               # Plot every 2 iterations
for m in arange(m_min, m_max, step):
    y = 0.5
    for i in range(1,201,1):            # Avoid transients
        y = m*y*(1-y)
    for i in range(201,402,1):
```

```
              y = m*y*( 1 - y)
         for i in range(201, 402, 1):        # Avoid transients
              oldy=int(1000*y)
18            y = m*y*(1 - y)
              inty = int(1000 * y)
              if  inty != lasty and count%2 == 0:
                   pts.plot(pos=(m,y))        # Avoid repeats
22            lasty = inty
              count  += 1
```

Listing 15.2 LyapLog.py Computes Lyapunov coefficient for the bifurcation plot of the logistic map as a function of growth rate. Note the fineness of the μ grid.

```
  # LyapLog.py:                  Lyapunov coef for logistic map

3 from visual.graph import *

  m_min = 3.5;          m_max = 4.5;          step = 0.25
  graph1 = gdisplay( title = 'Lyapunov coef (blue) for LogisticMap (red)',
7                    xtitle = 'm', ytitle = 'x , Lyap',
                     xmax=5.0, xmin=0, ymax = 1.0, ymin = - 0.6)
  funct1 = gdots(color = color.red)
  funct2 = gcurve(color = color.yellow)
11 for m in arange(m_min, m_max, step):                              # m loop
       y = 0.5
       suma = 0.0
       for i in range(1, 401, 1):   y = m*y*(1 - y)        # Skip transients
15     for i in range(402, 601, 1):
           y = m*y*(1 - y)
           funct1.plot(pos = (m, y) )
           suma = suma  + log(abs(m*(1. - 2.*y) ))             # Lyapunov
19     funct2.plot(pos = (m, suma/401) )                         # Normalize
```

Listing 15.3 Entropy.py Computes the Shannon entropy for the logistic map as a function of growth parameter μ.

```
  # Entropy.py Shannon Entropy with Logistic map using Tkinter

3 try:      from tkinter import *
  except:      from Tkinter import *
  import math
  from numpy import zeros, arange

7
  global Xwidth, Yheight
  root = Tk( );     root.title('Entropy versus mu ')
  mumin = 3.5;  mumax = 4.0;  dmu = 0.25; nbin = 1000;  nmax = 100000
11 prob = zeros( (1000), float)
  minx=mumin; maxx=mumax; miny=0; maxy=2.5; Xwidth=500; Yheight=500
  c = Canvas(root, width = Xwidth, height = Yheight)      # Init canvas
  c.pack()                                            # Pack canvas
15 Button(root, text = 'Quit', command = root.quit).pack()  # To quit

  def world2sc(xl, yt, xr, yb):  # x - left, y - top, x - right, y - bottom
      maxx = Xwidth     # canvas width
19    maxy = Yheight    # canvas height
      lm = 0.10*maxx    # left margin
      rm = 0.90*maxx    # right margin
      bm = 0.85*maxy    # bottom margin
23    tm = 0.10*maxy    # top margin
      mx = (lm - rm)/(xl - xr) #
      bx = (xl*rm - xr*lm)/(xl - xr) #
      my = (tm - bm)/(yt - yb) #
27    by = (yb*tm - yt*bm)/(yb - yt) #
      linearTr = [mx, bx, my, by]
      return linearTr                    # returns a list with 4 element
```

```
31  # Plot y, x, axes; world coord converted to canvas coordinates
    def xyaxis(mx, bx, my, by):           # to be called after call workd2sc
        x1 = (int)(mx*minx + bx)            # minima and maxima converted to
        x2 = (int)(mx*maxx + bx)              # canvas coordinades
35      y1 = (int)(my*maxy + by)
        y2 = (int)(my*miny + by)
        yc = (int)(my*0.0 + by)
        c.create_line(x1, yc, x2, yc, fill = "red")        # x axis
39      c.create_line(x1, y1, x1, y2, fill = 'red')        # y - axis
        for i in range (7):                                # x tics
            x = minx + (i - 1)*0.1              # world coordinates
            x1 = (int)(mx*x + bx)              # canvas coord
43          x2 = (int)(mx*minx + bx)
            y = miny + i*0.5                     # real coordinates
            y2 = (int)(my*y + by)              # canvas coords
            c.create_line(x1, yc - 4, x1, yc + 4, fill = 'red')    # tics x
47          c.create_line(x2 - 4, y2, x2 + 4, y2, fill = 'red')    # tics y
            c.create_text(x1 + 10, yc + 10, text = '%5.2f'% (x),\
                fill = 'red', anchor = E)             # x axis
            c.create_text(x2 + 30, y2, text = '%5.2f'% (y), fill = 'red',\
51              anchor = E)    # y axis
        c.create_text(70, 30, text = 'Entropy', fill = 'red', anchor = E)
        c.create_text(420, yc - 10, text = 'mu', fill = 'red', anchor = E)

55  mx, bx, my, by = world2sc(minx, maxy, maxx, miny)      # returns list
    xyaxis(mx, bx, my, by)                                 # axes values
    mu0 = mumin*mx + bx
    entr0 = my*0.0 + by
59  for mu in arange(mumin, mumax, dmu):                   # mu loop
        print(mu)
        for j in range(1, nbin):
            prob[j] = 0
63      y  = 0.5
        for n in range(1, nmax + 1):
            y = mu*y*(1.0 - y)                # Logistic map, Skip transients
            if (n > 30000):
67              ibin = int(y*nbin)  + 1
                prob[ibin]  += 1
        entropy = 0.
        for ibin in range(1, nbin):
71          if (prob[ibin]>0):
                entropy = entropy - (prob[ibin]/nmax)*math.log10(prob[ibin]/nmax)
        entrpc = my*entropy + by              # entropy to canvas coords
        muc = mx*mu + bx                      # mu to canvas coords
75      c.create_line(mu0, entr0, muc, entrpc, width = 1, fill = 'blue')
        mu0 = muc                             # begin values for next line
        entr0 = entrpc
    root.mainloop()                           # makes effective events
```

Listing 15.4   PredatorPrey.py Computes population dynamics for a group of interacting predators and prey.

```
# PredatorPrey.py:       Lotka-Volterra models

from visual import *
4  from visual.graph import *

Tmin = 0.0
Tmax = 500.0
8  y = zeros( (2), float)
Ntimes = 1000
y[0] = 2.0
y[1] = 1.3
12  h = (Tmax - Tmin)/Ntimes
    t = Tmin

    def f( t, y, F):              # Modify this function for your problem
```

```
16        F[0] = 0.2*y[0]*(1 - (y[0]/(20.0) )) - 0.1*y[0]*y[1]
          F[1]  = - 0.1*y[1] + 0.1*y[0]*y[1];

   def rk4(t, y, h, Neqs):                    # rk4 method,  DO NOT modify
20      F = zeros((Neqs), float)
        ydumb = zeros((Neqs), float)
        k1    = zeros((Neqs), float)
        k2    = zeros((Neqs), float)
24      k3    = zeros((Neqs), float)
        k4    = zeros((Neqs), float)
        f(t, y, F)
        for i in range(0, Neqs):
28          k1[i] = h*F[i]
            ydumb[i] = y[i] + k1[i]/2.
        f(t + h/2., ydumb, F)
        for i in range(0, Neqs):
32          k2[i] = h*F[i]
            ydumb[i] = y[i] + k2[i]/2.
        f(t + h/2., ydumb, F)
        for i in range(0, Neqs):
36          k3[i] = h*F[i]
            ydumb[i] = y[i] + k3[i]
        f(t + h, ydumb, F)
        for i in range(0, Neqs):
40          k4[i] = h*F[i]
            y[i] = y[i] + (k1[i] + 2.*(k2[i] + k3[i]) + k4[i])/6.

   graph1 = gdisplay(x= 0,y= 0, width = 500, height = 400, \
44        title = 'Prey p(green) and predator P(yellow) vs time',xtitle = 't', \
          ytitle = 'P, p',xmin=0,xmax=500,ymin=0,ymax=3.5)
   funct1 = gcurve(color = color.yellow)
   funct2 = gcurve(color = color.green)
48 graph2 = gdisplay(x= 0,y= 400, width = 500, height = 400,
                    title = 'Predator P vs prey p',
                    xtitle = 'P', ytitle = 'p',xmin=0,xmax=2.5,ymin=0,ymax=3.5)
   funct3 = gcurve(color = color.red)
52
   for t in arange(Tmin, Tmax + 1, h):
       funct1.plot(pos = (t, y[0]) )
       funct2.plot(pos = (t, y[1]) )
56     funct3.plot(pos = (y[0], y[1]) )
       rate(60)
       rk4(t, y, h, 2)
```

16

Nonlinear Dynamics of Continuous Systems

> *In Chapter 15 we explored the complex dynamics and chaos that occur in population models.*
> *In this chapter we explore physical systems that exhibit chaos, and, in particular, the driven*
> *realistic pendulum. The focus is on understanding chaos and using phase space to uncover*
> *the simplicity underlying complex behaviors.*

16.1 The Chaotic Pendulum

Problem Compute the motion of a driven pendulum with no restrictions on the magnitude of the displacements.

Although the plane pendulum is a classic subject for physics, it is usually studied for small displacements. That may be okay for large grandfather clocks, but not for this chapter. We will look at a *chaotic pendulum*, i.e., one with friction and a driving torque (Figure 16.1 left) and with no restriction to small displacements [Rasband, 1990]. Newton's laws of rotational motion tell us that the sum of the gravitational torque $-mgl \sin \theta$, the frictional torque $-\beta \dot{\theta}$, and the external torque $\tau_0 \cos \omega t$ equals the moment of inertia of the pendulum times its angular acceleration:

$$-mgl \sin \theta - \beta \frac{d\theta}{dt} + \tau_0 \cos \omega t = I \frac{d^2\theta}{dt^2}, \tag{16.1}$$

$$\Rightarrow \quad -\omega_0^2 \sin \theta - \alpha \frac{d\theta}{dt} + f \cos \omega t = \frac{d^2\theta}{dt^2}, \tag{16.2}$$

$$\omega_0 = \frac{mgl}{I}, \qquad \alpha = \frac{\beta}{I}, \qquad f = \frac{\tau_0}{I}. \tag{16.3}$$

Equation (16.2) is a second-order, time-dependent, nonlinear differential equation. The nonlinearity arises from the $\sin \theta$ dependence of the gravitational torque. The constant ω_0 is the natural frequency of oscillations for small displacements, with only a gravitational torque. The parameter α is a measure of the strength of friction, and the parameter f is a measure of the strength of the external driving torque. In our standard ODE form, $d\mathbf{y}/dt = \mathbf{f}$ of Chapter 8, we have two, simultaneous, first-order equations:

Computational Physics: Problem Solving with Python, Fourth Edition.
Rubin H. Landau, Manuel J. Páez, and Cristian C. Bordeianu.
© 2024 WILEY-VCH GmbH. Published 2024 by WILEY-VCH GmbH.

$$\frac{dy^{(0)}}{dt} = y^{(1)},$$

(16.4)

$$\frac{dy^{(1)}}{dt} = -\omega_0^2 \sin y^{(0)} - \alpha y^{(1)} + f \cos \omega t,$$

$$y^{(0)} = \theta(t), \qquad y^{(1)} = \frac{d\theta(t)}{dt}.$$

(16.5)

16.1.1 Free Pendulum Oscillations

If we ignore friction and the external torque, Newton's law (16.2) takes the simple, yet non-linear form:

$$\frac{d^2\theta}{dt^2} = -\omega_0^2 \sin \theta.$$

(16.6)

If the displacements are small, we can approximate $\sin \theta$ by θ and obtain the familiar linear equation of simple harmonic motion with frequency ω_0:

$$\frac{d^2\theta}{dt^2} \simeq -\omega_0^2\theta \Rightarrow \theta(t) = \theta_0 \sin(\omega_0 t + \phi).$$

(16.7)

In Chapter 8, we studied how nonlinearities produce anharmonic oscillations, and, indeed, (16.6) is another good candidate for such studies. As in that chapter, we expect solutions of (16.6) to be periodic, but with a frequency ω that equals ω_0 only for small oscillations. Furthermore, because the restoring torque, $mgl \sin \theta \simeq mgl(\theta - \theta^3/3)$, is less than the $mgl\theta$ assumed in a harmonic oscillator, realistic pendulums swing slower (have longer periods) as their angular displacements are made larger.

16.1.2 Analytic Solution as Elliptic Integrals

The analytic solution to the realistic pendulum is a standard textbook problem [Landau and Lifshitz, 1976; Marion and Thornton, 2019; Scheck, 2010]. However, it is a rather limited solution as it is only for the period T, and it is in terms of integrals that must be evaluated numerically. This solution is based on energy being a constant (integral) of the motion, and is for the pendulum released from rest at its maximum displacement θ_m, where its energy is all potential (Figure 16.1):

$$E = PE(t=0) = mgl - mgl \cos \theta_m = 2mgl \sin^2\left(\frac{\theta_m}{2}\right).$$

(16.8)

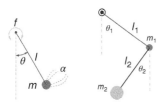

Figure 16.1 *Left:* A pendulum of length l driven through resistive air (dotted arcs) by an external sinusoidal torque (semicircle). The strength of the external torque is given by f and that of air resistance by α. *Right:* A double pendulum (Section 16.4.1) with neither air resistance nor a driving force. In both cases, there is a gravitational torque and the possibility of chaos. (The study of the single pendulum to follow can be replaced by the study of the double pendulum.)

Yet, because $E = KE + PE$ is a constant, we can express the energy for any value of θ, and then go on to solve for the period:

$$2mgl \sin^2 \tfrac{\theta_m}{2} = \tfrac{1}{2} I \left(\tfrac{d\theta}{dt} \right)^2 + 2mgl \sin^2 \tfrac{\theta}{2},$$

$$\Rightarrow \frac{d\theta}{dt} = 2\omega_0 \left[\sin^2 \tfrac{\theta_m}{2} - \sin^2 \tfrac{\theta}{2} \right]^{1/2} \Rightarrow \frac{dt}{d\theta} = \frac{T_0/\pi}{[\sin^2(\theta_m/2) - \sin^2(\theta/2)]^{1/2}},$$

$$\Rightarrow \frac{T}{4} = \frac{T_0}{4\pi} \int_0^{\theta_m} \frac{d\theta}{[\sin^2(\theta_m/2) - \sin^2(\theta/2)]^{1/2}}, \qquad (16.9)$$

$$\Rightarrow T \simeq T_0 \left[1 + \left(\tfrac{1}{2} \right)^2 \sin^2 \tfrac{\theta_m}{2} + \left(\tfrac{1 \cdot 3}{2 \cdot 4} \right)^2 \sin^4 \tfrac{\theta_m}{2} + \cdots \right]. \qquad (16.10)$$

Because the motion is periodic, we have assumed that it takes $T/4$ for the pendulum to travel from $\theta = 0$ to $\theta = \theta_m$. The integral in (16.9) can be expressed as an *elliptic integral of the first kind*. If you think of an elliptic integral as a generalization of a trigonometric function, then this is a closed-form solution; otherwise, it's an integral needing computation. The series expansion of the period (16.10) is obtained by expanding the denominator and integrating it term by term. It tells us, for example, that an amplitude of 80° leads to a period 10% longer than the small θ period. We will determine the period computationally without the need for any expansions.

16.1.3 Free Pendulum Implementation and Test

As a preliminary to the solution of the full equation (16.2), modify your rk4 program to solve (16.6) for the free oscillations of a realistic pendulum.

1) Start your pendulum at $\theta = 0$ with $\dot{\theta}(0) \neq 0$. Gradually increase $\dot{\theta}(0)$ to increase the importance of nonlinear effects.
2) Test your program for the linear case ($\sin \theta \rightarrow \theta$) and verify that:
   a) Your solution is harmonic with frequency $\omega_0 = 2\pi/T_0$, and that,
   b) The frequency of oscillation is independent of the amplitude.
3) Devise an algorithm to determine the period T of the oscillation by counting the time it takes for three successive passes of the amplitude through $\theta = 0$. (You need *three* passes to handle the case where the oscillation is not symmetric about the origin.) Test your algorithm for simple harmonic motion where you know T_0.
4) For the realistic pendulum, observe the change in period as a function of increasing initial energy. Plot your observations along with (16.10).
5) Verify that as the initial KE approaches $2mgl$, the motion remains oscillatory but not harmonic.
6) At $E = 2mgl$ (the *separatrix*), the motion transitions from oscillatory to rotational ("over the top" or "running"). See how close you can get to the separatrix and to its infinite period.
7) ⊙ Convert your numerical data to sound and listen to the difference between harmonic motion (boring) and anharmonic motion containing overtones (interesting). In Figure 16.2, we show some typical results.

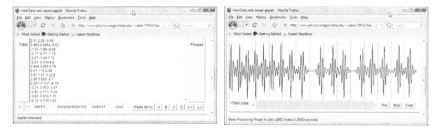

Figure 16.2 The data screen (*left*) and the output screen (*right*) of the applet `HearData` that converts data into sounds. Columns of $[t_i, x(t_i)]$ data are pasted into the data window, processed into the graph in the output window, and then converted to sound data that are played by Java. (Applets have now been outlawed.)

16.2 Phase Space

The conventional solution to an equation of motion is the position $x(t)$ and the velocity $v(t)$ as functions of time. Often behaviors that appear complicated as functions of time appear as familiar-looking geometric figures when viewed in the more abstract *phase space*. For the pendulum, the phase space ordinate is the velocity $v(t)$, and the abscissa is the position $x(t)$ (Figure 16.3, top). Furthermore, the motion of a complex system, when viewed in phase space over time, often displays a movement back-and-forth from one phase space structure to another, a behavior known as *strange attraction*. This is easy to understand. For example, the position and velocity of a free harmonic oscillator are given by the trigonometric functions

$$x(t) = A \sin(\omega t), \ v(t) = \frac{dx}{dt} = \omega A \cos(\omega t). \tag{16.11}$$

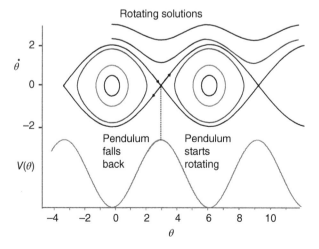

Figure 16.3 *Top:* Phase space trajectories for a pendulum including "over the top" motions. At the bottom of the figure is shown the corresponding θ dependence of the potential.

When substituted into the total energy, we obtain two important results:

$$E = KE + PE = \left(\tfrac{1}{2}m\right)v^2 + \left(\tfrac{1}{2}\omega^2 m^2\right)x^2 \tag{16.12}$$

$$= \tfrac{m\omega^2 A^2}{2}\cos^2(\omega t) + \tfrac{m\omega^2 A^2}{2}\sin^2(\omega t) = \tfrac{1}{2}m\omega^2 A^2. \tag{16.13}$$

Equation (16.12), being that of an ellipse in x-v space, indicates that the harmonic oscillator follows closed elliptical orbits in phase space, with the size of the ellipse increasing with the system's energy. Equation (16.13) proves that the total energy is a constant of the motion. Different initial conditions having the same energy start at different places on the same ellipse, yet transverse the same orbits.

Here are some phase space structures and behaviors that you will be asked to observe in your simulations:

Ellipses: The orbits of anharmonic oscillations will still be ellipse-like, but with angular corners that become more distinct with increasing nonlinearity (Figure 16.4 right).

Closed figures: Like those in Figures 16.3 and 16.4, describe periodic (not necessarily harmonic) oscillations with the same (x, v) occurring again and again. The restoring force leads to clockwise motion.

Open orbits: Like those in Figures 16.3 and 16.4 left, they correspond to nonperiodic or "running" motion (a pendulum rotating like a propeller). Regions where the potential is repulsive also lead to open trajectories in phase space.

Separatrix: As seen on top of Figure 16.3, this is an orbit in phase space that separates open and closed orbits. Motion on the separatrix is indeterminate, as the pendulum may balance, or move either way at the maximum potential.

Non-crossing orbits: Because solutions for different initial conditions are unique, different orbits do not cross. Yet different initial conditions can correspond to different starting positions along a single orbit.

Hyperbolic points: Open orbits intersect at points of unstable equilibrium called *hyperbolic points* (Figure 16.4 left), at which point an indeterminacy results.

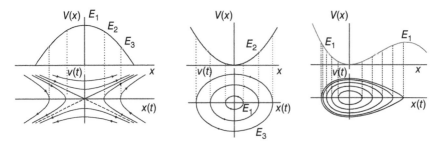

Figure 16.4 Three potentials and their behaviors in phase space. The different orbits below the potentials correspond to different energies, as indicated by the limits of maximum displacements within the potentials (dashed lines). *Left*: A repulsive potential leads to open orbits. The central crossing is an unstable hyperbolic point. *Middle*: The symmetric harmonic oscillator potential leads to symmetric ellipses. *Right*: A nonharmonic oscillator for small oscillations producing structures that are neither ellipses nor symmetric.

Figure 16.5 Position *versus* time and position *versus* velocity for two initial conditions of a chaotic pendulum that ends up with the same limit cycle. (W. Hager.)

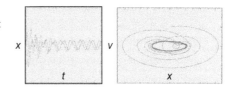

Fixed points: The inclusion of friction may cause the energy in a system to decrease with time, leading to phase-space orbits that spiral into a single *fixed-point*. However, if there is an external driving force, then the system would move away from the fixed point.

Limit cycle: If the parameters are just right, a closed ellipse-like figure called a *limit cycle* may occur (Figure 16.5 right). Here, the average energy put into the system during one period exactly balances the average energy dissipated by friction during that period:

$$\langle f \cos \omega t \rangle = \left\langle \alpha \frac{d\theta}{dt} \right\rangle = \left\langle \alpha \frac{d\theta(0)}{dt} \cos \omega t \right\rangle \Rightarrow f = \alpha \frac{d\theta(0)}{dt}. \qquad (16.14)$$

While a system may move into a limit cycle, it may also make sporadic jumps from one limit cycle to another.

Predictable attractors: The orbits, such as fixed points and limit cycles, into which the system settles or returns to often, and that are not particularly sensitive to initial conditions. If your location in phase space is near a predictable attractor, ensuing times will bring you to it.

Strange attractors: Well-defined, yet complicated, semiperiodic behaviors that appear to be uncorrelated with the motion at an earlier time. These are distinguished from *predictable attractors* by being fractal (Chapter 14) and highly sensitive to the initial conditions [José and Salatan, 1998]. Even after millions of oscillations, the motion remains *attracted* to them.

Mode locking: When the magnitude of the driving force is larger than that for a limit cycle (16.14), the driving force can overpower the natural oscillations, resulting in a steady-state motion at the frequency of the driver. While mode locking can occur for linear or nonlinear systems, for nonlinear systems the driving torque may lock onto an overtone, leading to a rational relation between the driving frequency and the natural frequency.

$$\frac{\omega}{\omega_0} = \frac{n}{m}, \ n, m = \text{integers.} \qquad (16.15)$$

Random motion: Appears in phase space as a diffuse cloud filling the entire energetically accessible region.

Chaotic paths: While periodic motion produces closed figures in phase space, and random motion a cloud, chaotic motion falls someplace in between, with dark or diffuse *bands* rather than single lines (Figure 16.7). The continuity of trajectories within bands means that the system flows continuously among the different trajectories within the band, which may well look very complicated or chaotic in normal space. The existence of these bands helps explain why the solutions are hypersensitive to the initial conditions and parameter values; the slightest change in values may cause the system to flow to nearby trajectories within the band.

Butterfly effect: The hypersensitivity of chaotic weather systems may, theoretically, lead to the weather pattern in North America being sensitive to the flapping of a butterfly's wings in South America. Although this appears to be counterintuitive because we know that systems with essentially identical initial conditions should behave the same, eventually the systems diverge. As seen on the right, in Figure 16.8, the initial conditions for both pendulums differ by only 1 part in 917, and so the initial paths in phase space are the same. Nonetheless, at just the time shown here, the pendulums balance in the vertical position, and then one falls before the other, leading to differing oscillations and differing phase-space plots from this time onward.

16.3 Chaotic Explorations

A challenge in understanding simulations of the chaotic pendulum (16.4) is that the 4D parameter space $(\omega_0, \alpha, f, \omega)$ is so immense that only sections of it can be studied systematically. We would expect that sweeping through the driving frequency ω should show resonances and beating; sweeping through the frictional force α should show underdamping, critical damping, and overdamping; and sweeping through the driving torque f might show resonances and mode locking. You should be able to observe all of these behaviors in your simulations, although they are somewhat mixed together.

Start by trying to reproduce the behavior shown in Figures 16.6 and 16.7. *Beware:*, because this is a potentially chaotic system, your solutions may be highly sensitive to the exact values of the initial conditions and to the details of your integration routine. We suggest that you

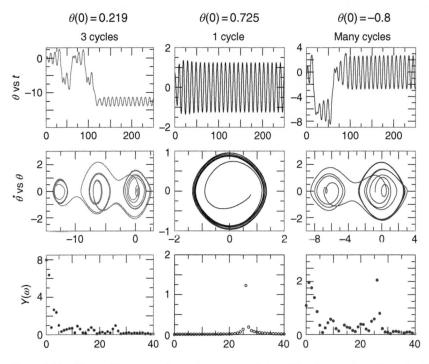

Figure 16.6 The position *versus* time, phase space plot, and Fourier spectrum for a chaotic pendulum with $\omega_0 = 1$, $\alpha = 0.2$, $f = 0.52$, and $\omega = 0.666$. The three differ only in initial conditions.

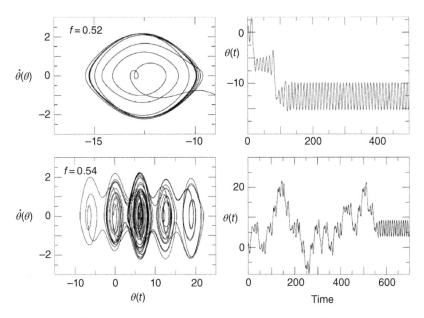

Figure 16.7   The behavior of a chaotic pendulum with slightly differing driving forces ($f = 0.52$, 0.54). On the left are the phase-space plots and on the right are plot position versus time. For $f = 0.54$, there occur the characteristic broadbands of chaos.

experiment; start with the parameter values we used to produce our plots, and then observe the effects of making *very small* changes in the parameters until you obtain different modes of behavior. Here are the recommended steps:

1) Take your solution to the realistic pendulum and include friction. Run it for a variety of initial conditions, including over-the-top ones. As no energy is fed to the system, you should see spirals in phase space. Note, if you plot the phase-space points at uniform time steps, without connecting them, then the spacing between the points gives an indication of the speed and direction of travel of the pendulum.
2) Try several small values for the driving torque, but without friction. Verify that you obtain distorted ellipses in phase space.
3) Turn friction back on and set the driving torque's frequency close to the natural frequency ω_0. Search for beats. Note, you may need to adjust the magnitude and phase of the driving torque to avoid an *impedance mismatch* between the pendulum and driver (being driven to the right while moving to the left).
4) Finally, scan through frequencies ω of the driving torque, and search for nonlinear resonance (it looks like beating).
5) **Explore chaos**: Start off with the initial conditions we used for Figure 16.6:

$$(x_0, v_0) = (-0.0885, 0.8), \quad (-0.0883, 0.8), \quad (-0.0888, 0.8). \tag{16.16}$$

To save time and storage, you may want to use a larger time step for plotting than the one used to solve the differential equations.
6) Identify which parts of the phase space plots correspond to transients.

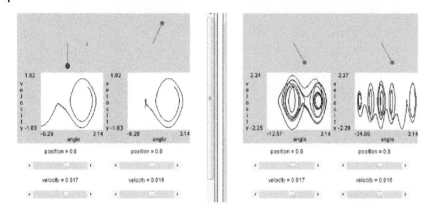

Figure 16.8 *Left*: The phase space plot for two pendulums with almost exactly the same initial conditions. Both arrive at the top (the separatrix), where one goes over the top while the other falls back down. *Right*: The long-term phase space plots for these same pendulums showing dark limit cycles.

7) Ensure that you find the following
   a) a period-3 limit cycle where the pendulum jumps between three major orbits,
   b) a running solution where the pendulum keeps going over the top,
   c) chaotic motion in which paths in the phase space are close enough together to appear as bands.
8) Look for the "butterfly effect" (Figure 16.8, left) by starting two pendulums off with identical positions, but with velocities that differ by 1 part in 1000. Notice that the initial motions are essentially identical, but eventually diverge.

16.3.1 Phase Space Without Velocities

Imagine that you have measured the displacement of a system as a function of time. Your measurements appear to indicate nonlinear behaviors, and you would like to view the system in phase space, but don't have data on the conjugate momenta or velocity. Amazingly enough, a plot of $x(t + \tau)$ *versus* $x(t)$ as a function of t also produces a phase space plot [Abarbanel *et al.*, 1993]. Here τ is a *lag time*, and should be chosen as some fraction of a characteristic time for the system. While this may not seem like a valid phase space plot, think of the forward difference approximation for the velocity,

$$v(t) = \frac{dx(t)}{dt} \simeq \frac{x(t + \tau) - x(t)}{\tau}. \tag{16.17}$$

Thus plotting $x(t + \tau)$ *versus* $x(t)$ is somewhat related to plotting $v(t)$ *versus* $x(t)$.

Exercise Create a phase space plot from the output of your chaotic pendulum by plotting $\theta(t + \tau)$ *versus* $\theta(t)$ for a large range of t values. Explore how the graphs change for different values of lag time τ. Compare your results to the conventional phase space plots you had obtained previously.

Figure 16.9 A bifurcation diagram for the damped pendulum with a vibrating pivot (see also the similar diagram for a double pendulum, Figure 16.11). The ordinate is $|d\theta/dt|$, the absolute value of the instantaneous angular velocity at the beginning of the period of the driver, and the abscissa is the magnitude of the driving force f. Note that the heavy line results from the overlapping of points, not from connecting the points (see enlargement in the inset).

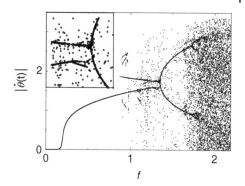

16.3.2 Chaotic Bifurcations

We have observed that a chaotic system contains a number of dominant frequencies, and that the system tends to "jump" from one oscillatory mode to another. This implies that the dominant frequencies occur sequentially, and not simultaneously, as they do in the Fourier analysis of linear systems.

Below, we outline the detailed steps to explore this possibility as a computer experiment. One starts by recording the instantaneous angular velocity $\dot\theta = d\theta/dt$ of the solution at various instances in time. By thinking of $d\theta/dt$ as a frequency, we obtain a series of frequencies, and, presumably, with the major Fourier components occurring most often, as if the system is *attracted* to them. Amazingly, when a scatterplot of the sampled $\dot\theta$'s is constructed as a function of the driving force, a bifurcation diagram similar to that of the logistic map (bugs) results!

Such a scatterplot is shown in Figure 16.9 for a chaotic pendulum with a vibrating pivot point (in contrast to our usual vibrating external torque):

$$\frac{d^2\theta}{dt^2} = -\alpha \frac{d\theta}{dt} - \left(\omega_0^2 + f \cos \omega t\right) \sin\theta. \tag{16.18}$$

Analytic and numerical studies of this system are present in the literature [Landau and Lifshitz, 1976; DeJong, 1992; Gould *et al.*, 2006]. To obtain the bifurcation diagram in Figure 16.9:

1) Use the initial conditions $\theta(0) = 1$ and $\dot\theta(0) = 1$.
2) Set $\alpha = 0.1$, $\omega_0 = 1$, $\omega = 2$, and vary $0 \le f \le 2.25$.
3) To permit transients to die off, wait 150 periods of the driver for each f value before sampling.
4) Sample $\dot\theta$ for 150 times at those instances at which the driving force passes through zero (or when the pendulum passes through its equilibrium position).
5) Plot 150 values of $|\dot\theta|$ *versus* f.

16.3.3 Fourier or Wavelet Analysis

We have seen that a realistic pendulum experiences a gravitational restoring torque

$$\tau_g \propto \sin\theta \simeq \theta - \frac{\theta^3}{3!} + + \frac{\theta^5}{5!} + \cdots . \tag{16.19}$$

The nonlinear terms lead to nonharmonic behavior in a free pendulum. When the pendulum is driven by an external sinusoidal torque, it may *mode lock* with the driver and oscillate at a frequency that is rationally related to the driver's frequency. Consequently, the behavior of the realistic pendulum is expected to be a combination of various periodic behaviors, with discrete jumps between them (as discussed in Section 16.3.2).

In this assessment, you are asked to determine the Fourier components present in some of the pendulum's complicated behaviors. You should find a three-cycle structure containing three major Fourier components, and a five-cycle structure with more frequencies. You should also notice that when the pendulum goes over the top, its spectrum contains a steady-state (DC) component.

1) Dust off your program for analyzing signal $y(t)$ into Fourier components.
2) Apply your analyzer to the solution of the chaotic pendulum for cases where there are one-, three-, and five-cycle structures in phase space. Deduce the major frequencies contained in these structures. Wait for the transients to die out before conducting your analysis.
3) Compare your results to those in Figure 16.6.
4) See if you can deduce a relation among the Fourier components, the natural frequency ω_0, and the driving frequency ω.
5) A signal of chaos is a broadband Fourier spectrum, though not necessarily a flat one. Examine your system for parameters that give chaotic behavior, and verify this statement by plotting the power spectrum on both linear and semi-logarithmic plots.

Wavelet Exploration We saw in Chapter 10 that a wavelet expansion is more appropriate than a Fourier expansion for signals containing components that occur for only finite periods of time. Chaotic oscillations are just such signals. Repeat the Fourier analysis of this section using wavelets instead of sines and cosines. Can you discern the temporal sequence of various components?

16.4 Other Chaotic Systems

16.4.1 The Double Pendulum

The study we outlined previously for the chaotic pendulum can be repeated, or replaced, by one for the realistic double pendulum. Figures 16.1 and 16.10 show such a system. As shown on the right of Figure 16.1, the double pendulum has a second pendulum connected to the first, but no external driving force. In this case, each pendulum acts as a driving force for the other, and so there are enough degrees of freedom for chaotic behavior to occur, even without an external driving force.

The equations of motions for the double pendulum are derived most directly from the Lagrangian formulation of mechanics. The Lagrangian is fairly simple, but with the θ_1 and θ_2 motions innately coupled:

$$L = KE - PE = \frac{1}{2}(m_1 + m_2)l_1^2\dot{\theta}_1^{\,2} + \frac{1}{2}m_2l_2^2\dot{\theta}_2^{\,2} \tag{16.20}$$
$$+ m_2l_1l_2\dot{\theta}_1\dot{\theta}_2\cos(\theta_1 - \theta_2) + (m_1 + m_2)gl_1\cos\theta_1 + m_2gl_2\cos\theta_2.$$

Textbooks usually approximate these equations for small oscillations, which diminish the nonlinear effects, and conclude that the system contains "slow" and "fast" modes. More

Figure 16.10 Photographs of a double pendulum built by a student after running his simulation (there is a video `DoublePend.mp4`). The upper pendulum consists of two separated shafts so that both pendulums can go over their tops. The first two frames show the pendulum released from rest and then moving quickly through various modes. The photographs with a faster shutter speed stop the motion in various stages (R. Landau (Author)).

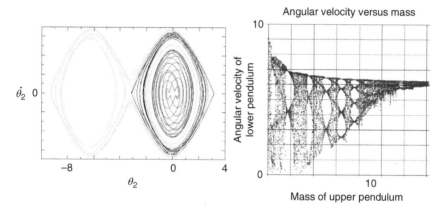

Figure 16.11 *Left:* Phase space trajectories for a double pendulum with $m_1 = 10m_2$ and with two dominant attractors. *Right:* A bifurcation diagram for the double pendulum displaying the instantaneous velocity of the lower pendulum as a function of the mass of the upper pendulum. (J. Danielson.)

interesting modes result when no small-angle approximations are made, and when the pendulums are given enough initial energy to go over the top. On the left of Figure 16.11, we see several phase-space plots of the lower pendulum's motion for $m_1 = 10m_2$. When given enough initial kinetic energy to go over the top, the trajectories are seen to flow between two major attractors, with energy being transferred back and forth between the pendulums.

On the right of Figure 16.11, is a bifurcation diagram for the double pendulum. This was created by sampling and plotting the instantaneous angular velocity $\dot{\theta}_2$ of the lower pendulum 70 times at instances when the pendulum passed through its equilibrium position. The mass of the upper pendulum (a convenient parameter) was then changed, and the process repeated. The resulting structure is fractal with bifurcations in the number of dominant frequencies in the motion. A plot of the Fourier or wavelet spectrum as a function of mass is expected to show similar characteristic frequencies.

16.4.2 Billiards

Deriving its name from the once-popular parlor game, a mathematical *billiard* is a dynamical system in which a particle moves freely in a straight line until it hits a boundary wall, at

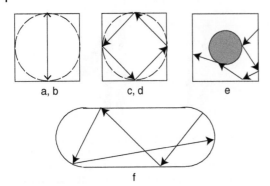

Figure 16.12 Square (a, c), circular (b, d), Sinai (e), and stadium billiards (f). The arrows are trajectories. The stadium billiard has two semicircles on the ends.

which point it undergoes specular reflection, and then continues on in a straight line until the next collision (Figure 16.12). The confining billiard table can be square, rectangular, circular, polygonal, a combination of the preceding, or even three-dimensional. Billiards are Hamiltonian systems in which there is no loss of energy, in which the motions continue endlessly, and which may display chaos.

In Figure 16.12, we show square (a, c), circular (b, d), Sinai (e), and stadium billiards (f), with the arrows indicating possible trajectories. Note how right-angle collisions lead to two-point periodic orbits for both square and circular billiards, while in (c) and (d) we see how 45° collisions leads to four-point periodic orbits. Figures (e) and (d) show nonperiodic trajectories that are ergodic, that is, orbits that will eventually pass through all points in the allowed space, and which can become chaotic.

1) Compute the trajectories for these four billiards, using a range of initial conditions. In Listing 16.1, we give a sample program for a square billiard that produces a VPython animation. (In Chapter 24, we do the quantum version of this problem.)
2) Plot the distance between successive collision points as a function of collision number. The plot should be simple for periodic motion, but should show irregular behavior as the motion becomes chaotic. Keep track of how many collisions occur before chaos sets in (typically 20–30). You need at least this many collisions to test hypersensitivity to initial conditions.
3) Since not all initial conditions lead to chaos, especially for circles, you may need to scan through various initial conditions.
4) For initial conditions that place you in the chaotic regime, explore the difference in behavior for a relatively slight ($\leq 10^{-3}$) variation in initial conditions.
5) Try initial conditions that differ at the machine precision level to gauge just how sensitive chaotic trajectories really are to initial conditions (be patient).

16.4.3 Multiple Scattering Centers

One expects the scattering of a projectile from a force center to be a continuous process. Nevertheless, when the potential has internal structure and the projectile undergoes multiple internal scatterings, complex behaviors may result. We have already seen some of this

Figure 16.13 One, two, and three stationary disks on a flat billiard table scatter point particles elastically, with some of the internal scattering leading to trapped, periodic orbits.

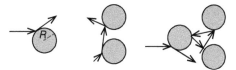

in Section 13.3.1 in our model for a pinball machine. If you have not solved that problem already, you may want to do so now and focus on how chaos shows itself there.

16.4.3.1 Hard Disk Scattering

Figure 16.13 shows one, two, and three hard disks attached to the surface of a 2D billiard table. In each case, the disks scatter point particles elastically (no energy loss). The disks all have radius R, center-to-center separations a, with the three-disk configuration forming an equilateral triangle. Also shown in the figure are imagined trajectories of particles scattered from the disks, with some of the internal scattering leading to trapped, periodic orbits in which the projectile bounces back and forth endlessly. For the two-disk case, there is just a single trapped orbit, but for the three-disk case, there are infinitely many, and that may lead to chaos. In Chapter 24, we explore the quantum mechanical version of this problem. Listing 24.2 3QMdisks.py in that chapter has a potential subroutine that can be used to model the present disks.

1) Modify the program already developed in Section 13.3.1 for the study of scattering from the four-peaked Gaussian (13.18) so that it can be applied to scattering from one, two, or three disks. Or write a new one.
2) Because infinite potentials cannot be handled numerically, pick instead a very large value for the potential. The exact value should not matter as long as it's large enough so that increasing its values has no significant effect on the scattering.
3) Plot a number of trajectories $[x(t), y(t)]$ that show both usual and unusual behaviors. In particular, plot those for which back-angle scattering occurs, and, consequently, for which there must have been significant number of multiple scatterings.
4) Plot a number of phase space trajectories $[x(t), \dot{x}(t)]$ and $[y(t), \dot{y}(t)]$. How do these differ from those of bound states?
5) Start with a projectile at $x \simeq -\infty$, that is, at some very large number, and at various distances (impact parameters) $y \equiv b$ from the center of the scattering region. Determine the scattering angle $\theta = $ atan2(Vx,Vy) as a function of b by determining the velocity components of the scattered particle after it has left the interaction region, that is, when $PE/E \leq 10^{-10}$.
6) Plot the discontinuities in $d\theta/db$ and $\sigma(\theta)$:

$$\sigma(\theta) = \left| \frac{d\theta}{db} \right| \frac{b}{\sin \theta(b)}. \tag{16.21}$$

7) Explore the different geometries in search for chaos. This should be near $a/R \simeq 6$ for the three disks.

16.4.4 Lorenz Attractors

In 1961, Edward Lorenz was using a simplified atmospheric convection model to predict weather patterns [Lorenz, 1963], and to save time, he entered the decimal 0.506 instead of

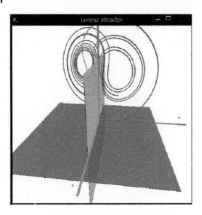

Figure 16.14 A 3D plot of a Lorenz attractor.

entering the full value 0.506127 for a parameter [Peitgen *et al.*, 1994; Motter and Campbell, 2013]. The results for the two numbers were so different that at first he thought it to be some kind of numerical error, but in time he realized that this was a nonlinear system with some unusual behavior that we now know as chaos. Here, we ask you to repeat his discovery.

   With simplified variables, the equation used by Lorenz are

$$\dot{x} = \sigma(y - x), \tag{16.22}$$

$$\dot{y} = x(\rho - z) - y, \tag{16.23}$$

$$\dot{z} = -\beta z + xy. \tag{16.24}$$

Here $x(t)$ is a measure of fluid velocity as a function of time t, $y(t)$ and $z(t)$ are measures of the temperature distributions in two directions, and σ, ρ, and β are parameters. Note that the xz and xy terms make these equations nonlinear.

1) Modify your ODE solver to handle the three, simultaneous Lorenz equations.
2) Start with the parameter values $\sigma = 10$, $\beta = 8/3$, and $\rho = 28$.
3) Make sure to use small enough step sizes so that good precision is obtained. You must have confidence that you are seeing chaos and not numerical error.
4) Make plots of x versus t, y versus t, and z versus t, and compare them to Figure 16.14.
5) The initial behavior in these plots are *transients* and are not considered dynamically interesting. Leave off these transients in the plots to follow.
6) Make a "phase space" plot of $z(t)$ versus $x(t)$ (the independent variable t does not appear in such a plot). The distorted, number eight-like figures you obtain are the Lorenz attractors, to which even chaotic solutions have an affinity.
7) Make phase space plots of $y(t)$ versus $x(t)$ and versus $z(t)$.
8) Make a 3D plot of $x(t)$ versus $y(t)$ and versus $z(t)$.
9) The parameters given to you should lead to chaotic solutions. Check this claim by seeing how small a change you can make in a parameter value and still, eventually, obtain different answers.

16.4.5 van der Pool Oscillator

The van der Pool equation describes the nonlinear behavior in once-common objects such as vacuum tubes and metronomes:

$$\frac{d^2x}{dt^2} + \mu(x^2 - x_0^2)\frac{dx}{dt} + \omega_0^2 x = 0. \tag{16.25}$$

1) Explain why (16.25) describes an oscillator with x-dependent damping.
2) Create phase space plots $\dot{x}(t)$ versus $x(t)$.
3) Verify that this equation produces a limit cycle with orbits internal to the cycle spiraling out to the limit cycle, and those external to it spiraling in to it.

16.4.6 The Duffing Oscillator

The Duffing Oscillator is another example of a damped, driven, nonlinear oscillator. It is described by a differential equation:

$$\frac{d^2x}{dt^2} = -2\gamma\frac{dx}{dt} - \alpha x - \beta x^3 + F\cos\omega t. \tag{16.26}$$

Our code `rk4D uffing.py` solves a form of this equation.

1) Modify your ODE solver to solve (16.26).
2) Start with parameter values corresponding to a simple harmonic oscillator, and verify that you obtain sinusoidal behavior and an elliptical phase space plot.
3) Include a driving force, wait 100 cycles in order to eliminate transients, and then create a phase space plot. We used $\alpha = 1.0$, $\beta = 0.2$, $\gamma = 0.2$, $\omega = 1.$, $F = 4.0$, $x(0) = 0.009$, and $\dot{x}(0) = 0$.
4) Search for period-three solutions like those in Figure 16.15, where we used $\alpha = 0.0$, $\beta = 1.$, $\gamma = 0.04$, $\omega = 1.$, and $F = 0.2$.
5) Change your parameters to $\omega=1$ and $\alpha=0$ in order to model an *Ueda* oscillator.

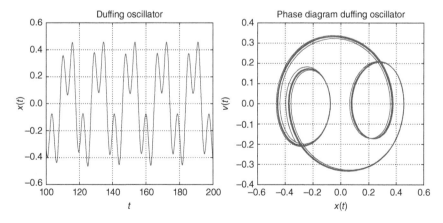

Figure 16.15 A period three solution for a forced Duffing oscillator. *Left*: $x(t)$ and *Right*: $v(t)$ versus $x(t)$.

16.5 Code Listings

Listing 16.1 **SqBilliardCM.py** Trajectories for motion on a square billiard table.

```
# SqBillardCM.py: Animated classical billiards on square table

from visual import *

dt = 0.01;   Xo = -90.;  Yo =  -5.4;  v = vector(13.,13.1)
r0 = r= vector(Xo,Yo); eps = 0.1;  Tmax = 500; tp = 0
scene = display(width=500, height=500, range=120,\
                background=color.white, foreground=color.black)
table = curve(pos=([(-100,-100,0),(100,-100,0),(100,100,0 ),\
                (-100,100,0),(-100,-100,0)]))
ball = sphere(pos=(Xo,Yo,0),color=color.red, radius=3,make_trail=True)

for t in arange(0,Tmax,dt):
    rate(5000)
    tp = tp + dt
    r = r0 + v*tp
    if(r.x>= 100 or r.x<=-100):          # Right and left walls
        v = vector(-v.x,v.y,0)
        r0 = vector(r.x,r.y,0)
        tp = 0
    if(r.y>= 100 or r.y<=-100):          # Top and bottom walls
        v = vector(v.x,-v.y,0)
        r0 = vector(r.x,r.y,0)
        tp = 0
    ball.pos = r
```

17

Thermodynamics Simulations and Feynman Path Integrals

The first part of this chapter extends the Monte-Carlo techniques studied in Chapter 4, now to the thermal behavior of a magnetic chain. The second part of this chapter applies the Metropolis algorithm, just used in the simulation of thermal behavior, now to Feynman's path integral formulation of quantum mechanics. The latter theory, while somewhat advanced for undergraduates, provides an unusual view of quantum mechanics, and is the basis for some of the most fundamental computations in physics.

17.1 An Ising Magnetic Chain

Ferromagnets contain finite size *domains* in which the spins of all the atoms point in the same direction. When an external magnetic field is applied to these materials, the different domains align, and the materials become "magnetized." Yet, as the temperature is raised, the degree of magnetism decreases, until a *phase transition* occurs at the Curie temperature, and all magnetization vanishes.

Problem Develop a model that exhibits the thermal behavior of ferromagnets.

Consider N magnetic dipoles fixed in place on the links of a linear chain (Figure 17.1). Because the particles are fixed, we need not worry about the symmetry of their wave function, or their positions, or their momenta. We assume that the particle at site i has spin s_i, which is either up or down:

$$s_i \equiv s_{z,i} = \pm\tfrac{1}{2}. \tag{17.1}$$

A configuration of the N particles is described by a quantum state vector:

$$|\alpha_j\rangle = |s_1, s_2, \dots, s_N\rangle = \left\{\pm\tfrac{1}{2}, \pm\tfrac{1}{2}, \dots\right\}, \quad j = 1, \dots, 2^N. \tag{17.2}$$

Due to the spin of each particle assuming one of *two* values, there are 2^N different possible states for the N particles.

The energy of the system arises from the interaction of the spins with each other and with an external magnetic field B. We know from quantum mechanics that an electron's spin and magnetic moment are proportional to each other, so a *spin–spin* interaction is equivalent

Computational Physics: Problem Solving with Python, Fourth Edition.
Rubin H. Landau, Manuel J. Páez, and Cristian C. Bordeianu.
© 2024 WILEY-VCH GmbH. Published 2024 by WILEY-VCH GmbH.

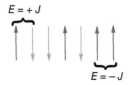

$E = + J$

$E = - J$

Figure 17.1 The 1D lattice of N spins used in the Ising model of magnetism. The interaction energy between nearest-neighbor pairs $E = \pm J$ is shown for aligned and opposing spins.

to a magnetic *dipole–dipole* interaction. In the simplest version of the model, we assume that each dipole interacts with the external magnetic field, and with its nearest neighbor, through the potential:

$$V_i = -J\,\mathbf{s}_i \cdot \mathbf{s}_{i+1} - g\mu_b\,\mathbf{s}_i \cdot \mathbf{B}. \tag{17.3}$$

Here, the constants are J, the *exchange energy*, g, the gyromagnetic ratio, and $\mu_b = e\hbar/(2m_e c)$, the Bohr magneton.

Even for a small numbers of particles, the 2^N possible spin configurations can get to be very large ($2^{20} > 10^6$), and it is computationally expensive to examine them all. Realistic chains of $\sim 10^{23}$ particles are beyond imagination. Consequently, statistical approaches are usually assumed, even for moderate values of N. Just how large N must be for statistics to be valid, is something you will be asked to explore with your simulations.

The energy of this system in state α_k is the expectation value of the sum of the potential V, over the spins of all the particles:

$$E_{\alpha_k} = \left\langle \alpha_k \left| \sum_i V_i \right| \alpha_k \right\rangle = -J \sum_{i=1}^{N-1} s_i s_{i+1} - B\mu_b \sum_{i=1}^{N} s_i. \tag{17.4}$$

An apparent paradox in the Ising model occurs if we turn off the external magnetic field B, since then there would be no preferred direction in space. This would imply that the average magnetization should vanish, despite our expectation that the lowest energy state would have all spins aligned. The resolution to the paradox is that the system with $B = 0$ is unstable; even if all the spins are aligned, there is nothing to stop the spontaneous reversal of all the spins. This instability leads to *Bloch-wall transitions* in which regions of different spin orientations change size spontaneously. Indeed, natural magnetic materials have multiple domains with all the spins aligned, but with the different domains pointing in different directions.

To proceed, we assume $B = 0$, which leaves just spin–spin interactions. However, be cognizant of the fact that this means there is no preferred direction in space, and so, some care may be needed when averaging observables over domains. For example, you may need to take an absolute value of the total spin when calculating the magnetization, that is, to calculate $\langle |\Sigma_i s_i| \rangle$ rather than $\langle \Sigma_i s_i \rangle$.

The equilibrium alignment of the spins depends on the sign of the exchange energy J. If $J > 0$, the lowest energy state will tend to have neighboring spins aligned, and if the temperature is low enough, the ground state will be a *ferromagnet*. Yet if $J < 0$, the lowest energy state will tend to have neighbors with opposite spins, and if the temperature is low enough, the ground state will be a *antiferromagnet* with alternating spins.

The solution to the 1D Ising model has its limitations. Although the model is accurate in describing a system in thermal equilibrium, it is not accurate in describing the *approach* to thermal equilibrium, or in predicting a phase transition at the Curie temperature. Also, we have postulated that only one spin is flipped at a time, whereas real magnetic materials tend to flip many spins at a time. Other limitations are straightforward to correct. For example, the addition of long-range interactions rather than just nearest neighbors, the motion of the centers, higher-multiplicity spin states, and extensions to two and three dimensions. In fact, the 2D and 3D models do support phase transitions (see Figure 17.4) [Yang, 1952].

17.1.1 Statistical Mechanics

Statistical mechanics starts with elementary interactions among a system's particles, and constructs the macroscopic thermodynamic properties, such as specific heat and magnetization. The essential assumption is that all configurations of the system consistent with the constraints are possible. In some simulations, such as the molecular dynamic ones in Chapter 18, the problem is set up such that the *energy* of the system is fixed. The states of that type of system are described by a *microcanonical ensemble*. In contrast, for the thermodynamic simulations we study in this chapter, the temperature, volume, and number of particles remain fixed, and so we have a *canonical ensemble*.

When we say that an object is *at* temperature T, we mean that the object's atoms are in thermodynamic equilibrium, and have an average kinetic energy proportional to T. Although this may be an equilibrium state, it is also a dynamic one in which the object's energy fluctuates as it exchanges energy with its environment. Indeed, one of the most illuminating aspects of the simulations to follow is its visualization of the continual and random interchange of energy that occurs at equilibrium.

The energy E_{α_j} of state α_j in a canonical ensemble is not constant, but rather is distributed with probabilities $P(\alpha_j)$ given by the Boltzmann distribution:

$$P(E_{\alpha_j}, T) = \frac{e^{-E_{\alpha_j}/k_B T}}{Z(T)}, \qquad Z(T) = \sum_{\alpha_j} e^{-E_{\alpha_j}/k_B T}. \tag{17.5}$$

Here k_B is Boltzmann's constant, T is the temperature, and $Z(T)$ is the partition function, a weighted sum over the individual *states* or *configurations* of the system. Another formulation, Wang–Landau sampling (WLS) as discussed in Section 17.3, instead sums over the *energies* of the states of the system with a density-of-states factor $g(E_i)$ [Landau and Wang, 2001].

17.1.1.1 Analytic Solution

For very large numbers of particles, one can solve for the internal energy $U = \langle E \rangle$ of the 1D Ising model [Plischke and Bergersen, 1994]:

$$\frac{U}{J} = -N \tanh \frac{J}{k_B T}, \tag{17.6}$$

$$= -N \frac{e^{J/k_B T} - e^{-J/k_B T}}{e^{J/k_B T} + e^{-J/k_B T}} = \begin{cases} N, & k_B T \to 0, \\ 0, & k_B T \to \infty. \end{cases} \tag{17.7}$$

The analytic results for the specific heat per particle and the magnetization are:

$$C(k_B T) = \frac{1}{N} \frac{dU}{dT} = \frac{(J/k_B T)^2}{\cosh^2(J/k_B T)}, \tag{17.8}$$

$$M(k_B T) = \frac{N e^{J/k_B T} \sinh(B/k_B T)}{\sqrt{e^{2J/k_B T} \sinh^2(B/k_B T) + e^{-2J/k_B T}}}. \tag{17.9}$$

The **2D Ising model** also has an analytic solution, which is not easy to derive [Yang, 1952; Huang, 1987]. Whereas internal energy and heat capacity are expressed in terms of elliptic integrals, the spontaneous magnetization per particle has the simple form:

$$\mathcal{M}(T) = \begin{cases} 0, & T > T_c, \\ \frac{(1+z^2)^{1/4}(1-6z^2+z^4)^{1/8}}{\sqrt{1-z^2}}, & T < T_c, \end{cases} \tag{17.10}$$

$$kT_c \simeq 2.269185J, \quad z = e^{-2J/k_BT}, \tag{17.11}$$

where the temperature is measured in units of the Curie temperature T_c.

17.2 Metropolis Algorithm

When trying to understand an algorithm that simulates thermal equilibrium, it is important to keep in mind that the Boltzmann distribution (17.5) does not require a system to always proceed to its lowest energy state; instead, it just requires it to be less likely to be found in a higher energy state than a lower energy one. Of course, as $T \to 0$, only the lowest energy state will be populated, but at finite temperatures, we expect the energy to fluctuate on the order of k_BT about the equilibrium energy, with the system sometimes moving to a higher-energy state.

In their simulation of neutron transmission through matter, Metropolis *et al.* [1953] devised an algorithm to improve the Monte Carlo calculation of averages. Because the sequence of configurations that their *Metropolis algorithm* produces accurately simulates the fluctuations occurring during thermal equilibrium, the algorithm has become a cornerstone of computational physics.

One can view the Metropolis algorithm as a combination of the variance reduction technique discussed in Section 5.7, and the Von Neumann rejection technique (stone throwing) discussed in Section 5.8. There, we showed how to make Monte Carlo integration more efficient by sampling random points, predominantly where the integrand is large, and how to generate random points weighted by an arbitrary probability distribution [now to be the Boltzmann function].

We want an approach that flips spins randomly, that equilibrates rapidly, and that produces a Boltzmann distribution of energies in the end:

$$\mathcal{P}(E_{\alpha_j}, T) \propto e^{-E_{\alpha_j}/k_BT}. \tag{17.12}$$

The procedure starts with the system at a fixed temperature and an arbitrary initial spin configuration. It then applies the Metropolis algorithm until a thermal equilibrium is reached. The continued application of the algorithm (typically $10\,N$ times for N particles) generates statistical fluctuations about the equilibrium, from which the thermodynamic quantities are deduced. Then, in order to deduce the temperature dependence of the thermodynamic quantities, the temperature is changed and the process is repeated. Explicitly:

1) Start with an arbitrary spin configuration $\alpha_k = \{s_1, s_2, \ldots, s_N\}$. (The equilibrium configuration should be independent of the initial distribution.)
   - A "hot" start has random values for the spins.
   - A "cold" start has all spins parallel ($J > 0$), or antiparallel ($J < 0$).

2) Generate a trial configuration α_{tr} by:
   (a) picking a particle i randomly, and
   (b) flipping its spin.
3) Calculate the energy $E_{\alpha_{tr}}$ of the trial configuration.
4) If $E_{\alpha_{tr}} \leq E_{\alpha_k}$, accept the trial by setting $\alpha_{k+1} = \alpha_{tr}$.
5) If $E_{\alpha_{tr}} > E_{\alpha_k}$, accept with probability $\mathcal{R} = \mathcal{P}_{tr}/\mathcal{P}_i = \exp(-\Delta E/k_B T)$. To do that:
   (a) Choose another uniform random number $0 \leq r_i \leq 1$.
   (b) Set $\alpha_{k+1} = \begin{cases} \alpha_{tr}, & \text{if } \mathcal{R} \geq r_j \text{ (accept)}, \\ \alpha_k, & \text{if } \mathcal{R} < r_j \text{ (reject)}. \end{cases}$
   (c) If the trial configuration is rejected, the next configuration is identical to the preceding one.

17.2.1 Metropolis Exercise

1) Write a program that implements the Metropolis algorithm, that is, that produces a new configuration α_{k+1} from the present configuration α_k. (Alternatively, use the program IsingViz.py given in Listing 17.1.)
2) Make the key data structure in your program an array s [N] containing the values of the spins s_i. For debugging, print out + and − (or o and blank) to give the spin at each lattice point (as we show in Figure 17.2). Examine the pattern for different trial lengths.
3) The value for the exchange energy J fixes the energy scale. Keep it fixed at $J = 1$. (You may also wish to study antiferromagnets with $J = -1$, but first examine ferromagnets whose domains are easier to understand.)
4) The thermal energy $k_B T$ is in units of J and is an independent variable for the model. Use $k_B T = 1$ for debugging.
5) Use periodic boundary conditions on your chain to minimize end effects. This means that the chain is a circle with the first and last spins adjacent to each other.
6) Try $N \simeq 20$ for debugging, and larger values for production runs.

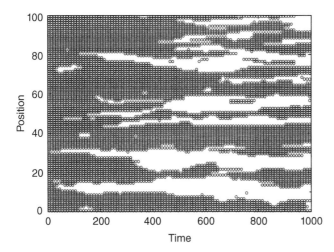

Figure 17.2 An Ising model simulation on a 1D lattice of 100 initially aligned spins (on the left). Up spins are indicated by circles, and down spins by blanks. Although the system starts with all up spins (a "cold" start), the system is seen to form domains of up and down spins as time progresses.

17.2.2 Equilibration and Thermodynamic Properties

1) Watch a chain of N atoms attain thermal equilibrium. At high temperatures, or for small number of atoms, you should see large fluctuations, while at lower temperatures, you should see smaller fluctuations.
2) Look for evidence of instabilities in which there is a spontaneous flipping of a large number of spins. This becomes more likely for larger $k_B T$ values.
3) Note how at thermal equilibrium, the system is still quite dynamic, with spins flipping all the time. It is this energy exchange that determines the thermodynamic properties.
4) You may well find that simulations at small $k_B T$ (say, $k_B T \simeq 0.1$ for $N = 200$) are slow to equilibrate. Higher $k_B T$ values equilibrate faster, yet have larger fluctuations.
5) Observe the formation of domains and the effect they have on the total energy. Regardless of the direction of spin within a domain, the atom–atom interactions are attractive, and so, they contribute negative amounts to the energy of the system when aligned. However, the ↑↓ or ↓↑ interactions between domains contribute positive energy. Therefore, you should expect a more negative energy at lower temperatures where there are larger and fewer domains.
6) Make a graph of average domain size *versus* temperature.

Thermodynamic Properties For a given spin configuration α_j, the energy and magnetization are given by:

$$E_{\alpha_j} = -J \sum_{i=1}^{N-1} s_i s_{i+1}, \qquad \mathcal{M}_j = \sum_{i=1}^{N} s_i. \tag{17.13}$$

The internal energy $U(T)$ is just the average value of the energy,

$$U(T) = \langle E \rangle, \tag{17.14}$$

where the average is taken over a system in equilibrium. At high temperatures, we expect a random assortment of spins, and so a vanishing magnetization. At low temperatures, when most the spins are aligned, we expect $\mathcal{M}$ to approach $N/2$. Although the specific heat can be computed from the elementary definition,

$$C = \frac{1}{N} \frac{dU}{dT}, \tag{17.15}$$

the numerical differentiation may be inaccurate due to the statistical fluctuations of U. A better approach is to first calculate the fluctuations in energy, occurring during M trials, and then determine the specific heat from the fluctuations:

$$U_2 = \frac{1}{M} \sum_{t=1}^{M} (E_t)^2, \tag{17.16}$$

$$C = \frac{1}{N^2} \frac{U_2 - (U)^2}{k_B T^2} = \frac{1}{N^2} \frac{\langle E^2 \rangle - \langle E \rangle^2}{k_B T^2}. \tag{17.17}$$

1) Extend your program to calculate the internal energy U and the magnetization $\mathcal{M}$ for the chain. Do not recalculate entire sums when only one spin changes.
2) Make sure to wait for your system to equilibrate before calculating thermodynamic quantities. (If equilibrated, U will fluctuate about its average.) Your results should resemble Figure 17.3.

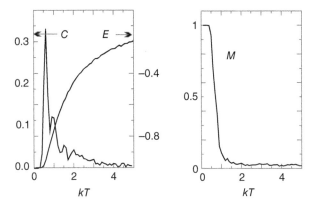

Figure 17.3 Simulation results from a 1D Ising model of 100 spins. *Left*: Energy and specific heat as functions of temperature; *Right*: Magnetization as a function of temperature.

3) Reduce the statistical fluctuations by running the simulation a number of times with different seeds, and by taking the average of the results.
4) The simulations you run for small N may be realistic, but may not agree with statistical mechanics, which assumes $N \simeq \infty$. (You may assume that $N \simeq 2000$ is close to infinity.) Check if that agreement with the analytic results for the thermodynamic limit is better for large N rather than small N.
5) Check that the simulated thermodynamic quantities are independent of initial conditions (within statistical uncertainties). This means that your cold and hot start results should agree.
6) Make a plot of the internal energy U as a function of $k_B T$, and compare it to the analytic result (17.6).
7) Make a plot of the magnetization $\mathcal{M}$ as a function of $k_B T$, and compare it to the analytic result. Does this agree with how you expect a heated magnet to behave?
8) Compute the energy fluctuations U_2 (17.16) and the specific heat C (17.17). Compare the simulated specific heat to the analytic result (17.8).

17.2.3 Explorations

1) Extend the model so that the spin–spin interaction (17.3) extends to next-nearest neighbors as well as nearest neighbors. For the ferromagnetic case, this should lead to more binding and less fluctuation because we have increased the couplings among spins, and thus increased the thermal inertia.
2) Extend the model so that the ferromagnetic spin–spin interaction (17.3) extends to nearest neighbors in two dimensions, and for the truly ambitious, three dimensions. Continue using periodic boundary conditions and keep the number of particles small, at least to start with [Gould *et al.*, 2006].
   (a) Form a square lattice and place $\sqrt{N}$ spins on each side.
   (b) Examine mean energy and magnetization as the system equilibrates.
   (c) Is the temperature dependence of the average energy qualitatively different from that of the 1D model?
   (d) Make a print out of the spin configuration for small N, and identify domains.

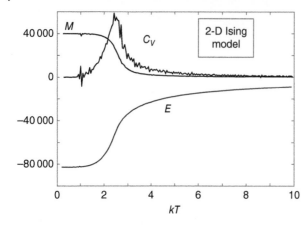

Figure 17.4 The energy, specific heat, and magnetization as a function of temperature from a 2D Ising model simulation with 40 000 spins. Evidence of a phase transition at the Curie temperature $kT = \simeq 2.5$ is seen in all three functions. The values of C and E have been scaled to fit on the same plot as M. (Courtesy of J. Wetzel.)

(e) Once your system appears to be behaving properly, calculate the heat capacity and magnetization of the 2D Ising model with the same technique used for the 1D model. Use a total number of particles of $100 \leq N \leq 2000$.

(f) Look for a phase transition from ordered to unordered configurations by examining the heat capacity and magnetization as functions of temperature. The former should diverge, while the latter should vanish at the phase transition (Figure 17.4).

17.3 Fast Equilibration via Wang–Landau Sampling ⊙

Although the Metropolis algorithm has been providing excellent service for more than 70 years, WLS [Landau and Wang, 2004; Clark University, 2011], with its shorter simulation time, has been showing increasing utility in research literature.[1] Our simulation with the Metropolis algorithm, which we have just described, used a Boltzmann distribution and focused on its temperature dependence. The WLS algorithm also uses a Boltzmann distribution, but focuses on its energy dependence. It starts with the probability that a system at a temperature T will contain the distribution of energy:

$$P(E_i, T) = g(E_i) \frac{e^{-E_i/k_B T}}{Z(T)}, \quad Z(T) = \sum_{E_i} g(E_i) e^{-E_i/k_B T}. \tag{17.18}$$

Here, $g(E_i)$ is the number, or density, of states of energy E_i, and $Z(T)$ is the partition function. The sum in Z is over all states of the system, but with states of the same energy entering just once, owing to $g(E_i)$ accounting for their degeneracy. Because the density-of-states $g(E)$ is a function of energy, but not temperature, once it has been computed, $Z(T)$ and all thermodynamic quantities, can be calculated without having to repeat the simulation for each temperature. For example, the internal energy and the entropy are:

$$U(T) \stackrel{\text{def}}{=} \langle E \rangle = \frac{\sum_{E_i} E_i g(E_i) e^{-E_i/k_B T}}{\sum_{E_i} g(E_i) e^{-E_i/k_B T}}, \quad S = k_B \ln g(E_i). \tag{17.19}$$

The density of states $g(E_i)$ is determined by taking the equivalent of a random walk in *energy* space. We flip a randomly chosen spin, record the energy of the new configuration,

---

1 We thank Oscar A. Restrepo of the Universidad de Antioquia for letting us use some of his material.

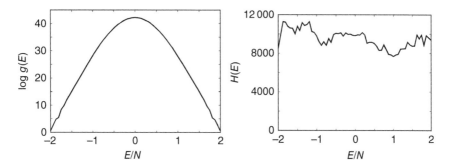

Figure 17.5 Wang–Landau sampling used in the 2D Ising model on an 8 × 8 lattice.
Left: Logarithm of the density of states log $g(E)$ *versus* the energy per particle. *Right:* The histogram
$H(E)$ showing the number of states visited as a function of the energy per particle. The aim of WLS
is to make $H(E)$ flat.

and keep on flipping spins and recording energies. When done, we have a *histogram* $H(E_i)$ of
the number of times each energy E_i is attained (Figure 17.5 right). If the flipping were continued for a very long time, the histogram $H(E_i)$ would eventually converge to the density
of states $g(E_i)$. Yet, because the walk would only rarely move away from the most probable
energies, even for small systems, some 10^{19} to 10^{30} steps might be required.

WLS increases the likelihood of sampling less probable configurations by increasing their
acceptance, while simultaneously decreasing the acceptance of more likely ones. To accomplish this trick, WLS accepts a new energy E_i with a probability inversely proportional to
the (initially unknown) density of states,

$$P(E_i) = \frac{1}{g(E_i)}, \tag{17.20}$$

and then builds up a histogram of visited states as the walk continues.

An apparent problem with WLS is that $g(E_i)$ is unknown. This is overcome by determining
$g(E_i)$ simultaneously with the execution of the random walk. One starts with an arbitrary
$g(E_i)$ function, and then multiplies $g(E_i)$ by an empirical factor $f > 1$, which increases the
likelihood of reaching states with small $g(E_i)$ values. As the histogram $H(E_i)$ gets flatter,
the multiplicative factor f is decreased until it is close to 1. At that point, we have a flat
histogram and a determination of $g(E_i)$ in which all energies have been visited equally.

17.3.1 WLS Implementation

Our implementation of WLS, `WangLandau.py`, is given in Listing 17.2. It assumes an Ising
model with $J = 1$, and nearest neighbor interactions. Rather than recalculating the energy
each time a spin is flipped, only the differences in energies are computed. For example, for
eight spins in a line:

$$-E_k = \sigma_0\sigma_1 + \sigma_1\sigma_2 + \sigma_2\sigma_3 + \sigma_3\sigma_4 + \sigma_4\sigma_5 + \sigma_5\sigma_6 + \sigma_6\sigma_7 + \sigma_7\sigma_0, \tag{17.21}$$

where we have assumed periodic boundary conditions. If spin 5 is flipped,

$$-E_{k+1} = \sigma_0\sigma_1 + \sigma_1\sigma_2 + \sigma_2\sigma_3 + \sigma_3\sigma_4 - \sigma_4\sigma_5 - \sigma_5\sigma_6 + \sigma_6\sigma_7 + \sigma_7\sigma_0, \tag{17.22}$$

and the difference in energies is:

$$\Delta E = E_{k+1} - E_k = 2(\sigma_4 + \sigma_6)\sigma_5. \tag{17.23}$$

For the 2D problem with spins on a lattice, the change in energy when spin $\sigma_{i,j}$ on site (i,j) is flipped is:

$$\Delta E = 2\sigma_{i,j}(\sigma_{i+1,j} + \sigma_{i-1,j} + \sigma_{i,j+1} + \sigma_{i,j-1}). \tag{17.24}$$

For N spins there are two states of minimum energy $E = -2N$, ones with all spins pointing in the same direction, either all up or all down. The maximum energy $2N$ corresponds to alternating spin directions on neighboring sites. Each spin flip on the lattice changes the energy by four units between these limits:

$$E_i = -2N, \quad -2N + 4, \quad -2N + 8, \ \dots, \quad 2N - 8, \ 2N - 4, \quad 2N. \tag{17.25}$$

The produced histogram $H(E_i)$ and entropy $S(T)$ are given in Figure 17.5.

17.4 Path Integral Quantum Mechanics ☉

Problem In classical mechanics, a particle's motion is described by it space-time trajectory $x(t)$. For a particle in a harmonic oscillator potential, relate the classical trajectory to the quantum mechanical wave function $\psi(x, t)$.

As the story goes, Feynman was looking for a formulation of quantum mechanics that had a more direct connection to classical mechanics than the Schrödinger theory does, and that also incorporated the statistical nature of quantum mechanics from the start. He followed a suggestion by Dirac that *Hamilton's principle of least action*, which can be used to derive classical dynamics, may be the $\hbar \to 0$ limit of a quantum least-action principle. Seeing that Hamilton's principle deals with the paths of particles through space-time, Feynman postulated [Feynman and Hibbs, 1965; Mannheim, 1983] that the quantum-mechanical wave function describing the propagation of a free particle from the space-time point $a = (x_a, t_a)$ to the point $b = (x_b, t_b)$, are related by:

$$\psi(x_b, t_b) = \int dx_a \, G(x_b, t_b; x_a, t_a)\psi(x_a, t_a), \tag{17.26}$$

where G is the *Green's function* or *propagator*:

$$G(x_b, t_b; x_a, t_a) \equiv G(b, a) = \sqrt{\frac{m}{2\pi i(t_b - t_a)}} \exp\left[i\frac{m(x_b - x_a)^2}{2(t_b - t_a)}\right]. \tag{17.27}$$

Equation (17.26) can be viewed as a form of Huygens's wavelet principle in which each point on the wavefront $\psi(x_a, t_a)$ emits a spherical wavelet $G(b; a)$ that propagates forward in space and time. Accordingly, the new wavefront $\psi(x_b, t_b)$ is created by summation over, and interference among, all of the emitted wavelets.

Feynman imagined that another way of viewing (17.26) is as a form of Hamilton's principle in which the probability amplitude ψ for a particle to be at B is equal to the sum over all *paths* through space-time originating at time A and ending at B (Figure 17.6). This view incorporates the statistical nature of quantum mechanics by assigning different probabilities for travel along different paths, with all paths possible, but with some more likely than others. The values for the probabilities of the paths derive from *Hamilton's classical principle of least action*:

The most general motion of a physical particle moving along the classical trajectory $\bar{x}(t)$ from time t_a to t_b is along a path such that the action $S[\bar{x}(t)]$ is an extremum:

$$\delta S[\bar{x}(t)] = S[\bar{x}(t) + \delta x(t)] - S[\bar{x}(t)] = 0, \tag{17.28}$$

with the paths constrained to pass through the endpoints:

$$\delta(x_a) = \delta(x_b) = 0.$$

This formulation of classical mechanics, which is based on the calculus of variations, is equivalent to Newton's differential equations if the action S is taken as the line integral of the Lagrangian along the classical trajectory:

$$S[\bar{x}(t)] = \int_{t_a}^{t_b} dt\, L\,[x(t), \dot{x}(t)]\,, \quad L = T\,[x, \dot{x}] - V[x]. \tag{17.29}$$

Here, T is the kinetic energy, V is the potential energy, $\dot{x} = dx/dt$, and the square brackets indicate a *functional*[2] of the function $x(t)$ and $\dot{x}(t)$.

Feynman observed that the classical action for a free ($V = 0$) particle,

$$S[b, a] = \frac{m}{2}(\dot{x})^2(t_b - t_a) = \frac{m}{2}\frac{(x_b - x_a)^2}{t_b - t_a}, \tag{17.30}$$

is related to the free-particle propagator (17.27) by:

$$G(b, a) = \sqrt{\frac{m}{2\pi i(t_b - t_a)}} e^{iS[b,a]/\hbar}. \tag{17.31}$$

Equation (17.31) is the much sought-after connection between quantum mechanics (LHS) and Hamilton's principle (RHS). Feynman went on to postulate a reformulation of quantum mechanics that incorporates its statistical aspects by postulating $G(b, a)$ to be a weighted sum of exponentials, each with an exponent that is the action for a *path* connecting a to b:

$$G(b, a) = \sum_{\text{paths}} e^{iS[b,a]/\hbar} \quad \text{(A path integral)}. \tag{17.32}$$

Figure 17.6 In the Feynman path-integral formulation of quantum mechanics, a collection of paths connect the initial space-time point A to the final point B. The solid line is the classical trajectory that minimizes the action S. The dashed lines are paths that are also sampled by a quantum particle.

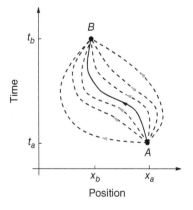

---

2 A *functional* is a number whose value depends on the complete behavior of some function and not just on its behavior at one point.

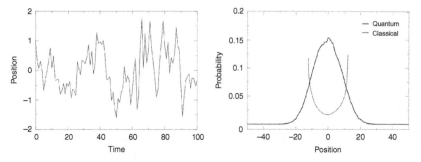

Figure 17.7 *Left:* A space-time quantum path resulting from applying the Metropolis algorithm. *Right:* The probability distribution for the harmonic oscillator ground state as determined by a path-integral calculation (the classical result has maxima at the two turning points).

The sum (17.32) is called a *path integral* because it sums the exponential of the classical action $S[b,a]$ over infinitely many paths (Figure 17.6), with each action itself being a classical line integral (in time) along a path.

The correspondence principle connecting classical and quantum mechanics applies here via the realization that because $\hbar \simeq 10^{-34}$ Js is a very small number, $S/\hbar \sim 10^{20}$ is a very large number. Accordingly, even though an infinity of paths may enter into the sum (17.32), the main contributions come from those paths that are adjacent to the classical trajectory $\bar{x}$. In fact, because S is an extremum for the classical trajectory, it remains constant to first order in the variation of paths, and so nearby paths have values (phases in the exponentials) that vary smoothly and relatively slowly. In contrast, paths far from the classical trajectory are weighted by a rapidly oscillating $\exp(iS/\hbar)$, and when many are summed over, they tend to cancel each other out. In the classical limit $\hbar \to 0$, only the single classical trajectory contributes, and (17.32) becomes Hamilton's principle of least action! In Figure 17.7 left, we show an example of an actual trajectory used in path-integral calculations.

17.4.1 Bound-State Wave Function

Although you may be thinking that you have already seen enough expressions for Green's function, there is yet another one we need for our computation. We start by assuming that the Hamiltonian operator $\tilde{H}$ supports a spectrum of eigenfunctions,

$$\tilde{H}\psi_n = E_n\psi_n, \tag{17.33}$$

each labeled by the index n. Because $\tilde{H}$ is Hermitian, its wave functions forms a complete orthonormal set in which we may expand a general solution:

$$\psi(x,t) = \sum_{n=0}^{\infty} c_n e^{-iE_n t}\psi_n(x), \tag{17.34}$$

$$c_n = \int_{-\infty}^{+\infty} dx\, \psi_n^*(x)\psi(x, t=0), \tag{17.35}$$

where the value for the expansion coefficients c_n follows from the orthonormality of ψ_n's. If we substitute this c_n back into the wave function expansion (17.34), we obtain the identity:

$$\psi(x,t) = \int_{-\infty}^{+\infty} dx_0 \sum_n \psi_n^*(x_0)\psi_n(x)e^{-iE_n t}\psi(x_0, t=0). \tag{17.36}$$

Comparison with (17.26) yields the eigenfunction expansion for G:

$$G(x, t; x_0, t_0 = 0) = \sum_n \psi_n^*(x_0)\psi_n(x)e^{-iE_n t}.$$ (17.37)

We relate this to the bound-state wave function (*recall that our* **problem** *is to calculate that*) by first requiring all paths to start and end at the space position $x_0 = x$, by then taking $t_0 = 0$, and, finally, by making an analytic continuation of (17.37) to negative imaginary time (permissible for analytic functions):

$$G(x, -i\tau; x, 0) = \sum_n |\psi_n(x)|^2 e^{-E_n \tau} = |\psi_0|^2 e^{-E_0 \tau} + |\psi_1|^2 e^{-E_1 \tau} + \cdots,$$

$$\Rightarrow \quad |\psi_0(x)|^2 = \lim_{\tau \to \infty} e^{E_0 \tau} G(x, -i\tau; x, 0).$$ (17.38)

The limit here corresponds to long imaginary times τ, after which the parts of ψ with higher energies decay more quickly, leaving only the ground state ψ_0.

Equation (17.38) provides a closed-form solution for the ground-state wave function directly in terms of the Green's function G. Although we will soon describe how to compute this function, for now look at Figure 17.7 right, showing some results of the computation. Although we start with a probability distribution that peaks near the classical turning points at the edges of the well, after a large number of iterations, we end up with a distribution that resembles the expected Gaussian. So, maybe the new formulation does work. On the left of Figure 17.7, we see a trajectory that has been generated via statistical variations about the classical trajectory $x(t) = A \sin(\omega_0 t + \phi)$.

17.5 Lattice Path Integration

Because both time and space need to be integrated over when evaluating a path integral, our simulation starts with a lattice of discrete space-time points [Potvin, 1993]. We visualize a particle's trajectory as a series of straight lines connecting one time to the next (Figure 17.8). We divide the time between the space-time points A and B into N equal time steps of size ε, and label them with the index j:

$$\varepsilon \overset{\text{def}}{=} \frac{t_b - t_a}{N} \Rightarrow t_j = t_a + j\varepsilon, \qquad (j = 0, N).$$ (17.39)

Although it is more precise to use the actual positions $x(t_j)$ of the trajectory at times t_j to determine the x_j's (as in Figure 17.8), we also discretize space uniformly with the links ending at the nearest lattice points. Seeing that we have a lattice, it is easy to evaluate derivatives or integrals on a link[3]:

$$\frac{dx_j}{dt} \simeq \frac{x_j - x_{j-1}}{t_j - t_{j-1}} = \frac{x_j - x_{j-1}}{\varepsilon},$$ (17.40)

$$S_j \simeq L_j \Delta t \simeq \frac{1}{2} m \frac{(x_j - x_{j-1})^2}{\varepsilon} - V(x_j)\varepsilon,$$ (17.41)

where we have assumed that the Lagrangian is constant over each link.

---

3 Although Euler's rule has a large error, it is often used in lattice calculations because of its simplicity. However, if the Lagrangian contains second derivatives, then the more precise central-difference method is needed to avoid singularities.

Lattice path integration is based on the *composition theorem* for propagators:

$$G(b,a) = \int dx_j\, G(x_b, t_b; x_j, t_j) G(x_j, t_j; x_a, t_a) \quad (t_a < t_j, t_j < t_b). \tag{17.42}$$

For a free particle this yields:

$$G(b,a) = \sqrt{\frac{m}{2\pi i(t_b - t_j)}} \sqrt{\frac{m}{2\pi i(t_j - t_a)}} \int dx_j\, e^{i(S[b,j] + S[j,a])}$$

$$= \sqrt{\frac{m}{2\pi i(t_b - t_a)}} \int dx_j\, e^{iS[b,a]}, \tag{17.43}$$

where we have added the actions because line integrals combine as $S[b,j] + S[j,a] = S[b,a]$. For the N-linked path in Figure 17.8, equation (17.42) becomes:

$$G(b,a) = \int dx_1 \cdots dx_{N-1}\, e^{iS[b,a]}, \quad S[b,a] = \sum_{j=1}^{N} S_j, \tag{17.44}$$

where S_j is the value of the action for link j. At this point, the integral over the *single* path shown in Figure 17.8 has become an N-term sum that becomes an infinite sum as the time step ε approaches zero.

To summarize, Feynman's path-integral postulate (17.32) means that we sum over all paths connecting A to B to obtain the Green's function $G(b, a)$. This, in turn, means that we must sum, not only over the links in one path, but *also* over all the different paths, in order to produce the variation in paths required by Hamilton's principle. The sum is constrained such that paths must pass through A and B and cannot double back on themselves (causality requires that particles move only forward in time). This is the essence of *path integration*. Because we are integrating over functions as well as along paths, the technique is also known as *functional integration*.

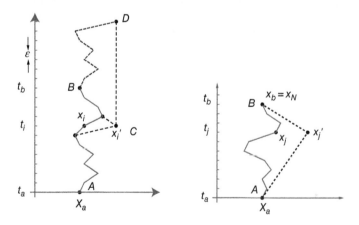

Figure 17.8 *Left:* A path through a space-time lattice that starts and ends at $x = x_a = x_b$. The action is an integral over this path, while the *path integral* is a sum of integrals over all paths. The dotted path BD is a transposed replica of path AC. *Right:* The dashed path joins the initial and final times in two equal time steps; the solid curve uses N steps each of size ε. The position of the curve at time t_j defines the position x_j.

The propagator (17.32) is the sum over all paths connecting A to B, with each path weighted by the exponential of the action along that path, explicitly:

$$G(x, t; x_0, t_0) = \oint dx_1\, dx_2 \cdots dx_{N-1}\, e^{iS[x,x_0]}, \tag{17.45}$$

$$S[x, x_0] = \sum_{j=1}^{N-1} S[x_{j+1}, x_j] \simeq \sum_{j=1}^{N-1} L\left(x_j, \dot{x}_j\right)\varepsilon, \tag{17.46}$$

where $L(x_j, \dot{x}_j)$ is the average value of the Lagrangian on link j at time $t = j\varepsilon$. The computation is made simpler by assuming that the potential $V(x)$ is independent of velocity and does not depend on other x values (local potential). Next, we observe that G is evaluated with a negative imaginary time in the expression (17.38) for the ground-state wave function. Accordingly, we evaluate the Lagrangian with $t = -i\tau$:

$$L(x, \dot{x}) = T - V(x) = +\frac{1}{2}m\left(\frac{dx}{dt}\right)^2 - V(x), \tag{17.47}$$

$$\Rightarrow L\left(x, i\frac{dx}{d\tau}\right) = -\frac{1}{2}m\left(\frac{dx}{d\tau}\right)^2 - V(x). \tag{17.48}$$

We see that the reversal of the sign of kinetic energy in L means that L now equals the negative of the Hamiltonian evaluated at a real positive time $t = \tau$:

$$H\left(x, \frac{dx}{d\tau}\right) = \frac{1}{2}m\left(\frac{dx}{d\tau}\right)^2 + V(x) = E, \tag{17.49}$$

$$\Rightarrow L\left(x, i\frac{dx}{d\tau}\right) = -H\left(x, \frac{dx}{d\tau}\right). \tag{17.50}$$

In this way, we rewrite the t-path integral of L as a τ-path integral of H, and so express the action and Green's function in terms of the Hamiltonian:

$$S[j+1, j] = \int_{t_j}^{t_{j+1}} L(x, t)\, dt = -i \int_{\tau_j}^{\tau_{j+1}} H(x, \tau)\, d\tau, \tag{17.51}$$

$$\Rightarrow G(x, -i\tau; x_0, 0) = \int dx_1 \ldots dx_{N-1}\, e^{-\int_0^\tau H(\tau')d\tau'}, \tag{17.52}$$

where the line integral of H is over an entire trajectory. Next, we express the path integral in terms of the average energy of the particle on each link, $E_j = T_j + V_j$, and then sum over the links to obtain the summed energy:

$$\int H(\tau)\, d\tau \simeq \sum_j \varepsilon E_j = \varepsilon \mathcal{E}(\{x_j\}), \tag{17.53}$$

$$\mathcal{E}(\{x_j\}) \overset{\text{def}}{=} \sum_{j=1}^{N} \left[\frac{m}{2}\left(\frac{x_j - x_{j-1}}{\varepsilon}\right)^2 + V\left(\frac{x_j + x_{j-1}}{2}\right)\right]. \tag{17.54}$$

In (17.54), we have approximated each link in the path as a *straight line*, used Euler's derivative rule to obtain the velocity, and evaluated the potential at the midpoint of the link. We now substitute this G into our expression (17.38) for the ground-state wave function, with identical initial and final points in space:

$$\lim_{\tau \to \infty} \frac{G(x, -i\tau, x_0 = x, 0)}{\int dx\, G(x, -i\tau, x_0 = x, 0)} = \frac{\int dx_1 \cdots dx_{N-1} \exp\left[-\int_0^\tau H d\tau'\right]}{\int dx\, dx_1 \cdots dx_{N-1} \exp\left[-\int_0^\tau H d\tau'\right]}$$

$$\Rightarrow \quad |\psi_0(x)|^2 = \frac{1}{Z} \lim_{\tau \to \infty} \int dx_1 \cdots dx_{N-1}\, e^{-\varepsilon \mathcal{E}}, \tag{17.55}$$

$$Z = \lim_{\tau \to \infty} \int dx\, dx_1 \cdots dx_{N-1} e^{-\varepsilon \mathcal{E}}. \tag{17.56}$$

Note the additional dx integrand in the expression for Z. The similarity of these expressions to thermodynamics, even with a partition function Z, is no accident. By making the time parameter imaginary, we have converted the time-dependent Schrödinger equation to the heat diffusion equation:

$$i \frac{\partial \psi}{\partial(-i\tau)} = \frac{-\nabla^2}{2m}\psi \quad \Rightarrow \quad \frac{\partial \psi}{\partial \tau} = \frac{\nabla^2}{2m}\psi. \tag{17.57}$$

It is not surprising then that the sum over paths in Green's function has each path weighted by the Boltzmann factor, $\mathcal{P} = e^{-\varepsilon \mathcal{E}}$, which is usually associated with thermodynamics. We make the connection complete by identifying the temperature with the inverse time step:

$$\mathcal{P} = e^{-\varepsilon \mathcal{E}} = e^{-\mathcal{E}/k_B T} \quad \Rightarrow \quad k_B T = \frac{1}{\varepsilon} \equiv \frac{\hbar}{\varepsilon}. \tag{17.58}$$

Consequently, the $\varepsilon \to 0$ limit, which makes time continuous, is equivalent to a high-temperature limit. The $\tau \to \infty$ limit, which is required to project the ground-state wave function, means that we must integrate over a path that is long in imaginary time, that is, long compared to typical time $\hbar/\Delta E$. Just as our simulation of the Ising model required us to wait for a long time for the system to equilibrate, so too does the present simulation require us to wait a long time, so that all but the ground-state wave function has decayed away. At last, we have the solution to our **problem** of finding the ground-state wave function via its connection to classical mechanics.

To summarize, we have expressed the Green's function as a path integral requiring integrations of the Hamiltonian along all paths (17.55). We evaluate this path integral as the sum over trajectories on a space-time lattice, with each path weighted by a probability based on the path's action. We use the Metropolis algorithm to perform the many, multi-dimensional integrations required as we examine all space-time paths. This is similar to what we did with the Ising model, however, rather than rejecting or accepting a *flip in spin*, we now reject or accept a *change in a link*, also based on the change in energy. The more iterations we let the algorithm run for, the more time the deduced wave function has to equilibrate to the ground state, and thus the more accurate the answer.

In general, these types of Monte Carlo Green's function techniques work best if we start with a good guess at the final answer, and then have the algorithm calculate variations on our guess. For the present problem, this means that if we start with a path in space-time close to the classical trajectory, the algorithm may be expected to do a good job at simulating the quantum fluctuations about that trajectory. However, it does not appear to be good at finding the classical trajectory from arbitrary locations in space-time.

17.5.1 A Time-Saving Trick

We have formulated the computation so that you pick a value of x and perform many computations of line integrals, over all space and time, to obtain $|\psi_0(x)|^2$. To obtain the wave function at another x, the entire simulation must be repeated from scratch. Rather than going through all that work again and again, we can compute the entire x dependence of the wave function in one fell swoop. The trick is to insert a delta function into the probability integral (17.55), thereby fixing the initial position to be x_0, and then to also integrate over all x_0s:

$$|\psi_0(x)|^2 = \int dx_1 \cdots dx_N \, e^{-\varepsilon \mathcal{E}(x,x_1,\dots)} \tag{17.59}$$

$$= \int dx_0 \cdots dx_N \delta(x - x_0) \, e^{-\varepsilon \mathcal{E}(x,x_1,\dots)}. \tag{17.60}$$

This equation expresses the wave function as an average of a delta function over all paths, a procedure that might seem totally inappropriate for numerical computation because one cannot compute singular functions. Yet, when we simulate the sum over all paths with (17.60), there will always be some x value for which the integral is nonzero, and so we accumulate the solution for whatever x value that is.

To understand how this works in practice, consider path AB in Figure 17.8, imagining that we have just calculated the summed energy. We form a new path by having one point on the chain jump to point C (which changes two links). If we replicate section AC, and use it as the extension AD to form the top path, we see that the path CBD has the same summed energy (action) as path ACB, and in this way it can be used to determine $|\psi(x'_j)|^2$. That being the case, once the system is equilibrated, we determine new values of the wave function at new locations x'_j by flipping links to new values and calculating new actions. The more frequently some x_j is accepted, the greater is the wave function at that point.

17.6 Implementation

The program QMC.py in Listing 17.3 evaluates the integral (17.32) by finding the average of the integrand $\delta(x_0 - x)$ with paths distributed according to the weighting function $\exp[-\varepsilon \mathcal{E}(x_0, x_1, \dots, x_N)]$. The physics enters via (17.62), the calculation of the summed energy $\mathcal{E}(x_0, x_1, \dots, x_N)$. We evaluate the action integral for the harmonic oscillator potential:

$$V(x) = \tfrac{1}{2}x^2, \tag{17.61}$$

and for a particle of mass $m = 1$. Using a convenient set of natural units, we measure lengths in $\sqrt{1/m\omega} \equiv \sqrt{\hbar/m\omega} = 1$, and times in $1/\omega = 1$. Correspondingly, the oscillator has a period $T = 2\pi$. Figure 17.7 shows results after the application of the Metropolis algorithm. In this computation, we started with an initial path close to the classical trajectory, and then examined half a million variations about this path. All paths were constrained to begin and end at $x = 1$.

When the time difference $t_b - t_a$ is small, say like $2T$, the system will not have enough time to equilibrate to its ground state, and so the computed wave function will look like the probability distribution of an excited state (nearly classical with the probability highest for the particle to be near its turning points, where its velocity vanishes). However, when $t_b - t_a$ equals a longer time, such as $20T$, the system will have enough time to decay to its ground state, and the wave function will look like the expected Gaussian. In either case (Figure 17.7 right), the trajectory through space-time fluctuates about the classical trajectory. This fluctuation is a consequence of the Metropolis algorithm occasionally going uphill in its search; if you modify the program so that searches go only downhill, the space-time trajectory will be a very smooth trigonometric function (the classical trajectory), but the wave function, which is a measure of the fluctuations about the classical trajectory, will vanish!

Here are the explicit steps [MacKeown, 1985; MacKeown and Newman, 1987]:

1) Construct a grid of N time steps each of length ε (Figure 17.8). Start at $t = 0$, and extend to time $\tau = N\varepsilon$ [N time intervals and $(N + 1)$ lattice points in time]. Note that time always increases monotonically along a path.
2) Construct a grid of M space points separated by steps of size δ. Start with $M \simeq N$, and use a range of x values several time larger than the characteristic size of the potential being used.
3) Any x or t value falling between lattice points should be assigned to the closest lattice point.
4) Associate a position x_j with each time τ_j, subject to the boundary conditions that the initial and final positions always remain at $x_N = x_0 = x$.
5) Choose a path consisting of straight-line links connecting the lattice points. This should correspond to the classical trajectory. Observe that the x values for the links of the path may have values that increase, decrease, or remain unchanged (in contrast to time, which always increases).
6) Starting at $j = 0$, evaluate the energy $\mathcal{E}$ by summing the kinetic and potential energies for each link of the path:

$$\mathcal{E}(x_0, x_1, \ldots, x_N) \simeq \sum_{j=1}^{N} \left[\frac{m}{2} \left(\frac{x_j - x_{j-1}}{\varepsilon} \right)^2 + V \left(\frac{x_j + x_{j-1}}{2} \right) \right]. \tag{17.62}$$

7) Begin a sequence of repetitive steps in which a random position x_j associated with time t_j is changed to the position x_j' (point C in Figure 17.8). This changes *two* links in the path.
8) Use the Metropolis algorithm to weigh the changed position with the Boltzmann factor.
9) For each lattice point, establish a running sum representing the squared modulus of the wave function at that point.
10) After each single-link change (or decision not to change), increase the running sum for the new x value by 1. After a sufficiently long running time, the sum divided by the number of steps is the simulated value for $|\psi(x_j)|^2$ at each lattice point x_j.
11) Repeat the entire link-changing simulation starting with a different seed. A wave function averaged over many intermediate length runs is better than one from a very long run.

17.6.1 Path Integration Exercise

1) Plot some of the actual space-time paths used in the simulation along with the classical trajectory.
2) For a more continuous picture of the wave function, make the x lattice spacing smaller; for a more precise value of the wave function at any particular lattice site, sample more points (run longer) and use a smaller time step ε.

3) Because there are no sign changes in a ground-state wave function, you can ignore the phase, assume $\psi(x) = \sqrt{\psi^2(x)}$, and then estimate the energy via:

$$E = \frac{\langle \psi | H | \psi \rangle}{\langle \psi | \psi \rangle} = \frac{\omega}{2\langle \psi | \psi \rangle} \int_{-\infty}^{+\infty} \psi^*(x) \left(-\frac{d^2}{dx^2} + x^2 \right) \psi(x)\, dx, \qquad (17.63)$$

where the space derivative is evaluated numerically.

4) Explore the effect of making $\hbar$ larger, and thus permitting greater fluctuations around the classical trajectory. Do this by decreasing the value of the exponent in the Boltzmann factor. Determine if this makes the calculation more or less robust in its ability to find the classical trajectory.

5) Test your ψ for the gravitational potential (see quantum bouncer below):

$$V(x) = mg|x|, \qquad x(t) = x_0 + v_0 t + \tfrac{1}{2} g t^2. \qquad (17.64)$$

17.6.2 Quantum Bouncer ⊙

Another problem for which the classical trajectory is well known is that of a *quantum bouncer*.[4] Here we have a particle dropped in a uniform gravitational field, hitting a hard floor, and then bouncing up. When treated quantum mechanically, quantized levels for the particle result [Gibbs, 1975; Goodings and Szeredi, 1992; Whineray, 1992; Vallée, 2000]. In 2002, an experiment to discern this gravitational effect at the quantum level was performed by Nesvizhevsky *et al.* [2002], and is described in Shaw [1992]. It consisted of dropping ultracold neutrons from a height of 14 μm unto a neutron mirror, and watching them bounce. It found a neutron ground state at 1.4 peV.

We start by determining the analytic solution to this problem for stationary states, and then generalizing it to include time-dependence. The time-independent Schrödinger equation for a particle in a uniform gravitation potential is:

$$-\frac{\hbar^2}{2m} \frac{d^2 \psi(x)}{dx^2} + mxg\, \psi(x) = E\, \psi(x), \qquad (17.65)$$

$$\psi(x \leq 0) = 0, \qquad \text{(boundary condition).} \qquad (17.66)$$

The boundary condition (17.66) is a consequence of the hard floor at $x = 0$. A change of variables converts (17.65) to a dimensionless form,

$$\frac{d^2 \psi}{dz^2} - (z - z_E)\, \psi = 0, \qquad (17.67)$$

$$z = x \left(\frac{2gm^2}{\hbar^2} \right)^{1/3}, \qquad z_E = E \left(\frac{2}{\hbar^2 mg^2} \right)^{1/3}. \qquad (17.68)$$

There is an analytic solution in terms of Airy functions Ai(z) [Press *et al.*, 2007]:

$$\psi(z) = N_n\, \text{Ai}(z - z_E), \qquad (17.69)$$

where N_n is a normalization constant. The boundary condition $\psi(0) = 0$ implies:

$$\psi(0) = N_E\, \text{Ai}(-z_E) = 0, \qquad (17.70)$$

which means that the allowed energies of the system correspond to the zeros z_n of Airy functions with negative arguments. To simplify the calculation, we take $\hbar = 1$, $g = 2$, and $m = \frac{1}{2}$, which leads to $z = x$ and $z_E = E$.

---

4 Oscar A. Restrepo assisted in the preparation of this section.

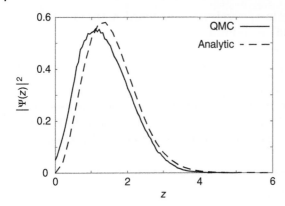

Figure 17.9 The analytic and quantum Monte Carlo solution for the quantum bouncer. The dashed line is the Airy function squared, and the solid line is $|\psi_0(z)|^2$ after a million trajectories.

The time-dependent solution for the quantum bouncer is an infinite sum over the eigenfunctions, each with a time-dependence determined by its energy:

$$\psi(z,t) = \sum_{n=1}^{\infty} C_n N_n \, \mathrm{Ai}(z - z_n) e^{-iE_n \, t/\hbar}, \tag{17.71}$$

where the C_ns are constants.

Figure 17.9 shows the results of solving for the quantum bouncer's ground-state probability $|\psi_0(z)|^2$ using Feynman's path integration, that is, quantum Monte Carlo. The time increment dt and the total time t were selected by trial and error in such a way as to satisfy the boundary condition $|\psi(0)|^2 \simeq 0$. To account for the potential being infinite for negative x values, we selected trajectories that have positive x values over all their links. This incorporates the fact that the particle can never penetrate the floor. Our program is given in Listing 17.4. The result after using 10^6 trajectories, and a time step $\varepsilon = d\tau = 0.05$, are shown in Figure 17.9. Both wave functions were normalized via a trapezoid integration. As can be seen, the agreement between the analytic and path integration wave function is satisfactory, though not perfect.

17.6.3 Path Integral Bouncer Exercises

1) You are given the fact that a particle falls at distance d in time $t = \sqrt{2D/g}$. Assume a quadratic dependence on distance and time,

$$d = \alpha t + \beta t^2, \tag{17.72}$$

and show, either analytically or numerically, that the action $S = \int_0^{t_0} L \, dt$ for the particle's trajectory is an extremum only when $\alpha = 0$ and $\beta = g/2$.

2) Consider a mass m attached to a harmonic oscillator with period $T = 1$ and frequency $\omega = 2\pi$:

$$x(t) = 10 \cos(\omega t). \tag{17.73}$$

(a) Propose a modification of (17.73) that agrees with it at $t = 0$ and $t = T$, though differs for intermediate values of t. Include an adjustable parameter in your modification.

(b) Compute the action for an entire range of values for the parameter in your proposed trajectory, and thereby verify that only the known analytic form yields a minimum action.

3) Consider a 1D harmonic oscillator with displacement q and momentum p. The energy:

$$E(p, q) = \frac{p^2}{2m} + \frac{m\omega^2 q^2}{2} \tag{17.74}$$

is an integral of the motion, and the area of the periodic orbit is:

$$A(E) = \oint p\, dq = 2 \int_{q_{min}}^{q_{max}} p\, dq. \tag{17.75}$$

(a) Use the analytic, or numeric, solution for simple harmonic motion to compute the area $A(E)$.

(b) Compute the derivative $T = dA(E)/dE$ via a central-difference approximation and compare to the analytic answer.

(c) Now repeat this problem using a nonlinear oscillator for which there is only a numerical solution. (Even oscillators of the form $V = kx^p$ with p should work just fine.) You can determine the period from the time dependence of your solution, and then use your solution to compute $A(E)$ for different initial conditions.

17.7 Code Listings

Listing 17.1 IsingViz.py A 1D Ising chain simulation with the Metropolis algorithm.

```
1  # IsingViz.py: Ising model

   from visual import *
   import random
5  from visual.graph import *

   # Display for the arrows
   scene = display(x=0,y=0,width=700,height=200, range=40,title='Spins')
9  engraph = gdisplay(y=200,width=700,height=300, title='E of Spin System',\
           xtitle='iteration', ytitle='E',xmax=500, xmin=0, ymax=5, ymin=-5)
   enplot = gcurve(color=color.yellow)
   N      = 30
13 B      = 1.
   mu     = .33                                      # g mu
   J      = .20
   k      = 1.                                       # Boltmann
17 T      = 100.
   state = zeros((N))                                # spins up(1), down (0)
   S     = zeros((N) ,float)
   test  = state
21 random.seed()                                     # Seed generator

   def energy ( S) :
       FirstTerm = 0.
25     SecondTerm = 0.
       for  i in range(0,N-2):   FirstTerm += S[i]*S[i + 1]
       FirstTerm *= -J
       for i in range(0,N-1):   SecondTerm += S[i]
29     SecondTerm *= -B*mu;
       return (FirstTerm + SecondTerm);

   ES = energy(state)
33
   def spstate(state):                               # Plots spins
       for obj in scene.objects: obj.visible=0       # Erase old arrows
       j=0
37     for i in range(-N,N,2):
           if state[j]==-1:  ypos = 5               # Spin down
```

```
               else :              ypos = 0
               if  5*state[j]<0:  arrowcol  = (1,1,1)        # White arrow if down
41             else :              arrowcol  =(0.7,0.8,0)
               arrow(pos=(i,ypos,0),axis=(0,5*state[j],0),color=arrowcol)
               j +=1

45  for   i in range(0 ,N):   state[i] = -1               # Initial spins all down

    for obj in scene.objects:    obj.visible=0
    spstate(state)
49  ES = energy(state)

    for   j in range (1,500):
          rate(3)
53        test = state
          r = int(N*random.random());    # Flip spin randomly
          test[r] *= -1
          ET = energy(test)
57        p = math.exp((ES-ET)/(k*T))    #  Boltzmann test
          enplot.plot(pos=(j,ES))        # Adds segment to curve
          if p >= random.random():
             state = test
61           spstate(state)
             ES = ET
```

Listing 17.2   WangLandau.py Wang Landau algorithm for 2-D spin system.

```
# WangLandau.py: Wang Landau algorithm for 2-D spin system

""" Author in Java: Oscar A. Restrepo,
4  Universidad de Antioquia, Medellin, Colombia
   Each time fac changes, a new histogrm is generated.
   Only the first Histogram plotted to reduce computational time"""
from visual import *
8  import random;
from visual.graph import *

L = 8;  N = (L*L)
12
# Set up graphics
entgr = gdisplay(x=0,y=0,width=500,height=250,title='Density of States',\
                         xtitle= 'E/N', ytitle='log g(E)', xmax=2.,
                               xmin=-2.,ymax=45,ymin=0)
16  entrp    = gcurve(color = color.yellow, display = entgr)
energygr = gdisplay(x=0, y=250, width=500, height=250, title='E vs T',\
           xtitle = 'T', ytitle='U(T)/N', xmax=8.,xmin=0, ymax =0.,ymin=-2.)
energ    = gcurve(color = color.cyan, display = energygr)
20  histogr  = display(x = 0, y = 500, width = 500, height = 300,\
            title = '1st histogram: H(E) vs. E/N, corresponds to log(f) = 1')
histo  = curve(x = list(range(0, N+1)), color=color.red, display=histogr)
xaxis    = curve(pos = [( - N,  - 10), (N,  - 10)])
24  minE     = label(text = ' - 2', pos = ( - N + 3,  - 15), box = 0)
maxE     = label(text = '2', pos = (N - 3,  - 15), box = 0)
zeroE    = label(text = '0', pos = (0,  - 15), box = 0)
ticm     = curve(pos = [( - N,  - 10), ( - N,  - 13)])
28  tic0     = curve(pos = [(0,  - 10), (0,  - 13)])
ticM     = curve(pos = [(N,  - 10), (N,  - 13)])
enr      = label(text = 'E/N', pos = (N/2,  - 15), box = 0)

32  sp       = zeros( (L, L) )                         # Grid size, spins
hist     = zeros( (N + 1) )
prhist = zeros( (N + 1) )                                   # Histograms
S        = zeros( (N + 1), float)              # Entropy = log g(E)
36
def iE(e):  return int((e + 2*N)/4)

def IntEnergy():
40     exponent = 0.0
       for T in arange (0.2, 8.2, 0.2 ):              # Select lambda max
```

```
                   Ener  = - 2*N
                   maxL = 0.0                                          # Initialize
44                 for i in range(0, N + 1):
                       if S[i]!= 0 and (S[i] - Ener/T)>maxL:
                           maxL = S[i] - Ener/T
                       Ener = Ener + 4
48                 sumdeno = 0
                   sumnume = 0
                   Ener    = -2*N
                   for i in range(0, N):
52                     if S[i] != 0:
                           exponent = S[i] - Ener/T - maxL
                       sumnume += Ener*exp(exponent)
                       sumdeno += exp(exponent)
56                     Ener    = Ener +  4.0
                   U = sumnume/sumdeno/N                     # internal energy U(T)/N
                   energ.plot(pos = (T, U) )

60  def WL():                                          # Wang - Landau sampling
        Hinf    = 1.e10                        # initial values for Histogram
        Hsup    = 0.
        tol     = 1.e-3                         # tolerance, stops the algorithm
64      ip      = zeros(L)
        im      = zeros(L)                              # BC R or down, L or up
        height = abs(Hsup - Hinf)/2.                    # Initialize histogram
        ave = (Hsup + Hinf)/2.                          # about average of histogram
68      percent = height / ave
        for i in range(0, L):
            for j in range(0, L): sp[i, j] = 1                  # Initial spins
        for i in range(0, L):
72          ip[i] = i + 1
            im[i] = i - 1                               # Case plus, minus
        ip[L - 1] = 0
        im[0]     = L - 1                                       # Borders
76      Eold = - 2*N                                    # Initialize energy
        for  j in range(0, N + 1): S[j] = 0             # Entropy initialized
        iter = 0
        fac = 1
80      while  fac > tol :

            i = int(N*random.random() )                 # Select random spin
            xg = i%L
84          # Must be i//L, not i/L for Python 3:
            yg      = i//L                              # Localize x, y, grid point
            Enew    = Eold + 2*(sp[ip[xg],yg] + sp[im[xg],yg] + sp[xg,ip[yg]]
                      + sp[xg, im[yg]] ) * sp[xg, yg]           # Change energy
88          deltaS = S[iE(Enew)]  - S[iE(Eold)]
            if  deltaS <= 0 or random.random() < exp( - deltaS):
                Eold = Enew;
                sp[xg, yg] *= - 1                                # Flip spin
92          S[iE(Eold)]   += fac;                       # Change entropy
            if iter%10000 == 0:                 # Check flatness every 10000 sweeps
                for j in range( 0, N + 1):
                    if  j == 0 :
96                      Hsup = 0
                        Hinf = 1e10             # Initialize new histogram
                    if  hist[j] == 0 : continue  # Energies never visited
                    if  hist[j] > Hsup: Hsup = hist[j]
100                 if  hist[j] < Hinf: Hinf = hist[j]
                height = Hsup - Hinf
                ave = Hsup + Hinf
                percent = 1.0* height/ave           # 1.0 to make it float number
104             if percent < 0.3 :                      # Histogram flat?
                    print(" iter ", iter, "  log(f) ", fac)
                    for j in range(0, N + 1):
                        prhist[j] = hist[j]                         # to plot
108                     hist[j]  = 0                                # Save hist
                    fac *= 0.5                   # Equivalent to log(sqrt(f))
            iter += 1
            hist[iE(Eold)]   += 1               # Change histogram, add 1, update
112         if fac >= 0.5:                       # just show the first histogram
```

```
                      # Speed up by using array calculations:
                      histo.x = 2.0*arange(0,N+1) - N
                      histo.y = 0.025*hist -10
116 deltaS = 0.0
    print("wait because iter > 13 000 000")              # not always the same
    WL()                                              # Call Wang Landau algorithm
    deltaS = 0.0
120 for j in range(0, N + 1):
        rate(150)
        order   = j*4  - 2*N
        deltaS  = S[j] -  S[0] + log(2)
124     if S[j] != 0 : entrp.plot(pos = (1.*order/N, deltaS))   # plot entropy
    IntEnergy();
    print("Done")
```

Listing 17.3   QMC.py Feynman path integration calculation of ground-state probability.

```
# QMC.py: Quantum MonteCarlo (Feynman path integration)

from visual import *;  from visual.graph import *;  import random
4
N = 100; Nsteps = 101; xscale = 10.                      # Initialize
path = zeros([Nsteps], float);  prob = zeros([Nsteps], float)

8 trajec = display(width = 300,height=500, title='Spacetime Paths')
  trplot = curve(y = range(0, 100), color=color.magenta, display = trajec)

  def PlotAxes():                                        # Axis
12    trax = curve(pos=[(-97,-100),(100,-100)],colo =color.cyan,display=trajec)
      label(pos = (0,-110), text = '0', box = 0, display = trajec)
      label(pos = (60,-110), text = 'x', box = 0, display = trajec)
  def WaveFunctionAxes():                                # Axes for probability
16    wvfax=curve(pos =[(-600,-155),(800,-155)],display=wvgraph,color=color.cyan)
      curve(pos = [(0,-150), (0,400)], display=wvgraph, color=color.cyan)
      label(pos = (-80,450), text='Probability', box = 0, display = wvgraph)
      label(pos = (600,-220), text='x', box=0, display=wvgraph)
20    label(pos = (0,-220), text='0', box=0, display=wvgraph)
  def Energy(path):                                       # HO Energy
      sums = 0.
      for i in range(0,N-2):sums += (path[i+1]-path[i])*(path[i+1]-path[i])
24    sums += path[i+1]*path[i+1];
      return sums
  def PlotPath(path):                                     # Plot trajectory
      for j in range (0, N):
28        trplot.x[j] = 20*path[j]
          trplot.y[j] = 2*j - 100
  def PlotWF(prob):                                       # Plot prob
      for i in range (0, 100):
32        wvplot.color = color.yellow
          wvplot.x[i] = 8*i - 400                         # Center fig

  wvgraph = display(x=340,y=150,width=500,height=300,title='Ground State')
36 wvplot = curve(x = range(0, 100), display = wvgraph)
  wvfax = curve(color = color.cyan)
  PlotAxes();   WaveFunctionAxes()                        # Plot axes
  oldE = Energy(path)
40 while True:                                            # Pick random element
      rate(10)                                            # Slow paintings
      element = int(N*random.random() )                   # Metropolis
      change = 2.0*(random.random() - 0.5)
44    path[element] += change                             # Change path
      newE = Energy(path);                                # Find new E
      if  newE > oldE and math.exp( - newE + oldE)<= random.random():
          path[element] -= change                         # Reject
48        PlotPath(path)                                  # Plot trajectory
      elem = int(path[element]*16 + 50)           # if path = 0, elem = 50

# elem = m *path[element] + b is the linear transformation
52 # if path=-3, elem=2 if path=3., elem=98 => b=50, m=16 linear TF.
```

```
                   # this way x = 0 correspond to prob[50]

          if elem < 0: elem = 0,
56        if elem > 100:  elem = 100                      # If exceed max
          prob[elem] += 1                            # increase probability
          PlotWF(prob)                                       # Plot prob
          oldE = newE
```

Listing 17.4   QMCbouncer.py Feynman path integration computation of a quantum particle in a gravitational field.

```
   # QMCbouncer.py:       g.s. wavefunction via path integration

 3 from visual import *
   import random
   from visual.graph import *

 7 N = 100;  dt = 0.05;         g = 2.0;        h = 0.00;        maxel = 0
   path = zeros([101], float);   arr = path;  prob = zeros([201],float)
   trajec = display(width = 300, height=500,title = 'Spacetime Trajectory')
   trplot = curve(y = range(0, 100), color=color.magenta, display = trajec)
11
   def trjaxs():                               # plot axis for trajectories
     trax=curve(pos=[(-97,-100),(100,-100)],color=color.cyan,display=trajec)
     curve(pos = [(-65, -100),(-65, 100)], color=color.cyan,display=trajec)
15   label(pos = (-65,110), text = 't', box = 0, display = trajec)
     label(pos = (-85, -110), text = '0', box = 0, display = trajec)
     label(pos = (60, -110), text = 'x', box = 0, display = trajec)
   wvgraph = display(x=350, y=80, width=500, height=300, title = 'GS Prob')
19 wvplot  = curve(x = range(0, 50), display = wvgraph)  # wave function plot
   wvfax   = curve(color = color.cyan)

   def wvfaxs():                               # plot axis for wavefunction
23   wvfax = curve(pos =[(-200,-155),(800,-155)],display=wvgraph,color=color.cyan)
     curve(pos = [(-200,-150),(-200,400)],display=wvgraph,color=color.cyan)
     label(pos = (-70, 420),text = 'Probability', box = 0, display=wvgraph)
     label(pos = (600, -220),text = 'x', box = 0, display = wvgraph)
27   label(pos = (-200, -220),text = '0', box = 0, display = wvgraph)
   trjaxs();  wvfaxs()                                         # plot axes

   def energy (arr):                                     # Energy of path
31   esum = 0.
     for i in range(0,N):
       esum += 0.5*((arr[i+1]-arr[i])/dt)**2+g*(arr[i]+arr[i+1])/2
     return esum
35
   def plotpath(path):                           # Plot xy trajectory
     for j in range (0, N):
       trplot.x[j] = 20*path[j] - 65
39     trplot.y[j] = 2*j - 100

   def plotwvf(prob):                           # Plot wave function
     for i in range (0, 50):
43     wvplot.color = color.yellow
       wvplot.x[i] = 20*i - 200
       wvplot.y[i] = 0.5*prob[i] - 150

47 oldE = energy(path)
   counter = 1
   norm = 0.                                    # Plot psi every 100
   maxx = 0.0
51 while 1:                                      # "Infinite" loop
     rate(100)
     element = int(N*random.random() )
     if element != 0 and element!= N:            # Ends not allowed
55     change = ( (random.random() - 0.5)*20.)/10.
       if  path[element] + change > 0.:          # No negative paths
         path[element]   += change
```

```
                newE = energy(path)                        # New trajectory E
59          if newE > oldE and exp( - newE + oldE) <= random.random() :
                path[element]    -= change                 # Link rejected
                plotpath(path)
            ele = int(path[element]*1250./100.)             # Scale changed
63          if  ele >= maxel:   maxel = ele           # Scale change 0 to N
            if  element != 0:  prob[ele]    += 1
            oldE = newE;
        if counter%100 == 0:                          # Plot psi every 100
67          for  i  in range(0, N):                        # Max x of path
                if  path[i] >= maxx:   maxx = path[i]
            h = maxx/maxel                                 # space step
            firstlast = h*0.5*(prob[0] +  prob[maxel])  # for trap. extremes
71          for  i  in range(0, maxel + 1):  norm = norm + prob[i]        # Norm
            norm = norm*h + firstlast                       # Trap rule
            plotwvf(prob)                               # Plot probability
        counter   += 1
```

18

Molecular Dynamics Simulations

You may recall from introductory chemistry that the ideal gas law can be derived from first principles by confining noninteracting molecules to a box. This chapter extends that model to molecules that interact with each other. Although the theory of Molecular Dynamics (MD) is straightforward, the simulations have proven to be powerful approaches for studying the physical and chemical properties of solids, liquids, amorphous materials, and biological molecules.

Problem Determine whether a collection of argon molecules placed in a box will coalesce into an ordered structure as the temperature is lowered.

Although we know that quantum mechanics is the proper theory for molecular interactions, MD uses Newton's laws as its basis and focuses on bulk properties, which are not particularly sensitive to the small r behaviors, where quantum effects may be important. Nevertheless, Car and Parrinello [1985] showed how MD can be extended to include quantum mechanics by using density functional theory to calculate the force between molecules. That technique, known as *quantum MD*, is an active area of research but is beyond the realm of the present chapter.[1] For those with further interests, there are full texts on MD [Rapaport, 1995; Hockney and Eastwood, 1988], fuller discussions in [Gould et al., 2006; Thijssen, 1999; Fosdick et al., 1996], as well as primers [Ercolessi, 1997], and codes available on-line [Nelson et al., 1996; Refson, 2000; Anderson et al., 2008].

Although MD's solution of Newton's laws is conceptually simple, when applied to a very large number of particles, it becomes the "high school physics problem from hell." Some approximations must be made in order to avoid solving the $\sim 10^{24}$ equations of motion of a realistic system, and so, present calculations tend to have an upper limit of $\sim 10^9$ particles confined to a finite region of space.

In a number of ways, MD simulations are similar to the thermal Monte Carlo simulations we studied in Chapter 17. Both typically involve a large number N of interacting particles that start out in some set configuration, and then equilibrate into a dynamic state. However, in MD we have a statistical mechanical *microcanonical ensemble* in which the energy E and volume V of the N particles are fixed. We then use Newton's laws to generate the dynamics

---

1 We thank Satoru S. Kano for pointing this out to us.

Computational Physics: Problem Solving with Python, Fourth Edition.
Rubin H. Landau, Manuel J. Páez, and Cristian C. Bordeianu.
© 2024 WILEY-VCH GmbH. Published 2024 by WILEY-VCH GmbH.

of the system. In contrast, Monte Carlo simulations do not start with first principles, but, instead, incorporate an element of chance and have the system remaining in contact with a heat bath at a fixed temperature, rather than keeping the energy E fixed. This is a *canonical ensemble*.

Since the molecules in an MD simulation are dynamic, their velocities and positions change continuously with time. After a simulation has run long enough to stabilize, we will compute time averages of the dynamic quantities in order to relate them to the thermodynamic properties. The simulations apply Newton's laws with the assumption that the net force on each molecule is the sum of the two-body forces with all of the other $(N-1)$ molecules:

$$m\frac{d^2\mathbf{r}_i}{dt^2} = \mathbf{F}_i(\mathbf{r}_0, \dots, \mathbf{r}_{N-1}), \tag{18.1}$$

$$m\frac{d^2\mathbf{r}_i}{dt^2} = \sum_{i<j=0}^{N-1} \mathbf{f}_{ij}, \quad i = 0, \dots, (N-1). \tag{18.2}$$

Here we have ignored the fact that an argon atom itself is a dynamic system composed of 18 electrons and a nucleus (Figure 18.1). Although it may be possible to ignore this internal structure when deducing the long-range properties of inert elements, it matters for systems such as polyatomic molecules that display rotational, vibrational, and electronic degrees of freedom as the temperature is raised.[2]

The force on molecule i derives from the sum of molecule–molecule potentials:

$$\mathbf{F}_i(\mathbf{r}_0, \mathbf{r}_1, \dots, \mathbf{r}_{N-1}) = -\nabla_{\mathbf{r}_i} U(\mathbf{r}_0, \mathbf{r}_1, \dots, \mathbf{r}_{N-1}), \tag{18.3}$$

$$U(\mathbf{r}_0, \mathbf{r}_1, \dots, \mathbf{r}_{N-1}) = \sum_{i<j} u(r_{ij}) = \sum_{i=0}^{N-2} \sum_{j=i+1}^{N-1} u(r_{ij}), \tag{18.4}$$

$$\Rightarrow \quad \mathbf{f}_{ij} = -\frac{du(r_{ij})}{dr_{ij}} \left(\frac{x_i - x_j}{r_{ij}}\hat{\mathbf{e}}_x + \frac{y_i - y_j}{r_{ij}}\hat{\mathbf{e}}_y + \frac{z_i - z_j}{r_{ij}}\hat{\mathbf{e}}_z \right). \tag{18.5}$$

Here $r_{ij} = |\mathbf{r}_i - \mathbf{r}_j| = r_{ji}$ is the distance between the centers of molecules i and j, and the limits on the sums assure that no interaction is counted twice. Because we have assumed a *conservative* potential, the total energy of the system, that is, the potential plus kinetic energies summed over all particles, should be conserved over time. Nonetheless, practical computations usually "cut the potential off" when the molecules are far apart $[u(r_{ij} > r_{cut}) = 0]$, which means that the derivative du/dr is infinite at r_{cut}, which is not what a conservative potential does, that, in turn, means that overall energy will no longer be precisely conserved. Yet because the cutoff radius is large, the cutoff occurs when the forces are minuscule, and so the violation of energy conservation should be small compared to other approximations and round-off errors.

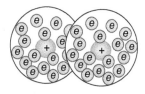

Figure 18.1 The molecule–molecule effective interaction arises from the many-body interaction of the electrons and nucleus in one molecule (circle) with the electrons and nucleus in another molecule (another circle). Note, the size of the nucleus at the center of each molecule is highly exaggerated, and real electrons have no size.

---

2 We thank Saturo Kano for clarifying this point.

In a true first-principles calculation, the potential between any two argon atoms would arise from the sum of approximately 1000 electron–electron and electron–nucleus Coulomb interactions. A more practical calculation would employ an effective potential derived from a many-body theory, such as Hartree–Fock or density functional theory. Our approach is simpler yet. We use the phenomenological Lennard–Jones potential,

$$u(r) = 4\epsilon \left[\left(\frac{\sigma}{r}\right)^{12} - \left(\frac{\sigma}{r}\right)^{6} \right], \tag{18.6}$$

$$\mathbf{f}(r) = -\frac{du}{dr}\frac{\mathbf{r}}{r} = \frac{48\epsilon}{r^2} \left[\left(\frac{\sigma}{r}\right)^{12} - \frac{1}{2}\left(\frac{\sigma}{r}\right)^{6} \right] \mathbf{r}. \tag{18.7}$$

Here the parameter ϵ governs the strength of the interaction, the parameter σ determines the length scale, and both are deduced by fits to data. Some typical values for the parameters and scales for the variables are given in Table 18.1. In order to make the simulation simpler and to avoid under- and overflows, it is helpful to measure all variables in the natural units of these constants. The interparticle potential and force then take the forms

$$u(r) = 4 \left[\frac{1}{r^{12}} - \frac{1}{r^6} \right], \qquad f(r) = \frac{48}{r} \left[\frac{1}{r^{12}} - \frac{1}{2r^6} \right]. \tag{18.8}$$

The Lennard–Jones potential is seen in Figure 18.2 to be the sum of a long-range attractive interaction $\propto 1/r^6$ and a short-range repulsive one $\propto 1/r^{12}$. The change from repulsion to attraction occurs at $r = \sigma$, with the potential's minimum at $r = 2^{1/6}\sigma = 1.1225\sigma$, which would be the atom–atom spacing in a solid bound by this potential. The $1/r^{12}$ term accounts for the Coulomb and Pauli principle repulsions that arise when the electron clouds from two atoms overlap. This $1/r^{12}$ term dominates at short distances and leads to atoms behaving like hard spheres. The precise value of 12 is not of theoretical significance (although it being large is) and may have been chosen because it is 2×6.

The $1/r^6$ term that dominates at large distances models the weak *van der Waals* induced dipole–dipole attraction between two molecules. This attraction arises from fluctuations in

Table 18.1 Parameter and scales for the Lennard–Jones potential.

Quantity	Mass	Length	Energy	Time	Temperature
Unit	m	σ	ϵ	$\sqrt{m\sigma^2/\epsilon}$	ϵ/k_B
Value	6.7×10^{-26} kg	3.4×10^{-10} m	1.65×10^{-21} J	4.5×10^{-12} s	119 K

Figure 18.2 The Lennard–Jones effective potential used in many MD simulations. Note the sign change at $r = 1$ and the minimum at $r \simeq 1.1225$ (natural units). Note too that because the r axis does not extend to $r = 0$, the infinitely high central repulsion is not shown.

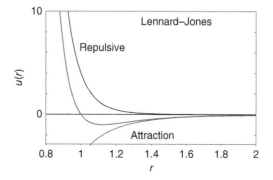

which, at some instant in time, a molecule on the right tends to be more positive on, say, the left side, like a dipole ⇐. This, in turn, attracts the negative charge in a molecule on its left, thereby inducing a dipole ⇐. As long as the molecules stay close to each other, the polarities continue to fluctuate in synchronization, ⇐⇐, ⇒⇒, so that the attraction is maintained. The resultant dipole–dipole attraction behaves like $1/r^6$, and although it's much weaker than a Coulomb force, it is responsible for the binding of neutral, inert elements, such as argon, for which the Coulomb force vanishes.

18.1   MD *Versus* Thermodynamics

Although an MD simulation is valid for any number of particles, if we assume that the number of particles is very large, then it becomes possible to use statistical mechanics to relate the results of a simulation to thermodynamic quantities. The equipartition theorem tells us that, on average, for molecules in thermal equilibrium at temperature T, each degree of freedom has an energy $k_B T/2$ associated with it, where $k_B = 1.38 \times 10^{-23}$ J/K is Boltzmann's constant. A simulation provides the kinetic energy of translation[3]:

$$KE = \frac{1}{2} \left\langle \sum_{i=0}^{N-1} v_i^2 \right\rangle. \tag{18.9}$$

The time average of KE (for three degrees of freedom) is related to temperature by

$$\langle KE \rangle = N\frac{3}{2}k_B T \Rightarrow T = \frac{2\langle KE \rangle}{3k_B N}. \tag{18.10}$$

The system's pressure P is determined by a version of the *Virial theorem*,

$$PV = Nk_B T + \frac{w}{3}, \quad w = \left\langle \sum_{i<j}^{N-1} \mathbf{r}_{ij} \cdot \mathbf{f}_{ij} \right\rangle, \tag{18.11}$$

where the Virial w is seen to be an average of force times interparticle distances. Note that because ideal gases have no intermolecular forces, their Virial vanishes, and we would obtain the ideal gas law. The pressure for the general case is

$$P = \frac{\rho}{3N} \left(2\langle KE \rangle + w \right), \tag{18.12}$$

where $\rho = N/V$ is the density of the particles.

18.2   Initial, Boundary, and Large *r* Conditions

Although we may start off an MD simulation with a velocity distribution characteristic of a definite temperature, this is not the true temperature of the system because it has not yet equilibrated. Eventually, there will be a redistribution of energy between KE and PE [Thijssen, 1999], and then the system will have a true temperature. It is interesting to note that this initial random distribution is the only place where chance enters into our MD simulation, and it is put there only to speed up the equilibration. Once started, the time evolution of the MD system is determined by Newton's laws, in contrast to Monte Carlo simulations, which are inherently stochastic.

---

3 Unless the temperature is very high, argon atoms, being inert spheres, have no rotational energy.

It is easy to believe that a simulation of 10^{23} molecules might predict bulk properties well, but with MD simulations employing only 10^6–10^9 particles, one must be clever to make less seem like more. Furthermore, because computers are finite, the molecules in the simulation are constrained to lie within a finite box, which inevitably introduces artificial *surface effects* arising from the walls. Surface effects are particularly significant when the number of particles is small because a large fraction of the molecules reside near the walls. For example, if 1000 particles are arranged in a $10 \times 10 \times 10$ cube, there are then 10^3–$8^3 =$ 488 particles one unit away from the surface, that is, 49% of the molecules. For 10^6 particles, this fraction falls to 6%.

The imposition of *periodic boundary conditions* (PBCs) strives to minimize the shortcomings of both the small numbers of particles and the artificial boundaries. Although we limit our simulation to an $L_x \times L_y \times L_z$ box, we imagine this box being replicated to infinity in all directions (Figure 18.3). Accordingly, after each time-integration step, we examine the position of each particle and check if it has left the simulation region. If it has, then we bring an *image* of the particle back through the opposite boundary (Figure 18.3):

$$x \Rightarrow \begin{cases} x + L_x, & \text{if } x \le 0, \\ x - L_x, & \text{if } x > L_x. \end{cases} \tag{18.13}$$

Consequently, each box looks the same and has continuous properties at the edges. As shown by the one-headed arrows in Figure 18.3, if a particle exits the simulation volume, its image enters from the other side, and thus balance is maintained.

In principle, a molecule interacts with all other molecules and all of their images, so despite the fact that there are a finite number of atoms in the interaction volume, there

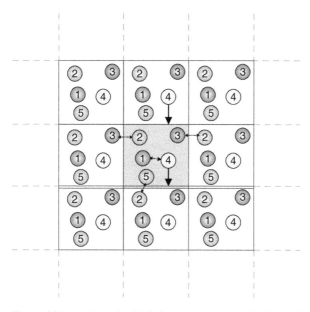

Figure 18.3 An imagined infinite space generated by imposing periodic boundary conditions on the particles within the simulation volume (shaded box). The two-headed arrows indicate how a particle interacts with the nearest version of another particle, be that within the simulation volume or an image. The vertical arrows indicate how the image of particle 4 enters when the actual particle 4 exits.

should be an infinite number of interactions [Ercolessi, 1997]. Nonetheless, because the Lennard–Jones potential falls off so rapidly for large r, $V(r = 3\sigma) \simeq V(1.13\sigma)/200$, far-off molecules do not contribute significantly to the motion of a molecule. An so we pick a value of the cutoff radius $r_{cut} \simeq 2.5\sigma$, beyond which we ignore the effect of the potential:

$$u(r) = \begin{cases} 4\left(r^{-12} - r^{-6}\right), & \text{for } r < r_{cut}, \\ 0, & \text{for } r > r_{cut}. \end{cases} \tag{18.14}$$

Accordingly, if the simulation region is large enough for $u(r > L_i/2) \simeq 0$, an atom interacts with only the *nearest image* of another atom.

As already indicated, a shortcoming with the cutoff potential (18.14) is that because the derivative du/dr is singular at $r = r_{cut}$, the potential is no longer conservative, and thus energy conservation is no longer ensured. However, because the forces are very small at r_{cut}, the violation is, presumably, very small.

18.3 Verlet Algorithms

A realistic, but small, MD simulation may require integration of the 3D equations of motion for 10^{10} time steps for each of 10^3–10^6 particles. Although we could use our standard rk4 ODE solver for this, time is saved by using a simpler rule. The Verlet algorithm uses the central-difference approximation (Chapter 5) for the second derivative to advance the solutions simultaneously by a single time step h for all N particles:

$$\mathbf{F}_i[\mathbf{r}(t), t] = \frac{d^2\mathbf{r}_i}{dt^2} \simeq \frac{\mathbf{r}_i(t+h) + \mathbf{r}_i(t-h) - 2\mathbf{r}_i(t)}{h^2}, \tag{18.15}$$

$$\Rightarrow \quad \mathbf{r}_i(t+h) \simeq 2\mathbf{r}_i(t) - \mathbf{r}_i(t-h) + h^2\mathbf{F}_i(t) + O(h^4), \tag{18.16}$$

where we have set $m = 1$. (Improved algorithms may vary the time step depending upon the speed of the particle.) Notice that although the atom–atom force does not have an explicit time dependence, we include an implicit t dependence in it as a way of indicating its dependence upon the other atoms' positions at that particular time.

Part of the efficiency of the Verlet algorithm (18.16) lies in its solving for the position of each particle without requiring a separate solution for the particle's velocity. However, once we have deduced the position for various times, we can use the central-difference approximation for the first derivative of $\mathbf{r}_i$ to obtain the velocity:

$$\mathbf{v}_i(t) = \frac{d\mathbf{r}_i}{dt} \simeq \frac{\mathbf{r}_i(t+h) - \mathbf{r}_i(t-h)}{2h} + O(h^2). \tag{18.17}$$

Finally, note that because the Verlet algorithm needs $\mathbf{r}$ from two previous steps, it is not self-starting, and so must be started with a forward difference derivative,

$$\mathbf{r}(t = -h) \simeq \mathbf{r}(0) - h\mathbf{v}(0) + \frac{h^2}{2}\mathbf{F}(0). \tag{18.18}$$

Velocity-Verlet Algorithm Another version of the Verlet algorithm, which we recommend because of its increased stability, uses a forward-difference approximation for the derivative to advance *both* the position and velocity simultaneously:

$$\mathbf{r}_i(t+h) \simeq \mathbf{r}_i(t) + h\mathbf{v}_i(t) + \frac{h^2}{2}\mathbf{F}_i(t) + O(h^3), \tag{18.19}$$

$$\mathbf{v}_i(t+h) \simeq \mathbf{v}_i(t) + h\,\overline{\mathbf{a}(t)} + O(h^2), \tag{18.20}$$

$$\simeq \mathbf{v}_i(t) + h\left[\frac{\mathbf{F}_i(t+h) + \mathbf{F}_i(t)}{2}\right] + O(h^2). \tag{18.21}$$

Although this algorithm appears to be of lower order than (18.16), the use of updated positions when calculating velocities, and the subsequent use of these velocities, give both algorithms similar precision.

Of interest is that (18.21) approximates the average force during a time step as $[\mathbf{F}_i(t+h) + \mathbf{F}_i(t)]/2$. Updating the velocity is a little tricky because we need the force at time $t+h$, which depends on the particle positions at $t+h$. Consequently, we must update all the particle positions and forces to $t+h$ before updating velocities, while saving the forces at an earlier time for use in (18.21). As soon as the positions are updated, we impose PBCs to establish that we have not lost any particles and then calculate the forces.

18.3.1 Implementation and Exercise

In the online materials for this book, you will find a number of animations (movies) of solutions to the MD equations. Some frames from these animations are shown in Figure 18.4. The program MD1D.py in Listing 18.1 implements a 1D simulation using the velocity-Verlet algorithm, MD2D.py in Listing 18.2 implements a 2D simulation, and MDPBC.py in Listing 18.3 implements a 2D simulation with PBCs. Use these as models for the following:

1) Establish that you can run and visualize MD2D.py.
2) Place the particles initially at the sites of a simple cubic lattice. The equilibrium configuration for a Lennard–Jones system at low temperature is a face-centered cubic, and if your simulation is running properly, then, as we show in Figure 18.4, the particles should migrate from simple cubic (SC) to face-centered cubic (FCC). An FCC lattice has four-quarters of a particle per unit cell, so an L^3 box with a lattice constant L/N contains (parts of) $4N^3 = 32, 108, 256, \dots$ particles.
3) To save computing time, assign initial particle velocities corresponding to a fixed-temperature Maxwellian distribution. (Recall, the sums of uniform random numbers follow a Gaussian distribution.)

 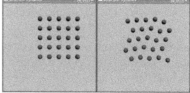

Figure 18.4 *Left*: Two frames from an animation of a 1D simulation. The spaces on the left and right differ slightly in these two frames due to an image atom moving off the frame to the right and then popping up on the left. *Right*: Two frames from the animation of a 2D simulation showing the initial and equilibrated states. Note how the atoms start off in a simple cubic arrangement, but then equilibrate to a face-centered cubic lattice. In both the 1D and 2D simulations, the atoms remain confined as a result of the interatomic forces.

4) Print the code and indicate on it which integration algorithm is used, where the PBCs are imposed, where the nearest image interaction is evaluated, and where the potential is cut off.

5) A typical time step is $\Delta t = 10^{-14}$ s, which in our natural units equals 0.004. You probably will need to make 10^4–10^5 such steps to equilibrate, which corresponds to a total time of only 10^{-9} s (a lot can happen to a speedy molecule in 10^{-9} s). Choose the *largest* time step that provides stability and gives results similar to Figure 18.5.

6) The PE and KE change with time as the system equilibrates. Even after that, there will be fluctuations in them because this is a dynamic system. Evaluate the time-averaged energies for an equilibrated system.

7) Compare the final temperature of your system to the initial temperature. Change the initial temperature and look for a simple relation between it and the final temperature (Figure 18.6).

18.3.2   Analysis

1) Modify your program so that it outputs the coordinates and velocities of a few particles throughout the simulation. Note that you do not need as many time steps to follow a trajectory as you do to compute it, and so you may want to use the *mod* operator `%100` for output.

2) Start your assessment with a 1D simulation at zero temperature. The particles should remain in place without vibration. Increase the temperature and note how the particles begin to move about and interact.

3) Try starting off all your particles at the minima in the Lennard–Jones potential. The particles should remain bound within the potential at low temperatures.

4) Repeat the simulations for a 2D system. The trajectories should resemble billiard ball-like collisions.

5) Create an animation of the time-dependent locations of several particles.

6) Calculate and plot the root-mean-square displacement of molecules as a function of temperature:

$$R_{\text{rms}} = \sqrt{\langle |\mathbf{r}(t + \Delta t) - \mathbf{r}(t)|^2 \rangle}, \tag{18.22}$$

where the average is over all the particles in the box. Determine the approximate time dependence of R_{rms}.

7) Test your system for time-reversal invariance. Stop it at a fixed time, reverse all velocities, and see if the system retraces its trajectories back to the initial configuration after this same fixed time.

8) **Diffusion**: It is well known that light molecules diffuse more quickly than heavier ones. See if you can simulate diffusion with your MD simulation using a Lennard–Jones potential and PBCs [Satoh, 2011].

   (a) Generalize the velocity-Verlet algorithm so that it can be used for molecules of different masses.

   (b) Modify the simulation code so that it can be used for five heavy molecules of mass $M = 10$ and five light molecules of mass $m = 1$.

   (c) Start with the molecules placed randomly near the center of the square simulation region.

(d) Assign random initial velocities to the molecules.
(e) Run the simulation several times and verify visually that the lighter molecules tend to diffuse more quickly than the heavier ones.
(f) For each ensemble of molecules, calculate the rms velocity at regular instances of time, and then plot the rms velocities as functions of time. Do the lighter particles have a greater rms velocity?

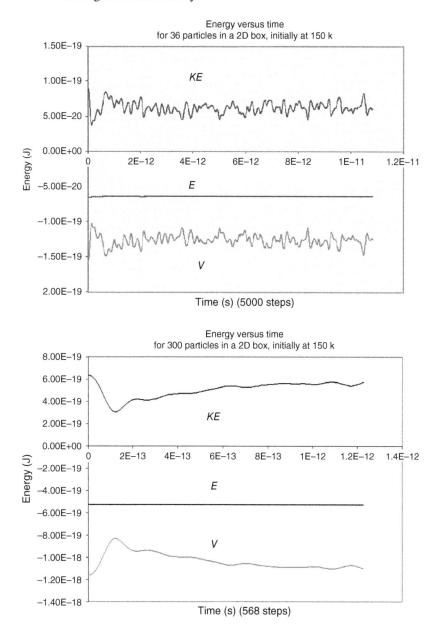

Figure 18.5 The kinetic, potential, and total energy for a 2D MD simulation with 36 particles (*top*), and 300 particles (*bottom*), both with an initial temperature of 150 K. The potential energy is negative, the kinetic energy is positive, and the total energy is seen to be conserved (flat).

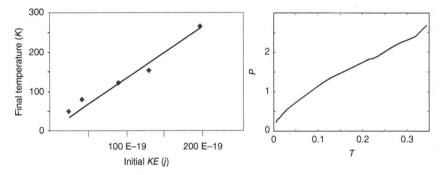

Figure 18.6 *Left*: The temperature after equilibration as a function of initial kinetic energy for a 2D MD simulation with 36 particles. *Right*: The pressure *versus* temperature for a simulation with several hundred particles. An ideal gas (noninteracting particles) would yield a straight line. (Courtesy of J. Wetzel.)

18.4 MD for 16 Particles

A small number of particles are placed in a box. The forces between the particles derive from the Lennard–Jones potential. A number of independent snapshots are taken of the particles in the box, and the number N_{rhs} of particles on the RHS of the box is recorded for each. If n is the number of frames that show N_{rhs} particles in the right-hand side, then the probability of finding N_{rhs} particles on the RHS is:

$$P(n) = \frac{C(n)}{2^{N_{rhs}}}. \tag{18.23}$$

Here $C(n)$ is the number of ways of placing n particles in the left half of the box.

1) Modify the previously developed MD program so that it runs for 16 particles inside a 2D box of side $L = 1$. Assume PBCs, and compute the positions and velocities of the particles using the velocity-Verlet algorithm (18.21).
   (a) Extend the program so that at the end of each time step, it counts the number of particles N_{rhs} on the RHS of the box.
   (b) Create, plot, and update continually a histogram containing the distribution of the number of times n that a N_{rhs} value occurs, as a function of N_{rhs}.
   (c) Make a histogram showing the probability (18.23) of finding N_{rhs} particles on the RHS, as a function of N_{rhs}.
   (d) Compare your plots to those in Figure 18.7, created by MDpBC.py.
2) Even though an MD simulation is deterministic, the particles do tend to equilibrate after a rather small number of collisions, in which case the system resembles a thermal one. This is consistent with the result from ergodic theory that, after a long time, a dynamical system tends to forget its initial state. Test this hypothesis by randomly assigning several different sets of initial positions and velocities to the 16 particles, and then determining the distributions for each initial condition. If the hypothesis is valid, the distributions should be much the same.
3) Use your simulation to determine the velocity distribution of the particles.
   (a) Create a histogram by plotting the number of particles with a velocity in the range v to $v + \Delta v$ versus v.
   (b) Start with random values for the initial positions of the particles.

(c) Start all particles off with the same speed v_0, though with random values for directions.

(d) Update the histogram after each step and continue until it looks like a normal distribution.

4) Compute and plot the heat capacity at a constant volume, $C_V = \partial E/\partial T$, as a function of temperature for the 16 particles in a box.

(a) As before, start with random values for the initial positions of the particles.

(b) Start all particles off with the same speed v_0, though with random values for directions.

(c) Take an average of the temperature for 10 initial conditions, all with the same v_0. Relate the temperature to the total energy.

(d) Repeat the computations for increasing values of v_0. (We suggest $0.5 \leq v_0 \leq 20$ in steps of 1.)

(e) Plot the total energy as a function of the average temperature.

(f) Use numerical differentiation to evaluate the heat capacity at a constant volume, $C_V = \partial E/\partial T$. Unless you change the parameters of the system, you should expect C_V to be constant within statistical fluctuations. Results of our calculation are shown in Figure 18.8.

(g) Explore the effect of a projectile hitting a group of particles, as we do in Figure 18.9.

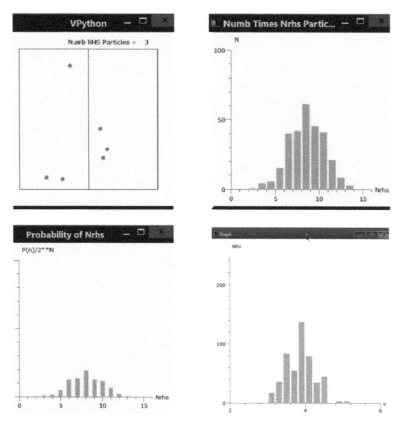

Figure 18.7 *Top Left*: Positions of particles at a single time. *Top Right*: Distribution showing the number of times N_{rhs} particles are present in the RHS of the box. *Lower Left*: The probability distribution for finding N_{rhs} particles in the RHS of the box. *Lower Right*: The velocity distribution for 16 particles in the box.

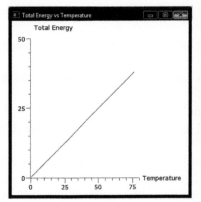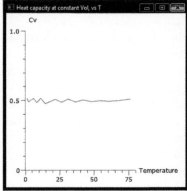

Figure 18.8 *Left*: The total energy versus temperature for 16 particles in a box. *Right*: The heat capacity at constant volume versus temperature for 16 particles in a box.

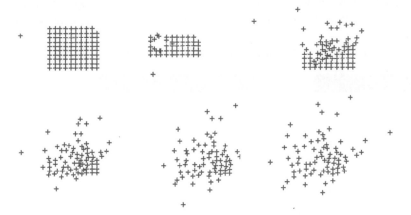

Figure 18.9 A simulation of a projectile shot into a group of particles. The energy introduced by the projectile is seen to lead to evaporation of the particles. (Courtesy of J. Wetzel.)

18.5 Code Listing

Listing 18.1 MD1D.py A 1D MD simulation with too small a number of too large time steps.

```
# MD1.py              Molecular dynamics in 1D

from visual import *
from visual.graph import *
import random

scene = display(x=0,y=0,width=700,height=350, title='Molecular Dynamics',
        range=12)                        # plot spheres
sceneK = gdisplay(x=0,y=350,width=600,height=150,title='Average KE',
        ymin=0.0,ymax=0.3,xmin=0,xmax=100,xtitle='time',ytitle='KE avg')
Kavegraph=gcurve(color= color.red)  # plot KE
scenePE = gdisplay(x=0,y=500,width=600,height=150,title='Pot Energy',
        ymin=-0.6,ymax=0.0,xmin=0,xmax=100,xtitle='time',ytitle='PE')
PEcurve = gcurve(color=color.cyan)
```

```
   Natom = 8
16 Nmax =   8
   Tinit = 10.0                                              # T initial
   t1 = 0
   x  = zeros( (Nmax), float)
20 vx = zeros( (Nmax), float)
   fx = zeros( (Nmax, 2), float)
   L = Natom                                       # Length of atom chain
   atoms = []
24
   def twelveran():                       # Gaussian as average 12 randoms
       s = 0.0
       for i in range (1,13):
28         s += random.random()
       return s/12.-0.5
   def initialposvel():                       # Initial positions, velocities
       i = −1
32     for ix in range(0, L):
           i = i + 1
           x[i] = ix
           vx[i] = twelveran()
36         vx[i] = vx[i]*sqrt(Tinit)
       for j in range(0,Natom):
           xc = 2*x[j] − 7            # Linear transform to place spheres
           atoms.append(sphere(pos=(xc,0), radius=0.5,color=color.red))
40 def sign(a, b):
       if (b >=  0.0):
           return abs(a)
       else:
44         return − abs(a)
   def Forces(t, PE):                                         # Forces
       r2cut = 9.                                             # Cutoff
       PE = 0.
48     for i in range(0, Natom):
           fx[i][t] = 0.0
       for i in range( 0, Natom−1 ):
           for j in range(i + 1, Natom):
52             dx = x[i] − x[j]
               if (abs(dx) > 0.50*L):
                   dx = dx − sign(L, dx)       # Interact with closer image
               r2 = dx*dx
56             if (r2 < r2cut):
                   if (r2 == 0.):                     # Avoid 0 denominator
                       r2 = 0.0001
                   invr2 = 1./r2
60                 wij =  48.*(invr2**3 − 0.5) *invr2**3
                   fijx = wij*invr2*dx
                   fx[i][t] = fx[i][t]  +  fijx
                   fx[j][t] = fx[j][t]  −  fijx    # opposite sense next i
64                 PE = PE  +  4.*(invr2**3)*((invr2**3) − 1.)
       return PE
   def timevolution():
       t1=0
68     t2 = 1
       h = 0.038                                      # Unstable if larger
       hover2 = h/2.0
       KE = 0.0
72     PE = 0.0
       initialposvel()
       PE = Forces(t1,PE)
       for i in range(0, Natom):                           # Kinetic energy
76         KE=KE+(vx[i]*vx[i])/2.0
       t = 0
       while t<100:                                          # Time loop
           rate(1)
80         for i in range(0,  Natom):
               PE = Forces(t1,  PE)
               x[i] = x[i] + h*(vx[i] + hover2*fx[i][t1])
               if x[i] <= 0.:
84                 x[i] = x[i] + L                  # Periodic boundary conditions
               if x[i] >= L :
```

```
                    x[i] = x[i] - L
                    xc = 2*x[i] - 8              # Linear transform to plot atoms
88                  atoms[i].pos=(xc,0)
            PE = 0.0
            PE = Forces(t2,  PE)
            KE = 0.
92          for i in range(0 , Natom):
                vx[i] = vx[i] + hover2*(fx[i][t1] + fx[i][t2])
                KE = KE + (vx[i]*vx[i] )/2
            T = 2*KE/(3*Natom)
96          Itemp = t1
            t1 = t2
            t2 = Itemp
            Kavegraph.plot(pos=(t,KE))                              # Plot KE
100         PEcurve.plot(pos=(t,PE),display=scenePE)               # Plot PE
            t += 1
    timevolution()
```

Listing 18.2   MD2D.py A 2D MD simulation with too small a number of too large time steps.

```
# MD2D.py:            Molecular dynamics in 2D

from visual import *
4 from visual.graph import *
import random

scene = display(x=0,y=0,width=350,height=350, title='Molecular Dynamics',
8               range=10)
sceneK = gdisplay(x=0,y=350,width=600,height=150,title='Average KE',
            ymin=0.0,ymax=5.0,xmin=0,xmax=500,xtitle='time',ytitle='KE avg')
Kavegraph=gcurve(color= color.red)
12 sceneT = gdisplay(x=0,y=500,width=600,height=150,title='Average PE',
            ymin=-60,ymax=0.,xmin=0,xmax=500,xtitle='time',ytitle='PE avg')
Tcurve = gcurve(color=color.cyan)
Natom = 25; Nmax =   25;  Tinit = 2.; dens = 1.;t1 = 0 # Den 1.20 for fcc
16 x  = zeros( (Nmax),    float)
y  = zeros( (Nmax),    float)
vx = zeros( (Nmax),    float)
vy = zeros( (Nmax),    float)
20 fx = zeros( (Nmax, 2), float)
fy = zeros( (Nmax, 2), float)
L = int(1.*Natom**0.5)                          # Side of lattice
atoms=[]
24
def twelveran():                          # Average 12 rands for Gaussian
    s=0.0
    for i in range (1,13):
28      s += random.random()
    return s/12.0-0.5
def initialposvel():                                     # Initialize
    i = -1
32  for ix in range(0, L):                  # x->   0  1  2  3  4
        for iy in range(0, L):              # y=0   0  5  10 15 20
            i = i + 1                       # y=1   1  6  11 16 21
            x[i]  = ix                      # y=2   2  7  12 17 22
36          y[i]  = iy                      # y=3   3  8  13 18 23
            vx[i] = twelveran()             # y=4   4  9  14 19 24
            vy[i] = twelveran()             # numbering of 25 atoms
            vx[i] = vx[i]*sqrt(Tinit)
40          vy[i] = vy[i]*sqrt(Tinit)
    for j in range(0,Natom):
        xc = 2*x[j] - 4
        yc = 2*y[j] - 4
44      atoms.append(sphere(pos=(xc,yc), radius=0.5,color=color.red))
def sign(a, b):
    if (b >=  0.0): return abs(a)
    else: return  - abs(a)
```

```
48  def Forces(t, w, PE, PEorW):                                    # Forces
        # invr2 = 0.
        r2cut = 9.                                     # Switch: PEorW = 1 for PE
        PE = 0.
52      for i in range(0, Natom):
            fx[i][t] = fy[i][t]  = 0.0
        for i in range( 0, Natom-1 ):
            for j in range(i + 1, Natom):
56              dx = x[i] - x[j]
                dy = y[i] - y[j]
                if (abs(dx) > 0.50*L):
                    dx = dx - sign(L, dx)       # Interact with closer image
60              if (abs(dy) > 0.50*L):
                    dy = dy - sign(L, dy)
                r2 = dx*dx + dy*dy
                if (r2 < r2cut):
64                  if (r2 == 0.):                       # To avoid 0 denominator
                        r2 = 0.0001
                    invr2 = 1./r2
                    wij =  48.*(invr2**3 - 0.5) *invr2**3
68                  fijx = wij*invr2*dx
                    fijy = wij*invr2*dy
                    fx[i][t] = fx[i][t] + fijx
                    fy[i][t] = fy[i][t] + fijy
72                  fx[j][t] = fx[j][t] - fijx
                    fy[j][t] = fy[j][t] - fijy
                    PE = PE + 4.*(invr2**3)*((invr2**3) - 1.)
                    w = w + wij
76      if (PEorW == 1):
            return PE
        else:
            return w
80  def timevolution():
        avT = 0.0
        avP = 0.0
        Pavg = 0.0
84      avKE = 0.0
        avPE = 0.0
        t1 = 0
        PE = 0.0
88      h = 0.031                                               # step
        hover2 = h/2.0
        KE = 0.0
        w = 0.0
92      initialposvel()
        for i in range(0, Natom):
            KE = KE+(vx[i]*vx[i]+vy[i]*vy[i])/2.0
     # System.out.println(""+t+" PE= "+PE+" KE = "+KE+" PE+KE = "+(PE+KE));
96      PE = Forces(t1,w,PE,1)
        time =1
        while 1:
            rate(100)
100         for i in range(0, Natom):
                PE = Forces(t1,w,PE,1)
                x[i] = x[i] + h*(vx[i] + hover2*fx[i][t1])
                y[i] = y[i] + h*(vy[i] + hover2*fy[i][t1]);
104             if x[i] <= 0.: x[i] = x[i] + L              # Periodic BC
                if x[i] >= L :  x[i] = x[i] - L
                if y[i] <= 0.: y[i] = y[i] + L
                if y[i] >= L:  y[i] = y[i] - L
108             xc = 2*x[i] - 4
                yc = 2*y[i] - 4
                atoms[i].pos=(xc,yc)
            PE = 0.
112         t2=1
            PE = Forces(t2, w, PE, 1)
            KE = 0.
            w = 0.
116         for i in range(0 , Natom):
                vx[i] = vx[i] + hover2*(fx[i][t1] + fx[i][t2])
                vy[i] = vy[i] + hover2*(fy[i][t1] + fy[i][t2])
```

```
                      KE = KE + (vx[i]*vx[i] + vy[i]*vy[i])/2
120            w = Forces(t2, w, PE, 2)
               P=dens*(KE+w)
               T=KE/(Natom)
               # increment averages
124            avT = avT + T
               avP = avP + P
               avKE = avKE + KE
               avPE = avPE + PE
128            time += 1
               t=time
               if (t==0):
                    t=1
132            Pavg  = avP /t
               eKavg = avKE /t
               ePavg = avPE /t
               Tavg  = avT /t
136            pre = (int)(Pavg*1000)
               Pavg = pre/1000.0
               kener = (int)(eKavg*1000)
               eKavg = kener/1000.0
140            Kavegraph.plot(pos=(t,eKavg))
               pener = (int)(ePavg*1000)
               ePavg = pener/1000.0
               tempe = (int)(Tavg*1000000)
144            Tavg = tempe/1000000.0
               Tcurve.plot(pos=(t,ePavg),display=sceneT)
       timevolution()
```

Listing 18.3   MDpBC.py A 2D MD simulation with periodic boundary conditions.

```
# MDpBC.py: 2-D MD with Periodic BC

from visual.graph import *
4 import random

L = 1; Natom = 16;  Nrhs = 0; dt = 1e-6
scene = display(width = 500,height = 500,range = (1.3)  )
8 ndist = gdisplay(x = 500, ymax = 200,
                width = 500, height = 500, xtitle = 'Nrhs', ytitle = 'N')
inside = label(pos = (0.4,1.1),text = 'PRatomticles here = ',box = 0)
inside2 = label(pos = (0.8,1.1),box = 0)
12 border = curve(pos = [(-L,-L),(L,-L),(L,L),(-L,L),(-L,-L)]) # Limits fig
half = curve(pos = [(0,-L),(0,L)],color = color.yellow)     # Middle
positions = []
vel = []
16 Atom = []                            # For Spheres
dN = []                              # Atoms in right half
fr = [0]*(Natom)                     # Atoms (spheres)
fr2 = [0]*(Natom)                    # second force
20 Ratom = 0.03                         # Radius of atom
pref = 4                             # Reference velocity
h = 0.01
factor = 1e-9                        # Lenn Jones
24 deltaN = 1                           # For histogram
distribution = ghistogram(bins=Ratomange(0.,Natom,deltaN),
                    accumulate=1, average=1, color=color.red)
for i in range (0,Natom):            # Initial r & v
28     col = (1.3*random.random(),1.3*random.random(),1.3*random.random())
       x =2.*(L-Ratom)*random.random()-L+Ratom    # Positons atoms
       y = 2.*(L-Ratom)*random.random()-L+Ratom   # Border forbidden
       Atom = Atom+[sphere(pos=(x,y),radius=Ratom,color=col)] # Add atoms
32     theta = 2*pi*random.random()          # Select angle  0<=theta<= 2pi
       vx = pref*cos(theta)                  # x component velocity
       vy = pref*sin(theta)
       positions.append((x,y))               # Add positions to list
36     vel.append((vx,vy))                    # Add momentum to list
       pos = Ratomray(positions)             # Ratomray with positions
       ddp = pos[i]
```

```
        if ddp[0] >=0 and ddp[0] <=L:            # count atoms right half
40          Nrhs+=1
        v = Ratomray(vel)
    def sign(a, b):                              # Sign function
        if (b >=  0.0): return abs(a)
44      else:   return  - abs(a)
    def forces(fr):
        fr =[0]*(Natom)
        for i in range( 0, Natom-1 ):
48          for j in range(i + 1, Natom):
                dr = pos[i]-pos[j]               # relative position
                if (abs(dr[0]) > L):             # smallest distance or image
                    dr[0] = dr[0]  - sign(2*L, dr[0])  # interact closer image
52              if (abs(dr[1]) > L):
                    dr[1] = dr[1]  - sign(2*L, dr[1])
                if i == 0 and j == 1:
                    curve(pos=[(pos[0]),(pos[0]-dr)])
56              r2 = mag2(dr)
                if (abs(r2) < Ratom):            # to avoid 0 denominator
                    r2 = Ratom
                invr2 = 1./r2
60              fij =invr2*factor*  48.*(invr2**3 - 0.5) *invr2**3
                fr[i] = ij*dr+ fr[i]
                fr[j]=-fij*dr +fr[j]
        return fr
64  for t in range (0,1000):
        Nrhs = 0                                 # begin 0 each time
        for i in range(0,Natom):
            fr = orces(fr)
68          dpos=pos[i]
            if dpos[0] <= -L:
                pos[i] = [dpos[0]+2*L,dpos[1]]   # x periodic BC
            if dpos[0] >= L:
72              pos[i] = [dpos[0]-2*L,dpos[1]]
            if dpos[1] <= -L:
                pos[i] = [dpos[0],dpos[1]+2*L]   # y periodic BC
            if dpos[1] >= L:
76              pos[i] = [dpos[0],dpos[1]-2*L]
            dpos=pos[i]
            if dpos[0] > 0 and dpos[0] <L :      # count at right
                Nrhs+=1
80          fr2 = orces(fr)
            fr2 = r
            v[i] = v[i]+0.5*h*h*(fr[i]+fr2[i])   # velocity Verlet
            pos[i] = pos[i]+h*v[i]+0.5*h*h*fr[i]
84          Atom[i].pos = pos[i]                 # plot new positions
        inside2.text = '%4s'%Nrhs                # RHS
        dN.append(Nrhs)                          # for histogram
        distribution.plot(data = dN)             # plot histogram
```

19

General Relativity

This chapter on general relativity (GR) is new for this 4th edition. It's here in response to requests, and also in response to recent developments in Einstein lensing, exoplanets, black holes, and computational GR. We review some GR theories, compute some GR tensors, and then apply the theory to the deflection of starlight by stars, to gravitational lensing, to corrections to planetary orbits, and to the visualization of wormholes as seen in movies.

Problem   GR tells us that the space we live in is curved, not flat, and that this curvature is the origin of the gravitational force. Your problem is to determine how much of a difference do GR effects make in observable phenomena.

19.1   Einstein's Field Equations

Einstein's   theory of GR postulates that the presence of matter or energy in a region of space distorts the spacetime there. The resulting local curvature of space is the origin of the gravitational force, and thereby provides a link between geometry and dynamics. The theory is expressed succinctly by the *Einstein field equations*, which in a simple form are: [Hartle, 2003]

$$R_{\mu\nu} - \tfrac{1}{2}R g_{\mu\nu} + \Lambda g_{\mu\nu} = \kappa T_{\mu\nu}, \tag{19.1}$$

where the standard is to have each term with dimension of $1/\text{length}^2$. Here the R's, about which we'll talk more about soon, describe the curvature of spacetime, $g_{\mu\nu}$ is the metric tensor ("the metric") that describes the path length, Λ is the cosmological constant, $T_{\mu\nu}$ is the energy-stress tensor, and κ is the Einstein *gravitational constant*,

$$\kappa = \frac{8\pi G}{c^4} \simeq 2.077 \times 10^{-43} \text{N}^{-1}, \tag{19.2}$$

where G is Newton's gravitational constant. Einstein added the *cosmological constant* Λ to the original form of his equations to explain a universe that neither contracts nor expands. It is now believed that Λ is needed to explain the accelerating expansion of the universe due to dark energy in the vacuum of space.

Computational Physics: Problem Solving with Python, Fourth Edition.
Rubin H. Landau, Manuel J. Páez, and Cristian C. Bordeianu.
© 2024 WILEY-VCH GmbH. Published 2024 by WILEY-VCH GmbH.

To unravel (19.1) a bit, we start with the metric $g_{\mu\nu}$, which, by defining how to compute the distance between two points, provides the basic description of the local spacetime:

$$ds^2 = g_{11}\,dx_1^2 + g_{12}\,dx_1 dx_2 + g_{22}\,dx_2^2 + \ldots \equiv \quad g_{\mu\nu}\,dx^\mu dx^\nu, \tag{19.3}$$

where we have used Einstein's convention of summing over repeated upper and lower Greek indices. As an example, in the Ellis extension of a spherical polar coordinates metric, which we will use in Section 19.4, the arc length is

$$ds^2 = -dt^2 + d\ell^2 + r^2(d\theta^2 + \sin^2\theta\,d\phi^2). \tag{19.4}$$

Exercise   Determine the metric tensor $g_{\mu\nu}$ for the Ellis metric.

Once we have a metric describing the arc length in space, we can connect that metric to surface measurements and curvature via the *Christoffel* symbols:

$$\Gamma^\mu_{\alpha\beta} = \tfrac{1}{2}g^{\mu\lambda}\left(\frac{\partial g_{\lambda\alpha}}{\partial x^\beta} + \frac{\partial g_{\lambda\beta}}{\partial x^\alpha} - \frac{\partial g_{\alpha\beta}}{\partial x^\lambda}\right). \tag{19.5}$$

Once we have the Christoffel symbols, we can calculate the *Ricci curvature tensor* $R_{\mu\nu}$ and the *scalar curvature R*:

$$R_{\mu\nu} = \partial_\nu\Gamma^\alpha_{\mu\alpha} - \partial_\alpha\Gamma^\alpha_{\mu\nu} + \Gamma^\alpha_{\nu\gamma}\Gamma^\gamma_{\mu\alpha} - \Gamma^\alpha_{\alpha\gamma}\Gamma^\gamma_{\mu\nu}, \tag{19.6}$$

$$R = g^{\mu\nu}R_{\mu\nu}. \tag{19.7}$$

The *stress-energy tensor* on the RHS of (19.1) is the source of the curvature of spacetime, and it arises from the presence of matter and energy. Specifically, the time-time component of $T_{\mu\nu}$ is the relativistic energy density due to mass and the EM field:

$$T_{00} = \frac{\rho_E}{c^2} + \frac{1}{c^2}\left(\tfrac{1}{2}\epsilon_0 E^2 + \tfrac{1}{2\mu_0}B^2\right). \tag{19.8}$$

The T_{kk} components are related to the stress or pressure in the k direction, while the T_{kl} components are related to shear stress due to momentum flux across a surface.

When all the pieces are assembled, we see that the Einstein field equations (19.1), while looking simple, are actually ten independent, nonlinear, partial differential equations, with 16 functions, two of which appear arbitrary. Except for some simple cases and assumed symmetries, they are generally too hard to solve analytically.

A *geodesic* is the shortest path between two points in spacetime. In addition to the field equations, a key element of GR is the *geodesic equation* that describes the motion of a freely falling particle in spacetime:

$$\frac{d^2x^\mu}{ds^2} = -\Gamma^\mu_{\alpha\beta}\frac{dx^\alpha}{ds}\frac{dx^\beta}{ds}, \tag{19.9}$$

where s is the scalar proper time and $\Gamma^\mu_{\alpha\beta}$ is the Christoffel symbol. Massive particles travel on time-like solutions of the geodesic equation (19.9), while light travels on space-like solutions. (We'll solve the equation in Section 19.2.) Of course "free" in GR, while not explicitly including the force of gravity, does account for gravity via the stress-energy tensor causing space to curve. After the use of the chain rule for derivatives, the geodesic equation can be written with an explicit time coordinate:

$$\frac{d^2x^\mu}{dt^2} = -\Gamma^\mu_{\alpha\beta}\frac{dx^\alpha}{dt}\frac{dx^\beta}{dt} + \Gamma^0_{\alpha\beta}\frac{dx^\alpha}{dt}\frac{dx^\beta}{dt}\frac{dx^\mu}{dt}. \tag{19.10}$$

This form of the geodesic equation, with its manifest nonlinearity, is the one used for numerical computations. Because d^2x^μ/ds^2 is an acceleration, the geodesic equation is analogous to Newton's second law of motion, with the force replaced by the geometry of spacetime on the RHS. For example, if a test particle's velocity is small (nonrelativistic), the terms quadratic and cubic in the velocity can be ignored, and we would be left with Galileo's hypothesis (although he would not have used these equations) that all particles, regardless of mass, have the same acceleration:

$$\frac{d^2x^i}{dt^2} \simeq -\Gamma^i_{00}, \qquad i = 1, 2, 3. \tag{19.11}$$

19.1.1 Calculating the Riemann and Ricci Tensors

Figure 19.1 shows two free particles moving along the two infinitesimally close geodesics $x^a(\tau)$ and $x^b(\tau)$. We consider the particle on x^a as a reference particle with $u^\mu = dx^\mu/d\tau$ its 4-velocity. The x^a and x^b trajectories start off parallel at time $\tau = 0$ and are connected by the vector $n(\tau)$:

$$x^a = x^b + n^\alpha(\tau). \tag{19.12}$$

For zero relative acceleration of the particles, the geodesics remain parallel, and so:

$$\frac{d^2n}{d\tau^2} = 0. \tag{19.13}$$

This derivative acts on the basis vectors, which in turn requires knowledge of the Christoffel symbols:

$$\left(\frac{d^2n}{d\tau^2}\right)^\alpha = \left(\partial_\sigma\Gamma^\alpha_{\mu\nu} - \partial_\nu\Gamma^\alpha_{\mu\sigma} + \Gamma^\alpha_{\sigma\gamma}\Gamma^\gamma_{\mu\nu} - \Gamma^\alpha_{\nu\gamma}\Gamma^\gamma_{\mu\sigma}\right)u^\sigma u^\mu u^\nu. \tag{19.14}$$

We recognize the quantity in parenthesis as the uncontracted version of the Riemann tensor:

$$R^\alpha_{\mu\nu\sigma} = \partial_\sigma\Gamma^\alpha_{\mu\nu} - \partial_\nu\Gamma^\alpha_{\mu\sigma} + \Gamma^\alpha_{\sigma\gamma}\Gamma^\gamma_{\mu\nu} - \Gamma^\alpha_{\nu\gamma}\Gamma^\gamma_{\mu\sigma}. \tag{19.15}$$

19.1.2 Riemann and Ricci Tensor Problems

The Schwarzschild metric,

$$ds^2 = \left(1 - \frac{2GM}{c^2r}\right)c^2dt^2 - \left(1 - \frac{2GM}{c^2r}\right)^{-1}dr^2 - r^2(d\theta^2 + \sin^2\theta d\phi^2), \tag{19.16}$$

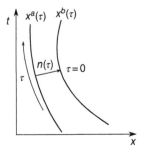

Figure 19.1 Two free particles move along the infinitesimally close geodesics $x^a(\tau)$ and $x^b(\tau)$. The particles start off parallel at time $\tau = 0$ and are connected by the vector $n(\tau)$.

permits a solution of the Einstein equation for a spherically symmetric geometry with zero cosmological constant Λ, and no matter present so $T_{\mu\nu} = 0$. The following problems can be solved with variations of the one same code. Our code `Ricci.py` in Listings 19.1 may help.

1) Create four matrices representing the Christoffel symbols, $\Gamma^0_{\mu\nu}, \Gamma^r_{\mu\nu}, \Gamma^\theta_{\mu\nu}, \Gamma^\phi_{\mu\nu}$, and use symPy or some other symbolic manipulation program to evaluate them for the Schwarzschild metric.
2) Use symPy to evaluate the Riemann tensor, $R^\alpha_{\mu\nu\sigma}$, for the Schwarzschild metric.
3) Use symPy to extract the Ricci curvature tensor, which is defined as the contraction

$$R_{\lambda\mu} \equiv R^\alpha_{\lambda\alpha\mu}. \tag{19.17}$$

4) The Ricci scalar gives a single numerical measure of the curvature at each point in space-time. If a spacetime is flat, then $R = 0$, and initially parallel geodesics remain so in time. If a spacetime is curved, then $R \neq 0$. Use symPy to extract the Ricci scalar from the Ricci curvature tensor:

$$R = g^{\mu\kappa} R_{\mu\kappa}. \tag{19.18}$$

19.1.3 Event Horizons

The curvature of spacetime and the speed of light create boundaries in space beyond which events cannot be observed. These are called *event horizons* and are important near black holes. Look again at the Schwarzschild metric (19.16), and imagine observing events simultaneously at one instant of time so that $dt = 0$. The proper distance, as would be measured by putting down metersticks, would then be the space-like distance:

$$-ds^2 = \left(1 - \frac{2GM}{c^2 r}\right)^{-1} dr^2 - r^2 d\theta^2, \tag{19.19}$$

where we have left off the ϕ dependence. This equation leads to the conclusion that distances, somehow, become singular at the Schwarzschild radius r_s:

$$r_h = r_s \stackrel{\text{def}}{=} \frac{2GM}{c^2}. \tag{19.20}$$

This is the event horizon. Although this singularity is a peculiarity of the Schwarzschild metric, physical black holes are believed to have event horizons.

Problem Calculate numerical values for the event horizons around the earth, the sun, and a black hole.

Black holes tend to rotate, and so the spacetime near one with angular momentum J is more appropriately described by the Kerr metric [Kerr, 1963]. For motion in the equatorial plane (no ϕ variation), the Kerr metric is:

$$ds^2 = -\left(1 - \frac{r_s r}{\Sigma}\right) c^2 dt^2 + \frac{\Sigma}{\Delta} dr^2 + \Sigma \, d\theta^2, \tag{19.21}$$

$$\Sigma = r^2 + a^2 \cos^2\theta, \qquad \Delta = r^2 - r_s r + a^2, \qquad a = \frac{J}{Mc}. \tag{19.22}$$

Again, we determine the horizon radius r_h by solving for the r value at which the distance becomes singular:

$$\Delta = 0 \Rightarrow r_h = \frac{r_s \pm \sqrt{r_s^2 - 4a^2}}{2}. \tag{19.23}$$

For real r_h, this limits the angular momentum of the black hole to

$$J \leq \frac{GM^2}{c}. \tag{19.24}$$

The full determination of the equations of motion for a particle near a Kerr metric black hole is a significant endeavor, and we suggest [Hanc and Taylor, 2004] and [Gould *et al.*, 2006] for the derivation.

19.2 Gravitational Deflection of Light

As we have said before, a *geodesic* is the shortest path between two points in spacetime. GR assumes that light travels on space-like geodesics, while massive particles travel on time-like geodesics. The geodesic path is the solution of the geodesic equation (introduced in Section 19.1):

$$\frac{d^2x^\beta}{d\lambda^2} + \Gamma^\beta_{\mu\nu} \frac{dx^\mu}{d\lambda} \frac{dx^\nu}{d\lambda} = 0. \tag{19.25}$$

We leave the application of this equation from first principles to the references and focus instead on what's been derived from it.

One of the early tests of GR was the prediction for the angle of deflection ϕ of light starting at an impact parameter $b = R$ relative to the sun's center and just grazing the sun's surface (Figure 19.2). Newtonian mechanics solved this problem by calculating the orbit of a massive particle around the sun, and then taking the $m \rightarrow 0$ limit for the particle. This yielded

$$\phi = 2\frac{GM}{Rc^2}, \tag{19.26}$$

where G is the gravitational constant, M is the mass of the sun, and R is the radius of the sun. Later, Einsteinian mechanics was used to solve the geodesic equation (19.9) approximately, and obtained twice as large a value,

$$\phi \simeq 4\frac{GM}{Rc^2}. \tag{19.27}$$

This agreed with measurements and helped to establish the validity of GR.

Now let's try to calculate some numerical values for the deflection. In 1916, Schwarzschild found an exact solution of the Einsteinian equations using (what else?) the Schwarzschild metric (19.16). For this metric and for light just grazing the sun ($b = R$), the orbit equation takes the simple form [Moore, 2013]:

$$\left(\frac{1}{r}\frac{dr}{d\phi}\right)^2 = \left(1 - \frac{2M}{R}\right)\frac{1}{R^2} - \left(1 - \frac{2M}{r}\right)\frac{1}{r^2}. \tag{19.28}$$

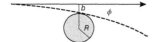

Figure 19.2 A light ray being bent by an angle ϕ due to the gravitational effect of the sun.

A change of variable to $u = R/r$ produces the numerically more robust equation:

$$\left(\frac{du}{d\phi}\right)^2 = 1 - u^2 - \frac{2M}{R}(1 - u^3).$$

$$(19.29)$$

1) Verify that an approximate solution to (19.29) is

$$\phi \simeq 4\frac{GM}{Rc^2}.$$

$$(19.30)$$

2) Evaluate this expression to determine a numerical value for the angle of deflection for light grazing the sun's surface (hint: It's small). Use parameters $M = 2 \times 10^{23}$ g, $R = 7 \times 10^{10}$ cm, and $G/c2 = 7.4 \times 10^{-29}$ cm/g.
3) Although the ODE (19.29) is nonlinear, that's not a problem for a numerical solution. Solve (19.29) numerically and compare your result with the value obtained from the approximate analytic expression.

19.2.1 Gravitational Lensing

In a different approach to the deflection of a light due to a very massive star, Moore [2013] assumes a Schwarzschild spacetime to describe the curved space outside of a spherically symmetric gravitational source (a star). In terms of the inverse variable $u = 1/r$, the geodesic equation is now

$$\frac{d^2u}{d\phi^2} = 3GM\,u^2 - u.$$

$$(19.31)$$

1) Modify your ODE solver to solve this equation. Employ units such that mass is measured in meters, $GM{=}1477.1$ m, and $M = 28M_\odot$ ($M_\odot$ is a solar mass). Our program `LensGravity.py` is given in Listing 19.2.
2) Equation (19.31) is quite sensitive to the initial conditions. Assume that initially the light is very distant, say $r \simeq 10^6$, and $u(\phi = 0) = du(\phi)/d\phi = 10^{-6}$.
3) Convert your solution for $r(\phi)$ into one for (x, y), and plot the photons' paths for $0 \le \phi \le \pi$. Our plots are shown on the left of Figure 19.3.

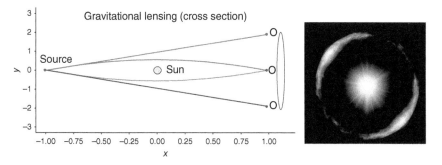

Figure 19.3 *Left*: Three trajectories showing the bending of light rays caused by the sun's mass. Note, the three images on the right would be seen as an Einstein ring. *Right*: A James Webb Telescope image showing an Einstein ring.

4) Employ the symmetry of this problem to rotate your solution about the $x = 0$ axis and thus create an Einstein ring. This is what an observer sees when viewing a distant light source lying behind a massive star and focusing on a point source (Figure 19.3 right).

19.3 Planetary Orbits in GR Gravity

19.3.1 Newton's Potential Corrected

The classical solution of Newton's laws, including his gravitational potential, is just fine for most everything here on or near the earth. However, there are corrections arising from GR, and while small, these corrections are actually critical to the accuracy of modern GPS devices. The usual approach is to determine an ODE with a GR correction to the familiar $1/r$ gravitational potential, and then to solve the ODE. We follow [Hartle, 2003; Moore, 2013; James *et al.*, 2015], who assume the Schwarzschild metric (19.16), and show that an effective potential for this metric is:

$$V_{\text{eff}}(r) = -\frac{GM}{r} + \frac{\ell^2}{2r^2} - \frac{GM\ell^2}{r^3}. \tag{19.32}$$

Here G is the gravitational constant, ℓ is the angular momentum per unit rest mass, M is the mass of the star, and the middle term is the usual angular momentum barrier. We see that (19.32) differs from the Newtonian potential by a $-GM\ell^2/r^3$ term that, in addition to the usual $-GM/r$ attraction, provides an additional strong attraction at short distances. We obtain a dimensionless, and simpler-to-compute, form of the potential by changing variables:

$$V_{\text{eff}}(r') = -\frac{G}{r'} + \frac{\ell'^2}{2r'^2} - \frac{G\ell'^2}{r'^3} \qquad r' = \frac{r}{M}, \quad \ell' = \frac{\ell}{M}. \tag{19.33}$$

1) Plot $V_{\text{eff}}(r')$ versus r' for $\ell = 4.3$ (like Figure 19.4).
2) Describe in words how the orbits within this potential change with energy.
3) At what values of r' does the effective potential have a maximum and a minimum?
4) At what value of r' does a circular orbit exist? (Hint: the small circles in Figure 19.4 correspond to circular orbits.)
5) Determine the range of r' values that occur for $\ell = 4.3$.
6) Indicate the above range on your plot by a horizontal line, and describe the orbits.
7) Describe the orbits for energies corresponding to the maximum in the potential.

19.3.2 Orbit Computation via Energy Conservation

A fairly simple way to determine the orbits of massive particles in the effective potential (19.33) derives from energy conservation. It starts with the energy per unit mass expressed as the sum of kinetic and potential terms:

$$E = \frac{1}{2}\left(\frac{dr}{d\phi}\right)^2 \frac{\ell^2}{r^4} - \frac{GM}{r} + \frac{\ell^2}{2r^2} - \frac{GM\ell^2}{r^3}, \tag{19.34}$$

where ϕ is the polar angle. We obtain an ODE for the orbit by differentiating both sides of the equation with respect to ϕ:

$$\frac{d^2r}{d\phi^2} = -\frac{GM}{r^2} + \frac{\ell^2}{r^3} - \frac{3GM\ell^2}{r^4}, \tag{19.35}$$

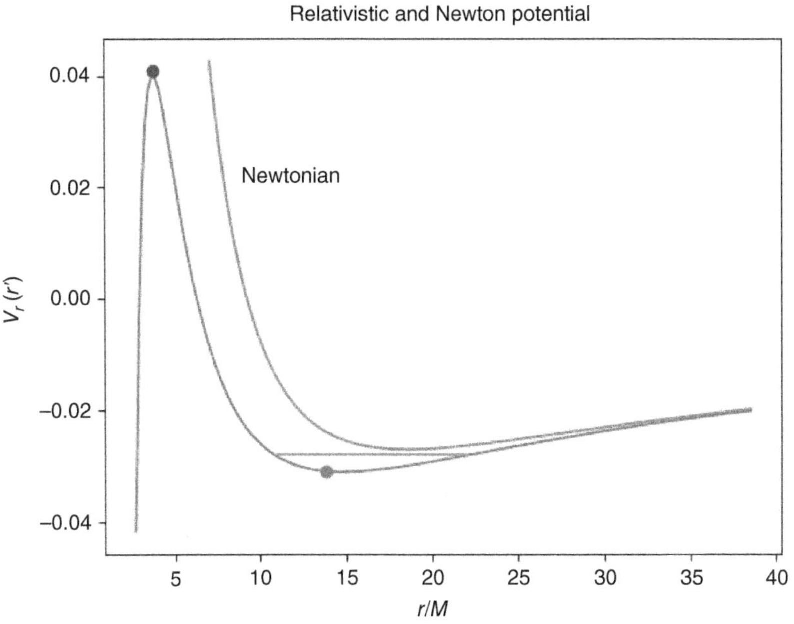

Relativistic and Newton potential

Figure 19.4 Relativistic and Newtonian potentials for $\ell/M = 4.3$. The two dots correspond to radii for circular orbits.

where a common $dr/d\phi$ factor has been canceled out. The ODE is simplified by a change of variables to:

$$\frac{d^2u}{d\phi^2} = -u + \frac{GM}{\ell^2} + 3GMu^2, \qquad u' = \frac{M}{r}, \qquad \ell' = \frac{\ell}{M}. \tag{19.36}$$

As with Newtonian orbits, the energy and angular momentum of the system determine the orbit characteristics. We used the energy integral (19.34) to determine the initial conditions for the ODE, and then solved for $du'/d\phi$:

$$\frac{du'}{d\phi} = \sqrt{\frac{2E}{\ell'^2} + 2\frac{Gu'}{\ell'^2} - u'^2 + 2Gu'^3}. \tag{19.37}$$

1) Use your ODE solver to explore numerically and graphically various orbits corresponding to various initial conditions and energies. Our program, `RelOrbits.py` is in Listing 19.3. Introduce some signals into your figures so that you can tell the direction of travel (or produce a time series of graphs).

(a) Set up your ODE solver appropriately for (19.37) using $G = 1$.

(b) Choose an energy corresponding to the maximum of the effective potential and compute your version of Figure 19.4. Pick an initial r value at which the potential is a maximum. As you may have deduced, this should lead to an unstable orbit, such as on the left of Figure 19.5.

(c) See if you can find initial conditions that lead to a circular orbit. Is it stable?

(d) Investigate the effect of gradually decreasing the angular momentum.

(e) Choose an energy that corresponds to the minimum in the effective potential and plot nearby orbits. Examine the sensitivity of these orbits to the choice of initial conditions.

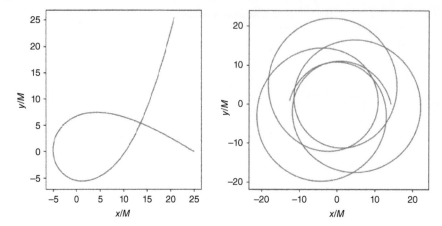

Figure 19.5 *Left*: An orbit corresponding to an energy at the maximum of the effective potential. *Right*: A rapidly precessing orbit.

(f) Determine the energy and initial conditions that produce a precessing perihelion, such as the one seen on the right of Figure 19.5. In this case, the massive particle moves between two turning points, as shown by the horizontal line in the potential well in Figure 19.4.

(g) Examine the orbits that occur if a particle is bound by the inner strong attraction. Can such a particle start at infinity and be captured?

19.3.3 Precession of the Perihelion of Mercury

Planets follow nearly perfect ellipses around the sun, with their major axes rotating very slowly, as shown in Figure 19.6. As viewed from the sun, the precession of Mercury is 9.55 minutes of arc per century [min = $(1/60)^{th}$ of a degree]. Mercury is the fastest of all the planets, and so its precession is the largest. All but about 0.01 of a degree of the precession can be explained with Newtonian mechanics as perturbations due to the other planets. This leaves a small mystery. The calculation of this correction to a small correction was one of the important early successes of GR. In this section, we present a first-principles calculation of that precession due to G. He.

The Schwarzschild metric with zero cosmological constant Λ describes a spacetime appropriate to a spherically symmetric geometry surrounding a mass M with no other matter present ($T_{\mu\nu} = 0$). As we now want to use actual mass and orbital values, we rewrite the metric (19.16) as

$$ds^2 = \left(1 - \frac{r_s}{r}\right) dt^2 - \frac{1}{1 - r_s/r} dr^2 - r^2 d\theta^2 - r^2 \sin^2\theta \, d\phi^2, \tag{19.38}$$

$$r_s \stackrel{\text{def}}{=} 2GM. \tag{19.39}$$

A time-like trajectory is a solution to the geodesic equation for a massive particle. If τ is the proper time, a time-like trajectory is described by

$$d\tau^2 = g_{\mu\nu} dx^\mu dx^\nu. \tag{19.40}$$

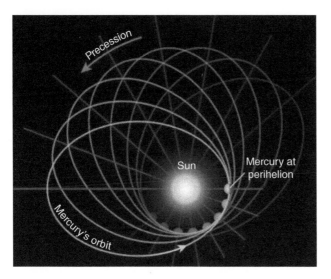

Figure 19.6 An artist's perception of the precession of the perihelion of Hg (www
.astronomicalreturns.com/2020/05/the-mystery-of-mercurys-missing).

For a planar orbit with $\theta = \pi/2$, this leads to

$$\left(\frac{d\tau}{dt}\right)^2 = \left(1 - \frac{r_s}{r}\right) - \frac{\dot{r}^2}{1 - r_s/r} - r^2\dot{\phi}^2. \tag{19.41}$$

We want an equation relating distance and angle. The derivatives can be rewritten in terms
of the constants of the motion,

$$\frac{d\tau}{dt} = \frac{1}{e}\left(1 - \frac{r_s}{r}\right), \qquad \frac{d\phi}{dt} = \frac{L}{er^2}\left(1 - \frac{r_s}{r}\right), \tag{19.42}$$

where $L = Rv/c$ is the angular momentum per unit mass, v is the linear velocity at the
apoapsis, and e is the energy per unit mass. Substitution leads to the differential equation

$$\left(\frac{dr}{dt}\right)^2 = \frac{1}{e^2}\left(1 - \frac{r_s}{r}\right)^2\left[(e^2 - 1) + \frac{r_s}{r} - \frac{L^2}{r^2} + \frac{L^2 r_s}{r^3}\right]. \tag{19.43}$$

Use of the chain rule,

$$\frac{dr}{dt} = \frac{dr}{d\phi}\frac{d\phi}{dt} \tag{19.44}$$

leads to the desired differential equation relating distance and angle:

$$\left(\frac{dr}{d\phi}\right)^2 = \frac{r^4}{L^2}\left[\left(1 - \frac{r_s}{R}\right)\left(1 + \frac{L^2}{R^2}\right) - \left(1 - \frac{r_s}{r}\right)\left(1 + \frac{L^2}{r^2}\right)\right], \tag{19.45}$$

Note that the mass of Mercury does not enter into the calculation, although its distance
from the sun does. This is the same as what happens with Newton's laws,

$$\frac{mMG}{r^2} = ma \quad \rightarrow \quad a = \frac{Mg}{r^2}, \tag{19.46}$$

where the m's cancel out.

Although (19.45) can be solved as it stands, the large differences in parameter values lead to numerical inaccuracies, and it is better to solve for the inverse distance $u = R/r$. This leads to the quadratic equation:

$$\left(\frac{du}{d\phi}\right)^2 = \frac{r_s}{R}(u-1)(u-u_+)(u-u_-),$$ (19.47)

$$u_\pm = \frac{-b \pm \sqrt{b^2 - 4ac}}{2a}, \quad a = \frac{r_s}{R}, \quad b = a - 1, \quad c = b + \frac{Rr_s}{L^2}.$$

1) Show that the perihelion precession per revolution can be written as

$$\Delta\phi = 2\sqrt{\frac{R}{r_s}} \int_1^{u_-} \frac{du}{\sqrt{(u-u_+)(u-u_-)(u-1)}} - 2\pi.$$ (19.48)

2) Compute $\Delta\phi$ and compare it to Landau and Lifshitz [1971]'s value of 5.02×10^{-7}. Here are some apoapsis numerical values:

$$r_s = 2950\,\text{m}, \quad r_a = 69.82 \times 10^9\,\text{m}, \quad r_p = 46.00 \times 10^9\,\text{m}.$$ (19.49)

The code PrecessHg.py, written by G. He is given online.

19.4 Visualizing Wormholes

Problem Create images of a wormhole of the type seen in the movie *Interstellar*. (As an alternative, you can reproduce some of the visualizations found online in Roman [1994].) Even better would be the creation of visualizations of travel through a wormhole [Nolan and Nolan, 2015].

During Christopher Nolan's direction of the science fiction movie *Interstellar*, Kip Thorne (a 2017 Noble laureate for GR) helped develop visualizations of rocket flight based on Einstein's field equations. The key element of the movie was that interstellar travel was possible in a single human lifetime if a spaceship passed through a *wormhole* (an *Einstein-Rosen bridge*). This wormhole would be a tunnel-like structure that connects one location in spacetime to another, or possibly to another universe [James *et al.*, 2015]. Figure 19.7 is a visualization of such a wormhole.

Although actual wormholes have never been observed, speculation is that they may occur as quantum fluctuations over distances on the Planck scale, $\sqrt{H\hbar/c^3} \sim 10^{-35}$ m (10^{-20} of the size of a proton). Further speculations imagine that the size of the wormhole might be enlarged to macroscopic size if there were some type of exotic matter with negative energy density at the throat of the wormhole. This might permit a rocket ship to pass through it [Roman, 1994].

However, if our 4D universe resided in a higher-dimensional space (called a *bulk*), such as the 5-D one imagined in *Interstellar*, then there might not be the need for exotic matter to hold open the wormhole. In any case, while unlikely, interstellar travel is not strictly forbidden (it is science fiction after all). Actually, as an exercise in GR, Morris and Thorne [1988] discuss the fundamentals of space travel using wormholes.

The equations that Thorne used to create the visualizations were expressed in units in which $G = 1$, $c = 1$, and time is measured in length $1\,\text{s} = c \times 1\,\text{s} = 2.998 \times 10^8$ m. Mass is

Figure 19.7 The Ellis wormhole connecting upper and lower (flatter) spaces. Note that this visualization has the wormhole's 4D bulk embedded within a 3D space. The throat diameter is 2ρ, and the proper distance traveled in a radial direction is ℓ.

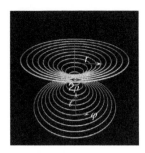

measured in length, $1\,\text{kg} = G/c^2 \times 1\,\text{kg}$, so that $1\,\text{kg} = 0.742 \times 10^{-27}$ m, in which case the sun's mass equals 1.476 km. The created wormhole connects two flat 3D spaces placed at the ends of a 4D cylinder. The 4D cylinder is of length $2a$, with cross sections that are spheres of radius ρ. In order to visualize the 4D wormhole, it is embedded in a 3D space, in which case the cross sections are circles of radius ρ (Figure 19.7).

Thorne used the Ellis extension of a spherical polar coordinate metric:

$$ds^2 = -dt^2 + d\ell^2 + r^2(d\theta^2 + \sin^2\theta\, d\phi^2). \tag{19.50}$$

Here, the radius coordinate r is a function of ℓ, and the physical distance (proper distance) traveled in a radial direction:

$$r(\ell) = \sqrt{\rho^2 + \ell^2}, \tag{19.51}$$

where ρ is the radius of the throat of the cylindrical wormhole. Note that the time coordinate t enters the metric (19.50) with a negative sign. This means that for fixed ℓ, θ, and ϕ, the time t increases in the timelike direction. Accordingly, t is the proper time as measured by a person at rest in the spatial (ℓ, θ, ϕ) coordinate system.

Because $r^2(d\theta^2 + \sin^2\theta\, d\phi^2)$ is the familiar metric describing the surface of a sphere of radius r, the created wormhole is spherically symmetric. This means that, as $\ell \to \pm\infty$, the radius of the sphere within the wormhole approaches the proper distance ℓ. This also means that as $\ell \to \pm\infty$, we would have two separate flat spaces connected by the wormhole. In the movie, the transition between the two flat spaces via the wormhole's throat is made to resemble the transition to an external space in which a nonspinning black hole resides. This is described by the Schwarzschild or hole metric [James *et al.*, 2015]:

$$ds^2 = -\left(1 - 2\frac{\mathcal{M}}{r}\right)dr^2 + \frac{dr^2}{1 - 2\mathcal{M}/r} + r^2(d\theta^2 + \sin^2\theta\, d\phi^2), \tag{19.52}$$

where $\mathcal{M}$ is the black hole's mass. With this metric, the radius r becomes the outward coordinate rather than the proper distance ℓ. The visualizations in the movie required a solution for $r(\ell)$, that is, a solution or an expression for the outward coordinate as a function of proper distance. To reduce the effort involved, the visualizations used an analytic expression for $r(\ell)$ outside the wormhole's cylindrical interior that is similar to the Schwarzschild $r(\ell)$:

$$r = \rho + \frac{2}{\pi}\int_0^{|\ell|-a}\arctan\left(\frac{2\xi}{\pi\mathcal{M}}\right)d\xi \tag{19.53}$$

$$= \rho + \mathcal{M}\left[x\arctan x - \frac{1}{2}\ln(1 + x^2)\right], \quad \text{for } |\ell| > a. \tag{19.54}$$

For cylindrical coordinates, the z coordinate is the height above the wormhole's midplane in the embedding space, and so the embedding space metric becomes

$$ds^2 = dz^2 + dr^2 + r^2\, d\phi^2. \tag{19.55}$$

In this case, the spatial metric of the wormhole's 2D equatorial surface is:

$$ds^2 = d\ell^2 + r^2(\ell)\, d\phi^2. \tag{19.56}$$

Combining these equations lets us solve for $z(\ell)$:

$$d\ell^2 = dz^2 + dr^2, \tag{19.57}$$

$$z(\ell) = \int_0^\ell \sqrt{1 - (dr/d\ell')^2}\, d\ell'. \tag{19.58}$$

One obtains the equations needed to visualize the wormhole by substituting (19.53) and (19.54) into (19.58).

19.5 Problems

1) In order to apply (19.58), we need to evaluate the derivative $dr/d\ell$. Use Python's symbolic algebra package sympy to show that

$$\frac{dr}{d\ell} = \frac{2}{\pi} \arctan \frac{2\ell - a}{\pi M}. \tag{19.59}$$

Our program WormHole.py in Listing 19.4 evaluates this derivative.

2) Insert this $dr/d\ell$ into (19.58) and evaluate the $z(\ell)$ integral numerically for

$$\rho = 1, \quad a = 1, \quad M = 0.5. \tag{19.60}$$

3) The contour lines or rings shown in Figure 19.7 correspond to different values of ℓ. They were obtained with the program visualWorm.ipynb in Listing 19.5.

4) Make your own plot of the wormhole for $\ell = 1, \cdots, 11$.

5) Create a cylindrical wormhole of length $2L$ with a spherical cross section of radius ρ. Visualize the wormhole with a 3D embedding diagram in which the missing dimension results in the cross sections appearing as circles rather than spheres. Follow the same steps as used for the Ellis wormhole, (19.50), but now with

$$r(\ell) = \begin{cases} \rho & |\ell| \le L \quad \text{(Wormhole interior)} \\ |\ell| - L + \rho, & |\ell| \ge L \quad \text{(Wormhole exterior)}. \end{cases} \tag{19.61}$$

19.6 Code Listings

Listing 19.1 **Ricci.py** uses SymPy to compute the Riemann and Ricci tensors and the Ricci scalar.

```
# Ricci.py: Riemann & Ricci tensors, Ricci scalar, uses Sympy

from sympy import * #symbolic python
import numpy as np

t,r,th, fi, rg = symbols('t r th fi rg')        # Schwarzschild metric
print("contravariant")                          # Upper indices

# Inverse matrix
gT = Matrix([[1/(-1 + rg/r),0,0,0], [0, 1-rg/r, 0,0],
    [0, 0, r**(-2), 0],[0, 0, 0, 1/(r**2*sin(th)**2)]])
```

```
     # 4-D array for alpha, beta, mu, nu
     Ri  =  [[[[[] for n in range(4)] for a in range(4)] for b in range(4)] for c in
            range(4)]
14   RT = [[[] for m in range(4)]for p in range(4)]                          # Ricci tensor

     # Christoffel symbols, upper t, r, theta, and phi
     Cht = Matrix([[0, 0.5*rg/(r*(r − rg)), 0, 0],
18       [0.5*rg/(r*(r − rg)), 0, 0, 0], [0, 0, 0, 0], [0, 0, 0, 0]])
     Chr = Matrix([[0.5*rg*(r − rg)/r**3,0,0,0], [0, −0.5*rg/(r*(r − rg)),0,0],
            [0,0, −1.0*r + 1.0*rg, 0], [0,0,0, (−1.0*r + rg)*sin(th)**2]])
     Chth = Matrix([[0, 0, 0, 0], [0, 0, 1.0/r, 0], [0, 1.0/r, 0, 0],
22              [0, 0, 0, −0.5*sin(2*th)]])
     Chfi = Matrix([[0, 0, 0, 0], [0, 0, 0, 1.0/r], [0, 0, 0, 1.0/tan(th)],
              [0, 1.0/r, 1.0/tan(th), 0]])
     for alpha in range(0,4):                        # Upper index in Christoffel
26       if alpha == 0:   Chalp = Cht
         elif alpha == 1: Chalp = Chr
         elif alpha == 2: Chalp = Chth
         else:  Chalp = Chfi
30       for be in range(0,4):                       # Beta
          for mu in range(0,4):
              if mu == 0:      der2 = t            # Derivative
              elif mu == 1:    der2 = r
34            elif mu == 2:    der2 = th
              elif mu == 3:    der2 = fi
              for nu in range(0,4):
                  if nu == 0:    der1 = t          # Other derivative
38                elif nu == 1: der1 = r
                  elif nu == 2: der1 = th
                  elif nu == 3: der1 = fi
                  a1 = diff(Chalp[be,nu],der2)     # Christoffel symbol
42                a2 = diff(Chalp[be,mu],der1)     # And derivative
                  sump = 0
                  sumn = 0
                  for gam in [t,r,th,fi]:
46                    if gam == t:
                         Chgam = Cht
                         gama = 0
                      elif gam == r:
50                       Chgam = Chr
                         gama = 1
                      elif gam == th:
                         Chgam = Chth
54                       gama = 2
                      elif gam == fi:
                         Chgam = Chfi
                         gama = 3
58                    sump = sump+Chalp[mu,gama]*Chgam[be,nu]
                      sumn = sumn+Chalp[nu,gama]*Chgam[be,mu]
                  R = simplify(a1−a2+sump−sumn) # Riemann tensor
                  if R == 0:   # Print nonzero components
62                    Ri[alpha][be][mu][nu] = 0
                  else :
                      Ri[alpha][be][mu][nu] = R
                      print("Ri[",alpha,"][",be,"][",mu,"][",nu,"] = ", Ri[alpha][be][mu][nu])
66   print("\n")
     print("Ricci Tensor\n")
     for ro in range(0,4):
         for de in range (0,4):
70           sum = 0
             for alp in range (0,4):
                 sum = sum+Ri[alp][ro][alp][de]
             RT[ro][de] = simplify(sum)
74           print("RT[",ro,"][",de,"] = ",RT[ro][de])          # Ricci's tensor
     sumR = 0  # Ricci Scalar
     for be in range(0,4):
         for nu in range(0,4): sumR = sumR+gT[be,nu]*RT[be][nu]
78       print(sumR)
     RS = (sumR)
     print("RS",RS)                                              # Ricci Scalar R
```

Listing 19.2 LensGravity.py Compute the deflection of light by the sun with Matplotlib.

```
# LensGravity.py:   Deflection of light by the sun wi Matplotlib

import numpy as np
import matplotlib.pyplot as plt

y = np.zeros((2),float)
ph = np.zeros((181),float)                    # Time
yy = np.zeros((181),float)
xx = np.zeros((181),float)
rx = np.zeros((181),float)
ry = np.zeros((181),float)
Gsun = 4477.1                                 # Meters,  sum massxG
GM = 28.*Gsun
y[0] = 1.e-6; y[1] = 1e-6                      # Initial condition u=1/r

def f(t,y):                                   # RHS, can modify
    rhs = np.zeros((2),float)
    rhs[0] = y[1]
    rhs[1] = 3*GM*(y[0]**2)-y[0]
    return rhs
def rk4Algor(t, h, N, y, f):                  # Do not modify
    k1=np.zeros(N); k2=np.zeros(N); k3=np.zeros(N); k4=np.zeros(N)
    k1 = h*f(t,y)
    k2 = h*f(t+h/2.,y+k1/2.)
    k3= h*f(t+h/2.,y+k2/2.)
    k4= h*f(t+h,y+k3)
    y = y+(k1+2*(k2+k3)+k4)/6.
    return y

f(0,y)                             # Initial conditions
dphi = np.pi/180.                  # 180 phi values
i = 0                        # counter
for phi in np.arange(0,np.pi+dphi,dphi):
    ph[i] = phi
    y = rk4Algor(phi,dphi,2,y,f)     # Call rk4
    xx[i] = np.cos(phi)/y[0]/1000000 # Scale for graph
    yy[i] = np.sin(phi)/y[0]/1000000
    i = i + 1
m = (yy[180] - yy[165])/(xx[180]-xx[165])      # Slope
b = yy[180]-m*xx[180]                           # Intercept
j = 0
for phi in np.arange(0,np.pi+dphi,dphi):
    ry[j] = m*xx[j]+b                # Straight line eqtn
    j = j+1
plt.figure(figsize=(12,6))
plt.plot(xx,yy)          # Light trajectory
plt.plot(xx,-yy)         # Symmetric for negative y
plt.plot(0,0,'ro')            # Mass at origin
plt.plot(0.98,0,'bo')         # Source
plt.plot(0.98,1.91,'go')      # Position source seen from O
plt.plot(0.98,-1.91,'go')
plt.text(1,0,'S')
plt.text(-1.04,-0.02,'O')
plt.text(1.02, 1.91,"S' ")
plt.text(1.02,-2,"S''")
plt.plot([0],[3.])    # Invisible poin
plt.plot([0],[-3.])  # Invisible point at -y
plt.plot(xx,ry)       # Upper straight
plt.plot(xx,-ry)      # Lower straight line
plt.xlabel('x')
plt.ylabel('y')
plt.show()
```

Listing 19.3   RelOrbits.py computes relativistic and Newtonian potentials and orbits.

```
# RelOrbits.py Reltv orbits in a gravitational potential, needs rk4

import matplotlib.pyplot as plt
import numpy as np

dh = 0.03
dt = dh
ell = 4.3 # is el/M
G = 1.0
N = 2
E = -0.028
phi = np.zeros((7000),float)
rr = np.zeros((7000),float)
y = np.zeros((2),float)
y[0] = 0.0692
y[1] = np.sqrt(2*E/ell**2 + 2*G*y[0]/ell**2-G*y[0]**2+2*G*y[0]**3)

def f(t,y):
    rhs = np.zeros(2)
    rhs[0] = y[1]
    rhs[1] = -y[0]+G/ell**2 +3*G*y[0]**2
    return rhs

f(0,y)
i = 0
for fi in np.arange(0,12.0*np.pi,dt):
    y = rk4(fi,dt,N,y,f)
    rr[i] = (1/y[0])*np.sin(fi)           # Note u = 1/r
    phi[i] = (1/y[0])*np.cos(fi)
    i = i+1
f1 = plt.figure()
plt.axes().set_aspect('equal')           # Equal aspect ratio
plt.plot(phi[:900],rr[:900])
plt.show()
```

Listing 19.4   WormHole.py computes the derivatives needed to construct the Ellis wormhole connecting an upper and lower space.

```
# WormHole.py: Sympy evaluation of derivative for Ellis Wormhole

from sympy import *
L, x, M, rho, a, r, I ,lp= symbols('L x M rho a r I lp')
x = (2*L-a)/(pi*M)
r = rho+M*(x*atan(x) -log(1+x*x)/2)
p = diff(r,L)
print(p)
print("hola")
n = simplify(p)
print(n)
v = integrate(sqrt(1-n*n),(L,0,lp))
print("integral",v)
# Result: 2*atan((2*L - a)/(pi*M))/pi
```

Listing 19.5   VisualWorm.ipynb visualizes the Ellis wormhole using Vpython in a Jupyter notebook.

```
# VisualWorm.ipynb

from vpython import *
escene = canvas(width=400,height=400, range= 15)
import numpy as np
import math
```

```
 8  a = 1                              # 2a = height inner cylinder
    ring(pos=vector(0,0,0),radius=1,axis=vector(0,1,0),color=color.yellow)
    def f(x):                          # function to be integrated
        M = 0.5                        # black hole mass
12      a = 1                          # 2a: cylinders height
        y = np.sqrt(1- (2*np.arctan(2*(x - a)/(np.pi*M))/np.pi)**2)
        return y

16  def trapezoid(Func,A,B,N):
        h = (B - A)/(N )               # step, A:initial, B:end
        sum = (Func(A)+Func(B))/2      # initialize, (first + last)/2
        for i in range(1, N):          # inside
20          sum += Func(A+i*h)         #
        return h*sum                   # sum times h
    def radiuss(L):                    # radius as function of L
        ro = 1                         # radius of cylinder (a/ro=1)
24      a = 1                          # 2a: height of inner cylinder
        M = 0.5                        # black hole (mass M/ro)=1
        xx = (2*(L-a))/(np.pi*M)
        p = M*(xx*np.arctan(xx))
28      q = -0.5*M*math.log(1+xx**2)
        r = ro+ p+q
        return r
    for i in range(1,12):              # Plot rings at z, -z
32      A = 0                          # limits of integration
        B = i
        N = 300                        # trapezoid rule points
        if i>6: N=600                  # more points
36      z = trapezoid(f,A,B,N)         # returns z
        L = i+1
        rr = radiuss(L)                # radius
        ring(pos=vector(0,z,0),radius=rr,axis=vector(0,1,0),color=color.yellow)
40      ring(pos=vector(0,-z,0),radius=rr,axis=vector(0,1,0),color=color.yellow)
```

20

Integral Equations

The power and accessibility of high-speed computers have changed the view about what kind of equations are solvable. We have seen how even nonlinear differential equations can be solved easily and can give new insight into the physical world. In this chapter, we examine how integral equations, even those containing singularities, can be solved as matrix equations. We first convert the quantum bound-state problem to a matrix eigenvalue problem and solve it. We then examine the quantum scattering problem, which leads to a singular integral equation, and solve it too.

20.1 Nonlocal Potential Binding

A particle undergoes an interaction with a many-body medium, as shown on the left of Figure 20.1. We replace this very difficult problem by an approximate, but easier one, in which the particle interacts with an effective, one-particle potential. Because the effective potential at **r** depends on the wave function at **r′**, where there are interactions with other particles, the effective potential at r also depends on r', that is, the potential $V(r, r')$ is *nonlocal*. This means that the $V(r)\psi(r)$ term in the Schrödinger equation gets replaced by:

$$V(r)\psi(r) \rightarrow \int dr' \, V(r, r')\psi(r'), \tag{20.1}$$

$$\Rightarrow \quad -\frac{1}{2m}\frac{d^2\psi(r)}{dr^2} + \int dr' \, V(r, r')\psi(r') = E\psi(r). \tag{20.2}$$

Your **problem** is to solve this *integrodifferential equation* for bound-state energies E_n and wave functions ψ_n. (These are natural units, $\hbar = 1$.)

20.2 Momentum-Space Schrödinger Equation

Although integrodifferential equations can be solved iteratively, a more direct approach is to solve the momentum-space version of (20.1) [Landau, 1996]:

$$\frac{k^2}{2m}\psi_n(k) + \frac{2}{\pi}\int_0^\infty dp\, p^2 V(k, p)\psi_n(p) = E_n\psi_n(k), \tag{20.3}$$

Computational Physics: Problem Solving with Python, Fourth Edition.
Rubin H. Landau, Manuel J. Páez, and Cristian C. Bordeianu.
© 2024 WILEY-VCH GmbH. Published 2024 by WILEY-VCH GmbH.

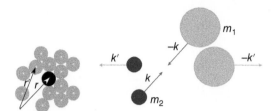

Figure 20.1 *Left*: A projectile (dark particle at r) scatters from a dense medium. *Right*: The same process viewed in the COM system where the projectile and target always have equal and opposite momenta.

where subscript n is used to enumerate the different bound-states. Here $V(k,p)$ is the momentum-space representation (double Fourier transform) of the coordinate-space potential, and if we restrict our solution to angular momentum $l = 0$ partial waves, it is simply related to $V(r)$:

$$V(k,p) = \frac{1}{kp} \int_0^\infty dr \sin(kr) V(r) \sin(pr). \tag{20.4}$$

In turn, the momentum-space wave function $\psi_n(k)$ is the probability amplitude for finding the particle with momentum k. It is the Fourier transform of $\psi_n(r)$:

$$\psi_n(k) = \int_0^\infty dr \, k \, r \, \psi_n(r) \sin(kr). \tag{20.5}$$

Equation (20.3) is an integral equation for $\psi_n(k)$. It differs from an integral representation of $\psi_n(k)$ in that the integral in it cannot be evaluated until the solution $\psi_n(p)$ is known. Although this may seem like a paradox, we will transform this equation into a matrix equation that can be solved with the matrix techniques discussed in Chapter 7.

20.2.1 Integral to Matrix Equations

We approximate the p integral over the potential in (20.4) as a weighted sum over N integration (usually Gauss quadrature) points $p = k_j$, $j = 1 \ldots N$:

$$\int_0^\infty dp \, p^2 \, V(k,p) \, \psi_n(p) \simeq \sum_{j=1}^N w_j \, k_j^2 \, V(k,k_j) \, \psi_n(k_j). \tag{20.6}$$

This converts the integral equation (20.3) to the algebraic equation:

$$\frac{k^2}{2m} \psi_n(k) + \frac{2}{\pi} \sum_{j=1}^N w_j \, k_j^2 \, V(k,k_j) \, \psi_n(k_j) = E_n. \tag{20.7}$$

Equation (20.7) contains N unknown function values $\psi_n(k_j)$, an unknown energy E_n, as well as the unknown functional dependence of $\psi_n(k)$. We eliminate the functional dependence of $\psi_n(k)$ by solving the equations for the same $k = k_i$ values as those used to approximate the integral. In other words, we solve the equations only for k values on the grid shown in Figure 20.2. This leads to a set of N coupled linear equations in $(N+1)$ unknowns:

$$\frac{k_i^2}{2m} \psi_n(k_i) + \frac{2}{\pi} \sum_{j=1}^N w_j \, k_j^2 \, V(k_i,k_j) \, \psi_n(k_j) = E_n \psi_n(k_i), \quad i = 1, N. \tag{20.8}$$

$k_1 \quad k_2 \quad k_3 \quad \bullet \quad \bullet \quad \bullet \quad k_N$

Figure 20.2 The grid of momentum values on which the integral equation is solved.

For example, for $N = 2$ we would have the two simultaneous linear equations:

$$\frac{k_1^2}{2m}\psi_n(k_1) + \frac{2}{\pi}w_1 k_1^2 V(k_1, k_1)\psi_n(k_1) + w_2 k_2^2 V(k_1, k_2)\psi_n(k_1) = E_n\psi_n(k_1),$$

$$\frac{k_2^2}{2m}\psi_n(k_2) + \frac{2}{\pi}w_1 k_1^2 V(k_2, k_1)\psi_n(k_1) + w_2 k_2^2 V(k_2, k_2)\psi_n(k_2) = E_n\psi_n(k_2).$$

Of course, a precise solution would require more than two integration points.

We write our coupled equations (20.8) in matrix form as:

$$[H][\psi_n] = E_n[\psi_n], \tag{20.9}$$

$$
\begin{bmatrix}
\frac{k_1^2}{2m} + \frac{2}{\pi}V(k_1, k_1)k_1^2 w_1 & \frac{2}{\pi}V(k_1, k_2)k_2^2 w_2 & \cdots & \frac{2}{\pi}V(k_1, k_N)k_N^2 w_N \\
\frac{2}{\pi}V(k_2, k_1)k_1^2 w_1 & \frac{2}{\pi}V(k_2, k_2)k_2^2 w_2 + \frac{k_2^2}{2m} & \cdots & \\
\ddots & & & \\
\cdots & & \cdots & \cdots \frac{k_N^2}{2m} + \frac{2}{\pi}V(k_N, k_N)k_N^2 w_N
\end{bmatrix}
$$

$$
\times
\begin{bmatrix}
\psi_n(k_1) \\
\psi_n(k_2) \\
\ddots \\
\psi_n(k_N)
\end{bmatrix}
= E_n
\begin{bmatrix}
\psi_n(k_1) \\
\psi_n(k_2) \\
\ddots \\
\psi_n(k_N)
\end{bmatrix}. \tag{20.10}
$$

Equation (20.9) is the matrix representation of the Schrödinger equation (20.3). The wave function $\psi_n(k)$ evaluated on the grid of integration point is the $N \times 1$ vector

$$
[\psi_n(k_i)] =
\begin{bmatrix}
\psi_n(k_1) \\
\psi_n(k_2) \\
\ddots \\
\psi_n(k_N)
\end{bmatrix}. \tag{20.11}
$$

The astute reader may be questioning the possibility of solving N equations for the $N + 1$ unknowns, $\psi_n(k_i)$ and E_n. Only sometimes, and only for certain values of E_n (the eigenvalues), will a solution exist. Let's start by trying to apply the matrix inversion technique by rewriting (20.9) as:

$$[H - E_n I][\psi_n] = [0]. \tag{20.12}$$

If we try to obtain a solution by multiplying both sides of (20.13) by the inverse of $[H - E_n I]$, we get something weird:

$$[\psi_n] = [H - E_n I]^{-1}[0]. \tag{20.13}$$

This equation tells us that if the inverse exists, then we have the *trivial* solution $\psi_n \equiv 0$, which is a solution, but is trivial. So maybe the assumption that the inverse exists is not valid, which would mean that the determinant of $[H - E_n I]$ vanishes:

$$\det[H - E_n I] = 0 \qquad \text{(bound-state condition)}. \tag{20.14}$$

Equation (20.14) is the $(N + 1)$th equation we need for a unique solution to the bound-state problem, in which E_n is the *eigenvalues* of (20.9).

20.2.2 Delta-Shell Potential

To keep things simple, and to have an analytic answer with which to compare, we consider the local, delta-shell potential:

$$V(r) = \frac{\lambda}{2m}\delta(r-b). \tag{20.15}$$

This might be a good model for an interaction that occurs when two particles are predominantly a fixed distance b apart. Equation (20.4) determines the momentum-space representation of the potential:

$$V(k',k) = \int_0^\infty \frac{\sin(k'r')}{k'k}\frac{\lambda}{2m}\delta(r-b)\sin(kr)\,dr = \frac{\lambda}{2m}\frac{\sin(k'b)\sin(kb)}{k'k}. \tag{20.16}$$

Beware: We have chosen this potential because it is easy to evaluate the momentum-space matrix elements. However, its singular nature in r space leads to (20.16) having a very slow fall off in k space, and this leads to poor numerical precision.

If the energy is parameterized in terms of a wave vector κ by $E_n = -\kappa^2/2m$, then for this potential there is, at most, one bound state, and it satisfies the transcendental equation [Gottfried and Yan, 2004]:

$$e^{-2\kappa b} - 1 = \frac{2\kappa}{\lambda}. \tag{20.17}$$

Due to b]ound states occurring only for attractive potentials, we must have $\lambda < 0$.

Exercise Pick some values of b and λ, and solve (20.17) for κ.

The numerical computation may follow two paths. One evaluates $\det[H - E_n I]$ in (20.14), and then *searches* for those values of energy at which the determinant vanishes. This provides E_n, but not wave functions. The other path solves the eigenvalue problem for all eigenvalues and eigenfunctions. In both cases, the solution must be searched for, and you may be required to guess starting values for the energy. We present our solution Bound.py in Listing 20.1.

Problems Write a program that solves the integral equation (20.9) for the delta-shell potential (20.16). Find either the E_n's for which the determinant vanishes *or*, the eigenvalues and eigenvectors for this H.

1) Set the scale by setting $2m = 1$ and $b = 10$.
2) Set up the potential and Hamiltonian matrices, $V(i,j)$ and $H(i,j)$, for Gaussian quadrature integration using at least $N = 16$ grid points.
3) Adjust the value and sign of λ for bound states. Start with a large negative value for λ and then make it progressively less negative. You should find that the eigenvalues move up in energy.
4) **Note**: Your eigenenergy solver may return several eigenenergies. The true bound state will be at negative energy and change little as the number of grid points changes. The others are numerical artifacts.
5) Try increasing the number of grid points in steps of 8, for example, 16, 24, 32, 64, ..., and see how the energy changes.

6) Extract the best value for the bound-state energy, and estimate its precision by seeing how it changes with the number of grid points.
7) If you are solving the eigenvalue problem, check your solution by comparing the RHS and LHS in the matrix multiplication $[H][\psi_n] = E_n[\psi_n]$.
8) Verify that, regardless of the potential's strength, there is only a single bound-state and that it gets deeper as the magnitude of λ increases. Compare with (20.17).

20.2.3 Wave Function (Exploration)

1) Determine the momentum-space wave function $\psi_n(k)$ using an eigenproblem solver. Does $\psi_n(k)$ fall off at $k \to \infty$? Does it oscillate? Is it well-behaved at the origin?
2) Using the same points and weights as used to evaluate the integral in the integral equation, determine how the coordinate-space wave function via the Bessel transforms.

$$\psi_n(r) = \int_0^\infty dk \psi_n(k) \frac{\sin(kr)}{kr} k^2. \tag{20.18}$$

Does $\psi_n(r)$ fall off as you would expect for a bound-state? Does it oscillate? Is it well-behaved at the origin?
3) Compare the r dependence of this $\psi_n(r)$ to the analytic wave function:

$$\psi_n(r) \propto \begin{cases} e^{-\kappa r} - e^{\kappa r}, & \text{for } r < b, \\ e^{-\kappa r}, & \text{for } r > b. \end{cases} \tag{20.19}$$

20.3 Scattering in Momentum Space ⊙

Again we have a particle interacting with the nonlocal potential, Figure 20.1 left, only now the particle has sufficiently high energy for it to scatter from the target particles and not be bound by them.

Problem Determine the scattering phase shift δ for this scattering.

20.3.1 Schrödinger to Lippmann–Schwinger Equation

Because scattering experiments measure scattering amplitudes, but not wave functions, it is more direct to have our theory calculate amplitudes. An integral form of the Schrödinger equation dealing with the scattering amplitude R is the *Lippmann–Schwinger equation*:

$$R(k', k) = V(k', k) + \frac{2}{\pi} P \int_0^\infty dp \frac{p^2 V(k', p) R(p, k)}{(k_0^2 - p^2)/2m}. \tag{20.20}$$

(R is actually the *reaction matrix*, but is related to the scattering amplitude, and is easier to calculate.) As in the bound-state problem, this equation is for partial wave $l = 0$ and $\hbar = 1$. In (20.20) the momentum k_0 is related to the energy E and the reduced mass m by:

$$E = \frac{k_0^2}{2m}, \qquad m = \frac{m_1 m_2}{m_1 + m_2}. \tag{20.21}$$

The initial and final COM momenta k and k' are the momentum-space variables. The experimental observable that results from a solution of (20.20) is the diagonal $(k = k' = k_0)$ matrix element $R(k_0, k_0)$, which is related to the scattering phase shift δ_0, and thus the cross section:

$$R(k_0, k_0) = -\frac{\tan \delta_l}{\rho}, \qquad \rho = 2mk_0. \tag{20.22}$$

Note that (20.20) is not just the evaluation of an integral, it is an integral equation in which $R(p, k)$ must be integrated over all p values. Yet because $R(p, k)$ is unknown, the integral cannot be evaluated until after the equation is solved! The symbol $\mathcal{P}$ in (20.20) indicates the *Cauchy principal-value* prescription for avoiding the singularity arising from the zero of the denominator at $p = k_0$.

20.3.2 Singular Integral Evaluations

A *singular* integral

$$\mathcal{G} = \int_a^b g(k)\, dk, \tag{20.23}$$

is one in which the integrand $g(k)$ is singular at the point k_0 within the integration interval, yet the integral $\mathcal{G}$ remains finite. (If the integral itself were infinite, we could not compute it.) Unfortunately, computers are notoriously incompetent at dealing with infinite numbers, and if an integration point gets too near to the singularity, overwhelming subtractive cancellation, or overflow, occurs. But we can deal with it.

In Figure 20.3 we show three ways to avoid the singularity at k_0. The paths in Figure 20.3a and b move the singularity slightly off the real k axis by giving k_0 a small imaginary part $\pm i\epsilon$. The Cauchy principal-value prescription $\mathcal{P}$ in Figure 20.3c says to integrate along a path that "pinches" both sides of the singularity at k_0, without integrating over it:

$$\mathcal{P} \int_{-\infty}^{+\infty} f(k)\, dk = \lim_{\epsilon \to 0} \left[\int_{-\infty}^{k_0 - \epsilon} f(k)\, dk + \int_{k_0 + \epsilon}^{+\infty} f(k)\, dk \right]. \tag{20.24}$$

The preceding three prescriptions are related by the identity

$$\int_{-\infty}^{+\infty} \frac{f(k)\, dk}{k - k_0 \pm i\epsilon} = \mathcal{P} \int_{-\infty}^{+\infty} \frac{f(k)\, dk'}{k - k_0} \mp i\pi f(k_0), \tag{20.25}$$

which follows from Cauchy's residue theorem.

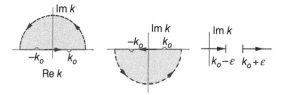

Figure 20.3 Three different paths in the complex k plane used to evaluate line integrals when there are singularities. Here the singularities are at $\pm k_0$, and the integration variable is k. In *Left* and *Center* the singularity is given a small imaginary part $k_0 \to k_0 \pm i\epsilon$ that moves it slightly off the real axis, while in *Right* the integration path "pinches" both sides of the singularity, without passing through it.

A numerical evaluation of the principal value limit (20.24) is troublesome because large cancellations will occur near the singularity. An accurate algorithm for evaluating the integral follows from the fact that

$$P \int_{-\infty}^{+\infty} \frac{dk}{k - k_0} = 0. \tag{20.26}$$

This equation says that a graph of $1/(k - k_0)$ versus k has equal and opposite areas on both sides of the singular point k_0:

$$P \int_{-\infty}^{+\infty} \frac{dk}{k - k_0} = \int_{-\infty}^{0} \frac{dk}{k - k_0} + \int_{0}^{+\infty} \frac{dk}{k - k_0} \tag{20.27}$$

$$= -\int_{0}^{+\infty} \frac{-dk}{-k - k_0} + \int_{0}^{+\infty} \frac{dk}{k - k_0}, \tag{20.28}$$

$$\Rightarrow P \int_{0}^{\infty} \frac{dk}{k^2 - k_0^2} = 0, \tag{20.29}$$

where we have broken the integral up into one over positive k and one over $-k$, and then changed variable $k \to -k$ in the first integral. We thus see that the principal-value exclusion of the singular point's contribution to the integral is equivalent to a simple subtraction of the zero integral (20.29):

$$P \int_{0}^{\infty} \frac{f(k)\, dk}{k^2 - k_0^2} = \int_{0}^{\infty} \frac{[f(k) - f(k_0)]\, dk}{k^2 - k_0^2}. \tag{20.30}$$

Notice that there is no P on the RHS of (20.30) because the integrand is no longer singular at $k = k_0$ (it is proportional to the df/dk). Therefore the integral on the RHS can be evaluated numerically using the usual rules. The integral (20.30) is called the *Hilbert transform* of f and also arises in subjects such as inverse problems.

20.3.3 Singular Integral Equations to Matrix Equations

Now that we have put the singularity out of the way, we go back to reducing the integral equation (20.20) to a set of linear equations. We rewrite the principal-value prescription as a definite integral [Haftel and Tabakin, 1970]:

$$R(k',k) = V(k',k) + \frac{2}{\pi} \int_{0}^{\infty} dp\, \frac{p^2 V(k',p)R(p,k) - k_0^2 V(k',k_0)R(k_0,k)}{(k_0^2 - p^2)/2m}. \tag{20.31}$$

We convert this integral equation to a set of simultaneous linear equations by approximating the integral as a sum over N Gaussian integration points k_j with weights w_j:

$$R(k,k_0) \simeq V(k,k_0) + \frac{2}{\pi} \sum_{j=1}^{N} \frac{k_j^2 V(k,k_j)R(k_j,k_0)w_j}{(k_0^2 - k_j^2)/2m}$$

$$- \frac{2}{\pi} k_0^2 V(k,k_0)R(k_0,k_0) \sum_{m=1}^{N} \frac{w_m}{(k_0^2 - k_m^2)/2m}. \tag{20.32}$$

We note that the last term in (20.32) implements the principal-value prescription and cancels the singular behavior of the previous term. This equation contains the $N + 1$ unknowns

$R(k_j, k_0)$ for $j = 0, N$. We turn it into $N + 1$ simultaneous equations by evaluating it for the N k values on the grid in Figure 20.2, and at the observable momentum k_0:

$$k = k_i = \begin{cases} k_j, & j = 1, N \quad \text{(quadrature points)}, \\ k_0, & i = 0 \quad \text{(observable point)}. \end{cases} \tag{20.33}$$

There are now $N + 1$ linear equations for the $N + 1$ unknowns $R_i \equiv R(k_i, k_0)$:

$$R_i = V_i + \frac{2}{\pi} \sum_{j=1}^{N} \frac{k_j^2 V_{ij} R_j w_j}{(k_0^2 - k_j^2)/2m} - \frac{2}{\pi} k_0^2 V_{i0} R_0 \sum_{m=1}^{N} \frac{w_m}{(k_0^2 - k_m^2)/2m}. \tag{20.34}$$

We express (20.34) in matrix form by combining the denominators and weights into a single denominator vector $\mathbf{D}$:

$$D_i = \begin{cases} +\dfrac{2}{\pi} \dfrac{w_i k_i^2}{(k_0^2 - k_i^2)/2m}, & \text{for } i = 1, N, \\[3ex] -\dfrac{2}{\pi} \sum\limits_{j=1}^{N} \dfrac{w_j k_0^2}{(k_0^2 - k_j^2)/2m}, & \text{for } i = 0. \end{cases} \tag{20.35}$$

The linear equations (20.34) now assume the matrix form

$$R - DVR = [1 - DV] R = V, \tag{20.36}$$

where R and V are the length $N + 1$ vectors:

$$[R] = \begin{bmatrix} R_{0,0} \\ R_{1,0} \\ \vdots \\ R_{N,0} \end{bmatrix}, \quad [V] = \begin{bmatrix} V_{0,0} \\ V_{1,0} \\ \vdots \\ V_{N,0} \end{bmatrix}. \tag{20.37}$$

We write our reduction of the integral equation as the matrix equation:

$$[F][R] = [V], \qquad F_{ij} = \delta_{ij} - D_j V_{ij}. \tag{20.38}$$

The F matrix is known as the *wave matrix*. With R the unknown vector, (20.38) is in the standard form $AX = B$, which can be solved by the mathematical subroutine libraries discussed in Chapter 7.

20.3.4 Solution

An elegant (but alas not most efficient) solution to (20.38) is by matrix inversion:

$$[R] = [F]^{-1}[V]. \tag{20.39}$$

Because the inversion of even complex matrices is a standard routine in linear algebra libraries, (20.39) is a *direct solution* for the R amplitude. Unless you need the inverse for other purposes (like calculating wave functions), a more efficient approach is *Gaussian elimination*, which is also contained in the linear algebra libraries.

Figure 20.4 The energy dependence of the cross section for angular momentum $l = 0$ scattering from an attractive delta-shell potential with $\lambda b = 15$. The dashed curve is the analytic solution (20.41), and the solid curve results from numerically solving the integral Schrödinger equation.

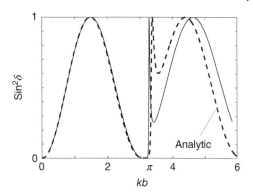

For this scattering problem we will use the same delta-shell potential (20.16) as we used in Section 20.2.2 for bound states:

$$V(k', k) = \frac{-|\lambda|}{2mk'k} \sin(k'b) \sin(kb). \tag{20.40}$$

This is one of the few potentials for which the Lippmann–Schwinger equation (20.20) has an analytic solution [Gottfried and Yan, 2004] with which to check:

$$\tan \delta_0 = \frac{\lambda b \sin^2(kb)}{kb - \lambda b \sin(kb) \cos(kb)}. \tag{20.41}$$

Our results were obtained with $2m = 1$, $\lambda b = 15$, and $b = 10$, the same as in Gottfried and Yan [2004]. In Figure 20.4, we give a plot of $\sin^2 \delta_0$ *versus* kb, which is proportional to the scattering cross section arising from the angular momentum $l = 0$ phase shift. Note that $sin^2\delta$ reaches its maximum values at energies corresponding to resonances. We present our solution, `scatt.py`, in Listing 20.2.

20.3.5 Exercises

1) Write a program for the matrices `v[]`, `D[]`, and `F[,]`. Use at least $N = 16$ Gaussian quadrature points for your grid.
2) Calculate the matrix F^{-1} using a library subroutine.
3) Calculate the vector R by matrix multiplication $R = F^{-1} V$.
4) Deduce the phase shift δ from $R(k_0, k_0)$:

$$R(k_0, k_0) = R_{0,0} = -\frac{\tan \delta}{\rho}, \quad \rho = 2mk_0. \tag{20.42}$$

5) Estimate the precision of your solution by increasing the number of grid point in steps of two (we found the best answer for $N = 26$). If your phase shift changes in the second or third decimal place, you probably have that much precision.
6) Plot $\sin^2 \delta$ *versus* energy $E = k_0^2/2m$ starting at zero energy and ending at energies where the phase shift is again small. Your results should be similar to those in Figure 20.4. Note that a *resonance* occurs when δ_l increases rapidly through $\pi/2$, that is, when $\sin^2 \delta_0 = 1$.
7) Check your answer against the analytic results (20.41).

20.3.6   Scattering Wave Function (Exploration)

The *wave matrix F^{-1}* in our solution to the integral equation

$$R = F^{-1}V = (1 - VG)^{-1}V \tag{20.43}$$

can be used to calculate the coordinate-space wave function:

$$u(r) = N_0 \sum_{i=1}^{N} \frac{\sin(k_i r)}{k_i r} F(k_i, k_0)^{-1}. \tag{20.44}$$

Here N_0 is a normalization constant, and the R amplitude is appropriate for standing-wave boundary conditions.

1) Plot $u(r)$ and compare it to a free wave.

20.4   Code Listings

Listing 20.1   Bound.py Solves the Lippmann–Schwinger integral equation for the quantum bounds states within a delta-shell potential.

```
# Bound.py: Bound state solutn of Lippmann-Schwinger equation in p space

from visual import *
from numpy import*
from numpy.linalg import*

min1 =0.;      max1 =200.;   u =0.5;    b =10.

def gauss(npts,a,b,x,w):
    pp = 0.;   m = (npts + 1)//2;   eps = 3.E-10         # Accuracy: ADJUST!

    for i in range(1,m+1):
        t = cos(math.pi*(float(i)-0.25)/(float(npts) + 0.5))
        t1 = 1
        while((abs(t-t1)) >= eps):
            p1 = 1. ;   p2 = 0.;
            for j in range(1,npts+1):
                p3 = p2
                p2 = p1
                p1=((2*j-1)*t*p2-(j-1)*p3)/j
            pp = npts*(t*p1-p2)/(t*t-1.)
            t1 = t; t = t1 - p1/pp
        x[i-1] = -t
        x[npts-i] = t
        w[i-1] = 2./((1.-t*t)*pp*pp)
        w[npts-i] = w[i-1]
    for i in range(0,npts):
        x[i] = x[i]*(b-a)/2. + (b + a)/2.
        w[i] = w[i]*(b-a)/2.

for M in range(16, 32, 8):
    z=[-1024, -512, -256, -128, -64, -32, -16, -8, -4, -2]
    for lmbda in z:
        A = zeros((M,M), float)                         # Hamiltonian
        WR = zeros((M), float)                  # Eigenvalues, potential
        k = zeros((M), float); w = zeros((M),float);          # Pts & wts
        gauss(M, min1, max1, k, w)                    # Call gauss points
        for i in range(0,M):                          # Set Hamiltonian
            for j in range(0,M):
                VR = lmbda/2/u*sin(k[i]*b)/k[i]*sin(k[j]*b)/k[j]
                A[i,j] = 2./math.pi*VR*k[j]*k[j]*w[j]
                if (i == j):
```

```
                        A[i,j] += k[i]*k[i]/2/u
              Es, evectors = eig(A)
              realev = Es.real                        # Real eigenvalues
46        for j in range(0,M):
              if (realev[j]<0):
                  print(" M (size), lmbda, ReE = ",M," ",lmbda," ",realev[j])
              break
```

Listing 20.2 Scatt.py Solves the Lippmann–Schwinger integral equation for quantum scattering from a delta-shell potential.

```
1 # Scatt.py:    Soln p space Lippmann Schwinger for scattering

  from visual import *
  from visual.graph import *
5 import numpy.linalg as lina                    # Numpy's LinearAlgebra

  def gauss(npts, job, a, b, x, w):
      m = i = j = t = t1 = pp = p1 = p2 = p3 = 0.
9     eps = 3.E-14                       # Accuracy: ******ADJUST THIS*******!
      m = (npts + 1)/2
      for i in arange(1, m + 1):
          t = cos(math.pi*(float(i) - 0.25)/(float(npts) + 0.5) )
13        t1 = 1
          while( (abs(t - t1) ) >= eps):
              p1 = 1. ;   p2 = 0.
              for j in range(1, npts + 1):
17                p3 = p2;   p2 = p1
                  p1 = ((2.*float(j)-1)*t*p2 - (float(j)-1.)*p3)/(float(j))
              pp = npts*(t*p1 - p2)/(t*t - 1.)
              t1 = t; t = t1  - p1/pp
21        x[i - 1] = - t;   x[npts - i] = t
          w[i - 1] = 2./( (1. - t*t)*pp*pp)
          w[npts - i] = w[i - 1]
      if (job == 0):
25        for i in range(0, npts):
              x[i] = x[i]*(b - a)/2. + (b + a)/2.
              w[i] = w[i]*(b - a)/2.
      if (job == 1):
29        for i in range(0, npts):
              xi  = x[i]
              x[i] = a*b*(1. + xi) / (b + a - (b - a)*xi)
              w[i] = w[i]*2.*a*b*b/( (b + a - (b-a)*xi)*(b + a - (b-a)*xi))
33    if (job == 2):
          for i in range(0, npts):
              xi = x[i]
              x[i] = (b*xi + b + a + a) / (1. - xi)
37            w[i] = w[i]*2.*(a + b)/( (1. - xi)*(1. - xi) )

  graphscatt = gdisplay(x=0, y=0, xmin=0, xmax=6,ymin=0, ymax=1, width=600,
          height=400,
      title='S Wave Cross Section vs E', xtitle='kb', ytitle=' [sin(delta)]**2')
41 sin2plot = gcurve(color=color.yellow)
  M = 27;                   b = 10.0;                n = 26
  k = zeros((M),float);       x = zeros((M),float);     w = zeros((M),float)
  Finv = zeros((M,M),float);  F = zeros((M,M), float);  D = zeros((M),float)
45 V = zeros((M), float);      Vvec = zeros((n+1,1),float)
  scale = n/2;                lambd = 1.5

  gauss(n, 2, 0., scale, k, w)                          # Set up points & wts
49 ko = 0.02
  for m in range(1,901):
      k[n] = ko
      for i in range (0, n):  D[i]=2/pi*w[i]*k[i]*k[i]/(k[i]*k[i]-ko*ko) #D
53    D[n] = 0.
      for j in range(0,n):   D[n]=D[n]+w[j]*ko*ko/(k[j]*k[j]-ko*ko)
      D[n] = D[n]*(-2./pi)
      for i in range(0,n+1):                            # Set up F & V
```

```
57        for j in range(0,n+1):
              pot = -b*b * lambd * sin(b*k[i])*sin(b*k[j])/(k[i]*b*k[j]*b)
              F[i][j] = pot*D[j]
              if i==j: F[i][j] = F[i][j] + 1.
61        V[i] = pot
      for  i in range(0,n+1):  Vvec[i][0]= V[i]
      Finv = lina.inv(F)                             # LinearAlgebra for inverse
      R = dot(Finv, Vvec)                                  # Matrix multiply
65    RN1 = R[n][0]
      shift = atan(-RN1*ko)
      sin2 = (sin(shift))**2
      sin2plot.plot(pos = (ko*b,sin2))                     # Plot sin**2(delta)
69    ko = ko + 0.2*pi/1000.
   print("Done")
```

Part IV

PDE Applications

21

PDE Review, Electrostatics and Relaxation

This chapter is the first of several dealing with partial differential equations (PDEs); several because PDEs are more complex than ODEs, and several because each type of PDE requires its own algorithm. We start with a review of the types of PDEs, and requirements for their unique solutions. Then we get down to business by examining the simple, but powerful, finite difference method for solving Poisson's and Laplace's equations. In Chapter 27, we introduce the more complicated, but computationally faster, finite element method (FEM) for solving the same equations.

21.1 Review

Physical quantities such as temperature and pressure vary continuously in both space and time. Such being our world, the function or *field* $U(x, y, z, t)$ used to describe these quantities must contain independent space and time variations. As time flows, the change in $U(x, y, z, t)$ at any one position affect the field at neighboring points. This means that the dynamic equations describing the dependence of U on four independent space-time variables must be written in terms of partial derivatives, and therefore, the equations must be *partial differential equations* (PDEs), in contrast to ordinary differential equations (ODEs).

The general form for a PDE with two independent variables is

$$A \frac{\partial^2 U}{\partial x^2} + 2B \frac{\partial^2 U}{\partial x \partial y} + C \frac{\partial^2 U}{\partial y^2} + D \frac{\partial U}{\partial x} + E \frac{\partial U}{\partial y} = F, \tag{21.1}$$

where A, B, C, and F are arbitrary functions of the variables x and y [Arfken and Weber, 2001]. In Table 21.1, we define the classes of PDEs by the value of the *discriminant* $d = AC - B^2$, and give examples there. We usually think of an *elliptic equation* as one containing second-order derivatives of all the variables, with all having the same sign when placed on the same side of the equal sign; a *parabolic equation* as one containing a first-order derivative in one variable and a second-order derivative in the other; and a *hyperbolic equation* as one containing second-order derivatives of all the variables, with opposite signs when placed on the same side of the equal sign.

After solving enough problems, one often develops some physical intuition as to whether one has sufficient *boundary conditions* for there to exist a unique solution for a given

Computational Physics: Problem Solving with Python, Fourth Edition.
Rubin H. Landau, Manuel J. Páez, and Cristian C. Bordeianu.
© 2024 WILEY-VCH GmbH. Published 2024 by WILEY-VCH GmbH.

Table 21.1 The types of PDE, their discriminants, and examples of each.

$d = AC - B^2 > 0$	$d = AC - B^2 = 0$	$d = AC - B^2 < 0$
$\nabla^2 U(x) = -4\pi\rho(x)$	$\nabla^2 U(\mathbf{x}, t) = a\,\partial U/\partial t$	$\nabla^2 U(\mathbf{x}, t) = c^{-2}\partial^2 U/\partial t^2$
Poisson's equation	Heat equation	Wave equation

Table 21.2 The relation between boundary conditions and uniqueness for PDEs.

Boundary condition	Elliptic (Poisson equation)	Hyperbolic (Wave equation)	Parabolic (Heat equation)
Dirichlet open surface	Underspecified	Underspecified	*Unique & stable (1D)*
Dirichlet closed surface	*Unique & stable*	Overspecified	Overspecified
Neumann open surface	Underspecified	Underspecified	*Unique & Stable (1D)*
Neumann closed surface	*Unique & stable*	Overspecified	Overspecified
Cauchy open surface	Nonphysical	*Unique & stable*	Overspecified
Cauchy closed surface	Overspecified	Overspecified	Overspecified

physical situation (this, of course, is in addition to requisite *initial conditions*). Table 21.2 gives the requisite boundary conditions for a unique solution to exist for each type of PDE. For instance, a string tied at both ends, or a heated bar placed in an infinite heat bath, are physical situations for which the boundary conditions are adequate. If the boundary condition is the value of the solution on a surrounding closed surface, we have a *Dirichlet boundary condition*. If the boundary condition is the value of the normal derivative on the surrounding surface, we have a *Neumann boundary condition*. If the value of both the solution and its derivative are specified on a closed boundary, we have a *Cauchy boundary condition*. Although having an adequate boundary condition is necessary for a unique solution, having too many boundary conditions, for instance, both Neumann and Dirichlet, may be an overspecification for which no solution exists.[1]

Solving PDEs numerically differs from solving ODEs in a number of ways. First, because we are able to write all ODEs in a standard form,

$$\frac{d\mathbf{y}(t)}{dt} = \mathbf{f}(\mathbf{y}, t), \tag{21.2}$$

with t the single independent variable, we are able to use a standard algorithm such as `rk4` to solve all such equations. Yet, because PDEs have several independent variables, for example $\rho(x, y, z, t)$, we would have to apply (21.2) simultaneously and independently to each variable, which would be very complicated. Second, because there are more equations to solve with PDEs than with ODEs, we need more information than just the two *initial conditions* $[x(0), \dot{x}(0)]$. In addition, because each PDE often has its own particular set of boundary conditions, we have to develop a special algorithm for each particular problem.

---

1 Although conclusions concerning uniqueness drawn for exact the PDEs may differ from those drawn for the finite difference equations we will use, they are usually the same [Jackson, 1988; Morse and Feshbach, 1953].

21.2 Laplace's Equation

Figure 21.1 shows a wire square in which the bottom and sides are "grounded" (kept at 0 V), while the top wire is connected to a voltage source that keeps it at a constant 100 V. There are no charges within the square.

Problem Find the electric potential for all points *inside* the square.

The voltages on the perimeter of the square in Figure 21.1 are the boundary conditions for this problem. (If you imagine there being infinitesimal insulators at the top corners of the box, then we have a closed boundary). Since the values of the potential are given on all sides, we have Neumann conditions on the boundary and, according to Table 21.2, a unique and stable solution exists.

It is known from classical electrodynamics that the electric potential $U(\mathbf{x})$ arising from static charges satisfies Poisson's PDE [Jackson, 1988]:

$$\nabla^2 U(\mathbf{x}) = -4\pi\rho(\mathbf{x}), \tag{21.3}$$

where $\rho(\mathbf{x})$ is the charge density at $\mathbf{x}$. In charge-free regions of space, that is, regions where $\rho(\mathbf{x}) = 0$, the potential satisfies *Laplace's equation*:

$$\nabla^2 U(\mathbf{x}) = 0. \tag{21.4}$$

Both these equations are elliptic PDEs of a form that occurs in various applications. We solve them in 2D rectangular coordinates:

$$\frac{\partial^2 U(x,y)}{\partial x^2} + \frac{\partial^2 U(x,y)}{\partial y^2} = 0, \qquad \text{Laplace's equation,} \tag{21.5}$$

$$\frac{\partial^2 U(x,y)}{\partial x^2} + \frac{\partial^2 U(x,y)}{\partial y^2} = -4\pi\rho(\mathbf{x}), \qquad \text{Poisson's equation.} \tag{21.6}$$

In both cases we see that the potential depends simultaneously on x and y. For Laplace's equation, the charges, which are the source of the field, enter indirectly by specifying the potential values in some region of space; for Poisson's equation they enter directly as well.

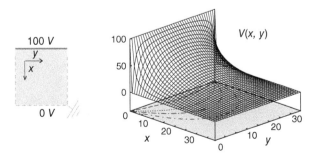

Figure 21.1 *Left*: The shaded region of space within a square in which we determine the electric potential by solving Laplace's equation. There is a wire at the top kept at a constant 100 V and a grounded wire (dashed) at the sides and bottom. *Right*: The computed electric potential as a function of x and y. The projections onto the shaded xy plane are equipotential contour lines.

21.2.1 Fourier Series Solution

For the simple geometry of Figure 21.1, an analytic solution of Laplace's equation (21.5) exists in the form of an infinite series. If we assume that the solution is the product of independent functions of x and y, and substitute the product into (21.5), we obtain:

$$U(x,y) = X(x)Y(y) \quad \Rightarrow \quad \frac{d^2X(x)/dx^2}{X(x)} + \frac{d^2Y(y)/dy^2}{Y(y)} = 0. \tag{21.7}$$

If $X(x)$ is a function of only x, and $Y(y)$ is a function of only y, the derivatives in (21.7) are *ordinary* as opposed to *partial* derivatives. Because $X(x)$ and $Y(y)$ are independent, the only way (21.7) can be valid for *all* values of x and y is for each term in (21.7) to be equal to a constant:

$$\frac{d^2Y(y)/dy^2}{Y(y)} = -\frac{d^2X(x)/dx^2}{X(x)} = k^2, \tag{21.8}$$

$$\Rightarrow \quad \frac{d^2X(x)}{dx^2} + k^2X(x) = 0, \qquad \frac{d^2Y(y)}{dy^2} - k^2Y(y) = 0. \tag{21.9}$$

We shall see that this choice of sign for the constant matches the boundary conditions and gives us periodic behavior in x. The other choice of sign would give periodic behavior in y, and that would not work with these boundary conditions.

The solutions for $X(x)$ are periodic, and those for $Y(y)$ are exponential:

$$X(x) = A \sin kx + B \cos kx, \qquad Y(y) = Ce^{ky} + De^{-ky}. \tag{21.10}$$

The $x = 0$ boundary condition $U(x = 0, y) = 0$ can be met only if $B = 0$. The $x = L$ boundary condition $U(x = L, y) = 0$ can be met only for:

$$kL = n\pi, \ n = 1, 2, \ldots. \tag{21.11}$$

Such being the case, for each value of n there is the solution:

$$X_n(x) = A_n \sin\left(\frac{n\pi}{L}x\right). \tag{21.12}$$

For each value of k_n, $Y(y)$ must satisfy the y boundary condition $U(x, 0) = 0$, which requires $D = -C$:

$$Y_n(y) = C(e^{k_n y} - e^{-k_n y}) \equiv 2C \sinh\left(\frac{n\pi}{L}y\right). \tag{21.13}$$

Because we are solving linear equations, the principle of linear superposition holds, which means that the most general solution is the sum of the products:

$$U(x,y) = \sum_{n=1}^{\infty} E_n \sin\left(\frac{n\pi}{L}x\right) \sinh\left(\frac{n\pi}{L}y\right). \tag{21.14}$$

The E_n values are arbitrary constants and are fixed by requiring the solution to satisfy the remaining boundary condition at $y = L$, $U(x, y = L) = 100$ V:

$$\sum_{n=1}^{\infty} E_n \sin\left(\frac{n\pi}{L}x\right) \sinh(n\pi) = 100\,\text{V}. \tag{21.15}$$

We determine the constants E_n by projection – Multiply both sides of the equation by $\sin(m\pi x/L)$, with m an integer, and integrate from 0 to L:

$$\sum_{n} E_n \sinh(n\pi) \int_0^L dx \sin\frac{n\pi}{L} x \sin\frac{m\pi}{L} x = \int_0^L dx\,100 \sin\frac{m\pi}{L} x. \tag{21.16}$$

The integral on the LHS is nonzero only for $n = m$, which yields

$$E_n = \begin{cases} 0, & \text{for } n \text{ even,} \\ \frac{4(100)}{n\pi \, \sinh(n\pi)}, & \text{for } n \text{ odd.} \end{cases}$$

(21.17)

Finally, we obtain an infinite series (analytic solution?) for the potential at any point (x, y):

$$U(x, y) = \sum_{n=1,3,5,\dots}^{\infty} \frac{400}{n\pi} \sin\left(\frac{n\pi x}{L}\right) \frac{\sinh(n\pi y/L)}{\sinh(n\pi)}.$$

(21.18)

21.2.2 Fourier Series as an Algorithm

If we try to use (21.18) as an algorithm, we must terminate the sum at some point. Yet, in practice, the convergence of the series is so painfully slow that many terms are needed for good accuracy, and so round-off errors may become a problem. In addition, the sinh functions in (21.18) overflows for large n, which can be avoided somewhat by expressing the quotient of the two sinh functions in terms of exponentials, and then taking a large n limit:

$$\frac{\sinh(n\pi y/L)}{\sinh(n\pi)} = \frac{e^{n\pi(y/L-1)} - e^{-n\pi(y/L+1)}}{1 - e^{-2n\pi}} \xrightarrow[n\to\infty]{} e^{n\pi(y/L-1)}.$$

(21.19)

A third problem with the "analytic" solution is that a Fourier series converges only in the *mean square* (Figure 21.2). This means that it converges to the *average* of the left- and right-hand limits in the regions where the solution is discontinuous, such as in the corners of the box [Kreyszig, 1998]. Explicitly, what you see in Figure 21.2 is a phenomenon known as the *Gibbs overshoot*, which occurs when a Fourier series with a finite number of terms is used to represent a discontinuous function. Rather than fall off abruptly, the series develops oscillations that tend to overshoot the function at the corner. To obtain a smooth solution, we had to sum 40,000 terms, where, in contrast, the numerical solution to follow required only several hundred evaluations.

Figure 21.2 The analytic (Fourier series) solution of Laplace's equation summing 21 terms. Gibbs-overshoot leads to the oscillations near $x = 0$, and persist even if a larger number of terms are summed over.

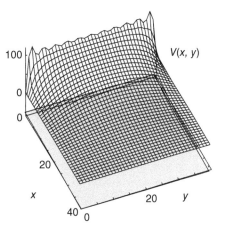

21.3 Finite-Difference Algorithm

To solve our 2D PDE numerically, we divide space into a lattice (Figure 21.3), and look for the solution U only on the lattice sites. Expressing derivatives in terms of the finite differences in the values of U at the lattice sites, is called a *finite-difference* method. A numerically more efficient method, but with more complicated set up, is the *finite-element* method (FEM), which solves the PDE for small geometric elements, and then matches the solutions from all of the elements. We discuss FEM in Chapter 27.

To derive the finite-difference algorithm for the numeric solution of (21.5), we take the same approach that we used in Section 5.1 to derive the forward-difference algorithm for differentiation. We start by adding the two Taylor expansions of the potential at the right and left of (x, y), and the two for above and below (x, y):

$$U(x + \Delta x, y) = U(x, y) + \frac{\partial U}{\partial x} \Delta x + \frac{1}{2} \frac{\partial^2 U}{\partial x^2} (\Delta x)^2 + \cdots, \tag{21.20}$$

$$U(x - \Delta x, y) = U(x, y) - \frac{\partial U}{\partial x} \Delta x + \frac{1}{2} \frac{\partial^2 U}{\partial x^2} (\Delta x)^2 - \cdots. \tag{21.21}$$

$$U(x, y + \Delta y) = U(x, y) + \frac{\partial U}{\partial y} \Delta y + \frac{1}{2} \frac{\partial^2 U}{\partial y^2} (\Delta y)^2 + \cdots, \tag{21.22}$$

$$U(x, y - \Delta y) = U(x, y) - \frac{\partial U}{\partial y} \Delta y + \frac{1}{2} \frac{\partial^2 U}{\partial y^2} (\Delta y)^2 - \cdots. \tag{21.23}$$

All odd terms cancel when we add these equations in pairs, and we obtain a central-difference approximation for the second partial derivative good to order Δ^4:

$$\frac{\partial^2 U(x, y)}{\partial x^2} \simeq \frac{U(x + \Delta x, y) + U(x - \Delta x, y) - 2U(x, y)}{(\Delta x)^2}, \tag{21.24}$$

$$\frac{\partial^2 U(x, y)}{\partial y^2} \simeq \frac{U(x, y + \Delta y) + U(x, y - \Delta y) - 2U(x, y)}{(\Delta y)^2}. \tag{21.25}$$

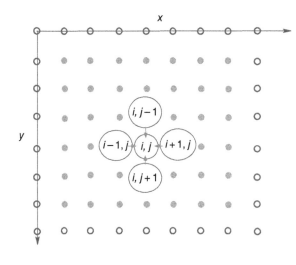

Figure 21.3 The lattice and algorithm for Laplace's equation. The potential at the point $(x, y) = (i, j)\Delta$ equals the average of the potential values at the four nearest neighbor points. The nodes with white centers correspond to fixed values of the potential along the boundaries.

Substitution of these approximations in Poisson's equation (21.6) produces the finite-difference form of the PDE:

$$\frac{U(x+\Delta x, y) + U(x-\Delta x, y) - 2U(x,y)}{(\Delta x)^2} \tag{21.26}$$

$$+ \frac{U(x, y+\Delta y) + U(x, y-\Delta y) - 2U(x,y)}{(\Delta y)^2} = -4\pi\rho. \tag{21.27}$$

If we take the x and y grids to be of equal spacings, $\Delta x = \Delta y = \Delta$, we obtain a simple form for the algorithm:

$$U(x+\Delta, y) + U(x-\Delta, y) + U(x, y+\Delta) + U(x, y-\Delta) - 4U(x,y) = -4\pi\rho. \tag{21.28}$$

The reader will notice that this equation shows a relation among the solutions at five points in space. When $U(x,y)$ is evaluated for the N_x x values on the lattice, and for the N_y y values, we obtain a set of $N_x \times N_y$ simultaneous linear algebraic equations to solve for $\mathtt{U[i,j]}$. One approach is to solve these equations explicitly as a (big) matrix problem. This is attractive as it is a direct solution, but it requires a great deal of memory and accounting.

The approach we use follows from the algebraic solution of (21.28) for $U(x,y)$:

$$4U(x,y) \simeq U(x+\Delta, y) + U(x-\Delta, y) + U(x, y+\Delta) + U(x, y-\Delta) + 4\pi\rho(x,y)\Delta^2, \tag{21.29}$$

where we would omit the $\rho(x)$ term for Laplace's equation. In terms of discrete locations on our lattice, the x and y variables are:

$$x = x_0 + i\Delta, \quad y = y_0 + j\Delta, \quad i, j = 0, \dots, N_{\max - 1}, \tag{21.30}$$

where we have placed our lattice in the square of side L. The finite-difference algorithm (21.29) becomes,

$$U_{i,j} = \tfrac{1}{4}\left[U_{i+1,j} + U_{i-1,j} + U_{i,j+1} + U_{i,j-1}\right] + \pi\rho(i\Delta, j\Delta)\Delta^2. \tag{21.31}$$

This equation says that when we have a proper solution, it will be the average of the potential at the four nearest neighbors in Figure 21.3, plus a contribution from the local charge density. As an algorithm, (21.31) does not provide a direct solution to Poisson's equation, but rather must be repeated many times to converge upon the solution. We start with an initial guess for the potential, improve it by sweeping through all space, taking the average over nearest neighbors at each node. We keep repeating the process until the solution no longer changes, at least to some level of precision, or until failure to converge is evident. When converged, the initial guess is said to have *relaxed* into the solution, and it does not matter what that guess may have been.

A reasonable question with this simple an approach is, "Does it always converge, and if so, does it converge fast enough to be useful?" In some sense the answer to the first question is not an issue; if the method does not converge, then we will know it; otherwise we have ended up with a solution, and the path we followed to get there is nobody's business! The answer to the question of speed is that relaxation methods may converge slowly (although still faster than a Fourier series), yet we will show you two clever tricks to accelerate the convergence.

At this point, it is important to remember that our algorithm arose from expressing the Laplacian ∇^2 in rectangular coordinates. While this does not restrict us from solving problems with circular symmetry, there may be geometries where it is better to develop an algorithm based on expressing the Laplacian in cylindrical or spherical coordinates in order to have grids that fit the geometry better.

21.3.1 Relaxation and Overrelaxation

There are a number of ways in which the algorithm (21.29) can be used to turn the boundary conditions into a solution. The most basic approach is the *Jacobi method*, in which the potential values are not changed until (21.29) is applied at each point on the lattice. This maintains the symmetry of the initial guess and boundary conditions. A rather obvious improvement on the Jacobi method is the *Gauss-Seidel method*, in which the updated guesses for the potential in (21.29) are used as soon as they have been computed. As a case in point, if the sweep starts in the upper-left-hand corner of Figure 21.3, then the leftmost `u([-1,j]` and topmost `u[i,j-1]` values of the potential used will be from the present generation of guesses, while the other two values of the potential will be from the previous generation:

$$U_{i,j}^{(\text{new})} = \tfrac{1}{4}\left[U_{i+1,j}^{(\text{old})} + U_{i-1,j}^{(\text{new})} + U_{i,j+1}^{(\text{old})} + U_{i,j-1}^{(\text{new})}\right]. \tag{21.32}$$

The *Gauss-Seidel method* usually leads to accelerated convergence, which, in turn, leads to less round-off errors. It also uses less memory as there is no need to store two generations of guesses. However, it does distort the symmetry of the boundary conditions, which one hopes is insignificant when convergence is reached.

A less obvious improvement in the relaxation technique, known as *successive overrelaxation* (SOR), starts by writing the algorithm (21.29) in a form that determines the new values of the potential $U^{(\text{new})}$ as the old values $U^{(\text{old})}$ plus a correction, or residual r:

$$U_{i,j}^{(\text{new})} = U_{i,j}^{(\text{old})} + r_{i,j}. \tag{21.33}$$

We rewrite the Gauss-Seidel technique here in the general form:

$$
\begin{aligned}
r_{i,j} &\overset{\text{def}}{=} U_{i,j}^{(\text{new})} - U_{i,j}^{(\text{old})} \\
&= \tfrac{1}{4}\left[U_{i+1,j}^{(\text{old})} + U_{i-1,j}^{(\text{new})} + U_{i,j+1}^{(\text{old})} + U_{i,j-1}^{(\text{new})}\right] - U_{i,j}^{(\text{old})}.
\end{aligned} \tag{21.34}
$$

The successive overrelaxation technique supposes that if convergence is obtained by adding r to U, then even more rapid convergence might be obtained by adding more or less of r [Press *et al.*, 2007; Garcia, 2000]:

$$U_{i,j}^{(\text{new})} = U_{i,j}^{(\text{old})} + \omega r_{i,j}, \quad (\text{SOR}), \tag{21.35}$$

where ω is a parameter that amplifies or reduces the residual. The nonaccelerated relaxation algorithm (21.32) corresponds to $\omega = 1$, accelerated convergence (overrelaxation) to $\omega \geq 1$, and underrelaxation to $\omega < 1$. Values of $1 \leq \omega \leq 2$ often works well, with $\omega > 2$ sometimes leading to numerical instabilities. Although a detailed analysis of the algorithm is needed to predict the optimal value for ω for a particular problem, we suggest a trial-and-error approach to see what works best.

21.4 Alternate Capacitor Problems

We give you a choice now. You can carry out the assessment using our wire-plus-grounded-box problem, or you can replace that problem with a more interesting one, involving a realistic capacitor, or nonplanar capacitors.

Elementary textbooks solve the capacitor problem for the uniform field confined between two infinite parallel plates. However, the field in a realistic (finite) capacitor varies near the edges (edge effects) and extends beyond the edges (fringe fields). We model the realistic capacitor in a grounded box (Figure 21.4) as two conducting plates (or wires) of finite length and width. Write your simulation such that it is convenient to vary the grid spacing Δ and the geometry of the box and plate. We pose three versions of this problem, each displaying somewhat different physics. In each case, the boundary condition $U = 0$ on the surrounding box must be imposed in order to obtain a unique solution.

1) For the simplest version, assume that the plates are very thin conductive sheets, with the top sheet maintained at 100 V, and the bottom at -100 V. Because the sheets are conductors, they must be equipotential surfaces, and so a battery could maintain them at these constant voltages. Write or modify the given program to solve Laplace's equation with fixed voltage plates.

2) For the next version of this problem, assume that the plates are composed of a line of dielectric material with uniform charge densities ρ on the top, and $-\rho$ on the bottom. Solve Poisson's equation (21.3) in the region including the plates, and Laplace's equation elsewhere. Experiment until you find a numerical value for ρ that gives a potential similar to that shown in Figure 21.5 for plates with fixed voltages.

3) For the final version of this problem, investigate how the charges on a capacitor with finite-thickness conducting plates (Figure 21.6) distribute themselves. Because the plates are conductors, they are still equipotential surfaces at 100 and -100 V, only now you should make them have a thickness of at least 2Δ (so we can see the difference between the potential near the top and the bottom surfaces of the plates). Such being the case, solve Laplace's equation (21.4) to determine $U(x,y)$. Once we have $U(x,y)$, substitute it into Poisson's equation (21.3), and determine how the charge density

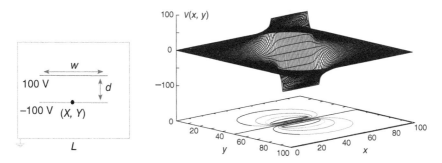

Figure 21.4 *Left*: A simple model of a parallel-plate capacitor within a box. A realistic model would have the plates close together, in order to condense the field, and the enclosing grounded box would be so large that it has no effect on the field near the capacitor. *Right*: A numerical solution for the electric potential for this geometry. The projection on the xy plane gives the equipotential lines.

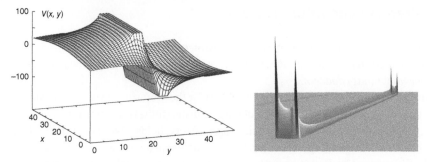

Figure 21.5 *Left*: A visualization of the computed electric potential for a capacitor with finite width plates. *Right*: A visualization of the charge distribution along one plate, determined by evaluating $\nabla^2 U(x, y)$ (courtesy of J. Wetzel). Note the "lightening rod" effect of charge accumulating at corners and points.

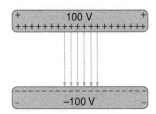

Figure 21.6 A guess as to how charge may rearrange itself on finite conducting plates.

distributes itself along the top and bottom surfaces of the plates. *Hint*: As the electric field is no longer uniform, we know that the charge distribution also will no longer be uniform. In addition, because the electric field now extends beyond the ends of the capacitor, and because field lines begin and end on charge, some charge may end up on the edges and outer surfaces of the plates (Figure 21.6).

4) The numerical solution to our PDE can be applied to arbitrary boundary conditions. The two boundary conditions to explore are triangular and sinusoidal:

$$U(x) = \begin{cases} 200x/w, & x \leq w/2, \\ 100(1 - x/w), & x \geq w/2, \end{cases} \quad \text{or} \quad U(x) = 100 \sin\left(\frac{2\pi x}{w}\right).$$

5) **Square conductors:** You have designed a piece of equipment consisting of a small metal box at 100 V within a larger grounded one (Figure 21.7). You find that sparking occurs between the boxes, which means that the electric field is too large. You need to determine where the field is greatest so that you can change the geometry and eliminate the sparking. Modify the program to satisfy these boundary conditions, and determine the field between the boxes. Gauss's law tells us that the field vanishes within the inner box because it contains no charge. Plot the potential and equipotential surfaces, and sketch in the electric field lines. Deduce where the electric field is most intense, and then redesign the equipment to reduce the field.

6) **Cracked cylindrical capacitor:** You have designed the cylindrical capacitor containing a long, outer cylinder surrounding a thin, inner cylinder (Figure 21.7 right). The cylinders have a small crack in them in order to connect them to the battery that maintains the inner cylinder at −100 V, and the outer cylinder at 100 V. Determine how this small crack affects the field configuration. In order for a unique solution to exist, place both cylinders within a large, grounded box.

Figure 21.7 *Left*: The geometry of a capacitor formed by placing two long, square cylinders within each other. *Right*: The geometry of a capacitor formed by placing two long, circular cylinders within each other. The cylinders are cracked on the side so that wires can enter the region.

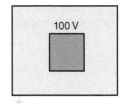 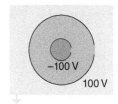

21.4.1 Implementation

In Listing 21.1, we present the code `LaplaceLine.py` that solves the square-wire problem (Figure 21.1) and produces the visualization there. Here, we have kept the code simple by setting the length of the box $L = N_{max} \Delta = 100$, and by taking $\Delta = 1$:

$$U(i, N_{max}) = 99 \quad \text{(top)}, \qquad U(1, j) = 0 \quad \text{(left)},$$
$$U(N_{max}, j) = 0 \quad \text{(right)}, \qquad U(i, 1) = 0 \quad \text{(bottom)}. \tag{21.36}$$

1) Write or modify `LaplaceLine.py` to find the electric potential for your choice of capacitor.
2) Start by having your program undertake 1000 iterations. Examine how the potential changes in some key locations as you iterate toward a solution.
3) Repeat the process for different step sizes Δ, and draw conclusions regarding the stability and accuracy of the solution. Keep in mind that this is a simple algorithm, and so may require many iterations for high precision.
4) Once your program produces reliable solutions, modify it so that it stops iterating once convergence is reached, or if the number of iterations becomes too large. Rather than trying to discern small changes in highly compressed surface plots, use a numerical measure of precision, for example, `trace` $= \sum_i |U[i,i]|$, that samples the solution along the diagonal. You should be able to obtain changes in the trace that are less than 1 part in 10^4. The `break` command or a `while` loop is useful for this type of test.
5) Equation (21.35) expresses the successive overrelaxation technique in which convergence is accelerated by using a judicious choice of ω. Determine by trial and error the best value of ω. This should let you double the speed of the algorithm.
6) Now that your code is accurate, modify it to simulate a more realistic capacitor in which the plate separation is approximately $\frac{1}{10}$ of the plate length. You should find the field more condensed and more uniform between the plates.
7) If you are working with the wire-in-the-box problem, compare your numerical solution to the analytic one (21.18). Do not be surprised if you need to sum thousands of terms before the analytic solution converges!

21.5 Electric Field Visualization

Create a 2D plot of the equipotential surfaces. You may want to start with a crude, hand- (or mouse-) drawn sketch of the electric field as curves orthogonal to the equipotential lines. We can do better than that! Because

$$\mathbf{E} = -\nabla U(x, y) = -\frac{\partial U(x, y)}{\partial x} \hat{\epsilon}_x - \frac{\partial U(x, y)}{\partial y} \hat{\epsilon}_y, \tag{21.37}$$

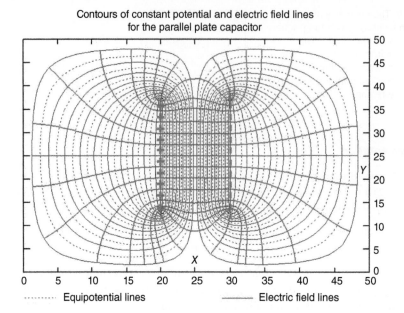

Contours of constant potential and electric field lines
for the parallel plate capacitor

········· Equipotential lines ——— Electric field lines

Figure 21.8 Computed equipotential surfaces and electric field lines for a realistic capacitor.

it is simple to calculate the field. Therefore, just use the central-difference approximation for the derivative to determine the field, for example:

$$E_x \simeq \frac{U(x+\Delta,y) - U(x-\Delta,y)}{2\Delta} = \frac{U_{i+1,j} - U_{i-1,j}}{2\Delta}. \qquad (21.38)$$

Once you have a data file containing the vector field, it can be visualized by plotting arrows of varying lengths and directions, or with just lines as in Figure 21.8.

21.6 Code Listings

Listing 21.1 LaplaceLine.py solves Laplace's equation via relaxation. Various parameters should be adjusted for an accurate solution.

```
# LaplaceLine.py: Matplotlib, Solve Laplace's eqtn in square

import matplotlib.pylab as p, numpy
from mpl_toolkits.mplot3d import Axes3D;   from numpy import *;

Nmax = 100; Niter = 50
V = zeros((Nmax, Nmax), float)
print ("Working hard, wait for the figure while I count to 60")

for k in range(0, Nmax-1):  V[0,k] = 100.0       # Line at 100V
for iter in range(Niter):
    if iter%10 == 0: print(iter)
    for i in range(1, Nmax-2):
        for j in range(1,Nmax-2):
            V[i,j] = 0.25*(V[i+1,j]+V[i-1,j]+V[i,j+1]+V[i,j-1])
    print ("iter, V[Nmax/5,Nmax/5]", iter, V[Nmax/5,Nmax/5])
```

```
17  x = range(0, 50, 2);  y = range(0, 50, 2)
    X, Y = p.meshgrid(x,y)

    def functz(V):                                          # V(x, y)
21      z = V[X,Y]
        return z

    Z = functz(V)
25  fig = p.figure()                            # Create figure
    ax = Axes3D(fig)                                # Plot axes
    ax.plot_wireframe(X, Y, Z, color = 'r')      # Red wireframe
    ax.set_xlabel('X');  ax.set_ylabel('Y');  ax.set_zlabel('V(x,y)')
29  ax.set_title('Potential within Square V(x=0)=100V (Rotatable)')
    p.show()                                        # Show fig
```

22

Heat Flow and Leapfrogging

This chapter introduces the time-stepping (leapfrog) method for solving a PDE on a space-time lattice. We use it, and the more precise Crank–Nicolson algorithm, to solve the heat equation. Time-stepping is simple, yet powerful, and we will use it again and again when solving wave equations.

Problem   You are given an aluminum bar of length $L = 1$ m and width w that is initially at 100 °C (Figure 22.1). Determine how the temperature varies along the length of the bar, after the ends are placed in ice water at 0 °C. Assume that the length of the bar is insulated, but not its ends.

22.1   The Parabolic Heat Equation

It's a basic fact of nature that heat flows from hot to cold, that is, from regions of high temperature to regions of low temperature. We express this mathematically by stating that the rate of the vector heat flow $\mathbf{H}$, through a material, is proportional to the gradient of the temperature T across the material:

$$\mathbf{H} = -K\,\nabla T(\mathbf{x}, t), \qquad (22.1)$$

where K is the thermal conductivity of the material. The total amount of heat $Q(t)$ in the material, at any one time, is proportional to the integral of the temperature over the material's volume:

$$Q(t) = \int d\mathbf{x}\, C\, \rho(\mathbf{x})\, T(\mathbf{x}, t), \qquad (22.2)$$

where C is the specific heat of the material, and ρ is its density. Because energy is conserved, the rate of decrease of Q, with time, must equal the amount of heat flowing out of the material. Applying energy balance leads to the *heat equation*:

$$\frac{\partial T(\mathbf{x}, t)}{\partial t} = \frac{K}{C\rho}\,\nabla^2 T(\mathbf{x}, t). \qquad (22.3)$$

Computational Physics: Problem Solving with Python, Fourth Edition.
Rubin H. Landau, Manuel J. Páez, and Cristian C. Bordeianu.
© 2024 WILEY-VCH GmbH. Published 2024 by WILEY-VCH GmbH.

Figure 22.1 A metallic bar insulated along its length with its ends in contact with ice. The bar is dark and the insulation is of lighter color.

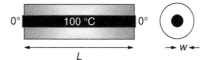

The heat equation (22.3) is a *parabolic* PDE with space and time as independent variables. The specification of this problem implies that there is no temperature variation in directions perpendicular to the bar, and so for our problem, we need to only consider one spatial coordinate, x, along the length of the bar:

$$\frac{\partial T(x,t)}{\partial t} = \frac{K}{C\rho} \frac{\partial^2 T(x,t)}{\partial x^2}. \tag{22.4}$$

As given, the initial temperature of the bar and the boundary conditions are:

$$T(x, t = 0) = 100\,^\circ\text{C}, \quad T(x = 0, t) = T(x = L, t) \equiv 0\,^\circ\text{C}. \tag{22.5}$$

22.1.1 Solution as Analytic Expansion

Analogous to Laplace's equation, the analytic solution starts with the assumption that the solution separates into the product of functions of space and time:

$$T(x, t) = X(x)\mathcal{T}(t). \tag{22.6}$$

When (22.6) is substituted into the heat equation (22.4), and the resulting equation is divided by $X(x)\mathcal{T}(t)$, two noncoupled ODE's result:

$$\frac{d^2X(x)}{dx^2} + k^2 X(x) = 0, \quad \frac{d\mathcal{T}(t)}{dt} + k^2 \frac{C}{C\rho} \mathcal{T}(t) = 0, \tag{22.7}$$

where k is a constant still to be determined. The boundary condition that the temperature equals zero at $x = 0$ requires a sine function for X:

$$X(x) = A \sin kx. \tag{22.8}$$

The boundary condition that the temperature equals zero at $x = L$ requires the sine function to vanish there, and so:

$$\sin kL = 0 \Rightarrow k = k_n = n\pi/L, \quad n = 1, 2, \ldots. \tag{22.9}$$

The function of time can be a growing or decaying exponential. To be physically reasonable (not blow up), it must be a decaying exponential with k in the exponent:

$$\mathcal{T}(t) = e^{-k_n^2 t/(C\rho)}, \tag{22.10}$$

$$\Rightarrow T(x, t) = A_n \sin k_n x e^{-k_n^2 t/(C\rho)}, \tag{22.11}$$

where n may be any integer, and A_n is an arbitrary constant. Since (22.4) is a linear equation, the most general solution is a linear superposition of $X_n(x)T_n(t)$ products for all values of n:

$$T(x, t) = \sum_{n=1}^{\infty} A_n \sin k_n x\, e^{-k_n^2 t/(C\rho)}. \tag{22.12}$$

The coefficients A_n are determined by the initial condition that at time $t = 0$, the entire bar has temperature $T = 100\,°C$:

$$T(x, t = 0) = 100 \Rightarrow \sum_{n=1}^{\infty} A_n \sin k_n x = 100. \tag{22.13}$$

Projecting the sine functions determines $A_n = 4T_0/n\pi$ for n odd, and so:

$$T(x, t) = \sum_{n=1,3,\dots}^{\infty} \frac{4T_0}{n\pi} \sin k_n x e^{-k_n^2 Kt/(C\rho)}. \tag{22.14}$$

22.2 Time Stepping (Leapfrog) Algorithm

As we did with Laplace's equation, the numerical solution is based on converting the differential equation to a finite difference (or just "difference") equation. We discretize space and time on the lattice in Figure 22.2, and seek solutions on the lattice sites. The horizontal nodes with white centers correspond to the known values of the temperature for the initial time, while the vertical white nodes correspond to the fixed temperature along the boundaries. If we *also* knew the temperature for times along the bottom row, then we could use a relaxation algorithm, as we did for Laplace's equation. However, with only the top row known, we shall end up with an algorithm that steps forward in time one row at a time, as in the children's game *leapfrog*.

As is often the case with PDEs, the algorithm is customized for the equation being solved, and for the constraints imposed by the particular set of initial and boundary conditions. With only one row of times to start with, we use a forward-difference approximation for the time derivative of the temperature:

$$\frac{\partial T(x, t)}{\partial t} \simeq \frac{T(x, t + \Delta t) - T(x, t)}{\Delta t}. \tag{22.15}$$

Because we know the spatial variation of the temperature along the entire top row, and the left and right sides, we are less constrained with the space derivative than with the time derivative. Consequently, as we did with the Laplace equation, we use the more accurate central-difference approximation for the space derivative:

$$\frac{\partial^2 T(x, t)}{\partial x^2} \simeq \frac{T(x + \Delta x, t) + T(x - \Delta x, t) - 2T(x, t)}{(\Delta x)^2}. \tag{22.16}$$

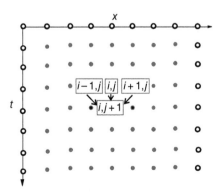

Figure 22.2 The algorithm for the heat equation in which the temperature, at the location $x = i\Delta x$ and time $t = (j + 1)\Delta t$, is computed from the temperature values at three points from an earlier time. The nodes with white centers correspond to known initial and boundary conditions. (The boundaries are placed artificially close for illustrative purposes.)

Figure 22.3 A visualization of a numerical calculation of the temperature *versus* position and *versus* time.

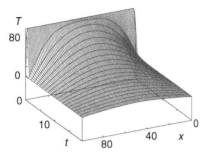

Substitution of these approximations into (22.4) yields the heat difference equation:

$$\frac{T(x, t + \Delta t) - T(x, t)}{\Delta t} = \frac{K}{C\rho} \frac{T(x + \Delta x, t) + T(x - \Delta x, t) - 2T(x, t)}{\Delta x^2}. \tag{22.17}$$

We reorder (22.17) into a form in which T can be stepped forward in t:

$$T_{i,j+1} = T_{i,j} + \eta \left[T_{i+1,j} + T_{i-1,j} - 2T_{i,j} \right], \qquad \eta = \frac{K\Delta t}{C\rho \Delta x^2}. \tag{22.18}$$

Here, $x = i\Delta x$ and $t = j\Delta t$. This algorithm is *explicit* because it provides a solution in terms of known values of the temperature. If we wanted to solve for the temperature at all lattice sites in Figure 22.2 simultaneously, then we would be using an *implicit* algorithm. With a time stepping algorithm, we need to keep track of only four temperatures. As indicated in Figure 22.2, the temperature at space-time point $(i, j + 1)$ is computed from the three temperature values at an earlier time j, and at adjacent space values $i \pm 1, i$. We start the solution at the top row, moving it forward in time for as long as we want, always keeping the temperature at the end fixed at 0 K. Figure 22.3 shows of the time and space dependence of the solution.

22.2.1 Von Neumann Stability Condition

When the difference-equation version of a PDE is solved, the hope is that its solution is a good approximation to the solution of the PDE. If the solution to the difference-equation diverges, then we don't have any solution; but if it does converge, then we should feel confident that we have a good solution to the PDE.[1] The *von Neumann stability analysis* tell us what we need do to get a good solution [Press *et al.*, 2007; Courant *et al.*, 1928]. The analysis is based on the assumption that after the jth time-step, the approximate solution has the form:

$$T_{i,j} = \xi(k)^j \, e^{Iki\Delta x}, \tag{22.19}$$

where $x = i\Delta x$, $t = j\Delta t$, and $I = \sqrt{-1}$ is the imaginary number. The constant k in (23.24) is an unknown wave vector $(2\pi/\lambda)$, and $\xi(k)$ is an unknown complex function. We view (22.19) as a function that oscillates in space (the exponential), with the amplitude or *amplification factor* $\xi(k)^j$ that gets multiplied by the power of ξ for each time step. Stability of the solution then requires $|\xi(k)| < 1$, else the solution would grow in time [Press *et al.*, 2007;

---

1 Well, in the case of shock waves, as in Chapter 25, divergences may not be such a bad thing.

Ancona, 2002]. To solve for the amplitude, we substitute (22.19) into the difference equation (22.18):

$$\xi^{j+1}e^{ikm\Delta x} = \xi^j e^{ikm\Delta x} + \eta\left[\xi^j e^{ik(m+1)\Delta x} + \xi^j e^{ik(m-1)\Delta x} - 2\xi^j e^{ikm\Delta x}\right].$$

After canceling a common factor, it is easy to solve for $\xi(k)$:

$$\xi(k) = 1 + 2\eta[\cos(k\Delta x) - 1].\tag{22.20}$$

In order for $|\xi(k)| < 1$ for all possible k values, we must have:

$$\eta = \frac{K\,\Delta t}{C\rho\,\Delta x^2} < \frac{1}{2}.\tag{22.21}$$

This equation tells us that if we make the time step Δt smaller, we will always improve the stability, as one would expect. But if we make the space step Δx smaller, without a concordant quadratic *increase* in the time step, we will worsen the stability.

22.2.2 Implementation

Recall that we want to solve for the temperature distribution within an aluminum bar of length $L = 1$ m, subject to the boundary and initial conditions:

$$T(x = 0, t) = T(x = L, t) = 0\,^\circ\text{C}, \qquad T(x, t = 0) = 100\,^\circ\text{C}.\tag{22.22}$$

The thermal conductivity, specific heat, and density for Al are:

$$K = 237\,\text{W/(mK)}, \qquad C = 900\,\text{J/(kg K)}, \qquad \rho = 2700\,\text{kg/m}^3.\tag{22.23}$$

1) Write a program, or modify `EqHeat.py` in Listing 22.1, to solve the heat equation.
2) Define a 2D array `T[101,2]` for the temperature as a function of space and time. The first index is for the 100 space divisions of the bar, and the second index is for present and past times (because you may have to make thousands of time steps, you save memory by saving only two times).
3) For time $t = 0$ ($j = 1$), initialize `T` so that all points on the bar, except the ends, are at 100. Set the temperature of the ends to 0.
4) Apply (22.15) to obtain the temperature at the next time step.
5) Assign the present time values of the temperature to the past values: `T[i,1] = T[i,2]`, `i = 1,..., 101`.
6) Start with 50 time steps. Once you are confident the program is running properly, use thousands of steps to see the bar cool smoothly with time. For approximately every 500 time steps, print the time and temperature along the bar.

22.2.3 Assessment and Visualization

1) Check that your program gives a temperature distribution that varies smoothly with time, that satisfies the boundary conditions, and that reaches equilibrium. You may have to vary the time and space steps to obtain stable solutions.
2) Compare the analytic and numeric solutions (and the wall times needed to compute them). If the solutions differ, suspect the one that does not appear smooth and continuous.
3) Make a surface plot of temperature *versus* position and *versus* time.

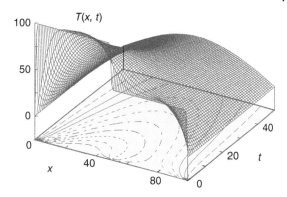

Figure 22.4 Temperature *versus* position and time when two bars at differing temperatures are placed in contact at $t = 0$. The projected contours show the isotherms.

4) Plot the *isotherms* (contours of constant temperature).
5) Create an animation that shows the temperature of the entire bar as a function of time. Our code `EqHeatAnimate.py` in the online codes directory does that with the Visual package.
6) **Stability test:** Verify the stability condition (22.21) by observing how the temperature distribution diverges if $\eta > \frac{1}{4}$.
7) **Material dependence:** Repeat the calculation for iron. Note that the stability condition requires you to change the size of the time step.
8) **Two bars in contact:** Two identical bars, 0.25 m long, are placed in end-to-end contact, with their other ends kept at $0\,°C$. One is initially at $100\,°C$, and the other at $50\,°C$. Determine how the temperature varies with time and location (Figure 22.4).

22.3 Newton's Radiative Cooling

Imagine now, a bar in contact with an environment at a temperature T_e. Newton's law of cooling gives the rate of temperature change, as a result of radiation, as:

$$\frac{\partial T}{\partial t} = -h(T - T_e), \tag{22.24}$$

with h a positive constant. Including radiative cooling and leaving off the constant modifies the heat equation to:

$$\frac{\partial T(x, t)}{\partial t} = \frac{K}{C\rho} \frac{\partial^2 T}{\partial^2 x} - hT(x, t). \tag{22.25}$$

1) Modify the heat equation algorithm to include Newton's cooling.
2) Compare the cooling of a radiating bar with that of the insulated bar.
3) Solve for conductive and radiative heat flow within a 2D rectangular iron plate which is initially at a temperature of $100\,°C$. Three sides are maintained at $100\,°C$, while the top is placed in contact with ice water at $0\,°C$.
   a) Make separate surface plots of $T(x, y = \text{fixed}, t)$ and $T(x = \text{fixed}, y, t)$, or a series of plots of $T(x, y, t = \text{fixed})$ for various t values.
   b) After an equilibrium is reached, make a plot of the isotherms $T(x, y, t = \infty)$.

4) Sometimes, an equilibrium is reached in which the temperature no longer changes as a function of time. In this case, the heat equation (22.3) takes the form:

$$\nabla^2 T(\mathbf{x}, t) = 0. \tag{22.26}$$

This is the same as Laplace's equation, which is studied in Section 21.4.1, and can be solved using the same *relaxation* algorithm developed there. And so, solve for isotherms within a 2D rectangular iron plate in which three sides are maintained at $100\,°C$, while the fourth side, remaining in contact with ice water, at $0\,°C$.

5) Compute the rate of heat flow from the center to the surface of a sphere of constant thermal conductivity. The center of the sphere is kept at $0\,°C$ and the surface is kept at $100\,°C$.

6) A sphere with constant thermal conductivity is initially at $0\,°C$, and is then placed in a heat bath of $100\,°C$. Compute the temperature profile of the sphere as a function of time.

7) A composite sphere is composed of a material of high thermal conductivity up to its middle, and low conductivity from its middle to its outside. Compute the rate of heat flow from the center to the surface of a sphere, if the center is kept at $0\,°C$ and the surface is kept at $100\,°C$.

22.4 The Crank–Nicolson Algorithm

The Crank–Nicolson algorithm provides a higher degree of precision for the heat equation (22.3) than the simple leapfrog method [Crank and Nicolson, 1946]. The algorithm calculates the time derivative with a central-difference approximation, in contrast to the forward-difference approximation used previously. In order to avoid introducing error for the initial time step, where only a single time value is known, the method uses a *split time step*,[2] so that time is advanced from time t to $t + \Delta t/2$:

$$\frac{\partial T}{\partial t}\left(x, t + \tfrac{\Delta t}{2}\right) \simeq \frac{T(x, t + \Delta t) - T(x, t)}{\Delta t} + O(\Delta t^2). \tag{22.27}$$

Yes, we know that this looks just like the forward-difference approximation for the derivative at time $t + \Delta t$, for which it would be a bad approximation; regardless, it is a better, but more complicated, approximation for the derivative at time $t + \Delta t/2$. Likewise, in (22.15), we gave the central-difference approximation for the second space derivative for time t. For $t = t + \Delta t/2$, that becomes:

$$2(\Delta x)^2 \frac{\partial^2 T}{\partial x^2}\left(x, t + \frac{\Delta t}{2}\right) \simeq [T(x - \Delta x,\ t) - 2T(x,\ t) + T(x + \Delta x,\ t)]$$
$$+ [T(x - \Delta x,\ t + \Delta t) - 2T(x,\ t + \Delta t) + T(x + \Delta x,\ t + \Delta t)] + O(\Delta x^2).$$

In terms of these expressions, the heat difference equation is:

$$T_{i,j+1} - T_{i,j} = \tfrac{\eta}{2}\left[T_{i-1,j+1} - 2T_{i,j+1} + T_{i+1,j+1} + T_{i-1,j} - 2T_{i,j} + T_{i+1,j}\right],$$
$$x = i\Delta x,\ t = j\Delta t,\ \eta = \tfrac{K\Delta t}{C\rho\,\Delta x^2}. \tag{22.28}$$

---

2 In Chapter 24, we develop split-time algorithms for solution to the Schrödinger equation and Maxwell's equations.

We group together terms involving the same temperature, to obtain an equation with future times on the LHS and present times on the RHS:

$$-T_{i-1,j+1} + \left(\frac{2}{\eta} + 2\right) T_{i,j+1} - T_{i+1,j+1} = T_{i-1,j} + \left(\frac{2}{\eta} - 2\right) T_{i,j} + T_{i+1,j}. \tag{22.29}$$

This equation represents an *implicit* scheme for the temperature $T_{i,j}$, where "implicit" means that we must solve simultaneous equations to obtain the solution at current (j) and future ($j+1$) times. In contrast, an *explicit* scheme uses the solution at current and past times to obtain it at future times. We start with the initial temperature distribution throughout all of space, the boundary conditions at the ends of the bar for all times, and the approximate values from the first derivative:

$$\begin{array}{llll} T_{i,0}, & \text{(known)}, & T_{0,j}, & \text{(known)}, & T_{N,j}, & \text{(known)}, \\ T_{0,j+1} = T_{0,j} = 0, & T_{N,j+1} = 0, & T_{N,j} = 0. \end{array} \tag{22.30}$$

We rearrange (22.29) so that we can use these known values of T to step the $j = 0$ solution forward in time, and express it as a set of simultaneous linear equations:

$$
\begin{bmatrix}
\left(\frac{2}{\eta}+2\right) & -1 & & & & \\
-1 & \left(\frac{2}{\eta}+2\right) & -1 & & & \\
& -1 & \left(\frac{2}{\eta}+2\right) & -1 & & \\
& & \ddots & \ddots & \ddots & \\
& & & -1 & \left(\frac{2}{\eta}+2\right) & -1 \\
& & & & -1 & \left(\frac{2}{\eta}+2\right)
\end{bmatrix}
\begin{bmatrix}
T_{1,j+1} \\
T_{2,j+1} \\
T_{3,j+1} \\
\vdots \\
T_{n-2,j+1} \\
T_{n-1,j+1}
\end{bmatrix}
$$

$$
=
\begin{bmatrix}
T_{0,j+1} + T_{0,j} + \left(\frac{2}{\eta} - 2\right) T_{1,j} + T_{2,j} \\
T_{1,j} + \left(\frac{2}{\eta} - 2\right) T_{2,j} + T_{3,j} \\
T_{2,j} + \left(\frac{2}{\eta} - 2\right) T_{3,j} + T_{4,j} \\
\vdots \\
T_{n-3,j} + \left(\frac{2}{\eta} - 2\right) T_{n-2,j} + T_{n-1,j} \\
T_{n-2,j} + \left(\frac{2}{\eta} - 2\right) T_{n-1,j} + T_{n,j} + T_{n,j+1}
\end{bmatrix}. \tag{22.31}
$$

Observe that the T's on the RHS are all at the present time j for various positions, and at future time $j + 1$ for the two ends (whose Ts are known for all times via the boundary conditions). We start the algorithm with the $T_{i,j=0}$ values of the initial conditions, then solve a matrix equation to obtain $T_{i,j=1}$. Once we have that solution, we know all the terms on the RHS of the equations ($j = 1$ throughout the bar, and $j = 2$ at the ends), and so can repeat the solution of the matrix equations to obtain the temperature throughout the bar for $j = 2$. So again, we time-step forward, only now we solve matrix equations at each step. That gives us the spatial solution at all locations simultaneously.

Not only is the Crank-Nicolson method more precise than the low-order time-stepping method, but it also is stable for all values of Δt and Δx. To prove that, we apply the von Neumann stability analysis, discussed in Section 22.2.1 to the Crank-Nicolson algorithm by substituting (22.18) into (22.29). This determines the amplitude:

$$\xi(k) = \frac{1 - 2\eta \sin^2(k\Delta x/2)}{1 + 2\eta \sin^2(k\Delta x/2)}. \tag{22.32}$$

Because the numerator is always smaller than the denominator, $|\xi| \leq 1$ for all Δt, Δx, and k, and so we always have stability.

22.4.1 Solution via Tridiagonal Matrix ⊙

The Crank-Nicolson equations (22.31) are in the standard form, $[A]\mathbf{x} = \mathbf{b}$, for linear equations, and so we can use our matrix methods to solve them. However, the coefficient matrix $[A]$ is tridiagonal (zero elements except for the main diagonal and two diagonals on either side of it):

$$
\begin{pmatrix}
d_1 & c_1 & 0 & 0 & \cdots & \cdots & \cdots & 0 \\
a_2 & d_2 & c_2 & 0 & \cdots & \cdots & \cdots & 0 \\
0 & a_3 & d_3 & c_3 & \cdots & \cdots & \cdots & 0 \\
\cdots & \cdots & \cdots & \cdots & \cdots & \cdots & \cdots & \cdots \\
0 & 0 & 0 & 0 & \cdots & a_{N-1} & d_{N-1} & c_{N-1} \\
0 & 0 & 0 & 0 & \cdots & 0 & a_N & d_N
\end{pmatrix}
\begin{pmatrix}
x_1 \\ x_2 \\ x_3 \\ \ddots \\ x_{N-1} \\ x_N
\end{pmatrix}
=
\begin{pmatrix}
b_1 \\ b_2 \\ b_3 \\ \ddots \\ b_{N-1} \\ b_N
\end{pmatrix},
$$

Consequently, a more robust and faster solution exists, and it makes this implicit method as fast as the explicit ones. Seeing that tridiagonal systems occur frequently, we now outline the specialized technique for solving them [Press *et al.*, 2007]. If we store the matrix elements $a_{i,j}$ using both subscripts, then we will need N^2 locations for elements, and N^2 operations to access them. However, if the matrix is tridiagonal, we only need to store those elements along, above, and below the diagonals, $\{d_i\}_{i=1,N}$, $\{c_i\}_{i=1,N}$, and $\{a_i\}_{i=1,N}$. The single subscripts on a_i, d_i, and c_i reduce the processing from N^2 to $(3N - 2)$ elements.

The solution to the matrix equation manipulates the individual equations until the coefficient matrix is in *upper triangular* form, with all the elements of the main diagonal equal to 1. We start by divide the first equation by d_1, then subtract a_2 times the first equation,

$$
\begin{pmatrix}
1 & \frac{c_1}{d_1} & 0 & 0 & \cdots & \cdots & \cdots & 0 \\
0 & d_2 - \frac{a_2 c_1}{d_1} & c_2 & 0 & \cdots & \cdots & \cdots & 0 \\
0 & a_3 & d_3 & c_3 & \cdots & \cdots & \cdots & 0 \\
\cdots & \cdots & \cdots & \cdots & \cdots & \cdots & \cdots & \cdots \\
0 & 0 & 0 & 0 & \cdots & a_{N-1} & d_{N-1} & c_{N-1} \\
0 & 0 & 0 & 0 & \cdots & 0 & a_N & d_N
\end{pmatrix}
\begin{pmatrix}
x_1 \\ x_2 \\ x_3 \\ \ddots \\ \cdot \\ x_N
\end{pmatrix}
=
\begin{pmatrix}
\frac{b_1}{d_1} \\ b_2 - \frac{a_2 b_1}{d_1} \\ b_3 \\ \ddots \\ \cdot \\ b_N
\end{pmatrix},
$$

and then dividing the second equation by the second diagonal element,

$$
\begin{pmatrix}
1 & \frac{c_1}{d_1} & 0 & 0 & \cdots & \cdots & \cdots & 0 \\
0 & 1 & \frac{c_2}{d_2-a_2\frac{c_1}{a_1}} & 0 & \cdots & & \cdots & 0 \\
0 & a_3 & d_3 & c_3 & \cdots & \cdots & \cdots & 0 \\
\cdots & \cdots & & \cdots & \cdots & \cdots & \cdots & \cdots \\
0 & 0 & 0 & 0 & & a_{N-1} & d_{N-1} & c_{N-1} \\
0 & 0 & 0 & 0 & \cdots & 0 & a_N & d_N
\end{pmatrix}
\begin{pmatrix}
x_1 \\ x_2 \\ x_3 \\ \vdots \\ \cdot \\ x_N
\end{pmatrix}
=
\begin{pmatrix}
\frac{b_1}{d_1} \\ \frac{b_2-a_2\frac{b_1}{d_1}}{d_2-a_2\frac{c_1}{d_1}} \\ b_3 \\ \vdots \\ \cdot \\ b_N
\end{pmatrix}.
$$

Assuming that we can repeat these steps without ever dividing by zero, the system of equations will be reduced to upper triangular form,

$$
\begin{pmatrix}
1 & h_1 & 0 & 0 & \cdots & 0 \\
0 & 1 & h_2 & 0 & \cdots & 0 \\
0 & 0 & 1 & h_3 & \cdots & 0 \\
0 & \cdots & \cdots & \ddots & \ddots & \cdots \\
0 & 0 & 0 & 0 & \cdots & \cdots \\
0 & 0 & 0 & \cdots & 0 & 1
\end{pmatrix}
\begin{pmatrix}
x_1 \\ x_2 \\ x_3 \\ \vdots \\ \cdot \\ x_N
\end{pmatrix}
=
\begin{pmatrix}
p_1 \\ p_2 \\ p_3 \\ \vdots \\ \cdot \\ p_N
\end{pmatrix},
\tag{22.33}
$$

where $h_1 = c_1/d_1$ and $p_1 = b_1/d_1$. We then recur for the other elements:

$$
h_i = \frac{c_i}{d_i - a_i h_{i-1}}, \qquad p_i = \frac{b_i - a_i p_{i-1}}{d_i - a_i h_{i-1}}.
\tag{22.34}
$$

Finally, back substitution leads to the explicit solution for the unknowns:

$$
x_i = p_i - h_i x_{i-1}; \quad i = n-1, n-2, \ldots, 1, \quad x_N = p_N.
\tag{22.35}
$$

In Listing 22.2, we give the program HeatCNTridiag.py that solves the heat equation using the Crank–Nicolson algorithm via a triadiagonal reduction.

22.4.2 Crank–Nicolson Implementation

1) Write a program using the Crank–Nicolson method to solve for heat flow in the metal bar of Section 22.1.
2) Solve the linear system of equations (22.31) using either NumPy or a special tridiagonal algorithm.
3) Check the stability of your solution by choosing different values for the time and space steps.
4) Construct a contoured surface plot of temperature *versus* position and versus time.
5) Compare the implicit and explicit algorithms used in this chapter for relative precision and speed. You may assume that a stable answer that uses very small time steps is accurate.

22.5   Code Listings

Listing 22.1   EqHeat.py solves the single space dimension heat equation on a lattice by leapfrogging the initial conditions forward in time. The parameters should be adjusted.

```python
# EqHeat.py: solves heat equation via finite differences, 3-D plot

from numpy import *; import matplotlib.pylab as p
from mpl_toolkits.mplot3d import Axes3D

Nx = 101;         Nt = 3000;      Dx = 0.03;      Dt = 0.9
kappa = 210.; C = 900.; rho = 2700. # Conductivity, specf heat, density
T = zeros((Nx,2), float);   Tpl = zeros((Nx, 31), float)
print("Working, wait for figure after count to 10")

for ix in range (1, Nx - 1):  T[ix, 0] = 100.0;            # Initial T
T[0,0] = 0.0 ;    T[0,1] = 0.                  # 1st & last T = 0
T[Nx-1,0] = 0. ; T[Nx-1,1] = 0.0
cons = kappa/(C*rho)*Dt/(Dx*Dx);
m = 1                                               # Counter
for t in range (1, Nt):
    for ix in range (1, Nx - 1):
        T[ix, 1] = T[ix, 0] +  cons*(T[ix+1, 0] + T[ix-1, 0] - 2.*T[ix,0])
    if t%300 == 0 or t == 1:                   # Every 300 steps
        for ix in range (1, Nx - 1, 2): Tpl[ix, m] = T[ix, 1]
        print(m)
        m = m + 1
    for ix in range (1, Nx - 1):  T[ix, 0] = T[ix, 1]
x = list(range(1, Nx - 1, 2))                      # Plot alternate pts
y = list(range(1, 30))
X, Y = p.meshgrid(x, y)

def functz(Tpl):
    z = Tpl[X, Y]
    return z

Z = functz(Tpl)
fig = p.figure()                                     # Create figure
ax = Axes3D(fig)
ax.plot_wireframe(X, Y, Z, color = 'r')
ax.set_xlabel('Position')
ax.set_ylabel('time')
ax.set_zlabel('Temperature')
p.show()
print("finished")
```

Listing 22.2   HeatCNTridiag.py solves the heat equation via the Crank-Nicolson method and a tridiagonal matrix algorithm.

```python
# HeatCNTridiag.py:  solution of heat eqtn via CN method

""" Dirichlet boundary conditions surrounding four walls
    Domain dimensions: WxH, with 2 triangles per square
    Based on FEM2DL_Box Matlab program in Polycarpou, Intro to the Finite
    Element Method in Electromagnetics, Morgan & Claypool (2006) """

import matplotlib.pylab as p;
from mpl_toolkits.mplot3d import Axes3D ;
from numpy import *;
import numpy;

Max = 51; n   = 50;    m = 50
Ta  = zeros((Max),float); Tb =zeros((Max),float); Tc = zeros((Max),float)
Td  = zeros((Max),float); a = zeros((Max),float); b = zeros((Max),float)
c   = zeros((Max),float); d = zeros((Max),float); x = zeros((Max),float)
```

```
18 t    = zeros( (Max, Max),float)

   def Tridiag(a, d, c, b, Ta, Td, Tc, Tb, x, n):
       Max = 51
22     h = zeros( (Max), float )
       p = zeros( (Max), float )
       for i in range(1,n+1):
           a[i] = Ta[i]
26         b[i] = Tb[i]
           c[i] = Tc[i]
           d[i] = Td[i]
       h[1] = c[1]/d[1]
30     p[1] = b[1]/d[1]
       for i in range(2,n+1):
           h[i] = c[i] / (d[i]-a[i]*h[i-1])
           p[i] = (b[i] - a[i]*p[i-1]) / (d[i]-a[i]*h[i-1])
34     x[n] = p[n]
       for i in range( n - 1, 1,-1 ): x[i] = p[i] - h[i]*x[i+1]

   width = 1.0; height = 0.1; ct = 1.0
38 for i in range(0, n):    t[i,0]  = 0.0
   for i in range( 1, m):    t[0][i] = 0.0
   h  = width  / ( n - 1 )
   k  = height / ( m - 1 )
42 r  = ct * ct * k / ( h * h )

   for j in range(1,m+1):
       t[1,j] = 0.0
46     t[n,j] = 0.0                                    # BCs
   for i in range( 2, n):    t[i][1] = sin( pi * h *i)         # ICs
   for i in range(1, n+1):    Td[i] = 2. + 2./r
   Td[1] = 1.; Td[n] = 1.
50 for i in range(1,n ): Ta[i] = -1.0;      Tc[i] = -1.0;     # Off diagonal
   Ta[n-1] = 0.0;    Tc[1] = 0.0; Tb[1] = 0.0; Tb[n] = 0.0
   print("I'm working hard, wait for fig while I count to 50")

54 for j in range(2,m+1):
       print(j)
       for i in range(2,n): Tb[i] = t[i-1][j-1] + t[i+1][j-1] \
               + (2/r-2) * t[i][j-1]
58     Tridiag(a, d, c, b, Ta, Td, Tc, Tb, x, n)            # Solve system
       for i in range(1, n+1):    t[i][j] = x[i]
   print("Finished")
   x = list(range(1, m+1))                              # Plot every other x
62 y = list(range(1, n+1))                              # every other y
   X, Y = p.meshgrid(x,y)

   def functz(t):                                       # Potential
66     z = t[X, Y]
       return z

   Z = functz(t)
70 fig = p.figure()
   ax = Axes3D(fig)
   ax.plot_wireframe(X, Y, Z, color= 'r')
   ax.set_xlabel('t')
74 ax.set_ylabel('x')
   ax.set_zlabel('T')
   p.show()                                             # Display figure
```

23

String and Membrane Waves

In this chapter, and in Chapters 24–26, we explore PDE's with wave-like solutions. Here we deal with 1D waves on strings, and 2D waves on membranes. In Chapter 24 we examine quantum wave packets and E&M waves, and in Chapter 25 we look at shock waves and solitary waves. The basic technique is the leapfrog algorithm that propagates the initial conditions forward in time, step by step. The numerical solutions let us include more physics than is possible with the familiar analytic treatments.

23.1   A Vibrating String's Hyperbolic Wave Equation

Recall the elementary physics demonstration in which a string, tied down at both ends, is plucked gently, and released, resulting in a pulse that travels along the string.

Problem   Develop an accurate model for wave propagation on a string, and see if it can produce both traveling and standing waves.

Consider a string of length L tied down at both ends (Figure 23.1 left). The string has a constant density ρ per unit length, no frictional forces acting on it, and a tension T that is high enough to let us ignore any sagging as a result of gravity. We assume that the displacement of the string from its rest position, $y(x, t)$, is only in the vertical direction, and that the displacement is a function of the horizontal location along the string x, and the time t.

To derive a linear equation of motion, we assume that the string's relative displacement $y(x, t)/L$ and slope $\partial y/\partial x$ are both small. In Figure 23.1 right, we isolate an infinitesimal section Δx of the string. We see there that the difference in the vertical components of the tension, at either end of the string, produces the restoring force that accelerates this section of the string up or down. By applying Newton's laws to this section, we obtain the familiar wave equation:

Computational Physics: Problem Solving with Python, Fourth Edition.
Rubin H. Landau, Manuel J. Páez, and Cristian C. Bordeianu.
© 2024 WILEY-VCH GmbH. Published 2024 by WILEY-VCH GmbH.

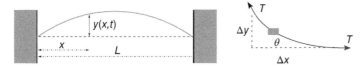

Figure 23.1 *Left*: A stretched string of length L tied down at both ends. The vertical disturbance of the string from its equilibrium position is $y(x, t)$. *Right*: A differential element of the string showing how the string's displacement leads to the restoring force.

$$\sum F_y = \rho \, \Delta x \, \frac{\partial^2 y}{\partial t^2}, \tag{23.1}$$

$$= T \sin \theta(x + \Delta x) - T \sin \theta(x) \tag{23.2}$$

$$= T \frac{\partial y}{\partial x}\bigg|_{x+\Delta x} - T \frac{\partial y}{\partial x}\bigg|_x \simeq T \frac{\partial^2 y}{\partial x^2}, \tag{23.3}$$

$$\Rightarrow \quad \frac{\partial^2 y(x, t)}{\partial x^2} = \frac{1}{c^2} \frac{\partial^2 y(x, t)}{\partial t^2}, \qquad c = \sqrt{\frac{T}{\rho}}. \tag{23.4}$$

Here, we have assumed that θ is small enough for $\sin \theta \simeq \tan \theta = \partial y/\partial x$. The existence of the two independent variables, x and t, makes (23.4) a PDE. The constant c here is the velocity with which a disturbance travels along the wave, and is seen to decrease for increasing density, and increase for increasing tension. Note that this *signal velocity c* is *not* the same as the velocity of a string element $\partial y/\partial t$.

The initial condition for our problem is that the string is plucked gently and released. We assume that the pluck places the string in a triangular shape, with the center of triangle $\frac{8}{10}$ of the way down the string, and with a height of 1:

$$y(x, t = 0) = \begin{cases} 1.25x/L, & x \le 0.8L, \\ (5 - 5x/L), & x > 0.8L, \end{cases} \quad \text{(initial condition 1).} \tag{23.5}$$

Because (23.4) is second-order in time, a second initial condition is needed to determine the solution. We interpret the "gentleness" of the pluck to mean that the string is released from rest:

$$\frac{\partial y}{\partial t}(x, t = 0) = 0, \quad \text{(initial condition 2).} \tag{23.6}$$

The boundary conditions have both ends of the string tied down at all times:

$$y(0, t) \equiv 0, \quad y(L, t) \equiv 0, \quad \text{(boundary conditions).} \tag{23.7}$$

23.1.1 Solution as Normal-Mode Expansion

The analytic solution to (23.4) is obtained via the familiar separation-of-variables technique. We assume that the solution is the product of a function of space and a function of time:

$$y(x, t) = X(x)T(t). \tag{23.8}$$

We substitute (23.8) into (23.4), divide it by $y(x, t)$, and are left with an equation that has a solution only if there are solutions to the two ODEs:

$$\frac{d^2T(t)}{dt^2} + \omega^2 T(t) = 0, \quad \frac{d^2X(x)}{dx^2} + k^2 X(x) = 0, \quad k \stackrel{\text{def}}{=} \frac{\omega}{c}. \tag{23.9}$$

The angular frequency ω and the wave vector k are determined by demanding that the solutions satisfy the boundary conditions:

$$X(x = 0, t) = X(x = l, t) = 0 \tag{23.10}$$

$$\Rightarrow \quad X_n(x) = A_n \sin k_n x, \quad k_n = \frac{\pi(n+1)}{L}, \quad n = 0, 1, \dots. \tag{23.11}$$

The time solution is:

$$T_n(t) = C_n \sin \omega_n t + D_n \cos \omega_n t, \quad \omega_n = nck_0 = n\frac{2\pi c}{L}, \tag{23.12}$$

where ω_n is the frequency of the nth *normal mode*. The *initial condition* (23.5) of zero velocity, $\partial y/\partial t(t = 0) = 0$, requires the C_n values in (23.12) to be zero. Putting the pieces together, the normal-mode are:

$$y_n(x, t) = \sin k_n x \cos \omega_n t, \quad n = 0, 1, \dots. \tag{23.13}$$

Because the wave equation (23.4) is linear in y, the principle of linear superposition holds, and the most general solution for waves on a string with fixed ends can be written as the sum of normal modes:

$$y(x, t) = \sum_{n=0}^{\infty} B_n \sin k_n x \cos \omega_n t. \tag{23.14}$$

(We will lose linear superposition once we include nonlinear terms in the wave equation.) The Fourier coefficient B_n is determined by the second initial condition (23.5), which describes how the wave is plucked:

$$y(x, t = 0) = \sum_{n}^{\infty} B_n \sin nk_0 x. \tag{23.15}$$

We multiply both sides by $\sin mk_0 x$, substitute the value of $y(x, 0)$ from (23.5), and integrate from 0 to L to obtain:

$$B_m = 6.25 \frac{\sin(0.8m\pi)}{m^2 \pi^2}. \tag{23.16}$$

You will be asked to compare the Fourier series (23.14) to your numerical solution. While it is in the nature of the approximation that the precision of the numerical solution depends on the choice of step size, it is also revealing to realize that the precision of the so-called analytic solution depends on summing an infinite number of terms, which can be summed only approximately.

23.2 Time-Stepping Algorithm

As with the heat equation, we look for a solution $y(x, t)$ only for discrete values of the independent variables, in this case x and t (Figure 23.2):

$$x = i\Delta x, \quad i = 1, \dots, N_x, \quad t = j\Delta t, \quad j = 1, \dots, N_t, \tag{23.17}$$

$$y(x, t) = y(i\Delta x, i\Delta t) \stackrel{\text{def}}{=} y_{i,j}. \tag{23.18}$$

Figure 23.2 The solutions of the wave equation for four earlier space-time points are used to obtain the solution at the present time. The boundary and initial conditions are indicated by the white-centered dots.

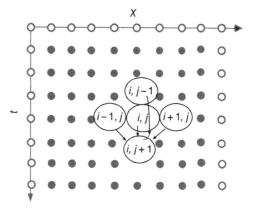

We seek solutions at the lattice sites of the space-time grid in Figure 23.2. That being the case, moving across a row corresponds to increasing x values along the string for a fixed time, while moving down a column corresponds to increasing time steps for a fixed position. Although the grid in Figure 23.2 may be square, we cannot use a relaxation technique like we did for the solution of Laplace's equation, because we do not know the solution on all four sides. The boundary conditions determine the solution along the right and left sides, while the initial time condition determines the solution along the top, but not the bottom.

As with the heat equation, we use the central-difference approximation to *discretize* the wave equation into a difference equation. First, we express the second derivatives in terms of finite differences:

$$\frac{\partial^2 y}{\partial t^2} \simeq \frac{y_{i,j+1} + y_{i,j-1} - 2y_{i,j}}{(\Delta t)^2}, \quad \frac{\partial^2 y}{\partial x^2} \simeq \frac{y_{i+1,j} + y_{i-1,j} - 2y_{i,j}}{(\Delta x)^2}. \tag{23.19}$$

Substituting (23.19) in the wave equation (23.4) yields a difference equation:

$$\frac{y_{i,j+1} + y_{i,j-1} - 2y_{i,j}}{c^2 (\Delta t)^2} = \frac{y_{i+1,j} + y_{i-1,j} - 2y_{i,j}}{(\Delta x)^2}. \tag{23.20}$$

Notice that this equation contains three time values: $j + 1 =$ the future, $j =$ the present, and $j - 1 =$ the past. Consequently, we rearrange it into a form that permits us to predict the future solution from the present and past solutions:

$$y_{i,j+1} = 2y_{i,j} - y_{i,j-1} + \frac{c^2}{c'^2} \left[y_{i+1,j} + y_{i-1,j} - 2y_{i,j} \right], \quad c' \stackrel{\text{def}}{=} \frac{\Delta x}{\Delta t}. \tag{23.21}$$

Here c' is a combination of numerical parameters, with the dimension of velocity, whose size relative to c determines the stability of the algorithm. As shown in Figure 23.2, the algorithm (23.21) propagates the wave from the two earlier times, j and $j - 1$, and from three nearby positions, $i - 1$, i, and $i + 1$, to a later time, $j + 1$, and a single space position, i.

We start the algorithm with the solution along the topmost row, and then move it down one step at a time. If we save the solution for present times, then we only need to store three time values on the computer. In fact, because the time steps must be quite small to obtain high precision, you may want to store the solution only for every fifth or tenth time for viewing.

Initializing the recurrence relation is a bit tricky because it requires displacements from two earlier times, whereas the initial conditions are for only one time. Nonetheless, the rest of the condition (23.5), when combined with the *central-difference* approximation, lets us extrapolate to negative time:

$$\frac{\partial y}{\partial t}(x, 0) \simeq \frac{y(x, \Delta t) - y(x, -\Delta t)}{2\Delta t} = 0, \ \Rightarrow \ y_{i,0} = y_{i,2}. \tag{23.22}$$

Here we take the initial time as $j = 1$, and so $j = 0$ corresponds to $t = -\Delta t$. Substituting this relation into (23.21) yields for the initial step:

$$y_{i,2} = y_{i,1} + \frac{c^2}{2c'^2}\left[y_{i+1,1} + y_{i-1,1} - 2y_{i,1}\right] \qquad \text{(for } j = 2 \text{ only)}. \tag{23.23}$$

Equation (23.23) uses the solution throughout all space at the initial time $t = 0$ to propagate (leapfrog) it forward to a time Δt. Subsequent time steps use (23.21), and are continued for as long as you like.

Exercise Devise a procedure for solving for the wave equation for all times in just one step. Estimate how much memory would be required.

23.3 von Neumann Stability Analysis

As discussed in Section 22.2.1, our approximation of converting a PDE into a difference equation will not be a good approach if the solution to the difference equation is unstable. The *von Neumann stability analysis* tell us what to do to get a good solution. The analysis is based on the assumption that eigenmodes of the difference equation have the form:

$$y_{i,j} = \xi(k)^j \, e^{Iki\Delta x}, \tag{23.24}$$

where $x = i\Delta x$, $t = j\Delta t$, and $I = \sqrt{-1}$ is the imaginary number. The constant k in (23.24) is an unknown wave vector $(2\pi/\lambda)$, and $\xi(k)$ is an unknown complex function. We view (23.24) as a function that oscillates in space (the exponential), with the amplitude or *amplification factor* $\xi(k)^j$ that gets multiplied by a power of ξ for each time step. Stability of the solution then requires $|\xi(k)| < 1$, else the solution would grow in time [Press *et al.*, 2007; Ancona, 2002].

Exercise Substitute the presumed form for the eigenmode (23.24) into difference equation (23.21) being used as the algorithm, and solve for the amplitude $\xi(k)$. *Hint*: in Section 22.2.1, we do this explicitly for the heat equation.

Even though the application of the stability condition may get complicated, [Press *et al.*, 2007; Courant *et al.*, 1928] show that $|\xi(k)| < 1$, and the difference equation solution will therefore be stable, for the general class of transport equations if:

$$c \leq c' = \Delta x/\Delta t, \qquad \text{(Courant condition)}. \tag{23.25}$$

Equation (23.25) means that the solution gets better with smaller *time* steps, but gets worse for smaller space steps (unless you simultaneously make the time step smaller too). Having different sensitivities to the time and space steps may appear surprising because the wave equation (23.4) is symmetric in x and t, yet the symmetry is broken by the nonsymmetric initial and boundary conditions.

In general, you should perform a stability analysis for every PDE you have to solve, although it can get complicated. Yet, even if you do not, the lesson here is that you may want to try different *combinations* of Δx and Δt until a stable and reliable solution is obtained. You may expect, nonetheless, that there are choices for Δx and Δt for which the numerical solution fails, and that simply decreasing an individual Δx or Δt, in the hope that this will increase precision, may not improve the solution.

23.3.1 Implementation and Assessment

The program `EqStringMat.py` in Listing 23.1 solves the wave equation for a string of length $L = 1$ m, with its ends fixed, and with the gently plucked initial condition. Note that although our use of $L = 1$ violates the assumption of small displacement, $y/L \ll 1$; you may want to use $L = 1000$ to be realistic. The values of density and tension are $\rho = 0.01$ kg/m, and $T = 40$ N, with the space grid set at 101 points, corresponding to $\Delta = 0.01$cm.

1) Solve the wave equation, and make a surface plot of displacement *versus* time and position.
2) Explore a number of space step and time step combinations. In particular, try steps that satisfy, and that do not satisfy, the Courant condition (23.25). Does your exploration conform with the stability condition?
3) Compare the analytic and numeric solutions, summing at least 200 terms in the analytic solution.
4) Use the plotted time dependence to estimate the peak's propagation velocity c. Compare the deduced c to (23.4).
5) Our solution of the wave equation for a plucked string leads to the formation of a wave packet that corresponds to the sum of multiple normal modes of the string. Solve the wave equation for a string initially placed in the single normal mode (standing wave):

$$y(x, t = 0) = 0.001 \sin 2\pi x, \quad \frac{\partial y}{\partial t}(x, t = 0) = 0. \tag{23.26}$$

See if a normal mode results and remains.
6) Observe the motion of a wave for initial conditions corresponding to the sum of two adjacent normal modes. Does beating occur?

23.4 Beyond The Simple Wave Equation

The string problem we have investigated so far can be handled by both numerical and analytic techniques. We now wish to extend the theory to include some more realistic physics. These extensions have only numerical solutions.

23.4.1 Including Friction

Plucked strings do not vibrate forever because the real world contains friction. Consider again the elements of a string between x and $x + dx$ in Figure 23.1 right, but now imagine that this element is moving in a viscous fluid such as air. An approximate model for the frictional force has it pointing in a direction opposite to the vertical velocity of the string,

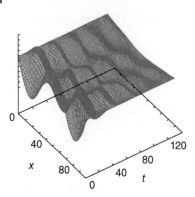

Figure 23.3 The vertical displacement as a function of position x and time t for waves on a string when friction is included.

proportional to that velocity, and also proportional to the length of the string element:

$$F_f \simeq -2\kappa \, \Delta x \, \frac{\partial y}{\partial t}. \tag{23.27}$$

Here, κ is a constant that is proportional to the viscosity of the medium. Including this force in the equation of motion changes the wave equation to:

$$\frac{\partial^2 y}{\partial t^2} = c^2 \frac{\partial^2 y}{\partial x^2} - \frac{2\kappa}{\rho} \frac{\partial y}{\partial t}. \tag{23.28}$$

In Figure 23.3, we show the resulting motion of a string plucked in the middle when friction is included. Observe how the initial pluck breaks up into waves traveling to the right and to the left, that then get reflected and inverted by the fixed ends. Since those parts of the wave with the higher velocity experience greater friction, the peak tends to get smoothed out the most as time progresses.

Exercise   Generalize the algorithm used to solve the wave equation to now include friction. Pick a value of κ large enough to cause a noticeable dampening, but not so large as to stop the oscillations. As a check, reverse the sign of κ and see if the wave grows in time.

23.4.2   Including Variable Tension and Density

We have derived the propagation velocity for waves on a string as $c = \sqrt{T/\rho}$. This says that waves move slower in regions of high density, and faster in regions of high tension. If the density of a string varies, for instance, by having the ends thicker in order to support the weight of the middle, then c will no longer be a constant, and our wave equation will need to be extended. In addition, if the density increases, then so will the tension, because it takes greater tension to accelerate a greater mass. If gravity acts, then we will also expect the tension at the ends of the string to be higher than in the middle, because the ends must support the entire weight of the string.

To derive the equation for wave motion with variable density and tension, we again consider the element of a string shown in Figure 23.1 right. If we do not assume the tension T is constant, then Newton's second law gives:

$$F = ma, \tag{23.29}$$

$$\frac{\partial}{\partial x} \left[T(x) \frac{\partial y(x,t)}{\partial x} \right] \Delta x = \rho(x) \Delta x \frac{\partial^2 u(x,t)}{\partial t^2}, \tag{23.30}$$

$$\frac{\partial T(x)}{\partial x}\frac{\partial y(x,t)}{\partial x} + T(x)\frac{\partial^2 y(x,t)}{\partial x^2} = \rho(x)\frac{\partial^2 y(x,t)}{\partial t^2}. \tag{23.31}$$

If $\rho(x)$ and $T(x)$ are known functions, then these equations can be solved with a slight modification of our algorithm.

In Section 23.4.3, we will solve for the tension in a stationary hanging string when gravity is present. Those readers interested in an **alternate easier problem**, that still shows the new physics, may assume that density and tension are proportional:

$$\rho(x) = \rho_0 e^{\alpha x}, \quad T(x) = T_0 e^{\alpha x}. \tag{23.32}$$

While we would expect the tension to be greater in regions of higher density (more mass to move and support), being proportional is clearly just an approximation. Substitution of (23.32) into (23.31) yields the wave equation:

$$\frac{\partial^2 y(x,t)}{\partial x^2} + \alpha\frac{\partial y(x,t)}{\partial x} = \frac{1}{c^2}\frac{\partial^2 y(x,t)}{\partial t^2}, \quad c^2 = \frac{T_0}{\rho_0}. \tag{23.33}$$

Here, c is a constant that would be the wave velocity if $\alpha = 0$. This equation is similar to the wave equation with friction, only now the first derivative is with respect to x and not t. The corresponding difference equation follows from using central-difference approximations for the derivatives:

$$y_{i,j+1} = 2y_{i,j} - y_{i,j-1} + \frac{\alpha c^2 (\Delta t)^2}{2\Delta x}[y_{i+1,j} - y_{i,j}] + \frac{c^2}{c'^2}[y_{i+1,j} + y_{i-1,j} - 2y_{i,j}],$$

$$y_{i,2} = y_{i,1} + \frac{c^2}{c'^2}[y_{i+1,1} + y_{i-1,1} - 2y_{i,1}] + \frac{\alpha c^2 (\Delta t)^2}{2\Delta x}[y_{i+1,1} - y_{i,1}]. \tag{23.34}$$

23.4.3 Waves on Catenary

Up until this point, we have been ignoring the effect of gravity upon the string's shape and tension. This is a good approximation if there is little sag in the string, as might happen if the tension is very high and the string is light. Even if there is some sag, our solution for $y(x,t)$ could still be used as the disturbance about the equilibrium shape. However, if the string is massive, say, like a chain or heavy cable, then the sag in the middle could be quite large (Figure 23.4), and the resulting variations in shape and tension need to be incorporated into

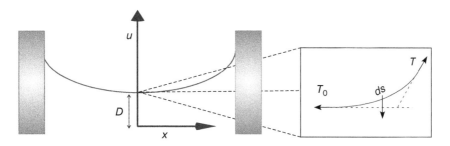

Figure 23.4 *Left*: A uniform string suspended from its ends in a gravitational field assumes a catenary shape. *Right*: A force diagram of a section of the catenary at its lowest point. Because the ends of the string must support the weight of the string, the tension now varies along the string.

the wave equation. Because the tension is no longer uniform, waves travel faster near the ends of the string, which are under greater tension.

The derivation of the catenary shape is straight forward. Consider a string of uniform density ρ acted upon by gravity. To avoid confusion with our use of $y(x)$ to describe a disturbance on a string, we call $u(x)$ the equilibrium shape of the string (Figure 23.4). The statics problem we need to solve is to determine the shape $u(x)$ and the tension $T(x)$. The inset in Figure 23.4 is a free-body diagram of the midpoint of the string and shows that the weight W of this section of arc length s is balanced by the vertical component of the tension T. The horizontal tension T_0 is balanced by the horizontal component of T:

$$T(x)\sin\theta = W = \rho g s, \quad T(x)\cos\theta = T_0, \tag{23.35}$$

$$\Rightarrow \tan\theta = \rho g s / T_0. \tag{23.36}$$

The trick is to convert (23.36) to a differential equation that we can solve. We do that by replacing the slope $\tan\theta$ by the derivative du/dx, and by taking the derivative with respect to x:

$$\frac{du}{dx} = \frac{\rho g}{T_0} s, \quad \Rightarrow \quad \frac{d^2 u}{dx^2} = \frac{\rho g}{T_0}\frac{ds}{dx}. \tag{23.37}$$

Yet, because $ds = \sqrt{dx^2 + du^2}$, we have our differential equation:

$$\frac{d^2 u}{dx^2} = \frac{1}{D}\frac{\sqrt{dx^2 + du^2}}{dx} \tag{23.38}$$

$$= \frac{1}{D}\sqrt{1 + (du/dx)^2}, \; D = T_0/\rho g. \tag{23.39}$$

Equation (23.39) is the equation for the *catenary*, and has the solution [Becker, 1954]

$$u(x) = D\cosh\frac{x}{D}. \tag{23.40}$$

Here, we have chosen the x axis to lie at distance D below the bottom of the catenary (Figure 23.4), so that $x = 0$ is at the center of the string where $y = D$ and $T = T_0$. Equation (23.37) tells us the arc length $s = D\,du/dx$, and so we can solve for $s(x)$ and for the tension $T(x)$ via (23.35):

$$s(x) = D\sinh\frac{x}{D}, \quad \Rightarrow \quad T(x) = T_0\frac{ds}{dx} = \rho g u(x) = T_0\cosh\frac{x}{D}. \tag{23.41}$$

It is this variation in tension that leads to an x dependence of the wave velocity.

23.4.4 Catenary Assessment

In Listing 23.1, we give the program `EqStringMat.py` that solves the wave equation. Modify it to solve for waves on a catenary including friction, or for the assumed density and tension given by (23.32) with $\alpha = 0.5$, $T_0 = 40$ N, and $\rho_0 = 0.01$ kg/m. Our code `CatFriction.py`, given in Listing 23.3, yields the solution shown in Figure 23.5.

1) Look for some interesting cases and create surface plots of the results.
2) Describe in words how the waves dampen, and how a wave's velocity appears to change.

Figure 23.5 The waves of a plucked catenary with friction at six different times.

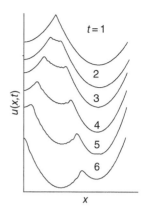

3) **Normal modes:** Search for normal mode solutions of the variable tension wave equation, that is, solutions that vary as:

$$u(x, t) = A \cos(\omega t) \sin(\gamma x). \tag{23.42}$$

Try using this form to start your program and see if you end up with standing waves. Use large values for ω.

4) When conducting physics demonstrations, we have set up standing wave patterns by continuously shaking one end of a string up and down. Try doing the same with your program; that is, build into your code the condition that for all times:

$$y(x = 0, t) = A \sin \omega t. \tag{23.43}$$

You may need to vary A and ω until a normal mode (standing wave) is obtained.

5) (For the exponential density case.) If you were able to find standing waves, then verify that this string acts like a high-frequency filter, that is, that there is a frequency below which no waves occur.

6) For the catenary problem, plot your results showing *both* the disturbance $u(x, t)$ about the catenary, and the actual height $y(x, t)$ above the horizontal for a plucked string initial condition.

7) Try the first two normal modes for a uniform string as the initial conditions for the catenary. These should be close to, but not exactly, normal modes.

8) We derived the normal modes for a uniform string after assuming that $k(x) = \omega/c(x)$ is a constant. For a catenary without too much x variation in the tension, we should be able to make the approximation:

$$c(x)^2 \simeq \frac{T(x)}{\rho} = \frac{T_0 \cosh(x/d)}{\rho}. \tag{23.44}$$

See if you get a better representation of the first two normal modes if you include some x dependence in k.

23.4.5 Including Nonlinear Terms

Show that an extension of the wave equation, including the next order in displacements, is [Taghipour *et al.*, 2014; Keller, 1959]:

$$c^2 \frac{\partial^2 y(x, t)}{\partial x^2} = \left[1 + \frac{\partial^2 y(x, t)}{\partial x^2} \right]^2 \frac{\partial^2 y(x, t)}{\partial t^2}. \tag{23.45}$$

1) Extend the leapfrog algorithm to solve this nonlinear equation, making whatever assumptions about initial conditions are needed.
2) Repeat some of the solutions to the wave equation studied previously, only now for large values of y/L, in which case nonlinear effects are important.
3) Examine what were normal modes for the linear problem, but now see if they persists for large amplitude oscillations.

23.5 Vibrating Membrane (2D Waves)

An elastic membrane is stretched securely across the top of a square box with sides of length π, and with the membrane in the asymmetrical shape [Kreyszig, 1998]:

$$u(x, y, t = 0) = \sin 2x \sin y, \quad 0 \le x \le \pi, \quad 0 \le y \le \pi, \tag{23.46}$$

where u is the vertical displacement from equilibrium.

Problem   Describe the motion of the membrane when it is released from rest.

The description of wave motion on a membrane is much the same as that of 1D waves on the string discussed in Section 23.1, only now with wave propagation in two directions. Consider Figure 23.6 showing a square section of the membrane under tension T. The membrane moves only vertically in the z direction, yet because the restoring force arising from the tension in the membrane varies in both the x and y directions, there is wave motion along the entire surface of the membrane.

Although the tension is constant over the small area in Figure 23.6, there will be a net vertical force on the displayed segment if the angle of incline of the membrane varies as we move through space. Accordingly, the net force on the membrane in the z direction, as a result of the change in y, is:

$$\sum F_z(x) = T\Delta x \sin \theta - T \Delta x \sin \phi, \tag{23.47}$$

where θ is the angle of incline at $y + \Delta y$, and ϕ the angle at y. If we assume that the displacements and the angles are small, then we can make the approximations:

$$\sin \theta \approx \tan \theta = \left. \frac{\partial u}{\partial y} \right|_{y+\Delta y}, \quad \sin \phi \approx \tan \phi = \left. \frac{\partial u}{\partial y} \right|_{y}, \tag{23.48}$$

$$\Rightarrow \quad \sum F_z(x_{\text{fixed}}) = T\Delta x \left(\left. \frac{\partial u}{\partial y} \right|_{y+\Delta y} - \left. \frac{\partial u}{\partial y} \right|_{y} \right) \approx T\Delta x \frac{\partial^2 u}{\partial y^2} \Delta y. \tag{23.49}$$

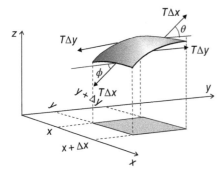

Figure 23.6   A small section of an oscillating membrane and the forces that act on it.

Similarly, the net force in the z direction, as a result of the variation in y, is:

$$\sum F_z(y_{\text{fixed}}) = T\Delta y \left(\frac{\partial u}{\partial x}\big|_{x+\Delta x} - \frac{\partial u}{\partial x}\big|_x \right) \approx T\Delta y \frac{\partial^2 u}{\partial x^2} \Delta x. \tag{23.50}$$

The membrane section has mass $\rho \Delta x \Delta y$, where ρ is the membrane's mass per unit area. We now apply Newton's second law to determine the acceleration of the membrane section in z direction, as a result of the sum of the net forces arising from both the x and y variations:

$$\rho \Delta x \Delta y \frac{\partial^2 u}{\partial t^2} = T\Delta x \frac{\partial^2 u}{\partial y^2} \Delta y + T\Delta y \frac{\partial^2 u}{\partial x^2} \Delta x, \tag{23.51}$$

$$\Rightarrow \qquad \frac{1}{c^2} \frac{\partial^2 u}{\partial t^2} = \frac{\partial^2 u}{\partial x^2} + \frac{\partial^2 u}{\partial y^2}, \qquad c = \sqrt{T/\rho}. \tag{23.52}$$

As one might have guessed, this is the 2D version of the wave equation (23.4) that we studied previously in one dimension. Here, c, the propagation velocity, is still the square root of tension over density, only now it is tension per unit length and mass per unit area.

23.6 Analytical Solution

The analytic or numerical solution of the partial differential equation (23.52) requires us to know both the boundary conditions and the initial conditions. The boundary conditions hold for all times, and were given when we were told that the membrane is attached securely to a square box of side π:

$$u(x = 0, y, t) = u(x = \pi, y, t) = 0, \tag{23.53}$$

$$u(x, y = 0, t) = u(x, y = \pi, t) = 0. \tag{23.54}$$

As required for a second-order equation, the initial conditions have two parts, the shape of the membrane at time $t = 0$, and the velocity of each point of the membrane. The first is initial configuration:

$$u(x, y, t = 0) = \sin 2x \sin y, \qquad 0 \le x \le \pi, \qquad 0 \le y \le \pi. \tag{23.55}$$

The second is the membrane being released from rest:

$$\frac{\partial u}{\partial t}\bigg|_{t=0} = 0, \tag{23.56}$$

where we write partial derivative because there are also spatial variations.

The analytic solution is based on the guess that because the wave equation (23.52) has separate derivatives with respect to each coordinate and time, the full solution $u(x, y, t)$ might be the product of separate functions of x, y, and t:

$$u(x, y, t) = X(x)\, Y(y)\, T(t). \tag{23.57}$$

After substituting into (23.52) and dividing by $X(x)Y(y)T(t)$, we obtain:

$$\frac{1}{c^2} \frac{1}{T(t)} \frac{d^2 T(t)}{dt^2} = \frac{1}{X(x)} \frac{d^2 X(x)}{dx^2} + \frac{1}{Y(y)} \frac{d^2 Y(y)}{dy^2}. \tag{23.58}$$

The only way that the LHS of (23.58) can be true for all times, while the RHS is also true for all coordinates, is if both sides are constant:

$$\frac{1}{c^2}\frac{1}{T(t)}\frac{d^2 T(t)}{dt^2} = -\xi^2 = \frac{1}{X(x)}\frac{d^2 X(x)}{dx^2} + \frac{1}{Y(y)}\frac{d^2 Y(y)}{dy^2}, \tag{23.59}$$

$$\Rightarrow \quad \frac{1}{X(x)}\frac{d^2 X(x)}{dx^2} = -k^2, \qquad \frac{1}{Y(y)}\frac{d^2 Y(y)}{dy^2} = -q^2, \tag{23.60}$$

where $q^2 = \xi^2 - k^2$. In (23.60), we have included the further realization that because each term on the RHS of (23.59) depends on either x or y, the only way for their sum to be constant, is if each term separately is a constant, in this case $-k^2$. The solutions of these equations are sinusoidal standing waves in the x and y directions:

$$X(x) = A \sin kx + B \cos kx, \tag{23.61}$$

$$Y(y) = C \sin qy + D \cos qy, \tag{23.62}$$

$$T(t) = E \sin c\xi t + F \cos c\xi t. \tag{23.63}$$

We now apply the boundary conditions:

$$u(x = 0, y, t) = u(x = \pi, y, z) = 0 \quad \Rightarrow \quad B = 0, \ k = 1, 2, \ldots,$$

$$u(x, y = 0, t) = u(x, y = \pi, t) = 0 \quad \Rightarrow \quad D = 0, \ q = 1, 2, \ldots,$$

$$\Rightarrow X(x) = A \sin kx, \quad Y(y) = C \sin qy. \tag{23.64}$$

The fixed values for the eigenvalues m and n, describing the modes for the x and y standing waves, are equivalent to fixed values for the constants q^2 and k^2. Yet because $q^2 + k^2 = \xi^2$, we must also have a fixed value for ξ^2:

$$\xi^2 = q^2 + k^2 \qquad \Rightarrow \qquad \xi_{kq} = \pi\sqrt{k^2 + q^2}. \tag{23.65}$$

The full space-time solution now takes the form:

$$u_{kq} = \left[G_{kq} \cos c\xi t + H_{kq} \sin c\xi t\right] \sin kx \sin qy, \tag{23.66}$$

where k and q are integers. Since the wave equation is linear in u, its most general solution is a linear combination of the eigenmodes (23.66):

$$u(x, y, t) = \sum_{k=1}^{\infty}\sum_{q=1}^{\infty} \left[G_{kq} \cos c\xi t + H_{kq} \sin c\xi t\right] \sin kx \sin qy. \tag{23.67}$$

While an infinite series is not a good algorithm, the initial and boundary conditions mean that only the $k = 2, q = 1$ term contributes, and we have a closed form solution:

$$u(x, y, t) = \cos c\sqrt{5} \sin 2x \sin y, \tag{23.68}$$

where c is the wave velocity.

23.7 Numerical Solution

The development of an algorithm for the solution of the 2D wave equation (23.52) follows that of the 1D equation in Section 23.2. We start by expressing the second derivatives in terms of central differences:

$$\frac{\partial^2 u(x,y,t)}{\partial t^2} = \frac{u(x,y,t+\Delta t) + u(x,y,t-\Delta t) - 2u(x,y,t)}{(\Delta t)^2}, \tag{23.69}$$

$$\frac{\partial^2 u(x,y,t)}{\partial x^2} = \frac{u(x+\Delta x,y,t) + u(x-\Delta x,y,t) - 2u(x,y,t)}{(\Delta x)^2}, \tag{23.70}$$

$$\frac{\partial^2 u(x,y,t)}{\partial y^2} = \frac{u(x,y+\Delta y,t) + u(x,y-\Delta y,t) - 2u(x,y,t)}{(\Delta y)^2}. \tag{23.71}$$

After discretizing the variables, $u(x = i\Delta, y = i\Delta y, t = k\Delta t) \equiv u_{i,j}^k$, we obtain our time-stepping algorithm by solving for the future solution in terms of the present and past ones:

$$u_{i,j}^{k+1} = 2u_{i,j}^k - u_{i,j}^{k-1} \frac{c^2}{c'^2}[u_{i+1,j}^k + u_{i-1,j}^k - 4u_{i,j}^k + u_{i,j+1}^k + u_{i,j-1}^k], \tag{23.72}$$

where, as before, $c' \stackrel{\text{def}}{=} \Delta x/\Delta t$. Whereas the present and past solutions, u^k and u^{k-1}, are known after the first step, to initiate the algorithm we need to know the solution at $t = -\Delta t$, that is, before the initial time. To find that, we use the fact that the membrane is released from rest, and so:

$$0 = \frac{\partial u(t=0)}{\partial t} \approx \frac{u_{i,j}^1 - u_{i,j}^{-1}}{2\Delta t}, \qquad \Rightarrow \qquad u_{i,j}^{-1} = u_{i,j}^1. \tag{23.73}$$

After substitution into (23.72), we obtain the algorithm for the first step:

$$u_{i,j}^1 = u_{i,j}^0 + \frac{c^2}{2c'^2}[u_{i+1,j}^0 + u_{i-1,j}^0 - 4u_{i,j}^0 + u_{i,j+1}^0 + u_{i,j-1}^0]. \tag{23.74}$$

Because the $t = 0$ displacement $u_{i,j}^0$ is known, we compute the solution for the first time step with (23.74), and for subsequent steps with (23.72).

The program Wave2D.py in Listing 23.2 solves the 2D wave equation using the leapfrog algorithm. The program Waves2Danal.py computes the analytic solution. The shape of the membrane at three different times is shown in Figure 23.7.

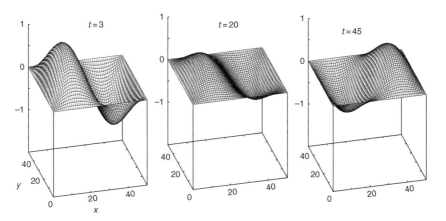

Figure 23.7 The standing wave pattern on a square box top at three different times.

23.8   Code Listings

Listing 23.1   EqStringMat.py Solves the wave equation for a gently plucked string.

```
# EqStringMat.py:   Animated leapfrog sol Vibrating string + MatPlotLib

from numpy import *
import numpy as np, matplotlib.pyplot as plt, matplotlib.animation as animation

rho = 0.01;    ten = 40.; c = sqrt(ten/rho)            # density, tension
c1 = c;          ratio = c*c/(c1*c1)                   # CFL criterium = 1
xi = np.zeros((101,3), float)                          # Declaration
k = range(0,101)

def Initialize():                                      # Initial conditions
    for i in range(0, 81):    xi[i, 0] = 0.00125*i
    for i in range (81, 101): xi[i, 0] = 0.1 - 0.005*(i - 80) # 2nd part
def animate(num):
    for i in range(1, 100):
        xi[i,2] = 2.*xi[i,1]-xi[i,0]+ ratio *(xi[i+1,1]+xi[i-1,1]-2*xi[i,1])
        line.set_data(k,xi[k,2])                       # Data to plot ,x,y
    for m in range (0,101):
        xi[m, 0] = xi[m, 1]                            # Recycle array
        xi[m, 1] = xi[m, 2]
    return line
Initialize()                                           # Plot initial string
fig = plt.figure()
ax = fig.add_subplot(111, autoscale_on=False, xlim=(0, 101), ylim=(-0.15, 0.15))
ax.grid()                                              # Plot  grid
plt.title("Vibrating String")
line, = ax.plot(k,  xi[k,0], lw=2)
for i in range(1,100):
    xi[i,1] = xi[i,0] + 0.5* ratio *(xi[i+1,0] + xi[i-1,0] -2*xi[i,0])
ani = animation.FuncAnimation(fig, animate,1)          # Dummy argument: 1
plt.show()
print("finished")
```

Listing 23.2   Waves2D.py Solves the wave equation for a vibrating membrane.

```
import matplotlib.pylab as p; from numpy import *
from mpl_toolkits.mplot3d import Axes3D

tim = 15;    N = 71
c = sqrt(180./390)                        # Speed = sqrt(ten[]/den[kg/m2;])
u = zeros((N,N,N),float);        v = zeros((N,N),float)
incrx = pi/N;               incry = pi/N
cprime = c;
covercp = c/cprime;     ratio = 0.5*covercp*covercp  # c/c' 0.5 for stable

def vibration(tim):
    y = 0.0
    for j in range(0,N):                              # Initial position
        x = 0.0
        for i in range(0,N):
            u[i][j][0] = 3*sin(2.0*x)*sin(y)          # Initial shape
            x += incrx
        y += incry
    for j in range(1,N-1):                            # First time step
        for i in range(1,N-1):
            u[i][j][1] = u[i][j][0] + 0.5*ratio *(u[i+1][j][0]+u[i-1][j][0]
                + u[i][j+1][0]+u[i][j-1][0]-4.*u[i][j][0])
    for k in range(1,tim):                            # Later time steps
        for j in range(1,N-1):
            for i in range(1,N-1):
                u[i][j][2] = 2.*u[i][j][1] - u[i][j][0] + ratio *(u[i+1][j][1]
                    + u[i-1][j][1] +u[i][j+1][1]+u[i][j-1][1] - 4.*u[i][j][1])
        u[:][:][0] = u[:][:][1]                       # Reset past
```

```
             u[:][:][1]  =  u[:][:][2]                        # Reset  present
30        for  j  in  range(0,N):
              for  i  in  range(0,N):
                  v[i][j]  =  u[i][j][2]                # Convert  to  2D  for  matplotlib
      return  v
34  v  =  vibration(tim)
    x1  =  range(0,  N)
    y1  =  range(0,  N)
    X,  Y  =  p.meshgrid(x1,y1)
38
    def  functz(v):       z  =  v[X,Y];  return  z

    Z  =  functz(v)
42  fig  =  p.figure()
    ax  =  Axes3D(fig)
    ax.plot_wireframe(X,  Y,  Z,  color  =  'r')
    ax.set_xlabel('x')
46  ax.set_ylabel('y')
    ax.set_zlabel('u(x,y)')
    p.show()
```

Listing 23.3   **CatFriction.py** Solves for waves on a catenary with friction.

```
    # CatFriction.py:  Solve  for  wave  on  catenary  with  friction

3   from  numpy  import  *

    dt  =  0.0001;  dx  =  0.01;  T  =  1;  rho  =  0.1;  maxtime  =  100;   kappa  =  30
    D  =  T/(rho*9.8)
7   x  =  zeros((512,3),float)
    q  =  open('CatFriction.dat','w');   rr  =  open('CatFunct.dat','w+t')

    for  i  in  range  (0,101):  x[i][0]  =  -0.08*sin(pi*i*dx)   # IC
11  for  i  in  range(1,100):  # First  step
        x[i][1]  =  (  dt*(T/rho)*((  x[i+1][0]-x[i][0]  )/dx*(  exp((i-50)*dx/D)
                      -exp(-(i-50)*dx/D))/D  +(exp((i-50)*dx/D)+exp(-(i-50)*dx/D))*
                      (  x[i+1][0]+x[i-1][0]-2.0*x[i][0]  )/(pow(dx,2))   )
15                    -2*kappa*x[i][0]+2*x[i][0]/dt   )/(2*kappa+(2/dt))
    for  k  in  range  (0,300):  # Other  steps
        for  i  in  range(1,100):
            x[i][2]  =  (dt*(T/rho)*((x[i+1][1]-x[i][1])/dx*(exp((i-50)*dx/D)  \
19                        -exp(-(i-50)*dx/D))/D  +(exp((i-50)*dx/D)+exp(-(i-50)*dx/D))  *\
                          (x[i+1][1]+x[i-1][1]-2.0*x[i][1])/(pow(dx,2)))\
                          -2*kappa*x[i][1]-(-2*x[i][1]+x[i][0])/dt)/(2*kappa+(1/dt))
        for  i  in  range(1,101):
23          x[i][0]  =  x[i][1]
            x[i][1]  =  x[i][2]
        if  (k%4==0  or  k==0):
            for  i  in  range(0,100):
27              a1=exp((i-50.)*dx/D)
                a2=exp(-(i-50.)*dx/D)
                rr.write("%7.3f"%(D*(a1+a2)))
                rr.write("\n")
31              q.write('%7.3f'%(x[i,2]))
                q.write("\n")
            q.write("\n");
            rr.write("\n");
35  rr.closed
    q.closed
    print("Data  stored  in  CatFrict.dat  and  CatFunct.dat")
```

24

Quantum Wave Packets and EM Waves

> This chapter extends the solution of wave equations that began in Chapter 23, to waves possessing multiple components. This requires algorithms with a bit more sophistication. First, we explore quantum wave packets, which have their real and imaginary parts solved for at slightly differing (split) times. Then, we explore electromagnetic waves, which have their interlaced E and H vector components also solved for at split times.

Problem An electron with a definite momentum is placed within an attractive potential that confines it to a 1D region the size of an atom. Determine the resulting electron's position in time and space.

24.1 Time-Dependent Schrödinger Equation

Because the region of confinement in this problem is the size of an atom, we need to solve it quantum mechanically. Because the particle started with both a definite momentum and position, we need to solve for both the spatial and time dependencies of the electron's wave function. Accordingly, we must solve the time-*dependent* Schrödinger equation. This is different from the time-*independent* eigenvalue problem of a particle bound in a box, which we considered in Chapters 6 and 13.

We model the electron being localized in space by a Gaussian with finite width centered at 5. Because we are told that the electron starts with a momentum, we multiply the Gaussian by a plane wave (a state of definite momentum k_o). And because this is the initial state at just the one time $t = 0$, we do not include any time dependence:

$$\psi(x, t = 0) = \exp\left[-\tfrac{1}{2}\left(\tfrac{x-5}{\sigma_0}\right)^2\right] e^{ik_o x}, \tag{24.1}$$

where we have set $\hbar = 1$. In spite of us setting up the wave function (24.1), as if it gives the electron a definite momentum and locations, (24.1) is an eigenstate of neither momentum nor position. Such are wave packets!

To solve the **problem**, we must propagate the initial wave function forward in time and through all of space (as we show in Figures 24.1 and 24.2). If (24.1) were an eigenstate of the Hamiltonian, it would have an exp($-i\omega t$) time dependence that can be factored out of the wave function (the usual in textbooks), and we would have a time-independent

Computational Physics: Problem Solving with Python, Fourth Edition.
Rubin H. Landau, Manuel J. Páez, and Cristian C. Bordeianu.

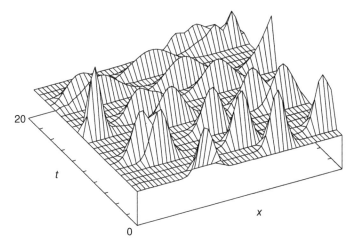

Figure 24.1 The probability density as a function of time and space for an electron confined to a square well. The electron is seen to start off on the left, spread out with time, and collide with the walls.

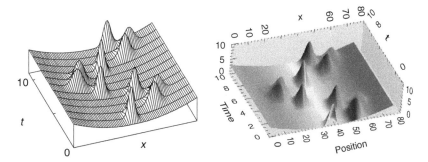

Figure 24.2 The probability density as a function of time for an electron confined to a 1D harmonic oscillator potential well. On the left is a conventional surface plot, while on the right is a color visualization.

Schrödinger equation. However, $\tilde{H}\psi \neq E\psi$ for this wave packet, and so we must solve the time-dependent Schrödinger equation:

$$i\frac{\partial \psi(x,t)}{\partial t} = \tilde{H}\psi(x,t) \tag{24.2}$$

$$i\frac{\partial \psi(x,t)}{\partial t} = -\frac{1}{2m}\frac{\partial^2 \psi(x,t)}{\partial x^2} + V(x)\psi(x,t), \tag{24.3}$$

where we have set $2m = 1$ and $\hbar = 1$. Because the initial wave function is complex (the plane wave), the wave function will be complex for all times. Accordingly, we have separate equations for the real and imaginary parts of the wave function:

$$\psi(x,t) = R(x,t) + iI(x,t), \tag{24.4}$$

$$\Rightarrow \quad \frac{\partial R(x,t)}{\partial t} = -\frac{\partial^2 I(x,t)}{\partial x^2} + V(x)I(x,t), \tag{24.5}$$

$$\frac{\partial I(x,t)}{\partial t} = +\frac{\partial^2 R(x,t)}{\partial x^2} - V(x)R(x,t), \tag{24.6}$$

where $V(x)$ is the confining potential.

24.2 Split-Time Algorithm

The time-dependent Schrödinger equation can be solved with both implicit (large-matrix) and explicit (leapfrog) methods. An extra challenge when solving the Schrödinger equation is the need to conserve probability, $\int_{-\infty}^{+\infty} dx\, \rho(x, t)$, to a high level of precision at all times. Here, we'll use an *explicit* method that, by design, provides a high level of probability conservation. It does so by solving for the real and imaginary parts of the wave function at slightly different, or "staggered", times [Askar and Cakmak, 1977; Visscher, 1991; Maestri *et al.*, 2000]. Explicitly, the real part R is determined at times $0, \Delta t, \ldots$, and the imaginary part I at times $\frac{1}{2}\Delta t, \frac{3}{2}\Delta t, \ldots$ The algorithm is based on (what else?) the Taylor expansions of R and I, for example,

$$R\left(x, t + \tfrac{1}{2}\Delta t\right) = R\left(x, t - \tfrac{1}{2}\Delta t\right) + [4\alpha + V(x)\,\Delta t]I(x, t)$$
$$- 2\alpha[I(x + \Delta x, t) + I(x - \Delta x, t)], \tag{24.7}$$

where $\alpha = \Delta t/[2(\Delta x)^2]$. In discrete form we have:

$$R_i^{n+1} = R_i^n - 2\left(\alpha[I_{i+1}^n + I_{i-1}^n] - 2[\alpha + V_i\,\Delta t]I_i^n\right), \tag{24.8}$$

$$I_i^{n+1} = I_i^n + 2\left(\alpha[R_{i+1}^n + R_{i-1}^n] - 2[\alpha + V_i\,\Delta t]R_i^n\right), \tag{24.9}$$

where the superscript n indicates the time, and the subscript i the position, e.g., $R_{x=i\Delta x}^{t=n\Delta t}$.

The probability density ρ is defined in terms of the wave function evaluated at three different times:

$$\rho(t) = \begin{cases} R^2(t) + I\left(t + \frac{\Delta t}{2}\right) I\left(t - \frac{\Delta t}{2}\right), & \text{for integer } t, \\[2mm] I^2(t) + R\left(t + \frac{\Delta t}{2}\right) R\left(t - \frac{\Delta t}{2}\right), & \text{for half-integer } t. \end{cases} \tag{24.10}$$

Although probability is not to be exactly conserved by the algorithm, the error is two orders higher than that in the wave function, and this is usually quite satisfactory. If it is not, then one must adjust the step sizes. While this definition of ρ may not seem intuitive, it reduces to the usual one for $\Delta t \to 0$, and so can be viewed as part of the art of numerical analysis. We will ask you to investigate just how well probability is conserved. We refer the reader to Koonin [1986] and Visscher [1991] for details on the stability of the algorithm.

24.2.1 Implementation

In Listing 24.1, you will find the program `HarmosAnimate.py` that solves for the motion of the wave packet (24.1) inside a harmonic oscillator potential. The program `slit.py` solves for the motion of a Gaussian wave packet as it passes through a slit (Figure 24.4). You should solve for a wave packet confined to an infinite square well:

$$V(x) = \begin{cases} \infty, & x < 0, \text{ or } x > 15, \\ 0, & 0 \leq x \leq 15. \end{cases}$$

1) Define arrays `psr[751,2]` and `psi[751,2]` for the real and imaginary parts of ψ, and `Rho[751]` for the probability. The first subscript refers to the x position on the grid, and the second to the present and future times.

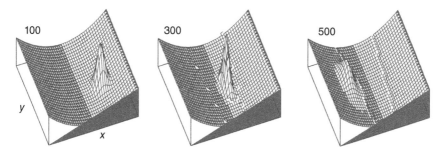

Figure 24.3 The probability density as a function of x and y at three different times for an electron confined to a 2D parabolic tube (infinitely long in y direction). The electron is initially a Gaussian wave packet in both the x and y directions.

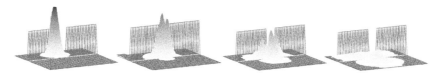

Figure 24.4 The probability density as a function of position and time for an electron incident upon and passing through a slit. A significant reflection is seen to be occurring.

2) Use the values $\sigma_0 = 0.5$, $\Delta x = 0.02$, $k_0 = 17\pi$, and $\Delta t = \frac{1}{2}\Delta x^2$.
3) Use equation (24.1) for the initial wave packet to define `psr[j,1]` for all j at $t = 0$, and to define `psi[j,1]` at $t = \frac{1}{2}\Delta t$.
4) Seeing that the wave function must vanish at the infinitely high well walls, set `Rho[1] = Rho[751] = 0.0`.
5) Increment time by $\frac{1}{2}\Delta t$. Use (24.8) to compute `psr[j,2]` in terms of `psr[j,1]`, and (24.9) to compute `psi[j,2]` in terms of `psi[j,1]`.
6) Repeat the steps through all of space, that is, for $i = 2–750$.
7) Throughout all of space, replace the present wave packet (second index 1) by the future wave packet (second index 2).
8) After you are sure that the program is running properly, repeat the time-stepping for ~ 5000 steps.

24.2.1.1 Animation

1) Output the probability density after every 200 steps.
2) Make a surface plot of probability *versus* position *versus* time. This should look like Figures 24.1 or 24.2.
3) Make an animation showing the wave function as a function of time.
4) Check how well probability is conserved for early and late times by determining the integral of the probability over all of space, $\int_{-\infty}^{+\infty} dx\, \rho(x)$, and seeing by how much it changes in time (its specific value doesn't matter as it's just normalization).
5) Explain why collisions with the walls cause the wave packet to broaden and break up. (*Hint*: The collisions do not appear so disruptive when a Gaussian wave packet is confined within a harmonic oscillator potential.)

24.2.2 Wave Packets in Other Wells

1D Well: Now confine the electron to a harmonic oscillator well:

$$V(x) = \tfrac{1}{2}x^2 \quad (-\infty \leq x \leq \infty). \tag{24.11}$$

Take the momentum as $k_0 = 3\pi$, the space step $\Delta x = 0.02$, and the time step $\Delta t = \tfrac{1}{4}\Delta x^2$. Note that the wave packet broadens in time, but then returns to its initial shape!

2D Well: Now confine the electron to a 2D parabolic tube (Figure 24.3):

$$V(x,y) = 0.9x^2, \quad -9.0 \leq x \leq 9.0, \quad 0 \leq y \leq 18.0. \tag{24.12}$$

The extra degree of freedom means that we must solve the 2D PDE:

$$i\frac{\partial \psi(x,y,t)}{\partial t} = -\left(\frac{\partial^2 \psi}{\partial x^2} + \frac{\partial^2 \psi}{\partial y^2}\right) + V(x,y)\psi. \tag{24.13}$$

Assume that the electron's initial localization is described by the 2D Gaussian wave packet:

$$\psi(x,y,t=0) = e^{ik_{0x}x}\, e^{ik_{0y}y}\, \exp\left[-\frac{(x-x_0)^2}{2\sigma_0^2}\right] \exp\left[-\frac{(y-y_0)^2}{2\sigma_0^2}\right]. \tag{24.14}$$

Note that you can solve the 2D equation by extending the method we just used in 1D, or you can look at Section 24.3 where we develop a special algorithm for the Schrödinger equation.

1) Determine the motion of a 2D Gaussian wave packet within a 2D harmonic oscillator potential:

$$V(x,y) = 0.3(x^2 + y^2), \quad -9.0 \leq x \leq 9.0, \quad -9.0 \leq y \leq 9.0. \tag{24.15}$$

   a) Center the initial wave packet at $(x,y) = (3.0, -3)$, and give it momentum $(k_{0x}, k_{0y}) = (3.0, 1.5)$.
   b) Young's single-slit experiment has a wave passing through a small slit with the transmitted wave showing interference effects. In quantum mechanics, where we represent a particle by a wave packet, this means that an interference pattern should be formed when a particle passes through a small slit. Pass a Gaussian wave packet of width 3 through a slit of width 5 (Figure 24.4), and observe the resulting quantum interference.

24.3 Special Schrödinger Algorithm

We have just developed an algorithm to solve the time-dependent Schrödinger equation in 2D by extending the 1D algorithm to another dimension. Another approach uses quantum theory to obtain a more powerful algorithm [Maestri *et al.*, 2000]. First, we note that equation (24.13) can be integrated formally to obtain the solution [Landau and Lifshitz, 1976]:

$$\psi(x,y,t) = U(t)\psi(x,y,t=0) \tag{24.16}$$

$$U(t) = e^{-i\tilde{H}t}, \qquad \tilde{H} = -\left(\frac{\partial^2}{\partial x^2} + \frac{\partial^2}{\partial y^2}\right) + V(x, y). \tag{24.17}$$

Here, $U(t)$ is an operator that translates a wave function forward in time, and $\tilde{H}$ is the Hamiltonian operator. From this formal solution, we deduce that a wave packet can be translated ahead of time Δt via:

$$\psi_{i,j}^{n+1} = U(\Delta t)\psi_{i,j}^n, \tag{24.18}$$

where the superscript denotes time, $t = n\Delta t$, and the subscripts denote the two spatial variables, $x = i\Delta x, y = j\Delta y$. Likewise, the inverse of the time evolution operator moves the solution back one time step:

$$\psi^{n-1} = U^{-1}(\Delta t)\psi^n = e^{+i\tilde{H}\Delta t}\psi^n. \tag{24.19}$$

While it would be nice to have an algorithm based on a direct application of (24.19), the references show that the resulting algorithm would not be stable [Askar and Cakmak, 1977]. That being so, we base our algorithm on an indirect application, namely, the relation between the difference in ψ^{n+1} and ψ^{n-1}:

$$\psi^{n+1} = \psi^{n-1} + (e^{-i\tilde{H}\Delta t} - e^{+i\tilde{H}\Delta t})\psi^n, \tag{24.20}$$

where the difference in sign of the exponents is to be noted. The algorithm derives from combining the $O(\Delta x^2)$ expression for the second derivative obtained from the Taylor expansion,

$$\frac{\partial^2\psi}{\partial x^2} \simeq -\frac{1}{2}(\psi_{i+1,j}^n + \psi_{i-1,j}^n - 2\psi_{i,j}^n), \tag{24.21}$$

with the corresponding-order expansion of the evolution equation (24.20). Substituting the resulting expression for the second derivative into the 2D time-dependent Schrödinger equation results in:[1]

$$\psi_{i,j}^{n+1} = \psi_{i,j}^{n-1} - 2i\left[\left(4\alpha + \frac{1}{2}\Delta t V_{i,j}\right)\psi_{i,j}^n - \alpha\left(\psi_{i+1,j}^n + \psi_{i-1,j}^n + \psi_{i,j+1}^n + \psi_{i,j-1}^n\right)\right],$$

where $\alpha = \Delta t/[2(\Delta x)^2]$. We convert this equation with its complex wave function to coupled real equations by substituting $\psi = R + iI$:

$$R_{i,j}^{n+1} = R_{i,j}^{n-1} + 2\left[\left(4\alpha + \frac{1}{2}\Delta t V_{i,j}\right)I_{i,j}^n - \alpha\left(I_{i+1,j}^n + I_{i-1,j}^n + I_{i,j+1}^n + I_{i,j-1}^n\right)\right],$$

$$I_{i,j}^{n+1} = I_{i,j}^{n-1} - 2\left[\left(4\alpha + \frac{1}{2}\Delta t V_{i,j}\right)R_{i,j}^n + \alpha\left(R_{i+1,j}^n + R_{i-1,j}^n + R_{i,j+1}^n + R_{i,j-1}^n\right)\right].$$

This is the algorithm for integrating the 2D Schrödinger equation. Probability is determined using the same expression (24.10) as used in 1D.

24.4 Quantum Chaos

Finding signals of chaos in quantum systems can be challenging. To avoid computation, much of the research has assumed the approximate, semiclassical formulation of quantum

---

1 The constants in the equation change as the dimension of the equation changes; that is, there will be different constants for the 3D equation, and therefore, our constants are different from those in the references!

mechanics. We won't do that. There are a number of ways in which chaotic behavior may occur in quantum systems, some rather obvious, and others more subtle. In this section, we look at the obvious; obvious because the classical versions of the systems display chaos, and so we know what to look for in the quantum systems.

24.4.1 Quantum Billiards

Here, we examine a quantum system whose classical analog has definite chaotic behavior, as seen in Figures 24.5 and 24.6. In the classical chaotic system, the billiard ball bounces around endlessly, filling all of the allowed space without ever repeating (Figure 24.6 left). The program for this is the same as for scattering from three disks, 3QMdisks.py in Listing 24.2, but with the disks removed. You must impose $\psi = 0$ at the boundaries to describe reflection from a hard wall. Some of our results after 200 time steps are shown in Figure 24.6 right. We see that the quantum system displays a signature of the endless classical reflections from the table's edges.

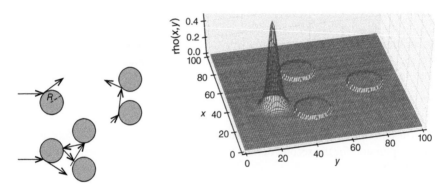

Figure 24.5 *Left:* Classical trajectories for scatterings from one, two, and three stationary disks. *Right*: A wave packet incident on three disks of radius 10 located at the vertices of an equilateral triangle.

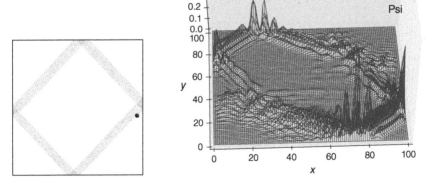

Figure 24.6 *Left*: The path followed by a classical ball bouncing elastically within a square billiard. *Right*: A quantum wave packet, initially Gaussian in shape, after 200 time steps confined within a square billiard table.

1) Compute the motion of an initial Gaussian wave packet confined to the top of a square billiards table. Match the initial conditions to those that lead to periodic orbits of the classical billiard.

    (a) Compute the classical motion of a square billiard for a series of times, ranging from those in which only a small number of reflections has occurred, to very large times.

    (b) Examine the wave function for the same range of times as used in the classical computation, and compare to the classical results.

    (c) How many reflections does it take for the wave packet to lose all traces of the classical trajectories?

2) Examine the motion of a quantum wave packet for the various billiards studied classically: by a) a circle, b) a stadium, c) a circle with a disk in the middle. In all cases, examine initial conditions that lead to classically periodic orbits.

24.4.2 Three Disks Scattering

Another system in which we might be able to see the quantum chaos happening is quantum scattering from three fixed hard disks (Figure 24.5 left). Here, we examine the scattering of a wave packet from several hard disk configurations (Figure 24.5 right), with quantum chaos possible for the three-disk scattering. To make your work easier, in Listing 24.2, we give our program 3QMdisks.py for quantum scattering from three fixed hard disks. The disks have radius R, a center-to-center separations of a, and are placed at the vertices of an equilateral triangle. As seen in Figure 24.5, a Gaussian wave packet,

$$\psi(x, y, t = 0) = e^{i(k_x x + k_y y)} e^{-A(x-x_0)^2 - A(y-y_0)^2}, \tag{24.22}$$

is incident upon the disks. *Note*: The program, as given, has the disks confined within a very small box. This may be appropriate for a bound state billiard problem, but not for scattering. You will need to enlarge the confining box in order to eliminate the effects of reflections from the boundary. And while you are free to make surface plots of Re $\psi(x, y)$ and Im $\psi(x, y)$, we find the interpretation easier with a surface plot of the probability density $z(x, y) = \rho(x, y)$.

1) Start with scattering from one disk–

    (a) Produce surface plots of $\rho(x, y)$ for times from 0 until the packet leaves the scattering region. Note, the edges of the billiard table will reflect the wave packet and, ergo, should be far from the disks in order to not interfere with the scattering from the disks.

    (b) Examine the qualitative effect of varying the size of the disk.

    (c) Examine the qualitative effect of varying the momentum of the wave packet.

    (d) Vary the initial position of the wave packet so that there are nearly head-on collisions as well as ones at glancing angles.

2) Next, repeat your investigation for the two disk system. Try to vary the parameters so that you obtain many *multiple* scatterings from the disks. In particular, see if you can find the analog of the classical case where there is a trapped orbit with unending back-and-forth scatterings. (*Hint*: Try starting the wave packet between the two disks.)

3) Next, extend and repeat the investigation to the three-disk system. Vary the parameters so that you obtain many multiple scatterings from the disks. In particular, see if you

can find the analog of the classical case where there are trapped orbits with unending back-and-forth scatterings.

(a) Develop an algorithm that determines the time delay of the wave packet, that is, the time it takes for most of the initial packet to leave the scattering region.

(b) Plot the time delay versus the wave packet momentum and look for indications of chaos, such as sharp peaks or rapid changes. The literature indicates that high degrees of multiple scattering occur when $a/R \simeq 6$.

24.5 E&M Waves: Finite Difference Time Domain

Simulations of electromagnetic waves are of tremendous practical importance. As with other wave equations, the basic technique is again calculating the solution on a spacetime lattice using finite difference and time stepping. When used for E&M simulations, the technique is known as the finite difference time domain (FDTD) method. What is new here is the coupling of the E and H fields, with the variations of one vector generating the other. More complete treatments can be found in Sullivan [2000] and Ward and Nelson [2004].

Problem You are given a region in space, $0 \leq z \leq 200$, in which the E and H fields are known to have a sinusoidal spatial variation:

$$E_x(z, t = 0) = 0.1 \sin\left(\frac{2\pi z}{100}\right), \qquad H_y(z, t = 0) = 0.1 \sin\left(\frac{2\pi z}{100}\right), \tag{24.23}$$

with all other components vanishing. Determine the fields for all subsequent times.

24.6 Maxwell's Equations

The description of electromagnetic (EM) waves via Maxwell's equations is covered in many textbooks. For propagation in just the z dimension, and for free space with no sinks or sources, there are four coupled PDE's:

$$\vec{\nabla} \cdot \mathbf{E} = 0 \Rightarrow \frac{\partial E_x(z, t)}{\partial x} = 0 \tag{24.24}$$

$$\vec{\nabla} \cdot \mathbf{H} = 0 \Rightarrow \frac{\partial H_y(z, t)}{\partial y} = 0, \tag{24.25}$$

$$\frac{\partial \mathbf{E}}{\partial t} = +\frac{1}{\epsilon_0}\vec{\nabla} \times \mathbf{H} \Rightarrow \frac{\partial E_x}{\partial t} = -\frac{1}{\epsilon_0}\frac{\partial H_y(z, t)}{\partial z}, \tag{24.26}$$

$$\frac{\partial \mathbf{H}}{\partial t} = -\frac{1}{\mu_0}\vec{\nabla} \times \mathbf{E} \Rightarrow \frac{\partial H_y}{\partial t} = -\frac{1}{\mu_0}\frac{\partial E_x(z, t)}{\partial z}. \tag{24.27}$$

As indicated in Figure 24.7, we have chosen the electric field $\mathbf{E}(z, t)$ to be polarized (oscillate) in the x direction, and the magnetic field $\mathbf{H}(z, t)$ to be polarized in the y direction. This is a transverse electromagnetic (TEM) wave. As indicated by the bold arrow in Figure 24.7, the direction of power flow is given by the right-hand rule applied to $\mathbf{E} \times \mathbf{H}$. Note that although we have set the initial conditions such that the EM wave is traveling in only the z dimension,

Figure 24.7 A single electromagnetic pulse traveling along the *z* axis. The coupled *E* and *H* pulses are indicated by solid and dashed curves, respectively, and the pulses at different *z* values correspond to different times.

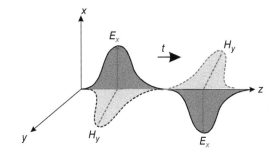

the electric field oscillates in the perpendicular *x* direction, and the magnetic field oscillates in the perpendicular *y* direction. So, while there may be propagation in just one direction, the vector nature of the fields means that the wave occurs in all three dimensions.

24.7 Split-Time FDTD

We need to solve the two coupled PDEs (24.26) and (24.27). As is usual for PDEs, we approximate the derivatives via the central-difference approximation, here in both time and space. For example,

$$\frac{\partial E(z,t)}{\partial t} \simeq \frac{E\left(z, t + \frac{\Delta t}{2}\right) - E\left(z, t - \frac{\Delta t}{2}\right)}{\Delta t}, \tag{24.28}$$

$$\frac{\partial E(z,t)}{\partial z} \simeq \frac{E\left(z + \frac{\Delta z}{2}, t\right) - E\left(z - \frac{\Delta z}{2}, t\right)}{\Delta z}. \tag{24.29}$$

We next substitute these approximations into Maxwell's equations, and rearrange the equations in the form of an algorithm that advances the solution in time. Because only first derivatives occur in Maxwell's equations, the equations are simple, although the electric and magnetic fields are intermixed.

As we have done with the time-dependent Schrodinger equation, we set up a space-time lattice (Figure 24.8), in which there are half-integer time steps, but now extended to half-integer space steps as well [Yee, 1966]. The magnetic field will be determined at integer time sites and half-integer space sites (open circles), while the electric field will be determined at half-integer time sites and integer space sites (filled circles). This extra level of complication of interlaced lattices leads to an accurate and robust algorithm. Because the fields already have subscripts indicating their vector nature, we indicate the lattice position as superscripts, for example,

$$E_x(z,t) \to E_x(k\Delta z, n\Delta t) \to E_x^{k,n}. \tag{24.30}$$

Maxwell's equations, (24.26) and (24.27), now become the discrete equations:

$$\frac{E_x^{k,n+1/2} - E_x^{k,n-1/2}}{\Delta t} = -\frac{H_y^{k+1/2,n} - H_y^{k-1/2,n}}{\epsilon_0 \Delta z},$$

$$\frac{H_y^{k+1/2,n+1} - H_y^{k+1/2,n}}{\Delta t} = -\frac{E_x^{k+1,n+1/2} - E_x^{k,n+1/2}}{\mu_0 \Delta z}.$$

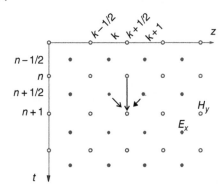

Figure 24.8 The algorithm for using the known values of E_x and H_y at two earlier times, and three different space positions, to advance the solution to the present time. The values of E_x are determined on the lattice of filled circles, corresponding to integer space indices and half-integer time indices. In contrast, the values of H_y are determined on the lattice of open circles, corresponding to half-integer space indices and integer time indices.

In summary, this formulation solves for the electric field at integer space steps, k, but half-integer time steps, n, while the magnetic field is solved for at half-integer space steps, but integer time steps.

We convert these equations into two simultaneous algorithms by solving for E_x at time $n + \frac{1}{2}$, and H_y at time n:

$$E_x^{k,n+1/2} = E_x^{k,n-1/2} - \frac{\Delta t}{\epsilon_0 \, \Delta z} \left(H_y^{k+1/2,n} - H_y^{k-1/2,n} \right), \tag{24.31}$$

$$H_y^{k+1/2,n+1} = H_y^{k+1/2,n} - \frac{\Delta t}{\mu_0 \Delta z} \left(E_x^{k+1,n+1/2} - E_x^{k,n+1/2} \right). \tag{24.32}$$

The very nature of Maxwell's equations requires these algorithms to be applied simultaneously: the space variation of H_y determines the time derivative of E_x, and the space variation of E_x determines the time derivative of H_y (Figure 24.8). These algorithms are more involved than our usual time-stepping ones in that the electric fields (filled circles in Figure 24.8) at future times, $t = n + \frac{1}{2}$, are determined from the electric fields at one time step earlier, $t = n - \frac{1}{2}$, and the magnetic fields at half a time step earlier, $t = n$. Likewise, the magnetic fields (open circles in Figure 24.8) at future times, $t = n + 1$, are determined from the magnetic fields at one time step earlier, $t = n$, and the electric field at half a time step earlier, $t = n + \frac{1}{2}$. In other words, it is as if we have two interleaved lattices, with the electric fields determined for half-integer times on lattice 1 and the magnetic fields at integer times on lattice 2.

Although these half-integer times are the norm for FDTD methods [Taflove and Hagness, 2000; Sullivan, 2000], it may be easier for some readers to understand them by doubling the index values, and referring to even and odd times instead:

$$E_x^{k,n} = E_x^{k,n-2} - \frac{\Delta t}{\epsilon_0 \Delta z} \left(H_y^{k+1,n-1} - H_y^{k-1,n-1} \right), \qquad k\text{ even}, n\text{ odd}, \tag{24.33}$$

$$H_y^{k,n} = H_y^{k,n-2} - \frac{\Delta t}{\mu_0 \Delta z} \left(E_x^{k+1,n-1} - E_x^{k-1,n-1} \right), \qquad k\text{odd}, n\text{ even}. \tag{24.34}$$

This makes it clear that E is determined for even space indices and odd times, while H is determined for odd space indices and even times.

We simplify the algorithm, and make its stability analysis simpler, by renormalizing the electric fields to have the same dimensions as the magnetic fields:

$$\tilde{E} = \sqrt{\frac{\epsilon_0}{\mu_0}} E. \tag{24.35}$$

The algorithms (24.31) and (24.32) now become:

$$\tilde{E}_x^{k,n+1/2} = \tilde{E}_x^{k,n-1/2} + \beta \left(H_y^{k-1/2,n} - H_y^{k+1/2,n} \right), \tag{24.36}$$

$$H_y^{k+1/2,n+1} = H_y^{k+1/2,n} + \beta \left(\tilde{E}_x^{k,n+1/2} - \tilde{E}_x^{k+1,n+1/2} \right), \tag{24.37}$$

$$\beta = \frac{c}{\Delta z/\Delta t}, \quad c = \frac{1}{\sqrt{\epsilon_0 \mu_0}}. \tag{24.38}$$

Here, c is the speed of light in a vacuum, and β is the ratio of the speed of light to the grid velocity $\Delta z/\Delta t$.

The space step Δz and the time step Δt must be chosen so that the algorithms are stable. The scales of the space and time dimensions are set by the wavelength and frequency, respectively, of the propagating wave. As a minimum, we want at least 10 grid points to fall within a wavelength:

$$\Delta z \leq \frac{\lambda}{10}. \tag{24.39}$$

The time step is then determined by the Courant stability condition [Taflove and Hagness, 2000; Sullivan, 2000] to be:

$$\beta = \frac{c}{\Delta z/\Delta t} \leq \frac{1}{2}. \tag{24.40}$$

As we have seen before, (24.40) implies that making the time step smaller improves precision and maintains stability, but making the space step smaller must be accompanied by a simultaneous decrease in the time step in order to maintain stability (you should check this).

24.7.1 Implementation and Assessment

In Listing 24.3, we provide a simple implementation of the FDTD algorithm for a z lattice of 200 sites. It produces the output shown in Figure 24.9. The initial conditions correspond to a sinusoidal variation of the E and H fields for $0 \leq z \leq 200$:

$$E_x(z, t = 0) = 0.1 \sin\left(\frac{2\pi z}{100}\right), \quad H_y(z, t = 0) = 0.1 \sin\left(\frac{2\pi z}{100}\right), \tag{24.41}$$

The algorithm then steps out in time for as long as the user desires. The discrete form of Maxwell equations used are:

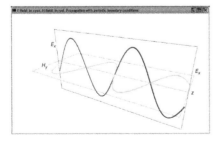 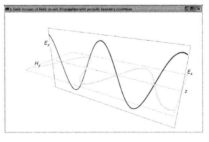

Figure 24.9 The E field (light colored) and the H field (dark) at the initial time (left), and at a later time (right). Periodic boundary conditions are used at the ends of the spatial region, which means that the large z wave continues into the $z = 0$ wave.

```
1 Ex[k, 1] = Ex[k, 0]  + beta * (Hy[k-1, 0] - Hy[k+1, 0])
  Hy[k, 1] = Hy[k, 0]  + beta * (Ex[k-1, 0] - Ex[k+1, 0])
```

The second index takes the values 0 and 1, with 0 being the old time and 1 the new. At the end of each iteration, the new field throughout all of space becomes the old one, and a newer one is computed. As the spatial endpoints, `k=0` and `k=xmax-1`, are not defined, we assume periodic boundary conditions.

1) Impose boundary conditions that make all fields vanish on the boundaries.
2) Test the Courant condition (24.40) by examining the stability of the solution for different values of Δz and Δt.
3) The direction of propagation of the pulse is in the $\mathbf{E} \times \mathbf{H}$ direction, which depends on the relative phase between the $\mathbf{E}$ and $\mathbf{H}$ fields. Verify that with no initial $\mathbf{H}$ field, we obtain pulses both to the right and the left.
4) Modify the program so that there is an initial $\mathbf{H}$ pulse, as well as an initial $\mathbf{E}$ pulse, both with Gaussian times in a sinusoidal shape.
5) Verify that the direction of propagation changes if the $\mathbf{E}$ and $\mathbf{H}$ fields have relative phases of 0 or π.
6) Investigate the resonator modes of a wave guide by picking the initial conditions corresponding to plane waves with nodes at the boundaries.
7) Investigate standing waves with wavelengths longer than the size of the integration region.
8) Simulate unbounded propagation by building in periodic boundary conditions into the algorithm.
9) Place a medium with periodic permittivity in the integration volume. This should act as a frequency-dependent filter, which blocks certain frequencies.
10) Extend the algorithm to include the effect of entering, propagating through, and exiting a *dielectric* material placed within the z integration region.
    (a) Ensure that you see both transmission and reflection at the dielectric boundaries.
    (b) Investigate the effect of varying the dielectric's index of refraction.

24.8 More E&M Problems

24.8.1 Circularly Polarized Waves

We now extend our treatment to EM waves that are not restricted to linear polarizations. Accordingly, we add H_x and E_y variations to (24.26) and (24.27):

$$\frac{\partial H_x}{\partial t} = \frac{1}{\mu_0} \frac{\partial E_y}{\partial z}, \tag{24.42}$$

$$\frac{\partial E_y}{\partial t} = \frac{1}{\epsilon_0} \frac{\partial H_x}{\partial z}. \tag{24.43}$$

When discretized in the same way as (24.31) and (24.32), we obtain:

$$H_x^{k+1/2,n+1} = H_x^{k+1/2,n} + \frac{\Delta t}{\mu_0 \, \Delta z} \left(E_y^{k+1,n+1/2} - E_y^{k,n+1/2} \right), \tag{24.44}$$

$$E_y^{k,n+1/2} = E_y^{k,n-1/2} + \frac{\Delta t}{\epsilon_0 \Delta z}\left(H_y^{k+1/2,n} - H_y^{k-1/2,n}\right).$$ (24.45)

To produce a circularly polarized traveling wave, we set as initial conditions:

$$E_x = \cos\left(t - \frac{z}{c} + \phi_y\right), \qquad H_x = \sqrt{\frac{\epsilon_0}{\mu_0}}\cos\left(t - \frac{z}{c} + \phi_y\right),$$ (24.46)

$$E_y = \cos\left(t - \frac{z}{c} + \phi_x\right), \qquad H_y = \sqrt{\frac{\epsilon_0}{\mu_0}}\cos\left(t - \frac{z}{c} + \phi_x + \pi\right).$$ (24.47)

We take the phases to be $\phi_x = \pi/2$ and $\phi_y = 0$, so that their difference $\phi_x - \phi_y = \pi/2$, which leads to circular polarization. We include the initial conditions in the same manner as we did the Gaussian pulse, only now with these cosine functions.

Listing 24.5 gives our implementation `EMcirc.py` for waves with transverse two-component **E** and **H** fields. Some results of the simulation are shown in Figure 24.11, where you will note the difference in phase between **E** and **H**.

24.8.2 Wave Plates

Problem Develop a numerical model for a wave plate that convert a linearly polarized electromagnetic wave into a circularly polarized one.

As seen in Figure 24.10, a wave plate is an optical device that alters the polarization of light traveling through it by shifting the relative phase of the components of the polarization vector. A quarter wave plate introduces a relative phase of $\lambda/4$, where λ is the wavelength of the light. Physically, a wave plate is often a birefringent crystal in which the different propagation velocities of waves in two orthogonal directions lead to a phase change. The thickness of the plate determines the amount of phase change.

To simulate a wave plate, we start with a linearly polarized wave, with both E_x and E_y components propagating along the z direction. The wave enters the plate and emerges from it while still traveling in the z direction, but now with the relative phase of these fields

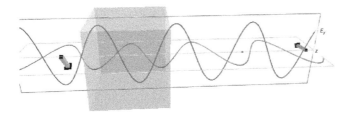

Figure 24.10 One frame from the program `quarterwave.py` in Listing 24.4, showing a linearly polarized electromagnetic wave entering a quarter wave plate from the left, and leaving as a circularly polarized wave on the right (the arrow on the left oscillates back and forth at 45° while the one on the right rotates).

Figure 24.11 *E* and *H* fields at $t = 100$ for a circularly polarized wave.

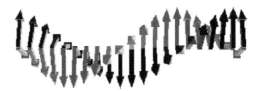

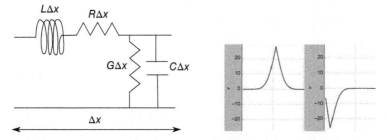

Figure 24.12 *Left*: A transmission line that repeats every Δx. *Right*: Two frames of an animation produced by `TeleMat.py` showing a wave transmitted along a telegraph line and being reflected from an end.

shifted. Of course, because this is an EM wave, there will also be coupled magnetic field components present, in this case H_x and H_y, and they too need be computed.

Maxwell equations for wave propagating along the z axis are:

$$\frac{\partial H_x}{\partial t} = +\frac{1}{\mu_0}\frac{\partial E_y}{\partial z}, \qquad \frac{\partial H_y}{\partial t} = -\frac{1}{\mu_0}\frac{\partial E_x}{\partial z}, \tag{24.48}$$

$$\frac{\partial E_x}{\partial t} = -\frac{1}{\epsilon_0}\frac{\partial H_y}{\partial z}, \qquad \frac{\partial E_y}{\partial t} = +\frac{1}{\epsilon_0}\frac{\partial H_x}{\partial z}. \tag{24.49}$$

We take as initial conditions a wave incident from the left along the z axis, linearly polarized (electric field direction of 45°), with corresponding, and perpendicular, $\boldsymbol{H}$ components:

$$E_x(t=0) = 0.1\cos\left(\tfrac{2\pi x}{\lambda}\right), \qquad E_y(t=0) = 0.1\cos\left(\tfrac{2\pi y}{\lambda}\right), \tag{24.50}$$

$$H_x(t=0) = 0.1\cos(\tfrac{2\pi x}{\lambda}), \qquad H_y(t=0) = 0.1\cos(\tfrac{2\pi y}{\lambda}). \tag{24.51}$$

Because only the relative phases matter, we simplify the calculation by assuming that the E_y and H_x components do not have their phases changed, but that the E_x and H_y components do (in this case by $\lambda/4$ when they leave the plate). Of course, after leaving the plate and traveling in free space, there are no further changes in relative phase.

24.8.3 Algorithm and Exercise

As in Section 24.7 and Figure 24.8, we follow the FDTD approach of using known values of E_x and H_y at three earlier times, and three different space positions, to obtain the solution at the present time. With the renormalized electric fields as in (24.35), this leads to the beautifully symmetric equations:

$$E_x^{k,n+1} = E_x^{k,n} + \beta\left(H_y^{k+1,n} - H_y^{k,n}\right), \tag{24.52}$$

$$E_y^{k,n+1} = E_y^{k,n} + \beta\left(H_x^{k+1,n} - H_x^{k,n}\right), \tag{24.53}$$

$$H_x^{k,n+1} = H_x^{k,n} + \beta\left(E_y^{k+1,n} - E_y^{k,n}\right), \tag{24.54}$$

$$H_y^{k,n+1} = H_y^{k,n} + \beta\left(E_x^{k+1,n} - E_x^{k,n}\right). \tag{24.55}$$

1) Modify the program FDTD.py in Listing 24.3 so that it solves the algorithm (24.52)–(24.55). Use $\beta = 0.01$.

   (a) After each time step, impose a gradual increment of the phase so that the total phase change will be one quarter of a wavelength. Our program for this is quarterplat.py.

   (b) Verify that the plate converts an initially linearly polarized wave into a circularly polarized one.

   (c) Verify that the plate converts an initially circularly polarized wave into a linearly polarized one.

   (d) What happens if you put two plates together? Three? Four? (Verify!)

24.8.4 Twin Lead Transmission Line

The model of a twin-lead transmission line consists of two parallel wires on which alternating current or pulses propagate [Sullivan, 2000]. The equivalent circuit for a segment of length Δx is shown on the left of Figure 24.12. There is an inductance $L\Delta x$, a resistance $R\Delta x$, a capacitance $C\Delta x$, and a conductance (inverse resistance of the dielectric material insulating the wires) $G\Delta x$. The telegrapher's equations describe the voltage and current:

$$\frac{\partial V(x,t)}{\partial x} = -RI - L\frac{\partial I(x,t)}{\partial t}, \tag{24.56}$$

$$\frac{\partial I(x,t)}{\partial x} = -GV - C\frac{\partial V(x,t)}{\partial t}. \tag{24.57}$$

For lossless transmission lines, that is those with $R = G = 0$, the equations become:

$$\frac{\partial V(x,t)}{\partial x} = -L\frac{\partial I(x,t)}{\partial t}, \qquad \frac{\partial I(x,t)}{\partial x} = -C\frac{\partial V(x,t)}{\partial t}. \tag{24.58}$$

Differentiation of these equations, and substitution into one another, leads to the familiar 1D wave equation:

$$\frac{\partial^2 V(x,t)}{c^2\partial t^2} - \frac{\partial^2 V(x,t)}{\partial x^2} = 0, \qquad c = \frac{1}{\sqrt{LC}}. \tag{24.59}$$

a) Apply the leapfrog algorithm to the telegrapher's equations (24.56), making sure to account for the stability condition:

$$c\frac{\Delta t}{\Delta x} \leq 1. \tag{24.60}$$

Experiment with different values for Δx and Δt in order to obtain better precision, or to speed up the computation.

b) Impose the boundary conditions $V(0,t) = V(L,t) = 0$, where L is the length of the transmission line.

c) Use as initial conditions a pulse with a constant voltage:

$$V(x,t=0) = 10\,e^{-x^2/0.1}, \qquad \frac{\partial V(x,t)}{\partial t} = 0. \tag{24.61}$$

d) Good values to try are $L = 0.1$, $C = 2.5$, $\Delta t = 0.025$, and $\Delta x = 0.05$.

e) Investigate the effect of a zero value for the conductance G and the resistance R. Do your results agree with what you might expect?

f) For nonzero R and G, investigate the distortion that occurs when a rectangular pulse is sent down the transmission line. At what point would you say the pulse shape becomes unrecognizable?

24.9   Code Listings

Listing 24.1   HarmosAnimate.py solves the time-dependent Schrödinger equation for
a Gaussian wave packet moving within a harmonic oscillator potential.

```
# HarmonsAnimate: Solve t-dependent Sch Eqt for HO wi animation

from visual import *

# Initialize wave function, probability, potential
dx = 0.04;     dx2 = dx*dx;   k0 = 5.5*pi;   dt = dx2/20.0;
xmax = 6.0; beta = dt/dx2
xs = arange(-xmax,xmax+dx/2,dx)                    # Array x values

g = display(width=500, height=250, title='Wave Packet in HO Well')
PlotObj = curve(x=xs, color=color.yellow, radius=0.1)
g.center = (0,2,0)                                 # Center of scene
                                                   # Initial wave packet
R = exp(-0.5*(xs/0.5)**2) * cos(k0*xs)             # Array Re I
I = exp(-0.5*(xs/0.5)**2) * sin(k0*xs)             # Array Im I
V   = 15.0*xs**2                                   # The potential

while True:
    rate(500)
    R[1:-1] = R[1:-1] - beta*(I[2:]+I[:-2]-2*I[1:-1])+dt*V[1:-1]*I[1:-1]
    I[1:-1] = I[1:-1] + beta*(R[2:]+R[:-2]-2*R[1:-1])-dt*V[1:-1]*R[1:-1]
    PlotObj.y = 4*(R**2 + I**2)
```

Listing 24.2   3QMdisks.py solves for a wave packet scattering from 3 disks.

```
# 3QMdisks.py: Wavepacket scattering from 3 disks wi MatPlot

import matplotlib.pylab as p,  numpy as np
from mpl_toolkits.mplot3d import Axes3D

r = 10;        N = 101;      x1 = 51; # 51 = 90.*sqrt 3/2-30
dx = 0.1;      dx2 = dx*dx; k0 = 20.;   k1 = 0.
dt  = 0.002; fc = dt/dx2; Xo = 40;     Yo = 25
    # Declare arrays
V    = np.zeros((N,N),float);   Rho  = np.zeros((N,N),float)
RePsi = np.zeros((N,N),float);  ImPsi = np.zeros((N,N),float)
ix = np.arange(0, 101);         iy = np.arange(0,101)
X, Y = np.meshgrid(ix, iy)
fig = p.figure(); ax = Axes3D(fig)                # Create figure

def Pot1Disk(xa,ya):                    # Potential single disk
    for i in range(ya-r,ya+r+1):
        for j in range(xa-r,xa+r+1):
            if np.sqrt((i-ya)**2+(j-xa)**2)<=r:  V[i,j] = 5.

def Pot3Disks():                        # Potential three disk
    Pot1Disk(30,45);   Pot1Disk(70,45); Pot1Disk(50,80)

def Psi_0(Xo,Yo):                          # Initial Psi
    for i in np.arange(0,N):
        for j in np.arange(0, N):
            Gaussian = np.exp(-0.03*(i-Yo)**2-0.03*(j-Xo)**2)
            RePsi[i,j] = Gaussian*np.cos(k0*i+k1*j)
            ImPsi[i,j] = Gaussian*np.sin(k0*i+k1*j)
            Rho[i,j] = RePsi[i,j]**2 + ImPsi[i,j]**2 + 0.01
Psi_0(Xo,Yo)   # Psi and Rho initial
Pot3Disks()                                     # Initial Psi
for t in range(0, 120):  # 120->30         # Compute Psi t < 120
    if t%5 == 0:  print ('t =', t )            # Print ea 5th t
    ImPsi[1:-1,1:-1] =  ImPsi[1:-1,1:-1] + fc*(RePsi[2: ,1:-1] \
        + RePsi[:-2 ,1:-1] - 4*RePsi[1:-1,1:-1] + RePsi[1:-1,2: ]\
```

```
                 + RePsi[1:-1, :-2]) + V[1:-1,1:-1]*dt*RePsi[1:-1,1:-1]
          RePsi[1:-1,1:-1] = RePsi[1:-1,1:-1] - fc*(ImPsi[2: ,1:-1]\
              +ImPsi[ :-2,1:-1] - 4*ImPsi[1:-1,1:-1] + ImPsi[1:-1,2: ]\
40            +ImPsi[1:-1, :-2]) + V[1:-1,1:-1]*dt*ImPsi[1:-1,1:-1]
          for i in range(1, N-1):                          # Compute Rho
              for j in range(1,N-1):
                  if V[i,j] !=0: RePsi[i,j] = 0; ImPsi[i,j] = 0 # Hard Disk
44                Rho[i,j] = 0.1*(RePsi[i,j]**2 + ImPsi[i,j]**2) + 0.0002*V[i,j]
     X, Y = np.meshgrid(ix, iy)
     Z = Rho[X,Y]
     ax.set_xlabel('y')
48   ax.set_ylabel('x')
     ax.set_zlabel('Rho(x,y)')
     ax.plot_wireframe(X, Y, Z, color = 'g')
     print("finito")
52   p.show()
```

Listing 24.3   FDTD.py solves Maxwell's equations via FDTD algorithm for linearly polarized wave propagation in the *z* direction.

```
    # FDTD.py   FDTD Maxwell's equations in 1-D wi Visual
2
    from visual import *

    Xm = 201;  Ym = 100; Zm = 100;  ts = 2;   beta = 0.01
6   Ex = zeros((Xm,ts),float);  Hy = zeros((Xm,ts),float)     # Declare arrays
    #              Set up 3-D Plots
    scene = display(x=0,y=0,width= 800, height= 500, \
             title= 'E: cyan, H: red. Periodic BC',forward=(-0.6,-0.5,-1))
10  Eplot = curve(x=list(range(0,Xm)),color=color.cyan,radius=1.5,display=scene)
    Hplot = curve(x=list(range(0,Xm)),color=color.red, radius=1.5,display=scene)
    vplane = curve(pos=[(-Xm,Ym),(Xm,Ym),(Xm,-Ym),(-Xm,-Ym),
                       (-Xm,Ym)],color=color.cyan)
14  zaxis = curve(pos=[(-Xm,0),(Xm,0)],color=color.magenta)
    hplane = curve(pos=[(-Xm,0,Zm),(Xm,0,Zm),(Xm,0,-Zm),(-Xm,0,-Zm),
                       (-Xm,0,Zm)],color=color.magenta)
    ball1 = sphere(pos = (Xm+30, 0,0), color = color.black, radius = 2)
18  ExLabel1 = label( text = 'Ex', pos = (-Xm-10, 50), box=0)
    ExLabel2 = label( text = 'Ex', pos = (Xm+10, 50), box=0)
    HyLabel  = label( text = 'Hy', pos = (-Xm-10, 0,50), box=0)
    zLabel   = label( text = 'Z', pos = (Xm+10, 0), box=0)
22
    def PlotFields():
        z = arange(Xm)
        Eplot.x = 2*z-Xm                          # World to screen coords
26      Eplot.y = 800*Ex[z,0]
        Hplot.x = 2*z-Xm
        Hplot.z = 800*Hy[z,0]

30  z = arange(Xm)
    Ex[:Xm,0] = 0.1*sin(2*pi*z/100.0)                         # Initial field
    Hy[:Xm,0] = 0.1*sin(2*pi*z/100.0)
    PlotFields()
34
    while True:
        rate(600)
        Ex[1:Xm-1,1] = Ex[1:Xm-1,0] + beta*(Hy[0:Xm-2,0]-Hy[2:Xm,0])
38      Hy[1:Xm-1,1] = Hy[1:Xm-1,0] + beta*(Ex[0:Xm-2,0]-Ex[2:Xm,0])
        Ex[0,1]      = Ex[0,0]      + beta*(Hy[Xm-2,0] -Hy[1,0])     # BC
        Ex[Xm-1,1]   = Ex[Xm-1,0]   + beta*(Hy[Xm-2,0] -Hy[1,0])
        Hy[0,1]      = Hy[0,0]      + beta*(Ex[Xm-2,0] -Ex[1,0])     # BC
42      Hy[Xm-1,1]   = Hy[Xm-1,0]   + beta*(Ex[Xm-2,0] - Ex[1,0])
        PlotFields()
        Ex[:Xm,0] = Ex[:Xm,1]                            # New -> old
        Hy[:Xm,0] = Hy[:Xm,1]
```

Listing 24.4   QuarterPlate.py Maxwell's equations solution via FDTD algorithm for quarter wave plate.

```
1  # QuarterPlate.py  FDTD solution of Maxwell's equations in 1–D

   from visual import *

5  xmax = 401; ymax = 100;  zmax = 100; ts = 2; beta = 0.01

   Ex = zeros((xmax,ts),float); Ey = zeros((xmax,ts),float); Hx =
       zeros((xmax,ts),float)
   Hy = zeros((xmax,ts),float); Hyy = zeros((xmax,ts),float)
9  Exx = zeros((xmax,ts),float);  Eyy = zeros((xmax,ts),float)
   Hxx = zeros((xmax,ts),float)

   scene = display(x=0,y=0,width= 800, height= 500,
13                 title= 'Ey : in cyan, Ex : in yellow. Propagation with periodic
                       boundary conditions',
                   forward=(-0.8,-0.3,-0.7))
   Exfield = curve(x=list(range(0,xmax)),color= color.yellow,radius=1.5,display=scene)
   Eyfield = curve(x=list(range(0,xmax)),color=color.cyan,radius=1.5,display=scene)
17 vplane= curve(pos=[(-xmax,ymax),(xmax,ymax),(xmax,-ymax),(-xmax,-ymax),
                     (-xmax,ymax)],color=color.cyan)
   zaxis = curve(pos=[(-xmax,0),(xmax,0)],color=color.magenta)
   hplane = curve(pos=[(-xmax,0,zmax),(xmax,0,zmax),(xmax,0,-zmax),(-xmax,0,-zmax),
21                   (-xmax,0,zmax)],color=color.magenta)
   ball1 = sphere(pos = (xmax+30, 0,0), color = color.black, radius = 2)
   ba2 = sphere(pos=(xmax-200,0),color=color.cyan,radius=3)
   plate = box(pos=(-100,0,0),height=2*zmax,width=2*ymax,length=0.5*xmax,
25             color=(1.0,0.6,0.0),opacity=0.4)
   Exlabel1 = label( text = 'Ey', pos = (-xmax-10, 50), box = 0 )
   Exlabel2 = label( text = 'Ey', pos = (xmax+10, 50), box = 0 )
   Eylabel = label( text = 'Ex', pos = (-xmax-10, 0,50), box = 0 )
29 zlabel  = label( text = 'Z', pos = (xmax+10, 0), box = 0 )
   polfield = arrow(display = scene)
   polfield2 = arrow(display = scene)
   ti = 0
33
   def InitField():
       kar = arange(xmax)
       phx = 0.5*pi
37     Hyy[:xmax,0] = 0.1*cos(-2*pi*kar/100)
       Exx[:xmax,0] = 0.1*cos(-2*pi*kar/100)
       Eyy[:xmax,0] = 0.1*cos(-2*pi*kar/100)
       Hxx[:xmax,0] = 0.1*cos(-2*pi*kar/100)
41     Ey[:xmax,0] = 0.1*cos(-2*pi*kar/100)
       Hx[:xmax,0] = 0.1*cos(-2*pi*kar/100)

   def InitExHy():
45     k  = arange(101)
       Ex[:101,0] = 0.1*cos(-2*pi*k/100)
       Hy[:101,0] = 0.1*cos(-2*pi*k/100)
       kk = arange(101,202)              # Inside plate, delay lambda/4
49     Ex[101:202,0]  = 0.1*cos(-2*pi*kk/100.0-0.005*pi*(kk-101)) # pi/2 phase
       Hy[101:202,0]  = 0.1*cos(-2*pi*kk/100.0-0.005*pi*(kk-101))
       kkk = arange(202,xmax)            # After plate, phase diff pi/2
       Ex[202:xmax,0]  = 0.1*cos(-2*pi*kkk/100-0.5*pi)
53     Hy[202:xmax,0]  = 0.1*cos(-2*pi*kkk/100-0.5*pi)

   def PlotFields(ti):                    # screen coordinates
       k  = arange(xmax)
57     Exfield.x = 2*k-xmax                # world to screen coords.
       Exfield.y = 800*Ey[k,ti]
       Eyfield.x = 2*k-xmax                # world to screen coords.
       Eyfield.z = 800*Ex[k,ti]
61
   InitField()
   InitExHy()
   PlotFields(ti)
65 j = 0
```

```
        end = 0
        while end < 5:
            rate(150)
69          Exx[1:xmax-1,1] = Exx[1:xmax-1,0] + beta*(Hyy[0:xmax-2,0] - Hyy[2:xmax,0])
            Eyy[1:xmax-1,1] = Eyy[1:xmax-1,0] + beta*(Hxx[0:xmax-2,0] - Hxx[2:xmax,0])
            Hyy[1:xmax-1,1] = Hyy[1:xmax-1,0] + beta*(Exx[0:xmax-2,0] - Exx[2:xmax,0])
            Hxx[1:xmax-1,1] = Hxx[1:xmax-1,0] + beta*(Eyy[0:xmax-2,0] - Eyy[2:xmax,0])
73          Ex[1:xmax-1,1] =  Ex[1:xmax-1,0]  + beta*(Hy[0:xmax-2,0]  - Hy[2:xmax,0])
            Ey[1:xmax-1,1] =  Ey[1:xmax-1,0]  + beta*(Hxx[0:xmax-2,0] - Hxx[2:xmax,0])
            Hy[1:xmax-1,1] =  Hy[1:xmax-1,0]  + beta*(Ex[0:xmax-2,0]  - Ex[2:xmax,0])
            Hx[1:xmax-1,1] =  Hx[1:xmax-1,0]  + beta*(Eyy[0:xmax-2,0] - Eyy[2:xmax,0])
77          polfield.pos = (-280,0,0)
            polfield.axis = (0,700*Exx[60,1],700*Eyy[60,1])
            polfield2.pos = (380,0,0)
            polfield2.axis = (0,700*Ex[360,1],-700*Ey[360,1])
81          Exx[0,1] = Exx[0,0] + beta*(Hyy[xmax-2,0] - Hyy[1,0])        # Periodic BC
            Eyy[0,1]=  Eyy[0,0] + beta*(Hxx[xmax-2,0] - Hxx[1,0])
            Hyy[0,1] = Hyy[0,0] + beta*(Exx[xmax-2,0] - Exx[1,0])       # Periodic BC, 0=100
            Hxx[0,1] = Hxx[0,0] + beta*(Eyy[xmax-2,0] - Eyy[1,0])
85          Hyy[xmax-1,1] = Hyy[xmax-1,0] + beta*(Exx[xmax-2,0] - Exx[1,0])
            Hxx[xmax-1,1] = Hxx[xmax-1,0] + beta*(Eyy[xmax-2,0] - Eyy[1,0])
            Exx[xmax-1,1] = Exx[xmax-1,0] + beta*(Hyy[xmax-2,0] - Hyy[1,0]) #conditions for
                first
            Eyy[xmax-1,1] = Eyy[xmax-1,0] + beta*(Hxx[xmax-2,0] - Hxx[1,0])
89          Ex[0,1] = Exx[0,0] + beta*(Hyy[xmax-2,0] - Hyy[1,0])         # Periodic BC
            Ey[0,1] = Eyy[0,0] + beta*(Hxx[xmax-2,0] - Hxx[1,0])
            Hy[0,1] = Hyy[0,0] + beta*(Exx[xmax-2,0] - Exx[1,0])        # Periodic BC 0=100
            Hx[0,1] = Hxx[0,0] + beta*(Eyy[xmax-2,0] - Eyy[1,0])
93          Hy[xmax-1,1] = Hy[xmax-1,0] + beta*(Ex[xmax-2,0] - Ex[xmax-100,0])
            Hx[xmax-1,1] = Hx[xmax-1,0] + beta*(Ey[xmax-2,0] - Ey[1,0])
            Ex[xmax-1,1] = Ex[xmax-1,0] + beta*(Hy[xmax-2,0] - Hy[xmax-100,0])
            Ey[xmax-1,1] = Ey[xmax-1,0] + beta*(Hxx[xmax-2,0] - Hxx[1,0])
97          PlotFields(ti)
            k = arange(101,202)
            Ex[101:202,1] = 0.1*cos(-2*pi*k/100-0.005*pi*(k-101)+2*pi*j/4996.004)
            Hy[101:202,1] = 0.1*cos(-2*pi*k/100-0.005*pi*(k-101)+2*pi*j/4996.004)
101         Exx[:xmax,0]  = Exx[:xmax,1]
            Eyy[:xmax,0]  = Eyy[:xmax,1]
            Hyy[:xmax,0]  = Hyy[:xmax,1]
            Hxx[:xmax,0]  = Hxx[:xmax,1]
105         Ex[:xmax,0] =     Ex[:xmax,1]
            Ey[:xmax,0] =     Ey[:xmax,1]
            Hx[:xmax,0] =     Hx[:xmax,1]
            Hy[:xmax,0]=      Hy[:xmax,1]
109         if j%4996 == 0:
                j = 0
                end += 1
            j = j+1
```

Listing 24.5 EMCirc.py solves Maxwell's equations via FDTD algorithm for circularly polarized wave propagation in the z direction.

```
    # EMcirc.py: Maxwell eqs. for circular polarization using FDTD
2
    from visual import *
    scene = display(x=0,y=0,width=600,height=400,range=200,
      title='Circular Polarized E (white) & H (yellow) Fields')
6   global phy, pyx
    max = 201; c = 0.01                                 # Stable if c < 0.1
    Ex = zeros((max+2,2),float);    Ey = zeros((max+2,2),float)
    Hy = zeros((max+2,2),float);    Hx = zeros((max+2,2),float)
10  arrowcol= color.white
    Earrows = [];                   Harrows = []
    for i in range(0,max,10):
        Earrows.append(arrow(pos=(0,i-100,0), axis=(0,0,0), color=arrowcol))
14      Harrows.append(arrow(pos=(0,i-100,0), axis=(0,0,0), color=color.yellow))

    def plotfields(Ex,Ey,Hx,Hy):
```

```
        for n, arr in enumerate(Earrows):
18            arr.axis = (35*Ey[10*n,1],0,35*Ex[10*n,1])
        for n, arr in enumerate(Harrows):
              arr.axis = (35*Hy[10*n,1],0,35*Hx[10*n,1])
    def inifields():                                    # Initial E & H
22      phx = 0.5*pi;  phy = 0.0
        z = arange(0,max)
        Ex[:-2,0] = cos(-2*pi*z/200+phx);        Ey[:-2,0] = cos(-2*pi*z/200+phy)
        Hx[:-2,0] = cos(-2*pi*z/200+phy+pi);     Hy[:-2,0] = cos(-2*pi*z/200+phx)
26  def newfields():
        while True:                                     # Time stepping
            rate(1000)
            Ex[1:max-1,1] =  Ex[1:max-1,0]+c*(Hy[:max-2,0]-Hy[2:max,0])
30          Ey[1:max-1,1] =  Ey[1:max-1,0] + c*(Hx[2:max,0]-Hx[:max-2,0])
            Hx[1:max-1,1] =  Hx[1:max-1,0] + c*(Ey[2:max,0]-Ey[:max-2,0])
            Hy[1:max-1,1] =  Hy[1:max-1,0] + c*(Ex[:max-2,0]-Ex[2:max,0])
            Ex[0,1]   = Ex[0,0]   + c*(Hy[200-1,0]-Hy[1,0])      # Periodic BC
34          Ex[200,1] = Ex[200,0] + c*(Hy[200-1,0]-Hy[1,0])
            Ey[0,1]   = Ey[0,0]   + c*(Hx[1,0]- Hx[200-1,0])
            Ey[200,1] = Ey[200,0] + c*(Hx[1,0]- Hx[200-1,0])
            Hx[0,1]   = Hx[0,0]   + c*(Ey[1,0]- Ey[200-1,0])
38          Hx[200,1] = Hx[200,0] + c*(Ey[1,0]- Ey[200-1,0])
            Hy[0,1]   = Hy[0,0]   + c*(Ex[200-1,0]-Ex[1,0])
            Hy[200,1] = Hy[200,0] + c*(Ex[200-1,0]-Ex[1,0])
            plotfields(Ex,Ey,Hx,Hy)
42          Ex[:max,0] = Ex[:max,1];   Ey[:max,0] = Ey[:max,1]   # Update fields
            Hx[:max,0] = Hx[:max,1];   Hy[:max,0] = Hy[:max,1]

    inifields()                                         # Initial fields
46  newfields()                                         # New fields
```

25

Shock and Soliton Waves

> *The first half of this chapter extends the discussion of waves to include nonlinearities, dispersion, and hydrodynamic effects. We end up with the Korteweg-de Vries (KDV) equation and shallow-water solitons. The second half of this chapter explores a model for solids as a chain of coupled, nonlinear oscillators. We end up with the Sine-Gordon equation (SGE) and solitons in solids.*

In 1844, J. Scott Russell reported an unusual occurrence on the Edinburgh-Glasgow canal [Russell, 1844] (Figure 25.1 left):

> *I was observing the motion of a boat which was rapidly drawn along a narrow channel by a pair of horses, when the boat suddenly stopped—not so the mass of water in the channel which it had put in motion; it accumulated round the prow of the vessel in a state of violent agitation, then suddenly leaving it behind, rolled forward with great velocity, assuming the form of a large solitary elevation, a rounded, smooth and well-defined heap of water, which continued its course along the channel, apparently without change of form or diminution of speed. I followed it on horseback, and overtook it still rolling on at the rate of some eight or nine miles an hour, preserving its original figure some thirty feet long and a foot to a foot and a half in height. Its height gradually diminished, and after a chase of one or two miles, I lost it in the windings of the channel. Such, in the month of August 1834, was my first chance interview with that singular and beautiful phenomenon....*

Russell also noticed that an initial, arbitrary waveform set in motion in the channel evolved into two or more waves that moved at different velocities, and that they progressively moved apart until they formed individual, solitary waves. In Figure 25.3 right, we see a single, step-like wave breaking up into approximately eight of these solitary waves (now called *solitons*). These eight solitons occur so frequently that some of the references consider them the equivalent of the normal modes for nonlinear systems. Russell went on to produce these solitary waves in a laboratory, and empirically deduced that their speed c is related to the depth h of the water in the canal, and to the amplitude A of the wave by,

$$c^2 = g(h + A), \tag{25.1}$$

where g is the acceleration due to gravity.

Computational Physics: Problem Solving with Python, Fourth Edition.
Rubin H. Landau, Manuel J. Páez, and Cristian C. Bordeianu.
© 2024 WILEY-VCH GmbH. Published 2024 by WILEY-VCH GmbH.

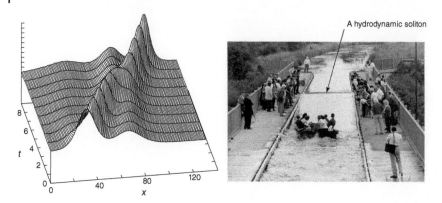

Figure 25.1 *Left*:Two computed shallow-water solitary waves crossing each other. The taller soliton on the left catches up with and overtakes the shorter one at $t \simeq 5$. The waves resume their original shapes after the collision. *Right*: A re-creation of Russel's soliton on the Union Canal near Edinburgh in July 1995. (Used with permission, Scott [2007].)

Equation (25.1) implies an effect not found in linear systems, namely, that waves with greater amplitudes A travel faster than those with smaller amplitudes. Observe that this is similar to the formation of shock waves, but different from *dispersion* in which waves of different wavelengths have different velocities. The dependence of c on the amplitude A is illustrated in Figure 25.1, where we see a taller soliton catching up with and passing through a shorter one. Indeed, these solitons appear to be related to the formation of *tsunamis*, ocean waves that form from sudden changes in the level of the ocean floor, and then travel over long distances without dispersion or attenuation.

Problem Explain Russell's observations.

25.1 The Continuity and Advection Equations

The motion of fluids is described by the continuity equation and the Navier–Stokes equation [Landau and Lifshitz, 1976]. We discuss the former here and the latter in Chapter 26. The continuity equation describes conservation of mass:

$$\frac{\partial \rho(\mathbf{x}, t)}{\partial t} + \vec{\nabla} \cdot \mathbf{j} = 0, \qquad \mathbf{j} \stackrel{\text{def}}{=} \rho \mathbf{v}(\mathbf{x}, t). \tag{25.2}$$

Here, $\rho(\mathbf{x}, t)$ is the mass density of the fluid, $\mathbf{v}(\mathbf{x}, t)$ is its velocity, and $\mathbf{j} = \rho \mathbf{v}$ is the mass current. As its name implies, the divergence $\vec{\nabla} \cdot \mathbf{j}$ describes the spreading of the current in a region of space, as might occur if there were a current source there. Physically, the continuity equation (25.2) states that changes in the density of the fluid within some region of space arises from the flow of current in and out of that region.

For 1D flow in the x direction, and for a fluid that is moving with a constant velocity $v = c$, the continuity equation (25.2) takes a simple form:

$$\frac{\partial \rho}{\partial t} + c\frac{\partial \rho}{\partial x} = 0. \tag{25.3}$$

This is the *advection equation*, where the term "advection" is used to describe the horizontal transport of a quantity from one region of space to another as a result of a flow's velocity field. For instance, advection describes the transportation of dissolved salt in water. The advection equation looks like a first-derivative form of the wave equation, and indeed, the two are related. A simple substitution proves that any function with the form of a traveling wave,

$$u(x, t) = f(x - ct),\tag{25.4}$$

will be a solution of the advection equation. If we consider a surfer riding along the crest of a traveling wave, that is, remaining at the same position relative to the wave's shape as time changes, then the surfer does not see the shape of the wave change in time, and that implies that:

$$x - ct = \text{constant} \quad \Rightarrow \quad x = ct + \text{constant}.\tag{25.5}$$

The speed of the surfer $dx/dt = c$, which is a constant. Any function $f(x - ct)$ is clearly a traveling wave solution in which an arbitrary pulse is carried along with the fluid at velocity c without changing shape.

Although the advection equation is simple, trying to solve it by a simple finite difference scheme (the leapfrog method) may lead to unstable numerical solutions. As we shall see when we look at the nonlinear version of this equation, there are better ways. Listing 25.1 presents our code `AdvecLax.py` for solving the advection equation using the Lax-Wendroff method (a better way).

25.2 Shock Waves via Burgers' Equation

In Section 25.4, we will examine the Korteweg-de Vries equation's description of solitary waves. In order to understand the physics contained in that equation, we study, one at a time, some of the terms in it. We start with two equivalent forms of Burgers' equation [Burgers, 1974]:

$$\frac{\partial u}{\partial t} + \epsilon u \frac{\partial u}{\partial x} = 0,\tag{25.6}$$

$$\frac{\partial u}{\partial t} + \epsilon \frac{\partial (u^2/2)}{\partial x} = 0,\tag{25.7}$$

where the second equation is the *conservative form*. This equation can be viewed as a variation on the advection equation (25.3), now with the wave speed $c = \epsilon u$ proportional to the amplitude of the wave, as Russell found for his waves. The second, nonlinear term in Burgers' equation leads to some unusual behaviors. Indeed, von Neumann studied this equation as a simple model for turbulence [Falkovich and Sreenivasan, 2006].

In the advection equation (25.3), all points on the wave move at the same speed c, and so the shape of the wave remains unchanged in time. In Burgers' equation (25.6), the points on the wave move ("advect") themselves such that the local speed depends on the local wave's amplitude, with the high parts of the wave moving progressively faster than the low parts. This changes the shape of the wave in time; if we start with a wave packet that has a smooth variation in height, the high parts will speed up and push their way to the front

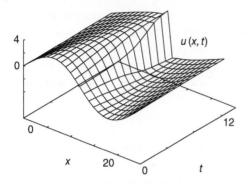

Figure 25.2 A visualization showing the formation of a shock wave (sharp edge) in a solution to Burgers' equation that started with a sine wave.

of the packet, thereby forming a sharp leading edge known as a *shock wave* [Tabor, 1989]. A shock wave solution to Burgers' equation with $\epsilon = 1$ is shown in Figure 25.2.

25.2.1 Lax–Wendroff Algorithm

We first solve Burgers' equation (25.3) via the usual approach in which we express the derivatives as central differences. This leads to a leapfrog scheme for the future solution in terms of present and past ones:

$$u(x, t + \Delta t) = u(x, t - \Delta t) - \frac{\beta}{2}[u^2(x + \Delta x, t) - u^2(x - \Delta x, t)], \tag{25.8}$$

$$u_{i,j+1} = u_{i,j-1} - \frac{\beta}{2}[u^2_{i+1,j} - u^2_{i-1,j}], \qquad \beta = \frac{\epsilon \Delta t}{\Delta x} \quad \text{(CFL number).}$$

Here u^2 is the square of u, not its second derivative, and β is a ratio of constants known as the *Courant–Friedrichs–Lewy* (CFL) *number*. As you should prove for yourself, $\beta < 1$ is required for stability.

While we have used a leapfrog method with success in our previous solution of PDEs, its low-order approximation for the derivative becomes inaccurate when the gradients can get large, as happens with shock waves, and this may cause the leapfrog algorithm to become unstable [Press *et al.*, 2007]. The *Lax–Wendroff method* attains better stability and accuracy by retaining second-order differences for the time derivative:

$$u(x, t + \Delta t) \simeq u(x, t) + \frac{\partial u}{\partial t} \Delta t + \frac{1}{2} \frac{\partial^2 u}{\partial t^2} \Delta t^2. \tag{25.9}$$

To covert (25.9) to an algorithm, we use Burgers' equation $\partial u / \partial t = -\epsilon \partial (u^2/2)/\partial x$ for the first-order time derivative. Likewise, we use Burger's equation to express the second-order time derivative in terms of space derivatives:

$$\frac{\partial^2 u}{\partial t^2} = \frac{\partial}{\partial t}\left[-\epsilon \frac{\partial}{\partial x}\left(\frac{u^2}{2}\right)\right] = -\epsilon \frac{\partial}{\partial x}\frac{\partial}{\partial t}\left(\frac{u^2}{2}\right) \tag{25.10}$$

$$= -\epsilon \frac{\partial}{\partial x}\left(u \frac{\partial u}{\partial t}\right) = \epsilon^2 \frac{\partial}{\partial x}\left[u \frac{\partial}{\partial x}\left(\frac{u^2}{2}\right)\right]. \tag{25.11}$$

We next substitute these derivatives into the Taylor expansion (25.9) to obtain:

$$u(x, t + \Delta t) = u(x, t) - \Delta t \epsilon \frac{\partial}{\partial x}(\frac{u^2}{2}) + \frac{(\Delta t)^2}{2}\epsilon^2 \frac{\partial}{\partial x}\left[u \frac{\partial}{\partial x}(\frac{u^2}{2})\right]. \tag{25.12}$$

We now replace the outer x derivatives by central differences of spacing $\Delta x/2$:

$$u(x, t + \Delta t) = u(x, t) - \frac{\Delta t \epsilon}{2} \frac{u^2(x+\Delta x,t) - u^2(x-\Delta x,t)}{2\Delta x} + \frac{(\Delta t)^2 \epsilon^2}{2} \tag{25.13}$$

$$\times \frac{1}{2\Delta x}\left[u(x+\tfrac{\Delta x}{2},t)\frac{\partial}{\partial x}u^2(x+\tfrac{\Delta x}{2},t) - u(x-\tfrac{\Delta x}{2},t)\frac{\partial}{\partial x}u^2(x-\tfrac{\Delta x}{2},t)\right].$$

Next we approximate $u(x \pm \Delta x/2, t)$ by the average of adjacent grid points,

$$u(x \pm \tfrac{\Delta x}{2}, t) \simeq \frac{u(x,t) + u(x \pm \Delta x, t)}{2}, \tag{25.14}$$

and apply a central-difference approximation to the second derivatives:

$$\frac{\partial u^2(x \pm \Delta x/2, t)}{\partial x} = \frac{u^2(x \pm \Delta x, t) - u^2(x, t)}{\pm \Delta x}. \tag{25.15}$$

Finally, putting all these derivatives together yields the algorithm:

$$u_{i,j+1} = u_{i,j} - \frac{\beta}{4}(u_{i+1,j}^2 - u_{i-1,j}^2) \tag{25.16}$$

$$+ \frac{\beta^2}{8}\left[(u_{i+1,j} + u_{i,j})(u_{i+1,j}^2 - u_{i,j}^2) - (u_{i,j} + u_{i-1,j})(u_{i,j}^2 - u_{i-1,j}^2)\right],$$

where we have substituted the CFL number β. This Lax–Wendroff scheme is explicit, centered upon the grid points, and stable for $\beta < 1$ (small nonlinearities).

25.2.2 Implementation and Assessment

Solve Burgers' equation (25.7) via the leapfrog method.

1) Define arrays `u0[100]` and `u[100]` for the old and new waves.
2) Take the initial wave to be sinusoidal, `u0[i]` $= 3\sin(3.2x)$, with speed $c = 1$.
3) Incorporate the boundary conditions `u[0]=0` and `u[100]=0`.
4) Keep the CFL number $\beta < 1$ for stability.
5) Save the initial wave and the solutions for a number of times in separate files, and plot them on the same graph in order to see the formation of a shock wave (like Figure 25.2).
6) Modify your program to solve Burgers' equation using the Lax–Wendroff method (25.16), and compare it with the leapfrog method. The leapfrog method should produce shock waves, but with ripples as the square edge develops. The ripples are numerical artifacts. The Lax–Wendroff method should give a better square edge, although some ripples may still occur.
7) Run the code for several increasingly large CFL values and check if the stability condition $\beta < 1$ is correct.

Our Lax–Wendroff code `AdvecLax.py` is given in Listing 25.1.

25.3 Including Dispersion

We have just seen that Burgers' equation has a solution in which an initially smooth wave transforms into a square-edged shock wave. A sort of inverse of this is *dispersion*, in which a waveform disperses, or broadens, as it travel through a medium. Dispersion does not cause waves to lose energy or attenuate, but it does lead to a loss of information with time. Physically, dispersion is important when the propagating medium has structures with a spatial regularity equal to some fraction of a wavelength. Mathematically, dispersion may arise from terms in the wave equation that contain higher-order space derivatives. For example, consider the waveform:

$$u(x, t) = e^{\pm i(kx - \omega t)}, \tag{25.17}$$

corresponding to a plane wave traveling to the right ("traveling" because the phase $kx - \omega t$ remains unchanged if you increase x with time). When this $u(x, t)$ is substituted into the advection equation (25.3), we obtain:

$$\omega = ck. \tag{25.18}$$

This equation is a *dispersion relation*, that is, a relation between frequency ω and wave vector k. Because *group velocity* of a wave is:

$$v_g = \frac{\partial \omega}{\partial k}, \tag{25.19}$$

the linear dispersion relation (25.18) implies that all frequencies have the same group velocity c. This is *dispersionless* propagation.

Let us now imagine that a wave is propagating with a small amount of *dispersion*, that is, with a frequency that has small correction to the linear increase with the wave number:

$$\omega \simeq ck - \beta k^3. \tag{25.20}$$

Note that because there are no even powers in (25.20), the group velocity,

$$v_g = \frac{d\omega}{dk} \simeq c - 3\beta k^2, \tag{25.21}$$

is the same for waves traveling to the left, or the right. Now we work backwards and deduce what might be the wave equation that produced this dispersion. If we have a plane-wave solution like (25.17), the ω term in the dispersion relation (25.20) would arise from a first-order time derivative. Likewise, the ck term would arise from a first-order space derivative, and the k^3 term from a third-order space derivative. And thus, the wave equation:

$$\frac{\partial u(x, t)}{\partial t} + c\frac{\partial u(x, t)}{\partial x} + \beta\frac{\partial^3 u(x, t)}{\partial x^3} = 0. \tag{25.22}$$

We leave it as an exercise to show that solutions to this equation do indeed have waveforms that disperse in time.

25.4 KdeV Solitons

Now we can put together all the pieces that are needed to understand the unusual water waves that occur in shallow, narrow channels such as canals [Abar 93, Tab 89]. The analytic description of this "heap of water" was given by Korteweg and de Vries [1895] with the *KdeV equation*:

$$\frac{\partial u(x, t)}{\partial t} + \varepsilon u(x, t)\frac{\partial u(x, t)}{\partial x} + \mu\frac{\partial^3 u(x, t)}{\partial x^3} = 0. \tag{25.23}$$

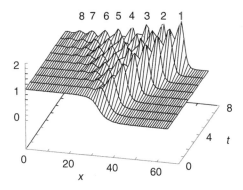

Figure 25.3 The formation of a tsunami. A single two-level waveform at time zero progressively breaks up into eight solitons (labeled) as time increases. The tallest soliton (1) is narrower and faster than the others in its motion to the right. You can generate an animation of this with the program **SolitonAnimate.py**.

As we discussed in Section 25.2 in our study of Burgers' equation, the nonlinear $\epsilon u \, \partial u / \partial t$ term leads to a sharpening of the wave, and ultimately a *shock* wave. In contrast, as we know from our discussion of dispersion, the $\partial^3 u / \partial x^3$ term tends to broaden a waveform, while the $\partial u / \partial t$ term produces a traveling wave. *For the proper parameters and initial conditions, the dispersive broadening exactly balances the nonlinear narrowing, and a stable traveling wave, a soliton, is formed.*

Korteweg and de Vries [1895] solved (25.23) analytically, and proved that the speed (25.1) given by Russell is, in fact, correct. Seventy years after its discovery, the KdeV equation was rediscovered by Zabusky and Kruskal [1965], who solved it numerically and found that a step-like initial condition broke up into eight solitary waves (Figure 25.3). They also found that the parts of the wave with larger amplitudes moved faster than those with smaller amplitudes. This is why, at later times, the higher peaks tend to be on the right in Figure 25.3. As if wonders never cease, Zabusky and Kruskal, who coined the name *soliton* for these solitary waves, also observed that a faster peak passed unscathed through a slower one (Figure 25.1).

25.4.1 Analytic Solution

The trick in analytic approaches to these types of nonlinear equations is to substitute a guessed solution that has the form of a traveling wave,

$$u(x, t) = u(\xi = x - ct). \tag{25.24}$$

This form means that if we move with a constant speed c, we will see a constant wave form. There is no guarantee that this form of a solution will exist, but it may lead you to one. Substituting (25.24) into the KdeV equation produces a solvable ODE:

$$-c \frac{\partial u}{\partial \xi} + \epsilon u \frac{\partial u}{\partial \xi} + \mu \frac{d^3 u}{d\xi^3} = 0, \tag{25.25}$$

$$\Rightarrow \quad u(x, t) = \frac{-c}{2} \operatorname{sech}^2 [\tfrac{1}{2} \sqrt{c}(x - ct - \xi_0)], \tag{25.26}$$

where ξ_0 is the initial phase. We see in the soliton waveform, (25.26), an amplitude that is proportional to the wave speed c, and a sech^2 function that gives a single lump-like wave.

25.4.2 Algorithm

The KdeV equation is solved numerically using a finite-difference scheme, with the time and space derivatives given by central-difference approximations:

$$\frac{\partial u}{\partial t} \simeq \frac{u_{i,j+1} - u_{i,j-1}}{2\Delta t}, \qquad \frac{\partial u}{\partial x} \simeq \frac{u_{i+1,j} - u_{i-1,j}}{2\Delta x}. \tag{25.27}$$

To approximate $\partial^3 u(x, t)/\partial x^3$, we expand $u(x, t)$ to $\mathcal{O}(\Delta t)^3$ about the four points $u(x \pm 2\Delta x, t)$ and $u(x \pm \Delta x, t)$:

$$u(x \pm \Delta x, t) \simeq u(x, t) \pm (\Delta x)\frac{\partial u}{\partial x} + \frac{(\Delta x)^2}{2!}\frac{\partial^2 u}{\partial^2 x} \pm \frac{(\Delta x)^3}{3!}\frac{\partial^3 u}{\partial x^3}. \tag{25.28}$$

We solve this for $\partial^3 u(x, t)/\partial x^3$. Finally, the factor $u(x, t)$ in the second term of (25.23) is taken as the average of three x values, all with the same t:

$$u(x, t) \simeq \frac{u_{i+1,j} + u_{i,j} + u_{i-1,j}}{3}. \tag{25.29}$$

We substitute these approximations to obtain the algorithm for the KdeV equation:

$$u_{i,j+1} \simeq u_{i,j-1} - \frac{\epsilon}{3}\frac{\Delta t}{\Delta x}\left[u_{i+1,j} + u_{i,j} + u_{i-1,j}\right]\left[u_{i+1,j} - u_{i-1,j}\right]$$

$$- \mu \frac{\Delta t}{(\Delta x)^3}\left[u_{i+2,j} + 2u_{i-1,j} - 2u_{i+1,j} - u_{i-2,j}\right]. \tag{25.30}$$

The algorithm predicts $u(x, t)$ at future times, given the solutions at the present and past times. The initial condition provides $u_{i,1}$ for all positions i. To find $u_{i,2}$, we use the forward-difference scheme in which we expand $u(x, t)$, keeping only two terms for the time derivative:

$$u_{i,2} \simeq u_{i,1} - \frac{\epsilon \, \Delta t}{6 \, \Delta x}\left[u_{i+1,1} + u_{i,1} + u_{i-1,1}\right]\left[u_{i+1,1} - u_{i-1,1}\right]$$

$$- \frac{\mu}{2}\frac{\Delta t}{(\Delta x)^3}\left[u_{i+2,1} + 2u_{i-1,1} - 2u_{i+1,1} - u_{i-2,1}\right]. \tag{25.31}$$

The keen observer will note that there are still some undefined columns of points, namely, $u_{1,j}$, $u_{2,j}$, $u_{N_{\max}-1,j}$, and $u_{N_{\max},j}$, where $N_{\max}$ is the total number of grid points. A simple technique for determining their value is to assume that $u_{1,2} = 1$ and $u_{N_{\max},2} = 0$. To obtain $u_{2,2}$ and $u_{N_{\max}-1,2}$, we assume that $u_{i+2,2} = u_{i+1,2}$, and $u_{i-2,2} = u_{i-1,2}$. (However, we avoid $u_{i+2,2}$ for $i = N_{\max} - 1$, and $u_{i-2,2}$ for $i = 2$). We carry out these steps, approximate (25.31), and thus simplify the relation to:

$$u_{i+2,2} + 2u_{i-1,2} - 2u_{i+1,2} - u_{i-2,2} \simeq u_{i-1,2} - u_{i+1,2}. \tag{25.32}$$

The truncation error and stability condition for this algorithm are related:

$$\mathcal{E}(u) = \mathcal{O}[(\Delta t)^3] + \mathcal{O}[\Delta t(\Delta x)^2] \qquad \text{(Error)}, \tag{25.33}$$

$$\tfrac{\Delta t}{\Delta x}[\epsilon|u| + 4\tfrac{\mu}{(\Delta x)^2}] \leq 1 \qquad \text{(Stability)}. \tag{25.34}$$

The first equation implies that smaller time and space steps lead to a smaller approximation error. Yet, as discussed in Chapter 3, if the steps made are very small, then you will need to take a large number of steps, and then the round-off error may get too large. Some balance is also indicated by the stability condition (25.34), where we see that making Δx too small leads to instability. Some experimentation is advised.

25.4.3 Implementation

Modify or run the program soliton.py in Listing 19.2 that solves the KdeV equation (25.23) for the initial condition:

$$u(x, t = 0) = \frac{1}{2}\left[1 - \tanh\left(\frac{x-25}{5}\right)\right], \tag{25.35}$$

with parameters $\epsilon = 0.2$ and $\mu = 0.1$. Start with $\Delta x = 0.4$ and $\Delta t = 0.1$. These constants satisfy (25.33) with $|u| = 1$. The code **SolitonAnimate.py** produces an animation.

1) Define a 2D array u[131,3], with the first index corresponding to the position x, and the second to the time t. With our choice of parameters, the maximum value for x is $130 \times 0.4 = 52$.
2) Initialize the time to $t = 0$, and assign values to u[i,1].
3) Assign values to u[i,2], $i = 3, 4, \ldots, 129$, corresponding to the next time interval. Use (25.31) to advance the time, but note that you cannot start at $i = 1$, or end at $i = 131$, because (25.31) would include u[132,2] and u[-1,1], which are beyond the limits of the array.
4) Increment time, and assume that u[1,2] $= 1$ and u[131,2] $= 0$. To obtain u[2,2] and u[130,2], assume that u[i+2,2] $=$ u[i+1,2] and u[i-2,2] $=$ u[i-1,2]. Avoid u[i+2,2] for i $= 130$, and u[i-2,2] for i $= 2$. To do this, approximate (25.31) so that (25.33) is satisfied.
5) Increment time, and compute u[i, j] for j $= 3$, and for i $= 3$, 4,1, 129, using equation (25.30). Again, follow the same procedures to obtain the missing array elements u[2, j] and u[130, j] (set u[1, j] $= 1$. and u[131, j] $= 0$). Print out the numbers during the iterations, and check that these are good choices.
6) Set u[i,1] $=$ u[i,2], and u[i,2] $=$ u[i,3] for all i. In this way, you are ready to find the next u[i,j] in terms of the previous two rows.
7) Repeat the previous two steps at least 2000 times. Write your solution in a file after approximately every 250 iterations.
8) Plot your results as a 3D graph of disturbance u versus position and time.
9) Observe the wave profile as a function of time, and try to confirm Russell's observation that a taller soliton travels faster than a smaller one.

25.4.4 Exploration: Phase Space Solitons and Soliton Crossings

1) Explore what happens when a tall soliton collides with a short one.
   (a) Start by placing a tall soliton of height 0.8 at $x = 12$ and a smaller soliton in front of it at $x = 26$:

$$u(x, t = 0) = 0.8\left[1 - \tanh^2\left(\frac{3x}{12} - 3\right)\right] + 0.3\left[1 - \tanh^2\left(\frac{4.5x}{26} - 4.5\right)\right]. \tag{25.36}$$

   (b) Do the two solitons reflect off each other? Do they go through each other? Do they interfere? Does the tall soliton still move faster than the short one after the collision (Figure 25.1)?
2) Construct phase space plots of $[\dot{u}(t)$ vs $u(t)]$ for all t's. Try out various parameter values. Note that only specific sets of parameters produce solitons. In particular, by correlating

the behavior of the solutions with your phase-space plots, show that the soliton solutions correspond to the *separatrix* solutions to the KdeV equation. In other words, the stability in time for solitons is analogous to the infinite period for a pendulum balanced straight up.

25.5 Pendulum Chain Solitons

In 1955, Fermi *et al.* [1955] published their investigation of how a 1D chain of coupled oscillators disperses waves. Since waves of differing frequencies traveled through the chain with differing speeds, they found that a pulse, which inherently includes a range of frequencies, broadens as time progresses. Surprisingly, when the oscillators were made more realistic by introducing a nonlinear term into Hooke's law,

$$F(x) \simeq -k(x + \alpha x^2),\tag{25.37}$$

the authors found that a sharp pulse could survive indefinitely, even in the presence of dispersion.

Problem Develop a model that explains how a combination of dispersion and nonlinearity can lead to a stable pulse in a chain of coupled oscillators.

We take a chain of realistic pendulums as our model for coupled oscillators, and in this way, extend our study of a nonlinear, single pendulum in Chapter 16. Figure 25.4 shows our chain of identical, equally-spaced pendulums, with the couplings between pendulums provided by torsion bars that twist as the pendulums swing. The angle θ_i measures the displacement of pendulum i from its equilibrium position, and the parameter a is the fixed distance between pivot points. If all the pendulums are set off swinging together, that is, with $\theta_i \equiv \theta_j$, the coupling torques would vanish and we would have our old friend, the equation for a realistic pendulum. We assume that three torques act on each pendulum, a gravitational torque trying to return the pendulum to its equilibrium position, and two torques from the twisting of the bar to the right and left of the pendulum. The equation of motion for pendulum j follows from Newton's law for rotational motion:

$$I\frac{d^2\theta_j(t)}{dt^2} = \sum_{j\neq i}\tau_{ji},\tag{25.38}$$

$$= -\kappa(\theta_j - \theta_{j-1}) - \kappa(\theta_j - \theta_{j+1}) - mgL\sin\theta_j,\tag{25.39}$$

$$\Rightarrow I\frac{d^2\theta_j(t)}{dt^2} = \kappa(\theta_{j+1} - 2\theta_j + \theta_{j-1}) - mgL\sin\theta_j.\tag{25.40}$$

Here, I is the moment of inertia of each pendulum, L is the length of each pendulum, and κ is the torque constant of the bar. As with our previous study of the realistic pendulum,

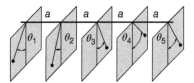

Figure 25.4 A 1D chain of pendulums, coupled via torsion bars between the pendulums. The pendulums swing in planes perpendicular to the length of the bar.

the nonlinearity in (25.40) arises from the $\sin\theta \simeq \theta - \theta^3/6 + \dots$ dependence of the gravitational torque. Equation (25.40) is a set of coupled nonlinear equations, with the number of equations equal to the number of oscillators, which would be large for model of a realistic solid. Now we want to include some dispersion in the motion of the pendulums.

25.5.1 Including Dispersion

Consider a surfer on the crest of a wave. Since she does not see the wave form change with time, her position is given by a function of the form $f(kx - \omega t)$. Consequently, to her, the wave has a constant phase:

$$kx - \omega t = \text{constant}. \tag{25.41}$$

The surfer's phase velocity is thus constant:

$$v_p = \frac{dx}{dt} = \frac{\omega}{k}. \tag{25.42}$$

In general, the frequency ω may have a nonlinear dependence on k, and this leads to the phase velocity varying with frequency, and, consequently, *dispersion*. If the wave was a pulse containing a range of Fourier components, then it would broaden and change shape in time, as each frequency moves with a different velocity. So, although dispersion does not lead to energy loss, it leads to a loss of information as pulses broaden and overlap.

The functional relation between frequency ω and the wave vector k is, of course, a *dispersion relation*. If the Fourier components in a wave packet are centered around a mean frequency ω_0, then the pulse's information travels, not with the phase velocity v_p, but with the *group velocity*:

$$v_g = \left. \frac{\partial\omega}{\partial k} \right|_{\omega=\omega_0}. \tag{25.43}$$

A comparison of (25.42) and (25.43) makes it clear that when there is dispersion, group and phase velocities may well differ.

To isolate the pure dispersive aspect of (25.40), we examine its linear version:

$$\frac{d^2\theta_j(t)}{dt^2} + \omega_0^2\theta_j(t) = \frac{\kappa}{I}(\theta_{j+1} - 2\theta_j + \theta_{j-1}), \tag{25.44}$$

where $\omega_0 = \sqrt{mgL/I}$ is the natural frequency of any one pendulum. We want to determine if a wave with a single frequency can propagate on this chain. To do that, we assume a traveling wave with frequency ω and wavelength λ,

$$\theta_j(t) = A e^{i(\omega t - kx_j)}, \quad k = \frac{2\pi}{\lambda}. \tag{25.45}$$

Substitution of (25.45) into the wave equation (25.44) produces the *dispersion relation* (Figure 25.5):

$$\omega^2 = \omega_0^2 - \frac{2\kappa}{I}(1 - \cos ka), \quad \text{(dispersion relation)}. \tag{25.46}$$

In dispersionless propagation, all frequencies propagate with the same velocity c. To have that, we need a linear relation between ω and k:

$$\lambda = c\frac{2\pi}{\omega} \quad \Rightarrow \quad \omega = ck, \quad \text{(dispersionless propagation)}. \tag{25.47}$$

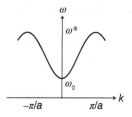

Figure 25.5 The dispersion relation for a linearized chain of pendulums.

This will be true for the chain only if ka is small enough for $\cos ka \simeq 1$, in which case $\omega \simeq \omega_0$.

Not only does the dispersion relation (25.46) change the speed of waves, it actually limits which frequencies can propagate on the chain. In order to have real k solutions, ω must lie in the range:

$$\omega_0 \leq \omega \leq \omega^* \qquad \text{(waves propagation)}. \qquad (25.48)$$

The minimum frequency ω_0 and the maximum frequency ω^* are related through the limits of $\cos ka$ in (25.46),

$$(\omega^*)^2 = \omega_0^2 + \tfrac{4\kappa}{I}. \qquad (25.49)$$

Waves with $\omega < \omega_0$ do not propagate, while those with $\omega > \omega^*$ are nonphysical because they correspond to wavelengths $\lambda < 2a$, that is, oscillations where there are no particles. These high and low ω cutoffs change the shape of a propagating pulse, that is, cause dispersion.

25.6 Continuum Limit, the Sine-Gordon Equation

If the wavelengths in a pulse are longer than the distance a between pendulums, then $ka \ll 1$, and the chain can be approximated as a continuous medium. In this limit, a becomes the continuous variable x, and the system of coupled *ordinary* differential equations becomes a single, *partial* differential equation:

$$\theta_{j+1} \simeq \theta_j + \frac{\partial\theta}{\partial x}\Delta x, \qquad (25.50)$$

$$\Rightarrow \quad (\theta_{j+1} - 2\theta_j + \theta_{j-1}) \simeq \frac{\partial^2\theta}{\partial x^2}\Delta x^2 \equiv \frac{\partial^2\theta}{\partial x^2}a^2, \qquad (25.51)$$

$$\Rightarrow \quad \frac{\partial^2\theta}{\partial t^2} - \frac{\kappa a^2}{I}\frac{\partial^2\theta}{\partial x^2} = \frac{mgL}{I}\sin\theta. \qquad (25.52)$$

If we measure time in units of $\sqrt{I/mgL}$ and distances in units of $\sqrt{\kappa a/(mgLb)}$, we obtain the standard form of the sine-Gordon equation (SGE):[1]

$$\frac{1}{c^2}\frac{\partial^2\theta}{\partial t^2} - \frac{\partial^2\theta}{\partial x^2} = \sin\theta \qquad \text{(SGE)}. \qquad (25.53)$$

The $\sin\theta$ on the RHS introduces nonlinear effects.

---

1 The name "sine-Gordon" is either a reminder that the SGE is like the Klein-Gordon equation of relativistic quantum mechanics with a sin u added to the RHS, or a reminder of how clever one can be in thinking up names.

25.6.1 Analytic Solution

Although simple looking, the nonlinearity of the sine-Gordon PDE (25.53) makes it hard to solve analytically. There is, however, a trick: as we did for solitons, guess a functional form of a traveling wave, substitute it into (25.53), and thereby convert the PDE into a solvable ODE:

$$\theta(x,t) \overset{?}{=} \theta(\xi = t \pm x/v), \qquad \Rightarrow \qquad \frac{d^2\theta}{d\xi^2} = \frac{v^2}{v^2 - 1}\sin\theta. \tag{25.54}$$

You should recognize (25.54) as our old friend, the equation of motion for a realistic pendulum with no driving force and no friction. The constant v is a velocity in natural units, and separates different regimes of the motion:

$$v < 1 : \text{ pendula initially down } \downarrow\downarrow\downarrow\downarrow\downarrow \text{ (stable)}, \tag{25.55}$$
$$v > 1 : \quad \text{pendula initially up } \uparrow\uparrow\uparrow\uparrow\uparrow \text{ (unstable)}.$$

Although the equation is familiar, we know that an analytic solution does not exists. However, for an energy $E = \pm 1$, we have separatrix motion, and a solution with characteristic *soliton* form,

$$\theta(x - vt) = \begin{cases} 4\tan^{-1}\left(\exp\left[+\frac{x-vt}{\sqrt{1-v^2}}\right]\right), & \text{for } E = 1, \\ 4\tan^{-1}\left(\exp\left[-\frac{x-vt}{\sqrt{1-v^2}}\right]\right) + \pi, & \text{for } E = -1. \end{cases} \tag{25.56}$$

This soliton corresponds to a solitary *kink*, traveling with velocity $v = -1$, that flips the pendulums around by 2π as it moves down the chain. There is also an *antikink* in which the initial $\theta = \pi$ values are flipped to final $\theta = -\pi$.

25.6.2 Numeric 2D Solitons (Pulsons)

It took a bit of manipulation, but we have already found how to solve the KdeV equation for 1D solitons. The elastic-wave solitons that arise from the SGE can be easily generalized to two dimensions, as we do here with the 2D generalization of the SGE equation (25.53):

$$\frac{1}{c^2}\frac{\partial^2 u}{\partial t^2} - \frac{\partial^2 u}{\partial x^2} - \frac{\partial^2 u}{\partial y^2} = \sin u \qquad \text{(2D SGE)}. \tag{25.57}$$

Whereas the 1D SGE describes wave propagation along a chain of connected pendulums, the 2D form might describe wave propagation in nonlinear elastic media. Interestingly enough, the same 2D SGE also occurs in quantum field theory, where the soliton solutions have been suggested as models for elementary particles [Christiansen and Lomdahl, 1981; Christiansen and Olsen, 1979]. The idea is that, like elementary particles, the solutions are confined to a region of space for a long period of time by nonlinear forces, and do not radiate away their energy.

We solve (25.57) in the finite region of 2D space and for positive times:

$$-x_0 < x < x_0, \quad -y_0 < y < y_0, \quad t \geq 0. \tag{25.58}$$

We take $x_0 = y_0 = 7$, and impose the *boundary conditions* that the derivative of the displacement vanishes at the ends of the region:

$$\frac{\partial u}{\partial x}(-x_0, y, t) = \frac{\partial u}{\partial x}(x_0, y, t) = \frac{\partial u}{\partial y}(x, -y_0, t) = \frac{\partial u}{\partial y}(x, y_0, t) = 0. \tag{25.59}$$

We also impose the *initial condition* that at time $t = 0$ the waveform is that of a pulse (Figure 25.6) with its surface at rest:

$$u(x, y, t = 0) = 4\tan^{-1}(\exp 3 - \sqrt{x^2 + y^2}), \qquad \frac{\partial u}{\partial t}(x, y, t = 0) = 0. \tag{25.60}$$

We discretize the equation by looking for solutions on a space-time lattice:

$$x = m\Delta x, \qquad y = l\Delta x, \qquad t = n\Delta t, \tag{25.61}$$

$$u_{m,l}^{(n)} \stackrel{\text{def}}{=} u(m\Delta x, \ l\Delta x, \ n\Delta t). \tag{25.62}$$

Next, we replace the derivatives in (25.57) by their finite-difference approximations:

$$u_{m,l}^{(n+1)} \simeq -u_{m,l}^{(n-1)} + 2\left[1 - 2\left(\frac{\Delta t}{\Delta x}\right)^2\right]u_{m,l}^{(n)} \tag{25.63}$$

$$+ \left(\frac{\Delta t}{\Delta x}\right)^2 (u_{m+1,l}^{(n)} + u_{m-1,l}^{(n)} + u_{m,l+1}^{(n)} + u_{m,l-1}^{(n)})$$

$$- \Delta t^2 \sin\left[\tfrac{1}{4}(u_{m+1,l}^{(n)} + u_{m-1,l}^{(n)} + u_{m,l+1}^{(n)} + u_{m,l-1}^{(n)})\right].$$

To make the algorithm simpler and establish stability, we assume that time and space steps are proportional, $\Delta t = \Delta x/\sqrt{2}$. This leads to all of the $u_{m,l}^{(n)}$ terms dropping out:

$$u_{m,l}^{(2)} \simeq \tfrac{1}{2}(u_{m+1,l}^{(1)} + u_{m-1,l}^{1} + u_{m,l+1}^{(1)} + u_{m,l-1}^{(1)})$$

$$- \frac{\Delta t^2}{2}\sin\left[\tfrac{1}{4}(u_{m+1,l}^{(1)} + u_{m-1,l}^{(1)} + u_{m,l+1}^{(1)} + u_{m,l-1}^{(1)})\right]. \tag{25.64}$$

Likewise, the discrete form of vanishing initial velocity (25.60) becomes:

$$\partial u(x, y, 0)/\partial t = 0 \qquad \Rightarrow \qquad u_{m,l}^{(2)} = u_{m,l}^{(0)}. \tag{25.65}$$

The values on lattice points on the edges and corners cannot be obtained from these relations, but, instead, are obtained by applying the boundary conditions (25.59):

$$\frac{\partial u}{\partial z}(x_0, y, t) = \frac{u(x + \Delta x, y, t) - u(x, y, t)}{\Delta x} = 0, \tag{25.66}$$

$$\Rightarrow \qquad u_{1,l}^{(n)} = u_{2,l}^{(n)}. \tag{25.67}$$

Similarly, the other derivatives in (25.59) give:

$$u_{N_{\max},l}^{(n)} = u_{N_{\max}-1,l}^{(n)}, \quad u_{m,2}^{(n)} = u_{m,1}^{(n)}, \quad u_{m,N_{\max}}^{(n)} = u_{m,N_{\max}-1}^{(n)}, \tag{25.68}$$

where $N_{\max}$ is the number of grid points used for one space dimension.

25.6.3 Implementation

1) Define an array $u[N_{\max}, N_{\max}, 3]$ with $N_{\max} = 201$ for the space points, and 3 for the time points.
2) Place the solution (25.60) for the initial time in $u[m, l, 1]$.
3) Place the solution for the second time Δt in $u[m, l, 2]$, and the solution for time $2\Delta t$ in $u[m, l, 3]$.

4) Assign values for the constants, $\Delta x = \Delta y = 7/100$, $\Delta t = \Delta x / \sqrt{2}$, and $y_0 = x_0 = 7$.
5) Start the solution $t = 0$ with the initial conditions, which, along with the boundary conditions defines it over the entire lattice.
6) For the second time step, use (25.64) for all points on the lattice increase time by Δt, but do not include the edges.
7) At the edges, for $i = 1, 2, \ldots, 200$, set:

$$u[i, 1, 2] = u[i, 2, 2], \quad u[i, N_{\max}, 2] = u[i, N_{\max -1}, 2]$$

$$u[1, i, 2] = u[2, i, 2], \quad u[N_{\max}, i, 2] = u[N_{\max -1}, i, 2].$$

8) To find values for the four points in the corners for the second time step, again use initial condition (25.64):

$$u[1, 1, 2] = u[2, 1, 2], \quad u[N_{\max}, N_{\max}, 2] = u[N_{\max -1}, N_{\max -1}, 2],$$

$$u[1, 1, N_{\max}] = u[2, N_{\max}, 2], \quad u[N_{\max}, 1, 2] = u[N_{\max -1}, 1, 2].$$

9) For the third time step (the future), use the full algorithm (25.66).
10) Continue the propagation forward in time, reassigning the future to the present, and determining a new future. In this way, you need to store the solutions for only three time steps.

We see in Figure 25.6, the time evolution of a circular ring soliton resulting from the stated initial conditions. We note that the ring at first shrinks in size, then expands, and then shrinks back into another (but not identical) ring soliton. A small amount of the particle does radiate away, and in the last frame we can notice some interference between the radiation and the boundary conditions. An animation of this sequence can be found online.

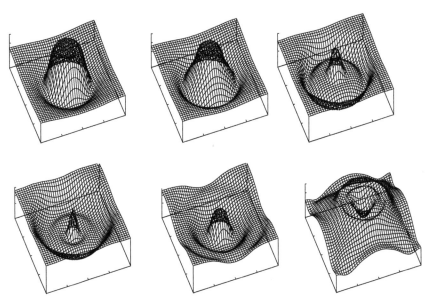

Figure 25.6 A circular ring soliton at times 8, 20, 40, 60, 80, and 120. This type of entity has been proposed as a model for an elementary particle. (Created with `TwoDsol.java`.)

25.7 Code Listings

Listing 25.1 AdvecLax.py solves the advection equation via the Lax–Wendroff scheme.

```
# AdvecLax.py:      Solve advection eqnt via Lax-Wendroff scheme
# du/dt+ c*d(u**2/2)/dx=0;   u(x,t=0)=exp(-300(x-0.12)**2)

from vpython import *
m = 100                                          # No steps in x
c = 1.;        dx = 1./m;      beta = 0.8         # beta = c*dt/dx
u = [0]*(m+1);                                    # Initial Numeric
u0 = [0]*(m+1);
uf = [0]*(m+1)
dt = beta*dx/c;
T_final = 0.5;
n = int(T_final/dt)                              # N time steps

graph1 = graph(width=600, height=500, xtitle = 'x', xmin=0,xmax=1,
        ymin=0, ymax=1, ytitle = 'u(x), Cyan=exact, Yellow=Numerical',
        title='Advect Eqn: Initial (red), Exact (cyan),Numerical (yellow)')
initfn = gcurve(color = color.red);
exactfn = gcurve(color = color.cyan)
numfn = gcurve(color = color.yellow)             # Numerical solution

def plotIniExac():                      # Plot initial & exact solution
  for i in range(0, m):
      x = i*dx
      u0[i] = exp(-300.* (x - 0.12)**2)          # Gaussian initial
      initfn.plot(pos = (0.01*i, u0[i]) )        # Initial function
      uf[i] = exp(-300.*(x - 0.12 - c*T_final)**2) # Exact = cyan
      exactfn.plot(pos = (0.01*i, uf[i]) )
      rate(50)
plotIniExac()

def numerical():                        # Finds Lax Wendroff solution
  for j in range(0, n+1):                          # Time loop
    for i in range(0, m- 1):                       # x loop
        u[i + 1] = (1.-beta*beta)*u0[i+1]-(0.5*beta)*(1.-beta)*u0[i+2]
        +(0.5*beta)*(1. + beta)*u0[i]    # Algorithm
        u[0] = 0.;      u[m-1] = 0.;     u0[i] = u[i]
numerical()
for j in range(0, m-1 ):
    rate(50)
    numfn.plot(pos = (0.01*j, u[j]) )            # Plot numeric soltn
```

Listing 25.2 Soliton.py solves the KdeV equation for 1D solitons corresponding to a "bore" initial conditions.

```
# Soliton.py:      Korteweg de Vries equation for a soliton

from visual import *
import matplotlib.pylab as p;
from mpl_toolkits.mplot3d import Axes3D ;
import numpy

ds = 0.4;    dt = 0.1;    max = 2000; mu = 0.1; eps = 0.2;    mx = 131
u  = zeros( (mx, 3), float); spl = zeros( (mx, 21), float); m = 1

for i in range(0, 131):                          # Initial wave
    u[i, 0] = 0.5*(1  -((math.exp(2*(0.2*ds*i-5.))-1)/(math.exp(2*(0.2*ds*i-5.))+1)))
u[0,1] = 1. ; u[0,2] = 1.; u[130,1] = 0. ; u[130,2] = 0.    # End points

for i in range (0, 131, 2): spl[i, 0] = u[i, 0]
fac = mu*dt/(ds**3)
print("Working. Please hold breath and wait while I count to 20")
for i in range (1, mx-1):                        # First time step
```

```
        a1 = eps*dt*(u[i + 1, 0] + u[i, 0] + u[i - 1, 0])/(ds*6.)
        if i > 1 and  i < 129: a2 = u[i+2,0]+2.*u[i-1,0]-2.*u[i+1,0]-u[i-2,0]
        else:  a2 = u[i-1, 0] - u[i+1, 0]
22      a3 = u[i+1, 0] - u[i-1, 0]
        u[i, 1] = u[i, 0] - a1*a3 - fac*a2/3.
    for j in range (1, max+1):                          # Next time steps
        for i in range(1, mx-2):
26          a1 = eps*dt*(u[i + 1, 1]  +  u[i, 1]  +  u[i - 1, 1])/(3.*ds)
            if i > 1 and i < mx-2:
                a2 = u[i+2,1] + 2.*u[i-1,1] - 2.*u[i+1,1] - u[i-2,1]
            else:  a2 = u[i-1, 1] - u[i+1, 1]
30          a3       = u[i+1, 1] - u[i-1, 1]
            u[i, 2] = u[i,0] - a1*a3 - 2.*fac*a2/3.
        if j%100 ==  0:                                 # Plot every 100 time steps
            for i in range (1, mx - 2): spl[i, m] = u[i, 2]
34          print(m)
            m = m + 1
        for k in range(0, mx):                          # Recycle array saves memory
            u[k, 0] = u[k, 1]
38          u[k, 1] = u[k, 2]

    x = list(range(0, mx, 2))                           # Plot every other point
    y = list(range(0, 21))                  # Plot 21 lines every 100 t steps
42  X, Y = p.meshgrid(x, y)

    def functz(spl):
        z = spl[X, Y]
46      return z

    fig  = p.figure()                                   # create figure
    ax = Axes3D(fig)                                        # plot axes
50  ax.plot_wireframe(X, Y, spl[X, Y], color = 'r')     # red wireframe
    ax.set_xlabel('Positon')                            # label axes
    ax.set_ylabel('Time')
    ax.set_zlabel('Disturbance')
54  p.show()                                    # Show figure, close Python shell
    print("That's all folks!")
```

26

Fluid Hydrodynamics

> *We have already covered some fluid dynamics in our discussion of shallow-water solitons in Chapter 25. This chapter examines the more general equations of fluid dynamics and their numerical solutions.[1] The equations are nonlinear, yet have striking similarities to those of E&M, and support elegant mathematical and computational treatments. Analytic solutions, however, are rare, which helps explain why computation fluid dynamics (CFD) is an important specialty (think airplanes). We recommend [Fetter and Walecka, 1980; Landau and Lifshitz, 1987] for those interested in the derivations, and Shaw [1992] for more details about the computations.*

In order for migrating salmon to have a place to rest during their arduous upstream journey, the Oregon Department of Environment wants to place objects in several deep, wide, fast-flowing streams. One such object is a long beam of rectangular cross section (Figure 26.1 left), and another is a set of plates (Figure 26.1 right). The objects are to be placed far enough below the water's surface so as not to disturb the surface flow, and also far enough from the bottom of the stream so as not to disturb the flow there.

Problem Determine how the size and location of the beam and plates affect the stream's velocity profile.

26.1 Navier–Stokes Equation

As with our study of shallow-water solitons, we assume that water is *incompressible* with constant density ρ. The problem description implies that we can assume a steady state, but not that we can ignore friction (*viscosity*). As before, the first equation of hydrodynamics is the continuity equation (25.2):

$$\frac{\partial \rho(\mathbf{x}, t)}{\partial t} + \nabla \cdot \mathbf{j} = 0, \qquad \mathbf{j} \overset{\text{def}}{=} \rho \mathbf{v}(\mathbf{x}, t). \tag{26.1}$$

---

1 We acknowledge some helpful reading and comments by Satoru S. Kano.

Computational Physics: Problem Solving with Python, Fourth Edition.
Rubin H. Landau, Manuel J. Páez, and Cristian C. Bordeianu.
© 2024 WILEY-VCH GmbH. Published 2024 by WILEY-VCH GmbH.

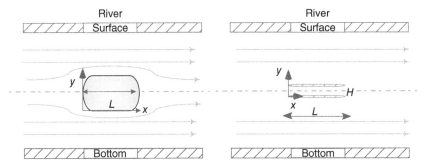

Figure 26.1 Side view of the flow of a stream around a submerged beam (*left*) and around two parallel plates (*right*). Both beam and plates have length L along the direction of flow. The flow is seen to be symmetric about the centerline and to be unaffected at the bottom and at the surface by the submerged object.

The second equation of hydrodynamics employs a special time derivative, the *hydrodynamic derivative* $D\mathbf{v}/Dt$ [Fetter and Walecka, 1980]:

$$\frac{D\mathbf{v}}{Dt} \stackrel{\text{def}}{=} (\mathbf{v} \cdot \nabla)\mathbf{v} + \frac{\partial \mathbf{v}}{\partial t}. \tag{26.2}$$

This derivative gives the rate of change, as viewed from a stationary frame, of the velocity of material within *an element of flowing fluid*. It thus incorporates changes as a result of the motion of the fluid (first term), as well as any explicit time dependence of the velocity (second term). It is particularly noteworthy that $D\mathbf{v}/Dt$ is second order in velocity, and as such it introduces nonlinearities into the theory. You may think of these nonlinearities as arising from fictitious (inertial) forces that occur when describing the motion of an element in the fluid's rest frame, which, in general, is an accelerating frame.

The material derivative is the leading term in the *Navier–Stokes equation*:

$$\frac{D\mathbf{v}}{Dt} = \nu \nabla^2 \mathbf{v} - \frac{1}{\rho} \nabla P(\rho, T, x), \tag{26.3}$$

where ν is the kinematic viscosity and P is the pressure. Though less elegant to look at, the computational solution is based on (26.3)'s Cartesian form:

$$\frac{\partial v_x}{\partial t} + \sum_{j=x}^{z} v_j \frac{\partial v_x}{\partial x_j} = \nu \sum_{j=x}^{z} \frac{\partial^2 v_x}{\partial x_j^2} - \frac{1}{\rho} \frac{\partial P}{\partial x},$$

$$\frac{\partial v_y}{\partial t} + \sum_{j=x}^{z} v_j \frac{\partial v_y}{\partial x_j} = \nu \sum_{j=x}^{z} \frac{\partial^2 v_y}{\partial x_j^2} - \frac{1}{\rho} \frac{\partial P}{\partial y}, \tag{26.4}$$

$$\frac{\partial v_z}{\partial t} + \sum_{j=x}^{z} v_j \frac{\partial v_z}{\partial x_j} = \nu \sum_{j=x}^{z} \frac{\partial^2 v_z}{\partial x_j^2} - \frac{1}{\rho} \frac{\partial P}{\partial z}.$$

The Navier–Stokes equation describes the transfer of the momentum of the fluid within some region of space as a result of (i) the forces on the fluid (think $d\mathbf{p}/dt = \mathbf{F}$), and (ii) the fluid flow. The $\mathbf{v} \cdot \nabla \mathbf{v}$ term in $D\mathbf{v}/Dt$ describes the transport of momentum in some region of space resulting from the fluid's flow, and it is often called the *convection* or *advection* term.[2] The ∇P term describes the velocity change resulting from pressure changes, and the $\nu\nabla^2\mathbf{v}$ term describes the velocity change resulting from viscous forces that tend to impede the flow.

The explicit functional dependence of the pressure on the fluid's density and temperature, $P(\rho, T, x)$, is known as the *equation of state of the fluid*, and it is assumed to be known before trying to solve the Navier–Stokes equation. To keep our problem simple, we assume that the pressure is independent of density and temperature. This leaves us with four simultaneous partial differential equations to solve, the continuity equation (26.1), and the Navier–Stokes equation (26.3). Because our problem is one with *steady state* flow, we will set all time derivatives of the velocity to zero. Because water is incompressible, the time derivative of the density also vanishes. Equations (26.1) and (26.3) then become:

$$\nabla \cdot \mathbf{v} \equiv \sum_i \frac{\partial v_i}{\partial x_i} = 0, \tag{26.5}$$

$$(\mathbf{v} \cdot \nabla)\mathbf{v} = \nu\nabla^2\mathbf{v} - \frac{1}{\rho}\nabla P. \tag{26.6}$$

The first equation expresses the equality of inflow and outflow, and is known as the *condition of incompressibility*. In as much as the stream in our problem is much wider than the width of the beam, and because we want a solution in the middle of the stream, and not near the banks, we will ignore the z dependence of the velocity. The explicit PDE's we need to solve then reduce to:

$$\frac{\partial v_x}{\partial x} + \frac{\partial v_y}{\partial y} = 0, \tag{26.7}$$

$$\nu\left(\frac{\partial^2 v_x}{\partial x^2} + \frac{\partial^2 v_x}{\partial y^2}\right) = v_x\frac{\partial v_x}{\partial x} + v_y\frac{\partial v_x}{\partial y} + \frac{1}{\rho}\frac{\partial P}{\partial x}, \tag{26.8}$$

$$\nu\left(\frac{\partial^2 v_y}{\partial x^2} + \frac{\partial^2 v_y}{\partial y^2}\right) = v_x\frac{\partial v_y}{\partial x} + v_y\frac{\partial v_y}{\partial y} + \frac{1}{\rho}\frac{\partial P}{\partial y}. \tag{26.9}$$

26.2 Flow Through Parallel Plates

The parallel plate problem is one of the few that have analytic solutions. Yet since we still have plenty of work to do in order to set up the numerical solution, we will skip the details and just give the result we need for comparison: there is a parabolic velocity profile:

$$\rho\nu\,v_x(y) = \frac{1}{2}\frac{\partial P}{\partial x}\left(y^2 - yH\right). \tag{26.10}$$

A unique solution to the PDEs (26.7)–(26.9) requires knowledge of appropriate boundary conditions. As far as we can tell, setting boundary conditions is somewhat of an acquired skill. We assume that the submerged parallel plates are placed in a stream that is flowing with a constant velocity V_0 in the horizontal direction (Figure 26.1 right). If the velocity V_0 is not too high, or if the kinematic viscosity ν is sufficiently large, then the flow should be

---

2 We discuss advection in Section 25.1. In oceanology or meteorology, convection implies the transfer of mass in the vertical direction where it overcomes gravity, whereas advection refers to transfer in the horizontal direction.

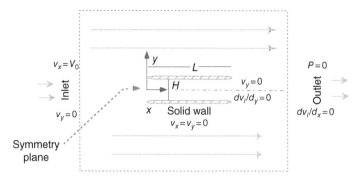

Figure 26.2 The boundary conditions for two thin submerged plates. The surrounding box is the integration volume within which we solve the PDE's, and upon whose surface we impose the boundary conditions. In practice the box would be much larger than *L* and *H*.

smooth and without turbulence. Such flow is called *laminar*. Typically, a fluid undergoing laminar flow moves in smooth paths that do not close on themselves, like the flow of water from a faucet. If we imagine attaching a vector to each element of the fluid, then the path swept out by that vector is called a *streamline*, or *line of motion*, of the fluid. These streamlines can be visualized experimentally by adding colored dye to the stream. We assume that the plates are so thin that the flow through and around them remains laminar.

If the plates are thin, then the flow far upstream of them will not be affected, and we can limit our solution space to the rectangular region in Figure 26.2. We assume that the length *L* and separation *H* of the plates are small compared to the size of the stream, so the flow returns to uniform as we get far downstream from the plates. As seen in Figure 26.2, there are boundary conditions at the *inlet*, where the fluid enters the solution space, at the *outlet*, where it leaves, and at the stationary plates, where it just passes through. In addition, because the plates are far from the stream's bottom and surface, we assume that the dotted-dashed centerline is a plane of symmetry, with identical flow above and below the plane. We thus have four different types of boundary conditions to impose on our solution:

Solid plates: In as much as there is friction (viscosity) between the fluid and the plate surface, the only way to have laminar flow is to have the fluid's velocity equal to the plate's velocity, which means both are zero:

$$v_x = v_y = 0. \tag{26.11}$$

Such being the case, we have smooth flow in which the negligibly thin plates lie along streamlines of the fluid (like a "streamlined" vehicle).

Inlet: The fluid enters the integration domain at the inlet with a horizontal velocity V_0. Because the inlet is far upstream from the plates, we assume that the fluid velocity at the inlet is unchanged by the presence of the plates:

$$v_x = V_0, \qquad v_y = 0. \tag{26.12}$$

Outlet: The fluid leaves the integration domain at the outlet. While it is totally reasonable to assume that the fluid returns to its unperturbed state there, we are not sure what that might be. So, instead, we assume that there is a physical outlet at the end with the water just shooting out of it. This means that the water pressure equals zero at the outlet (as at the end of a garden hose), and that the velocity does not change in a direction normal to the outlet:

$$P = 0, \qquad \frac{\partial v_x}{\partial x} = \frac{\partial v_y}{\partial x} = 0. \tag{26.13}$$

Symmetry plane: If the flow is symmetric about the $y = 0$ plane, then there cannot be flow through the plane, which means that the spatial derivatives of the velocity components normal to the plane must vanish:

$$v_y = 0, \qquad \frac{\partial v_y}{\partial y} = 0. \tag{26.14}$$

This condition follows from the assumption that the plates are along streamlines and that they are negligibly thin. It means that all the streamlines are parallel to the plates, as well as to the water surface, and so it must be that $v_y = 0$ everywhere. The fluid enters in the horizontal direction, the plates do not change the vertical y component of the velocity, and the flow remains symmetric about the centerline. There is a retardation of the flow around the plates as a result of the viscous nature of the flow, and as a result of the $\mathbf{v} = 0$ boundary layers formed on the plates, but there are no actual v_y components.

26.3 Navier–Stokes Difference Equation

Now we develop the difference equation forms of the Navier–Stokes and continuity PDEs. They will be solved with *successive overrelaxation*, a variation of the method used in Chapter 21 to solve Poisson's equation. We start by dividing space into a rectangular grid with spacing h in both the x and y directions:

$$x = ih, \qquad i = 0, \dots, N_x; \qquad y = jh, \qquad j = 0, \dots, N_y.$$

Next, we use the central-difference approximation to express the derivatives in (26.7)–(26.9) as finite differences of the values of the velocities at the grid points. For $v = 1\,\mathrm{m}^2/\mathrm{s}$ and $\rho = 1\,\mathrm{kg/m}^3$, this yields:

$$v_{i+1,j}^{(x)} - v_{i-1,j}^{(x)} + v_{i,j+1}^{(y)} - v_{i,j-1}^{(y)} = 0, \tag{26.15}$$

$$v_{i+1,j}^{(x)} + v_{i-1,j}^{(x)} + v_{i,j+1}^{(x)} + v_{i,j-1}^{(x)} - 4v_{i,j}^{(x)} \tag{26.16}$$

$$= \frac{h}{2} v_{i,j}^{(x)} [v_{i+1,j}^{(x)} - v_{i-1,j}^{(x)}] + \frac{h}{2} v_{i,j}^{(y)} [v_{i,j+1}^{(x)} - v_{i,j-1}^{(x)}] + \frac{h}{2} [P_{i+1,j} - P_{i-1,j}],$$

$$v_{i+1,j}^{(y)} + v_{i-1,j}^{(y)} + v_{i,j+1}^{(y)} + v_{i,j-1}^{(y)} - 4v_{i,j}^{(y)} \tag{26.17}$$

$$= \frac{h}{2} v_{i,j}^{(x)} [v_{i+1,j}^{(y)} - v_{i-1,j}^{(y)}] + \frac{h}{2} v_{i,j}^{(y)} [v_{i,j+1}^{(y)} - v_{i,j-1}^{(y)}] + \frac{h}{2} [P_{i,j+1} - P_{i,j-1}].$$

Because $v^{(y)} \equiv 0$, we can solve for $v^{(x)}$:

$$4v_{i,j}^{(x)} = v_{i+1,j}^{(x)} + v_{i-1,j}^{(x)} + v_{i,j+1}^{(x)} + v_{i,j-1}^{(x)} - \frac{h}{2} v_{i,j}^{(x)} [v_{i+1,j}^{(x)} - v_{i-1,j}^{(x)}]$$

$$- \frac{h}{2} v_{i,j}^{(y)} [v_{i,j+1}^{(x)} - v_{i,j-1}^{(x)}] - \frac{h}{2} [P_{i+1,j} - P_{i-1,j}]. \tag{26.18}$$

We recognize (26.18) as an algorithm similar to the one we used in solving Laplace's equation by relaxation. Indeed, as we did there, we can accelerate the convergence by writing the algorithm with the new value of $v^{(x)}$ given by the old value plus a correction (residual):

$$v_{i,j}^{(x)} = v_{i,j}^{(x)} + r_{i,j}, \qquad r \overset{\text{def}}{=} v_{i,j}^{x(\text{new})} - v_{i,j}^{x\text{old}} \tag{26.19}$$

$$\Rightarrow r = \frac{1}{4}\left\{ v_{i+1,j}^{(x)} + v_{i-1,j}^{(x)} + v_{i,j+1}^{(x)} + v_{i,j-1}^{(x)} - \frac{h}{2}v_{i,j}^{(x)}[v_{i+1,j}^{(x)} - v_{i-1,j}^{(x)}] \right.$$

$$\left. - \frac{h}{2}v_{i,j}^{(y)}[v_{i,j+1}^{(x)} - v_{i,j-1}^{(x)}] - \frac{h}{2}[P_{i+1,j} - P_{i-1,j}] \right\} - v_{i,j}^{(x)}. \tag{26.20}$$

As before, successive iterations sweep the interior of the grid, continuously adding in the residual (26.19) until the change becomes smaller than some set level of tolerance, $|r_{i,j}| < \varepsilon$.

A variation of this method, *successive overrelaxation* (SOR), increases the speed at which the residuals approach zero by including an amplifying factor ω:

$$v_{i,j}^{(x)} = v_{i,j}^{(x)} + \omega r_{i,j} \qquad \text{(SOR)}. \tag{26.21}$$

The standard relaxation algorithm (26.19) is obtained with $\omega = 1$, accelerated convergence (*overrelaxation*) is obtained with $\omega \geq 1$, and *underrelaxation* occurs for $\omega < 1$. Values $\omega > 2$ are found to lead to numerical instabilities. Although a detailed analysis of the algorithm is necessary to predict the optimal value for ω, we suggest that you test different values for ω to see which one provides the fastest, yet stable, convergence for the problem at hand.

26.3.1 Successive Overrelaxation Algorithm

1) Modify the program `Beam.py`, or write your own, to solve the Navier-Stokes equation for the velocity of a fluid in 2D flow. Represent the x and y components of the velocity by the arrays `vx[Nx,Ny]` and `vy[Nx,Ny]`.
2) Specialize your solution to the rectangular domain and boundary conditions indicated in Figure 26.2.
3) Use of the parameter values,

$$v = 1\,\text{m}^2/\text{s}, \quad \rho = 10^3\,\text{kg/m}^3, \quad \text{(flow parameters)},$$

$$N_x = 400, \quad N_y = 40, \ h = 1, \quad \text{(grid parameters)},$$

leads to the equations

$$\frac{\partial P}{\partial x} = -12, \quad \frac{\partial P}{\partial y} = 0, \quad v^{(x)} = \frac{3j}{20}\left(1 - \frac{j}{40}\right), \quad v^{(y)} = 0. \tag{26.22}$$

4) For the relaxation method, output the iteration number and the computed $v^{(x)}$.
5) Verify that your numerical solution agrees with the analytic result (26.10) for flow through parallel plates.
6) Repeat the calculation and see if SOR speeds up the convergence.

26.4 Vorticity Form of Navier–Stokes Equation

Now that the comparison with an analytic solution has shown that our CFD simulation works, we return to determining if the beam in Figure 26.1 might produce a good resting place for salmon. While we have no analytic solution with which to compare, our canoeing and fishing adventures have taught us that *standing waves* with fish in them are often formed behind rocks in streams, and so we will look for evidence of a standing wave forming behind the beam.

We have seen how to numerically solve the hydrodynamics equations:

$$\nabla \cdot \mathbf{v} = 0, \tag{26.23}$$

$$(\mathbf{v} \cdot \nabla)\mathbf{v} = -\frac{1}{\rho}\nabla P + \nu \nabla^2 \mathbf{v}. \tag{26.24}$$

These equations determine the components of a fluid's velocity, pressure, and density as functions of position. Recall how in electrostatics it is usually simpler to solve for a scalar potential, rather than a vector field, and then take the potential's gradient to determine the vector field. In analogy, we recast the hydrodynamic equation into forms that permit us to solve two simpler equations, from which the velocity is obtained via a gradient operation.[3]

We define a *stream function* $\mathbf{u}(\mathbf{x})$, from which the velocity is determined by the curl operator:

$$\mathbf{v} \stackrel{\text{def}}{=} \nabla \times \mathbf{u}(\mathbf{x}) = \hat{e}_x \left(\frac{\partial u_z}{\partial y} - \frac{\partial u_y}{\partial z} \right) + \hat{e}_y \left(\frac{\partial u_x}{\partial z} - \frac{\partial u_z}{\partial x} \right). \tag{26.25}$$

Note the absence of a z component of velocity for our problem. Since $\nabla \cdot (\nabla \times \mathbf{u}) \equiv 0$, we see that any $\mathbf{v}$ that can be a written as the curl of $\mathbf{u}$ automatically satisfies the continuity equation $\nabla \cdot \mathbf{v} = 0$. Furthermore, because the $\mathbf{v}$ for our problem has only x and y components, $\mathbf{u}(\mathbf{x})$ needs have only a z component:

$$u_z \equiv u \quad \Rightarrow \quad v_x = \frac{\partial u}{\partial y}, \quad v_y = -\frac{\partial u}{\partial x}. \tag{26.26}$$

Note that in 2D flows, the contour lines $u = $ constant are the *streamlines*.

The second simpler function is the *vorticity* field $\mathbf{w}(\mathbf{x})$, which is related physically, and alphabetically, to the angular velocity ω of the fluid. Vorticity is defined as the curl of the velocity (sometimes with a $-$ sign):

$$\mathbf{w} \stackrel{\text{def}}{=} \nabla \times \mathbf{v}(\mathbf{x}). \tag{26.27}$$

Because our problem's velocity does not change in the z direction, the z derivative vanishes:

$$w_z = \left(\frac{\partial v_y}{\partial x} - \frac{\partial v_x}{\partial y} \right). \tag{26.28}$$

Physically, we see that vorticity is a measure of how much the fluid's velocity curls or rotates, with the direction of the vorticity determined by the right-hand rule for rotations. In fact, if we could remove a small element of the fluid into space (so it would not feel the internal strain of the fluid), we would find that it is rotating like a solid with angular velocity $\omega \propto \mathbf{w}$ [Lamb, 1993]. That being the case, it is useful to think of vorticity as giving the local value of the fluid's angular velocity vector, with $\mathbf{w} = 0$ describing *irrotational* flow.

In general, the field lines of w are continuous and move as if attached to elements of the fluid. A uniformly flowing fluid would have vanishing curls, while a nonzero vorticity

---

3 If we had to solve only the simpler problem of *irrotational flow* (no turbulence), then we would be able to use a scalar velocity potential, in close analogy to electrostatics [Lamb, 1993]. For the more general *rotational flow*, two vector potentials are required.

indicates that the current rotates, or curls back on itself. From the definition of the stream function (26.25), we see that the vorticity $\mathbf{w}$ is related to it by:

$$\mathbf{w} = \nabla \times \mathbf{v} = \nabla \times (\nabla \times \mathbf{u}) = \nabla(\nabla \cdot \mathbf{u}) - \nabla^2\mathbf{u}, \tag{26.29}$$

where we have used a vector identity for $\nabla \times (\nabla \times \mathbf{u})$. But because $\mathbf{u}$ has only a z component that does not vary with z (or because there is no source for $\mathbf{u}$), the divergence $\nabla \cdot \mathbf{u} = 0$. We now have the basic relation between the stream function $\mathbf{u}$ and the vorticity $\mathbf{w}$:

$$\nabla^2\mathbf{u} = -\mathbf{w}. \tag{26.30}$$

Equation (26.30) is analogous to Poisson's equation of electrostatics, $\nabla^2\phi = -4\pi\rho$, only now each component of vorticity $\mathbf{w}$ acting as the source for the corresponding component of the stream function $\mathbf{u}$. If the flow is irrotational, that is, if $\mathbf{w} = 0$, then we need only solve Laplace's equation for each component of u. Rotational flow, with its coupled nonlinearities equations, leads to more interesting behavior.

As is to be expected from the definition of $\mathbf{w}$, the vorticity form of the Navier–Stokes equation is obtained by taking the curl of the velocity form, that is, by operating on both sides with $\nabla\times$. After significant manipulations one obtains

$$\nu\nabla^2\mathbf{w} = [(\nabla \times \mathbf{u}) \cdot \nabla]\mathbf{w}. \tag{26.31}$$

This and (26.30) are the two simultaneous PDE's that we need to solve. In 2D, with $\mathbf{u}$ and $\mathbf{w}$ having only z components, they are

$$\frac{\partial^2 u}{\partial x^2} + \frac{\partial^2 u}{\partial y^2} = -w, \tag{26.32}$$

$$\nu\left(\frac{\partial^2 w}{\partial x^2} + \frac{\partial^2 w}{\partial y^2}\right) = \frac{\partial u}{\partial y}\frac{\partial w}{\partial x} - \frac{\partial u}{\partial x}\frac{\partial w}{\partial y}. \tag{26.33}$$

So after all that work, we end up with two simultaneous, nonlinear, elliptic PDE's that look like a mixture of Poisson's equation with the wave equation. The equation for u is Poisson's equation with source w, and must be solved simultaneously with the second. It is this second equation that contains mixed products of the derivatives of u and w, and thus the nonlinearity.

26.4.1 Vorticity Difference Equation

We solve (26.32) and (26.33) on an $N_x \times N_y$ grid of uniform spacing h with:

$$x = i\Delta x = ih, \ i = 0, \dots, N_x, \ y = j\Delta y = jh, \ j = 0, \dots, N_y. \tag{26.34}$$

Because the beam is symmetric about its centerline (Figure 26.1 left), we need the solution only in the upper half-plane. We apply the now familiar central-difference approximation to the Laplacians of u and w to obtain the difference Laplacian:

$$\frac{\partial^2 u}{\partial x^2} + \frac{\partial^2 u}{\partial y^2} \simeq \frac{u_{i+1,j} + u_{i-1,j} + u_{i,j+1} + u_{i,j-1} - 4u_{i,j}}{h^2}. \tag{26.35}$$

Likewise, for the product of first derivatives,

$$\frac{\partial u}{\partial y}\frac{\partial w}{\partial x} \simeq \frac{u_{i,j+1} - u_{i,j-1}}{2h}\frac{w_{i+1,j} - w_{i-1,j}}{2h}. \tag{26.36}$$

The difference vorticity Navier–Stokes equation (26.32) is now:

$$u_{i,j} = \tfrac{1}{4}(u_{i+1,j} + u_{i-1,j} + u_{i,j+1} + u_{i,j-1} + h^2 w_{i,j}), \tag{26.37}$$

$$w_{i,j} = \tfrac{1}{4}(w_{i+1,j} + w_{i-1,j} + w_{i,j+1} + w_{i,j-1}) - \tfrac{R}{16}\left\{\left[u_{i,j+1} - u_{i,j-1}\right]\right.$$
$$\left. \times \left[w_{i+1,j} - w_{i-1,j}\right] - \left[u_{i+1,j} - u_{i-1,j}\right]\left[w_{i,j+1} - w_{i,j-1}\right]\right\}, \tag{26.38}$$

$$R = \tfrac{1}{\nu} \quad \left(\tfrac{V_0 h}{\nu} \text{ in normal units}\right). \tag{26.39}$$

Note, in order to obtain an algorithm appropriate for solution by relaxation, we have placed $u_{i,j}$ and $w_{i,j}$ on the LHS of the equations.

The parameter R in (26.39) is related to the *Reynolds number*. When we solve the problem in natural units, we measure distances in units of grid spacing h, velocities in units of initial velocity V_0, stream functions in units of $V_0 h$, and vorticity in units of V_0/h. The second form is in regular units and is dimensionless. This R is known as the *grid Reynolds number*, and differs from the physical R, which has a pipe diameter in place of the grid spacing h.

The grid Reynolds number is a measure of the strength of the coupling of the nonlinear terms in the equation. When the physical R is small, the viscosity acts as a frictional force that damps out fluctuations and keeps the flow smooth. When R is large ($R \simeq 2000$), physical fluids undergo phase transitions from laminar to turbulent flow in which turbulence occurs at a cascading set of smaller and smaller space scales. However, simulations that produce the onset of turbulence have been a research problem since Reynolds' first experiments in 1883 [Reynolds, 1883; Falkovich and Sreenivasan, 2006]. Possibly because laminar flow is stable against small perturbations, it may be that some large-scale "kick" is needed to change it from laminar to turbulent.

As discussed in Section 26.3, the finite difference algorithm can have its convergence accelerated by the use of successive overrelaxation (26.37):

$$u_{i,j} = u_{i,j} + \omega\, r_{i,j}^{(1)}, \qquad w_{i,j} = w_{i,j} + \omega\, r_{i,j}^{(2)} \quad \text{(SOR)}. \tag{26.40}$$

Here ω is the overrelaxation parameter, and should lie in the range $0 < \omega < 2$ for stability. The residuals are just the changes in a single step, $r^{(1)} = u^{\text{new}} - u^{\text{old}}$ and $r^{(2)} = w^{\text{new}} - w^{\text{old}}$:

$$r_{i,j}^{(1)} = \tfrac{1}{4}(u_{i+1,j} + u_{i-1,j} + u_{i,j+1} + u_{i,j-1} + w_{i,j}) - u_{i,j}, \tag{26.41}$$

$$r_{i,j}^{(2)} = \tfrac{1}{4}\left(w_{i+1,j} + w_{i-1,j} + w_{i,j+1} + w_{i,j-1} - \tfrac{R}{4}\left\{\left[u_{i,j+1} - u_{i,j-1}\right]\right.\right.$$
$$\left.\left. \times \left[w_{i+1,j} - w_{i-1,j}\right] - \left[u_{i+1,j} - u_{i-1,j}\right]\left[w_{i,j+1} - w_{i,j-1}\right]\right\}\right) - w_{i,j}.$$

26.4.2 Beam Boundary Conditions

A well-defined solution to these elliptic PDEs requires a combination of (less than obvious) boundary conditions on u and w. Consider Figure 26.3, based on the analysis of Koonin [1986]. The assumption is that the inlet, outlet, and surface are far from the beam (which

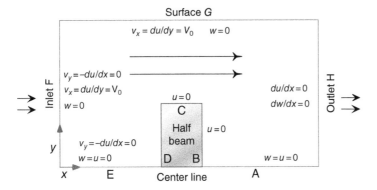

Figure 26.3 Boundary conditions for flow around the beam in Figure 26.1. The flow is symmetric about the centerline, and the beam has length L in the x direction (along flow).

may not be evident from the not-to-scale figure). We refer the interested reader to the references, and just give the **Boundary Conditions**:

$$u = 0; \qquad w = 0 \qquad\qquad\qquad \text{Centerline EA}$$

$$u = 0, \qquad w_{i,j} = -\frac{2}{h^2}(u_{i+1,j} - u_{i,j}) \quad \text{Beam back B}$$

$$u = 0, \qquad w_{i,j} = -\frac{2}{h^2}(u_{i,j+1} - u_{i,j}) \quad \text{Beam top C}$$

$$u = 0, \qquad w_{i,j} = -\frac{2}{h^2}(u_{i-1,j} - u_{i,j}) \quad \text{Beam front D}$$

$$\partial u/\partial x = 0, \qquad w = 0 \qquad\qquad\qquad \text{Inlet F}$$

$$\partial u/\partial y = V_0, \qquad w = 0 \qquad\qquad\qquad \text{Surface G}$$

$$\partial u/\partial x = 0, \quad \partial w/\partial x = 0 \qquad\qquad \text{Outlet H}$$

`Beam.py` in Listing 26.1 is our program for solution of the vorticity form of the Navier–Stokes equation. You will notice that while the relaxation algorithm is rather simple, some care is needed in implementing the many boundary conditions. Relaxation of the stream function and of the vorticity is performed by separate functions.

26.5 Assessment and Exploration

1) Use `Beam.py` as a basis for your solution for the stream function u and the vorticity w using the finite-differences algorithm (26.37). Figures 26.4 and 26.5 show some typical results.
2) A good place to start your simulation is with a beam of size $L = 8h$, $H = h$, Reynolds number $R = 0.1$, and intake velocity $V_0 = 1$. Keep your grid small during debugging, say, $N_x = 24$ and $N_y = 70$.
3) Explore the convergence of the algorithm.
   a) Print out the iteration number and u values upstream from, above, and downstream from the beam.
   b) Determine the number of iterations necessary to obtain three-place convergence for successive relaxation ($\omega = 1$).

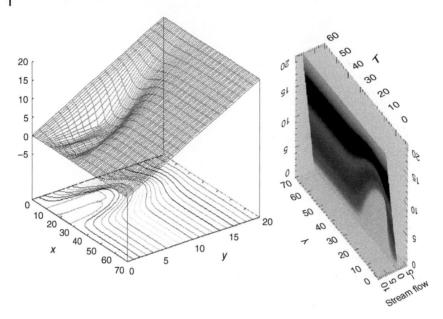

Figure 26.4 Two visualizations of the stream function *u* for Reynold's number $R = 5$.

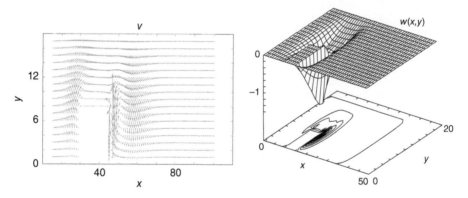

Figure 26.5 *Left*: The vorticity as a function of *x* and *y*. Rotation is seen to be largest behind the beam. *Right*: The velocity field around the beam as represented by vectors.

    c) Determine the number of iterations necessary to obtain three-place convergence for successive overrelaxation ($\omega \simeq 1.3$). Use this number for future calculations.

4) Change the beam's horizontal placement, so that you can see the undisturbed current entering from the left, and then developing into a standing wave. Note that you may need to increase the size of your simulation volume to see the effect of all the boundary conditions.

5) Make surface plots including contours of the stream function *u* and the vorticity *w*. Explain the behavior seen.

6) Is there a region where a big fish can rest behind the beam?

7) The results of the simulation (Figure 26.4) are for the one-component stream function *u*. Make several visualizations showing the fluid velocity throughout the simulation region.

Note that velocity is a vector with two components, and the individual components are interesting to visualize. A vector plot works well here.

8) Explore how increasing the Reynolds number *R* changes the flow pattern. Start at *R* = 0 and gradually increase *R* while watching for numeric instabilities. To overcome numerical instabilities, reduce the size of the relaxation parameter *ω* and continue to larger *R* values.

9) Verify that the flow around the beam is smooth for small *R* values, but that it separates from the back edge for large *R*, at which point a small vortex develops.

26.5.1 Explorations

1) Determine the flow behind a circular rock in the stream.
2) The boundary condition at an outlet far downstream should not have much effect on the simulation. Explore the use of other boundary conditions there.
3) Determine the pressure variation around the beam.

26.6 Code Lisitings

Listing 26.1 Beam.py Solves the Navier–Stokes equation for the flow over a plate.

```
# Beam.py: solves Navier-Stokes equation for flow around beam

import matplotlib.pylab as p;
from mpl_toolkits.mplot3d import Axes3D;
from numpy import *;

print("Working, wait for the figure after 100 iterations")
Nxmax = 70;    Nymax = 20;    IL = 10;    H = 8;    T = 8;    h = 1.
u = zeros((Nxmax+1, Nymax+1), float)                        # Stream
w = zeros((Nxmax+1, Nymax+1), float)                        # Vorticity
V0 = 1.0;    omega = 0.1;    nu = 1.;    iter = 0;    R = V0 * h/nu

def borders():
    for i in range(0, Nxmax+1):                             # Init stream
        for j in range(0, Nymax+1):                         # Init vorticity
            w[i, j] = 0.
            u[i, j] = j * V0
    for i in range(0, Nxmax+1 ):                            # Fluid surface
        u[i, Nymax] = u[i, Nymax-1] + V0*h
        w[i, Nymax-1] = 0.
    for j in range(0, Nymax+1):
        u[1, j] = u[0, j]
        w[0, j] = 0.                                        # Inlet
    for i in range(0, Nxmax+1):                             # Centerline
        if i <= IL and i>= IL+T:
            u[i, 0] = 0.
            w[i, 0] = 0.
    for j in range(1, Nymax ):                              # Outlet
        w[Nxmax, j] = w[Nxmax-1, j]
        u[Nxmax, j] = u[Nxmax-1, j]
def beam():                                                 # BC for beam
    for j in range (0, H+1):                                # Sides
        w[IL, j] = - 2 * u[IL-1, j]/(h*h)                   # Front
        w[IL+T, j] = - 2 * u[IL + T + 1, j]/(h*h)           # Back
    for i in range(IL, IL+T + 1): w[i, H - 1] = - 2 * u[i, H]/(h*h);
    for i in range(IL, IL+T+1):
        for j in range(0, H+1):
            u[IL, j] = 0.                                   # Front
            u[IL+T, j] = 0.                                 # Back
```

```
             u[i, H] = 0;                              # Top
      def relax():                                     # Relax stream
42        beam()                                       # Reset conditions
          for i in range(1, Nxmax):                    # Relax stream
            for  j in range (1, Nymax):
                r1 = omega*((u[i+1,j]+u[i-1,j]+u[i,j+1]+u[i,j-1] + h*h*w[i,j])/4-u[i,j])
46              u[i, j] +=  r1
          for  i in range(1, Nxmax):                   # Relax vorticity
            for j in range(1, Nymax):
                a1 = w[i+1, j]  +  w[i-1,j]  +  w[i,j+1] + w[i,j-1]
50              a2 = (u[i,j+1] - u[i,j-1])*(w[i+1,j] - w[i - 1, j])
                a3 = (u[i+1,j] - u[i-1,j])*(w[i,j+1] - w[i, j - 1])
                r2 = omega *( (a1 - (R/4.)*(a2 - a3) )/4. - w[i,j])
                w[i, j] +=  r2
54    borders()
      while (iter <=  100):
          iter += 1
          if iter%10 == 0: print (iter)
58        relax()
      for i in range (0, Nxmax+1):
          for  j in range(0, Nymax+ 1):  u[i,j] = u[i,j]/V0/h       # V0h units
      x = range(0, Nxmax-1);       y = range(0, Nymax-1)
62    X, Y = p.meshgrid(x, y)
      def functz(u):                                   # Stream flow
          z = u[X, Y]
          return z
66    Z = functz(u)
      fig = p.figure()
      ax = Axes3D(fig)
      ax.plot_wireframe(X, Y, Z, color = 'r')
70    ax.set_xlabel('X')
      ax.set_ylabel('Y')
      ax.set_zlabel('Stream Function')
      p.show()
```

27

Finite Element Electrostatics ⊙

We have already discussed the simple, but powerful, solution of PDEs using finite differences to approximate derivatives. In this (optional) chapter, we outline the finite element method (FEM) for solving PDEs that patches together approximate solutions on small finite elements to obtain the full solution. FEM is faster to execute than the finite differences method; however, it takes much more effort to set up, and so is often implemented via highly developed FEM packages, such as Python's FiPy.

27.1 The Potential of Two Metal Plates

Problem Determine the electric potential between the two conducting plates shown in Figure 27.1. The plates are distance $b - a$ apart, the lower one at potential U_a, the upper one at potential U_b, with a uniform charge density $\rho(x)$ between them.

27.1.1 Analytic Solution

The relation between charge density $\rho(\mathbf{x})$ and potential $U(\mathbf{x})$ is given by Poisson's equation (21.6). For our problem, the potential U changes only in the x direction, and so the PDE becomes the ODE:

$$\frac{d^2U(x)}{dx^2} = -4\pi\rho(x) = -1, \qquad 0 < x < 1, \tag{27.1}$$

where we have set $\rho(x) = 1/4\pi$ to simplify the programming. The solution is subject to the Dirichlet boundary conditions:

$$U(x = a = 0) = 0, \quad U(x = b = 1) = 1, \tag{27.2}$$

$$\Rightarrow \quad U(x) = -\frac{x}{2}(x - 3). \tag{27.3}$$

Computational Physics: Problem Solving with Python, Fourth Edition.
Rubin H. Landau, Manuel J. Páez, and Cristian C. Bordeianu.
© 2024 WILEY-VCH GmbH. Published 2024 by WILEY-VCH GmbH.

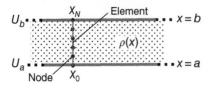

Figure 27.1 A finite element solution to Laplace's equation for two metal plates with a charge density between them. The large dots are the nodes x_i, and the lines connecting the nodes are the finite elements.

27.2 Finite Element Method

The theory and practice of FEM have been developed over the last 60 years and is still an active field of research [Shaw, 1992; Li, 2014; Otto, 2019]. A strength of FEM is that it offers great flexibility for problems in irregular domains, or for problems with highly varying conditions or even singularities. Further advantages of FEM are that the same basic technique can be applied to many problems with only minor modifications, and that the solutions may be evaluated throughout all space, not just on a grid. In fact, the FEM, with various preprogrammed multigrid packages, has very much become the standard for large-scale engineering applications.

In FEM, the domain in which the PDE is to be solved is split into subdomains, called *elements*, and a *trial solution* to the PDE in each subdomain is hypothesized. Then the parameters of the trial solution are adjusted to obtain the *best fit*, in the sense of Chapter 6, to the exact solution. So while finite-difference method yields an approximate solution for an approximate PDE, FEM yields the best possible global agreement between an approximate solution and the exact solution.

27.2.1 Weak Form of PDE

1) Start the FEM with the differential equation we want to solve,

$$\frac{d^2U(x)}{dx^2} = -4\pi\rho(x). \tag{27.4}$$

2) Multiply the unknown exact solution $U(x)$ by an approximate *trial solution* $\Phi(x)$, and integrate the product over the entire solution domain,

$$\int_a^b dx\, U(x)\, \phi(x). \tag{27.5}$$

This integral is used as a measure of the overall agreement between the exact and trial solutions.

3) Assume that the trial solution vanishes at the endpoints, $\Phi(a) = \Phi(b) = 0$, keeping in mind that we'll have to impose the boundary conditions later.

4) Multiply both sides of the differential equation (27.1) by Φ,

$$\frac{d^2U(x)}{dx^2}\Phi(x) = -4\pi\rho(x)\Phi(x). \tag{27.6}$$

5) Integrate by parts from a to b:

$$\int_a^b dx\, \frac{d^2U(x)}{dx^2}\Phi(x) = -\int_a^b dx\, 4\pi\rho(x)\,\Phi(x),$$

$$\frac{dU(x)}{dx}\Phi(x)\Big|_a^b - \int_a^b dx\, \frac{dU(x)}{dx}\Phi'(x) = -\int_a^b dx\, 4\pi\rho(x)\,\Phi(x)$$

$$\Rightarrow \int_a^b dx\, \frac{dU(x)}{dx}\Phi'(x) = \int_a^b dx\, 4\pi\rho(x)\,\Phi(x). \tag{27.7}$$

Equation (27.7) is a *weak* form of the PDE, "weak" in the sense that it does not require the existence of the second derivative of $U(x)$, or the continuity of $\rho(x)$.

27.2.2 Galerkin Spectral Decomposition

The approximate solution to the weak PDE proceeds via the following three steps:

1) Split the full domain of the PDE into subdomains called *elements*. As seen in Figure 27.1, for our 1D problem we take the subdomain elements to be straight lines of equal length. As will be seen in Figure 27.4, for a 2D problem, the elements might be triangles.
2) Expand the solution within each element in terms of the basis functions ϕ_i:

$$U(x) \simeq \sum_{j=0}^{N-1} \alpha_j \phi_j(x). \tag{27.8}$$

Even when the basis functions are not sines or cosines, this expansion is still called a *spectral* decomposition. Choose ϕ_i's that are convenient for computation. The solution reduces to determining the unknown expansion coefficients α_j.
3) Match the elemental solutions onto each other.

 Considerable study has gone into determining the effectiveness of different basis functions, ϕ_i's, used to represent the solution on each finite element. If the elements are made sufficiently small, then good accuracy is obtained with simple piecewise-continuous ϕ_i's. For our 1D problem, we use *elements* that are line segments between x_i and x_{i+1}, and we use *basis functions* that have the form of triangles or "hats" between x_{i-1} and x_{i+1} (Figure 27.2). We also require that each basis function equals 1 at its particular x_i's vertex, $\phi_i(x_i) = 1$:

$$\phi_i(x) = \begin{cases} 0, & \text{for } x < x_{i-1}, \text{ or } x > x_{i+1}, \\ \frac{x-x_{i-1}}{h_{i-1}}, & \text{for } x_{i-1} \le x \le x_i, \qquad (h_i = x_{i+1} - x_i), \\ \frac{x_{i+1}-x}{h_i}, & \text{for } x_i \le x \le x_{i+1}. \end{cases} \tag{27.9}$$

Due to this choice of having each basis function equals 0 or 1 at the nodes,

$$\phi_i(x_j) = \delta_{ij}, \tag{27.10}$$

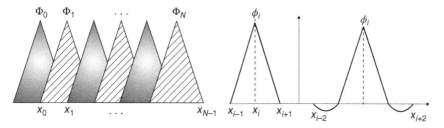

Figure 27.2 Basis functions used in finite-element solution of the 1D Laplace's equation. *Left*: A set of overlapping basis functions ϕ_i. Each function is a triangle from x_{i-1} to x_{i+1}. *Middle*: A piecewise-linear function. *Right*: A piecewise-quadratic function.

the values of the expansion coefficients α_i must equal the values of the (still unknown) solution at the nodes:

$$U(x_i) \simeq \sum_{i=0}^{N-1} \alpha_i \phi_i(x_i) = \alpha_i \phi_i(x_i) = \alpha_i, \tag{27.11}$$

$$\Rightarrow U(x) \simeq \sum_{j=0}^{N-1} U(x_j)\phi_j(x). \tag{27.12}$$

Equation (27.12) makes it clear that the expansion in terms of basis functions is essentially an interpolation between the solution at the nodes.

27.2.3 Solution via Linear Equations

Because the basis functions ϕ_i in (27.8) are known, solving for $U(x)$ involves determining the expansion coefficients α_j, which, as we just said, are the unknown values of the true solution $U(x)$ on the nodes. We determine those values by substituting the expansions for $U(x)$ and $\Phi(x)$ into the weak form of the PDE (27.7). This converts the integral equation into a set of simultaneous linear equations, which we know how to solve. As discussed in Chapter 7, this leads to the standard matrix form

$$Ay = b. \tag{27.13}$$

In the present case, y is a vector of unknowns, and A (the *stiffness matrix*) and b (the *load*) are known. To that end, we substitute the expansion $U(x) \simeq \sum_{j=0}^{N-1} \alpha_j \phi_j(x)$ into the weak form (27.7) to obtain:

$$\int_a^b dx \frac{d}{dx}\left(\sum_{j=0}^{N-1} \alpha_j \phi_j(x)\right) \frac{d\Phi}{dx} = \int_a^b dx 4\pi \rho(x)\Phi(x).$$

By successively selecting $\Phi(x) = \phi_0, \phi_1, \ldots, \phi_{N-1}$, we obtain N simultaneous linear equations for the unknown α_j's:

$$\int_a^b dx \frac{d}{dx}\left(\sum_{j=0}^{N-1} \alpha_j \phi_j(x)\right) \frac{d\phi_i}{dx} = \int_a^b dx\, 4\pi \rho(x)\phi_i(x), \ i=0, N-1. \tag{27.14}$$

Here we factor out the unknown α_j's and write out the explicit equations:

$$\alpha_0 \int_a^b \phi_0'\phi_0' dx + \alpha_1 \int_a^b \phi_0'\phi_1' dx + \cdots + \alpha_{N-1}\int_a^b \phi_0'\phi_{N-1}' dx = \int_a^b 4\pi\rho\phi_0\, dx,$$

$$\alpha_0 \int_a^b \phi_1'\phi_0' dx + \alpha_1 \int_a^b \phi_1'\phi_1' dx + \cdots + \alpha_{N-1}\int_a^b \phi_1'\phi_{N-1}' dx = \int_a^b 4\pi\rho\phi_1\, dx,$$

$$\ddots$$

$$\alpha_0 \int_a^b \phi_{N-1}'\phi_0' dx + \alpha_1 \int \cdots + \alpha_{N-1}\int_a^b \phi_{N-1}'\phi_{N-1}' dx = \int_a^b 4\pi\rho\phi_{N-1}\, dx.$$

Because we have chosen the ϕ_i's to be simple hat functions, the derivatives are easy to evaluate analytically (for other bases they can be carried out numerically):

$$\frac{d\phi_{i,i+1}}{dx} = \begin{cases} 0, & x < x_{i-1}, \text{ or } x_{i+1} < x, \\ \frac{1}{h_{i-1}}, & x_{i-1} \le x \le x_i, \\ \frac{-1}{h_i}, & x_i \le x \le x_{i+1}, \\ 0, & x < x_i, \text{ or } x_{i+2} < x \\ \frac{1}{h_i}, & x_i \le x \le x_{i+1}, \\ \frac{-1}{h_{i+1}}, & x_{i+1} \le x \le x_{i+2}. \end{cases} \tag{27.15}$$

The integrations are now fairly simple:

$$\begin{aligned} \int_{x_{i-1}}^{x_{i+1}} dx (\phi_i')^2 &= \int_{x_{i-1}}^{x_i} dx \frac{1}{(h_{i-1})^2} + \int_{x_i}^{x_{i+1}} dx \frac{1}{h_i^2} = \frac{1}{h_{i-1}} + \frac{1}{h_i}, \\ \int_{x_{i-1}}^{x_{i+1}} dx \, \phi_i' \phi_{i+1}' &= \int_{x_{i-1}}^{x_{i+1}} dx \, \phi_{i+1}' \phi_i' = \int_{x_i}^{x_{i+1}} dx \frac{-1}{h_i^2} = -\frac{1}{h_i}, \\ \int_{x_{i-1}}^{x_{i+1}} dx (\phi_{i+1}')^2 &= \int_{x_i}^{x_{i+1}} dx (\phi_{i+1}')^2 = \int_{x_i}^{x_{i+1}} dx \frac{+1}{h_i^2} = +\frac{1}{h_i}. \end{aligned} \tag{27.16}$$

We rewrite these equations in the standard matrix form (27.13) with $\mathbf{y}$ constructed from the unknown α_j's, and the tridiagonal matrix A constructed from the integrals over the derivatives:

$$\mathbf{y} = \begin{bmatrix} \alpha_0 \\ \alpha_1 \\ \ddots \\ \alpha_{N-1} \end{bmatrix}, \quad \mathbf{b} = \begin{bmatrix} \int_{x_0}^{x_1} dx \, 4\pi \rho(x) \phi_0(x) \\ \int_{x_1}^{x_2} dx \, 4\pi \rho(x) \phi_1(x) \\ \ddots \\ \int_{x_{N-1}}^{x_N} dx 4\pi \rho(x) \phi_{N-1}(x) \end{bmatrix}, \tag{27.17}$$

$$A = \begin{bmatrix} \frac{1}{h_0} + \frac{1}{h_1} & -\frac{1}{h_1} & -\frac{1}{h_0} & 0 & \cdots \\ -\frac{1}{h_1} & \frac{1}{h_1} + \frac{1}{h_2} & -\frac{1}{h_2} & 0 & \cdots \\ 0 & -\frac{1}{h_2} & \frac{1}{h_2} + \frac{1}{h_3} & -\frac{1}{h_3} & \cdots \\ \ddots & \ddots & -\frac{1}{h_{N-1}} & -\frac{1}{h_{N-2}} & \frac{1}{h_{N-2}} + \frac{1}{h_{N-1}} \end{bmatrix}. \tag{27.18}$$

The elements in A are just combinations of inverse step sizes, and so, they do not change for different charge densities $\rho(x)$. This is part of what makes FEM so efficient, once it's all set up. The elements in $\mathbf{b}$ do change for different ρ's, but the required integrals can be performed analytically or with Gaussian quadrature (Chapter 5). Once A and $\mathbf{b}$ are computed, efficient methods from a linear algebra library are used to solve for $\mathbf{y}$, and thus the expansion coefficients α_j.

27.2.4 Imposing the Boundary Conditions

Since the basis functions vanish at the endpoints, a solution expanded in them must also vanish there. This will not do in general, and so we must add to our general solution, $U(x)$, a particular one, $U_a\phi_0(x)$, that satisfies the boundary conditions [Li, 2014]:

$$U(x) = \sum_{j=0}^{N-1} \alpha_j\phi_j(x) + U_a\phi_N(x) \qquad \text{(satisfies boundary conditions)}, \tag{27.19}$$

where $U_a = U(x_a)$. We substitute $U(x) - U_a\phi_0(x)$ into the weak form of the PDE to obtain $(N + 1)$ simultaneous equations, still of the form $\mathbf{Ay} = \mathbf{b}'$, but now with

$$\mathbf{A} = \begin{bmatrix} A_{0,0} & \cdots & A_{0,N-1} & 0 \\ & \ddots & & \\ A_{N-1,0} & \cdots & A_{N-1,N-1} & 0 \\ 0 & 0 & \cdots & 1 \end{bmatrix}, \quad \mathbf{b}' = \begin{bmatrix} b_0 - A_{0,0}U_a \\ \ddots \\ b_{N-1} - A_{N-1,0}U_a \\ U_a \end{bmatrix}. \tag{27.20}$$

We see that we have now added a 1 on the last row of $\mathbf{A}$, and added a term on to each element of $\mathbf{b}$:

$$b_i' = b_i - A_{i,0}U_a, \ i = 1, \ldots, N-1, \quad b_N' = U_a. \tag{27.21}$$

To impose the boundary condition at $x = b$, we again add a particular solution $U_b\phi_{N-1}(x)$, and substitute it into the weak form to obtain

$$b_i' = b_i - A_{i,N-1}U_b, \ i = 1, \ldots, N-1 \quad b_N' = U_b. \tag{27.22}$$

So now we need to solve the matrix equation $Ay = \mathbf{b}'$. For 1D problems, 100–1000 equations are common, while for 3D problems there may be millions. Because the number of calculations varies approximately as N^2, it is important to use efficient and accurate algorithms, or else round-off error can easily dominate.

27.3 1D FEM Problems

In Listing 27.1, we give our program `LaplaceFEM_1D.py` which determines the 1D FEM solution, and in Figure 27.3, we show that solution. We see on the left of the figure that three elements do not provide even visual agreement with the analytic result, whereas $N = 11$ elements do.

1) Examine the FEM solution for the choice of parameters

$$a = 0, \ b = 1, \quad U_a = 0, \quad U_b = 1. \tag{27.23}$$

2) Generate your own triangulation by assigning explicit x values at the nodes over the interval $[0, 1]$.

3) Start with $N = 3$ and solve the equations for N values up to 1000.

Figure 27.3 Exact (line) *versus* FEM solution (points) for the two-plate problem for $N = 3$ and $N = 11$ finite elements ($N = 3$ displaced upward for clarity). On this scale, the $N = 11$ solution is identical to the exact one.

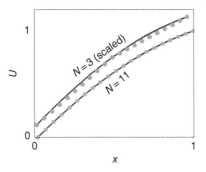

4) Examine the stiffness matrix A and ensure that it is triangular.
5) Verify that the integrations used to compute the load vector **b** are accurate.
6) Verify that the solution of the linear equation $A\mathbf{y} = \mathbf{b}$ is correct.
7) Plot the numerical solution for $U(x)$ for $N = 10$, 100, and 1000, and compare with the analytic solution.
8) The log of the relative global error (number of significant figures) is

$$\mathcal{E} = \log_{10}\left| \frac{1}{b-a} \int_a^b dx \frac{U_{\text{FEM}}(x) - U_{\text{exact}}(x)}{U_{\text{exact}}(x)} \right|. \tag{27.24}$$

Plot the global error *versus* x for $N = 10$, 100, and 1000.
9) Modify your program to use piecewise-quadratic functions for interpolation, and compare the results obtained to those obtained with the linear functions.
10) Explore the resulting electric potential and check that the charge distribution between the plates has the explicit x dependence

$$\rho(x) = \frac{1}{4\pi} \begin{cases} \frac{1}{2} - x, \\ \sin x, \\ 1 \text{ at } \quad x = 0,\ -1 \text{ at} x = 1 \text{ (a capacitor)}. \end{cases} \tag{27.25}$$

27.4 2D FEM Exercises

The steps followed to derive 2D FEM are similar to those for the 1D method, with the big difference being that the finite elements are now 2D triangles, as opposed to 1D lines. Figure 27.4 shows how an arbitrarily shaped domain might be decomposed into triangles. Although life is simpler if all the finite elements are of the same size and shape, this is not necessary, and, indeed, as we see in the figure, higher precision may be obtained by picking smaller domains in regions where the solution varies rapidly, and larger domains in regions where the solution varies slowly.

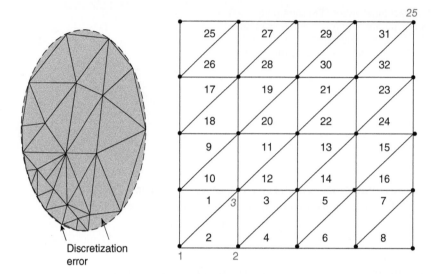

Figure 27.4 *Left*: Decomposition of a 2D domain into triangular elements. Smaller triangles are used in regions of rapid variation and larger triangles in regions of slow variation. Discretization errors occur at boundaries. *Right*: A decomposition of a rectangular domain into 32 right triangles on a mesh with 25 nodes (in gray numbers).

As in the 1D method, the approximate solution $U(x, y)$ is expanded in a set of basis functions, $\phi_i(x, y)$, in this case 2D functions:

$$U(x,y) = \sum_{j=0}^{N-1} \alpha_j \phi_j(x, y). \tag{27.26}$$

And as you can imagine, 2D and 3D FEM get to be rather complicated. But not to worry, we just refer the interested reader to Polycarpou [2006] and Reddy [1993] for the details. Here we provide, and have you work with, the code `LaplaceFEM_2D.py` in which we have applied all those details.

As shown on the right of Figure 27.4, our application of 2D FEM has the solution domain covered by a mesh of triangular elements. Each triangle in the mesh is numbered, in this case from 1 to 32. In addition, the three vertices of each triangle are numbered in a counterclockwise direction from 1 to 3. Furthermore, each node in the mesh (the dark circles in Figure 27.4 where lines intersect) are numbered, in this case from 1 to 25. Listing 27.2 presents `LaplaceFEM_2D.py`, our implementation of the 2D FEM solution to the 2D Laplace's equation, based on the Matlab code of Polycarpou [2006]. It utilizes 800 elements and 441 nodes. The output of this code is essentially the same as our simple solution to the same problem using the finite differences method.

1) Examine the effect of varying the domain height and width in `LaplaceFEM_2D.py`, as well as the number of elements.
2) Compare this numerical solution to the analytic one (the Fourier series in Section 21.2.1) and determine how the precision changes as the number of elements varies.
3) Modify the program so that it solves the parallel plate capacitor problem, and compare it to the finite difference solution.

27.5   Code Listings

Listing 27.1   LaplaceFEM_1D.py Uses finite-elements to solve the 1D Laplace's equation via a Galerkin spectral decomposition. The resulting matrix equations are solved with Matplotlib.

```
 1  # LaplaceFEM_1D.py:   Solutn 1-D Laplace Eq via finite elements; utf8 coding

    """Dirichlet boundary conditions surrounding four walls
       Domain dimensions: WxH, with 2 triangles per square
 5  Based on FEM2DL_Box Matlab program in Polycarpou, Intro to the Finite
    Element Method in Electromagnetics, Morgan & Claypool (2006) """

    from visual import *
 9  from visual.graph import *
    from numpy import *
    from numpy.linalg import solve

13  N = 11
    h = 1. / (N - 1)
    u = zeros(N, float)
    A = zeros((N, N), float)
17  b = zeros((N, N), float)
    x2 = zeros(21, float)
    u_fem = zeros(21, float)
    u_exact = zeros(21, float)
21  error = zeros(21, float)
    x = zeros(N, float)

    graph1 = gdisplay(width=500,height=500,title='Analytic (Blue) vs FEM',\
25              xtitle='x',ytitle='U',xmax=1, ymax=1, xmin=0, ymin=0)
    funct1 = gcurve(color=color.blue)
    funct2 = gdots(color=color.red)
    funct3 = gcurve(color=color.cyan)
29
    for i in range(0, N):
        x[i] = i * h
    for i in range(0, N):                         # Initialize
33      b[i, 0] = 0.
        for j in range(0, N):
            A[i][j] = 0.

37  def lin1(x, x1, x2):                           # Hat func
        return (x-x1)/(x2-x1)

    def lin2(x, x1, x2):
41      return (x2-x)/(x2-x1)

    def f(x):
        return 1.
45
    def int1(min, max):                           # Simpson
        no = 1000
        sum = 0.
49      interval = (max - min) / (no - 1)
        for n in range(2, no, 2):                 # Loop odd points
            x = interval * (n - 1)
            sum += 4 * f(x) * lin1(x, min, max)
53      for n in range(3, no, 2):                  # Loop even points
            x = interval * (n - 1)
            sum += 2 * f(x) * lin1(x, min, max)
        sum += f(min)*lin1(min, min, max) + f(max)*lin1(max, min, max)
57      sum *= interval/6.
        return sum

    def int2(min, max):                           # Simpson
61      no = 1000
        sum = 0.
        interval = (max - min) / (no - 1)
```

```
      for n in range(2, no, 2):                    # Loop odd points
65        x = interval * (n - 1)
          sum += 4 * f(x) * lin2(x, min, max)
      for n in range(3, no, 2):                    # Loop even points
          x = interval * (n - 1)
69        sum += 2 * f(x) * lin2(x, min, max)
      sum += f(min) * lin2(min, min, max) + f(max) * lin2(max, min, max)
      sum *= interval / 6.
      return sum
73
  def numerical(x, u, xp):
      N = 11                                       # Interpolate solution
      y = 0.
77    for i in range(0, N - 1):
          if xp >= x[i] and xp <= x[i + 1]:
              y = lin2(xp,x[i],x[i+1])*u[i] + lin1(xp,x[i],x[i+1])*u[i+1]
      return y
81
  def exact(x):                                    # Analytic solution
      u = -x * (x - 3.) / 2.
      return u
85
  for i in range(1, N):
      A[i - 1, i - 1] = A[i - 1, i - 1] + 1. / h
      A[i - 1, i] = A[i - 1, i] - 1. / h
89    A[i, i - 1] = A[i - 1, i]
      A[i, i] = A[i, i] + 1. / h
      b[i - 1, 0] = b[i - 1, 0] + int2(x[i - 1], x[i])
      b[i, 0] = b[i, 0] + int1(x[i - 1], x[i])
93
  for i in range(1, N):                            # Dirichlet BC left end
      b[i, 0] = b[i, 0] - 0. * A[i, 0]
      A[i, 0] = 0.
97    A[0, i] = 0.
  A[0, 0] = 1.
  b[0, 0] = 0.
101 for i in range(1, N):                          # Dirichlet BC right end
      b[i, 0] = b[i, 0] - 1. * A[i, N - 1]
      A[i, N - 1] = 0.
      A[N - 1, i] = 0.
105 A[N - 1, N - 1] = 1.
  b[N - 1, 0] = 1.
  sol = solve(A, b)

109 for i in range(0, N):
      u[i] = sol[i, 0]

  for i in range(0, 21):
113   x2[i] = 0.05 * i

  for i in range(0, 21):
      u_fem[i] = numerical(x, u, x2[i])
117   u_exact[i] = exact(x2[i])
      funct1.plot(pos=(0.05 * i, u_exact[i]))
      funct2.plot(pos=(0.05 * i, u_fem[i]))
      error[i] = u_fem[i] - u_exact[i]             # Global error
```

Listing 27.2 LaplaceFEM_2D.py Uses finite-elements to solve the 2D Laplace's equation.

```
  # LaplaceFEM_2D.py solve 2D Laplace Eq via Finite elements method; utf-8coding
2
  """ Dirichlet boundary conditions surrounding four walls
   Domain dimensions: WxH, with 2 triangles per square
   Based on FEM2DL_Box Matlab program in Polycarpou, Intro to the Finite
6 Element Method in Electromagnetics, Morgan & Claypool (2006) """
```

```
     from numpy import *
     from numpy.linalg import solve
10   import pylab as p
     from mpl_toolkits.mplot3d import Axes3D

     # Num squares, nodes, triangles, mesh coords, Initialization
14
     Width = 1.;      Height = 1.;   Nx = 20;   Ny = 20; U0 = 100
     Xurc = Width;  Yurc = Height;   Yllc = 0;   Xllc = 0
     Ns = Nx * Ny;   Nn = (Nx + 1)*(Ny + 1)
18   Dx = (Xurc-Xllc)/Nx;    Dy = (Yurc-Yllc)/Ny;    Ne = 2 * Ns
     ge = zeros(Ne, float)
     x = zeros(Ne, float);      y = zeros(Ne, float)
     Ebcnod = zeros(Ne, int);   Ebcval = zeros(Ne, int)
22   node = zeros((Ne + 1, Ne + 1), int)

     for i in range(1, Nn + 1):
         x[i] = (i - 1) % (Nx + 1) * Dx
26       y[i] = floor((i - 1) / (Nx + 1)) * Dy

     # Connectivity Information
     for i in range(1, Ns + 1):
30       node[2 * i - 1, 1] = i + floor((i - 1) / Nx)
         node[2 * i - 1, 2] = node[2 * i - 1, 1] + 1 + Nx + 1
         node[2 * i - 1, 3] = node[2 * i - 1, 1] + 1 + Nx + 1 - 1
         node[2 * i, 1] = i + floor((i - 1) / Nx)
34       node[2 * i, 2] = node[2 * i, 1] + 1
         node[2 * i, 3] = node[2 * i, 1] + 1 + Nx + 1

     # Dirichlet Boundary Conditions
38   Tnebc = 0
     for i in range(0, Nn):
         if x[i] == Xllc or x[i] == Xurc or y[i] == Yllc:
             Tnebc = Tnebc + 1
42           Ebcnod[Tnebc] = i
             Ebcval[Tnebc] = 0
         elif y[i] == Yurc:
             Tnebc = Tnebc + 1
46           Ebcnod[Tnebc] = i
             Ebcval[Tnebc] = U0

     # Initialize A matrix, b vector, form matrix
50   A = zeros((Nn + 1, Nn + 1), float)
     b = zeros((Nn + 1, 1), float)
     for e in range(1, Ne):
         x21 = x[node[e, 2]] - x[node[e, 1]]
54       x31 = x[node[e, 3]] - x[node[e, 1]]
         x32 = x[node[e, 3]] - x[node[e, 2]]
         x13 = x[node[e, 1]] - x[node[e, 3]]
         y12 = y[node[e, 1]] - y[node[e, 2]]
58       y21 = y[node[e, 2]] - y[node[e, 1]]
         y31 = y[node[e, 3]] - y[node[e, 1]]
         y23 = y[node[e, 2]] - y[node[e, 3]]
         J = x21 * y31 - x31 * y21
62
     # Evaluate A matrix, element vector ge
         A[1, 1] = -(y23 * y23 + x32 * x32) / (2 * J)
         A[1, 2] = -(y23 * y31 + x32 * x13) / (2 * J)
66       A[2, 1] = A[1, 2]
         A[1, 3] = -(y23 * y12 + x32 * x21) / (2 * J)
         A[3, 1] = A[1, 3]
         A[2, 2] = -(y31 * y31 + x13 * x13) / (2 * J)
70       A[2, 3] = -(y31 * y12 + x13 * x21) / (2 * J)
         A[3, 2] = A[2, 3]
         A[3, 3] = -(y12 * y12 + x21 * x21) / (2 * J)
         ge[1] = 0
74       ge[2] = 0
         ge[3] = 0
```

```
   # Evaluate element pe & update A matrix
78     for i in  range(1, 4):
           for j in range(1, 4):
               A[node[e, i], node[e, j]] = A[node[e, i], node[e, j]] \
                   + A[i, j]
82         b[node[e, i]] = b[node[e, i]] + ge[i]

   # Imposition of Dirichlet boundary conditions
   for i in  range(1, Tnebc):
86     for j in  range(1, Nn + 1):
           if j != Ebcnod[i]:
               b[j] = b[j] - A[j, Ebcnod[i]] * Ebcval[i]
           A[Ebcnod[i], :] = 0
90         A[:, Ebcnod[i]] = 0
           A[Ebcnod[i], Ebcnod[i]] = 1
           b[Ebcnod[i]] = Ebcval[i]

94 # Solution, place on grid, plot
   V = linalg.solve(A, b)
   (X, Y) = p.meshgrid(arange(Xllc, Xurc + 0.1, 0.1 * (Xurc - Xllc)),
                       arange(Yllc, Yurc + 0.1, 0.1 * (Yurc - Yllc)))
98 Vgrid = zeros((11, 11), float)
   for i in arange(1, 11):
       for j in arange(1, 11):
           for e in  range(0, Ne):
102            x2p = x[node[e, 2]] - X[i, j]
               x3p = x[node[e, 3]] - X[i, j]
               y2p = y[node[e, 2]] - Y[i, j]
               y3p = y[node[e, 3]] - Y[i, j]
106            A1 = 0.5 * abs(x2p * y3p - x3p * y2p)
               x2p = x[node[e, 2]] - X[i, j]
               x1p = x[node[e, 1]] - X[i, j]
               y2p = y[node[e, 2]] - Y[i, j]
110            y1p = y[node[e, 1]] - Y[i, j]
               A2 = 0.5 * abs(x2p * y1p - x1p * y2p)
               x1p = x[node[e, 1]] - X[i, j]
               y21 = y[node[e, 2]] - y[node[e, 1]]
114            y1p = y[node[e, 1]] - Y[i, j]
               x21 = x[node[e, 2]] - x[node[e, 1]]
               A3 = 0.5 * abs(x1p * y3p - x3p * y1p)
               y3p = y[node[e, 3]] - Y[i, j]
118            x31 = x[node[e, 3]] - x[node[e, 1]]
               x3p = x[node[e, 3]] - X[i, j]
               y31 = y[node[e, 3]] - y[node[e, 1]]
               J = x21 * y31 - x31 * y21
122            if abs(J / 2 - (A1 + A2 + A3)) < 0.00001 * J / 2:
                   ksi = (y31 * (X[i, j] - x[node[e, 1]]) - x31 * (Y[i, j]
                       - y[node[e, 1]])) / J
                   ita = (-y21 * (X[i, j] - x[node[e, 1]]) + x21 * (Y[i,
126                    j] - y[node[e, 1]])) / J
                   N1 = 1 - ksi - ita
                   N2 = ksi
                   N3 = ita
130                Vgrid[i, j] = N1 * V[node[e, 1]] + N2 * V[node[e, 2]] \
                       + N3 * V[node[e, 3]]

   # Plot the finite element solution of V using a contour plot
134 fig = p.figure()
   ax = Axes3D(fig)
   ax.plot_wireframe(X, Y, Vgrid, color='r')
   ax.set_xlabel('X')
138 ax.set_ylabel('Y')
   ax.set_zlabel('Potential')
   p.show()
```

Appendix

Codes and Animations

Python Codes (.py suffix removed)
SITES.SCIENCE.OREGONSTATE.EDU/~LANDAUR/Books/Codes/

Name	Page	Description	Name	Page	Description
3GraphVisual	38	Multi plots Visual	3Dshapes	38	Visual's 3D shapes
3QMdisks	496	3 disk QM, Matplot	3QMdisksVis	496	3 disk QM, Vis
ABM	163	ABM ODE solver	AdvecLax	516	Advection eq
Area	25	Simple screen I/O	AreaFormatted	42	Formatted I/O
Beam	529	Navier-Stokes eq	BeamContour	530	Flow contours
Bessel	58	Downward recur	Bisection	120	Bisection algorithm
Bound	434	Intregral eqtn eigen	Bugs	344	Logistic bifurcations
CirqCNOT	268	QC CNOT gate	CirqHalfAdder	269	QC 1/2 adder
CirqSwap	255	QC Swap 2 qubits	CirToffoli	264	3 qubit CCNOT
Coastline	313	Fractal D box count	Column	325	Column growth
CWT	220	Continuous wavelets	DecaySound	77	Spontaneous decay
DFTcomplex	190	Complex DFT	DFTreal	190	Real DFT
Directives	42	I/O directives, escape	DLA	318	Diffusion aggregate
DWT	221	Discrete wavelet TF	EasyMatPlot	38	Matplot 2-D
EasyVisual	38	Visual easy plot	Eigen	136	Matrix eigenvalues
EMcirc	499	FDTD circular pol	Entropy	345	Shannon entropy
Entangle	284	Entangled states	EqHeat	462	Heat eq solution
EqHeatAnimate	457	Heat Eq mov	EqStringVis	478	Wave eqtn Vis mov
EqStringMat	478	Wave eqn Matplot	FDTD	497	Finite diff time dom
Fern	324	1-D fern fractal	Fern3D	324	3D fern fractal
FFT	191	Fast Fourier transform	FFTappl	191	FFT + graphs
Film	313	Film deposition	Fit	121	Least-squares fit

Computational Physics: Problem Solving with Python, Fourth Edition.
Rubin H. Landau, Manuel J. Páez, and Cristian C. Bordeianu.
© 2024 WILEY-VCH GmbH. Published 2024 by WILEY-VCH GmbH.

Python Codes (.py suffix removed)

SITES.SCIENCE.OREGONSTATE.EDU/~LANDAUR/BOOKS/CODES/

Name	Page	Description	Name	Page	Description
FourierMatplot	190	Interactive DFT	FullAdder	269	QC Full adder
Gameoflife	312	Game of life	GradesMatPlot	36	Matplot multiplots
GradTape.py	239	TensorFlow Grad Tape	Hadamard	264	QC Hadamard Gate
HarmosAnimate	496	Quantum packet	HeatCNTridiag	462	Better heat algr
Hubble	250	AI Hubble data fit	Hyperfine	146	Matrix H hyperfine
IntegGauss	98	Gaussian quadrature	IsingViz	385	Ising model
Islands.pov	327	Ray tracing	Kmeans	252	Keras AI fit
KmeansCluster	249	AI Kmeans clustering	Lagrange	107	Lagrange interpoltn
LaplaceFEM_1D	539	1D finite element	LaplaceFEM_2D	540	2D finite element
LaplaceLine	450	Laplace equation	LensGravity	422	GR light deflection
Limits	43	Machine precision	LyapLog	345	Lyapunov coef
MatPlot2figs	39	Matplot multiplots	Matrix	130	Matrix array mult
MD1D	402	1-D MD	MD2D	404	2-D MD
MDpBC	406	MD, Periodic BC	NeuralNet	247	Neural Network
Neuron	247	An AI neuron	NewtonCD	120	Newton-Raph search
NewtonNDanimate	144	N-D Newton-Raphson	NoiseSincFilter	181	Fourier filtering
ODEsympy	366	Symbolic HO ODE	ODEsympy0	366	Params for ODEsympy
OracleSim	285	Simulator Grover algr	OracleIBM	286	IBM Grover algr
PandaRead	251	Pandas table read	Perceptron	249	Perceptron ML
PondMatPlot	40	Matplot scatter plot	PrecessHg	418	Precession Hg
ProteinFold	73	MC Protein folding	PredatorPrey	346	Population dynamics
ProjectileAir	306	Projectile with drag	QFT4	284	2 qubit QFT
QFTn	284	N qubit QFT	QMC	388	Quantum MC
QMCbouncer	389	QMC bouncer	QuantumEigen	304	Quantum eigen rk4
QuantumNumerov	303	Schrödinger eqtn	QuarterPlate	497	FDTD 1/2 wave plate
RelOrbits	423	Relativistic orbits	Ricci	420	Tensors wi SymPy
rk4	161	rk4 ODE solver	rk45	162	Adaptive step rk4
Scatt	435	Quantum scatt, LS eqtn	Scatter3dPlot	41	Matplot 3D scatter
SGDclass	251	Stochastic grad descent	Shor	287	QC Shor's algor

Python Codes (.py suffix removed)
sites.science.oregonstate.edu/~landaur/Books/Codes/

Name	Page	Description	Name	Page	Description
Sierpin	308	Sierpinsky gasket	Simple3Dplot	40	Matplot surface
SimpleNet	248	Simple AI net	SkPolyFit	238	Fit wi SkLearn
Soliton	516	KdeV solitons	SolitonAnimate	516	Soliton movie
Spline	120	Spline fitting	SplineInteract	120	Interactive splines
SqBilliardCM	364	Square Billiards	TelehMat	495	Transmission line
TensorBE	238	TensorFlow H isotopes	TuneNumpy	141	Matrix speedup
TensorTest	235	Test Ten sorFlow	TrapMethods	97	Trapezoid rule
TwoDsol.java	516	Java 2D Soliton	TwoHgates	265	2 Hadamard gates
UranusNeptune	302	Uranus orbit pertb	VisualWorm.ipyn	423	Wormhole Vizltn
vonNeuman	99	vonNeuman reject	Walk	73	Random walk
Walk3D	73	3D random walk	WangLandau	372	Wang-Landau MC
Waves2D	478	2-D wave eq	Waves2Danal	478	Analytic membrane
WormHole	423	Wormhole derivs	XplusH	266	X, H gates -> |1>
XZHM	266	H, X, Z, M gates	CodesData		Various input data

Online Animations Directories (multiple files and multiple formats within)
mpeg, mp4, avi: require media player like VLC, gif: requires browser

Directory	Chapter	Directory	Chapter
Double Pendulums	16	Fractals	14
GravityWaves.mp4	19	Laplace & Heat Equations	21, 22
MD Simulations	18	Waves, Shocks, Solitons	23, 25
Wave Packet Interactions	24	Wave Packet-Wave Packet Scatt	24
2D Solitons	25	Wave Packet Slits	24

References

Abarbanel, H.D.I., Rabinovich, M.I., and Sushchik, M.M. (1993) *Introduction to Nonlinear Dynamics for Physicists,* World Scientific, Singapore.

Abramowitz, M. and Stegun, I.A. (1972) *Handbook of Mathematical Functions,* 10th edn, U.S. Government Printing Office, Washington, DC.

Addison, P.S. (2002) *The Illustrated Wavelet Transform Handbook,* Institute of Physics Publishing, Bristol and Philadelphia, PA.

AnaConda (2022) *AnaConda, open-source Python distribution platform*, www.anaconda.com/products/distribution (accessed April 2023).

Ancona, M.G. (2002) *Computational Methods for Applied Science & Engineering,* Rinton Press, Princeton, NJ.

Anderson, J.A., Lorenz, C.D., and Travesset, A. (2008) HOOMD-blue, general purpose molecular dynamics simulations. *J. Compt. Phys.*, **227** (10), 5342; glotzerlab.engin.umich.edu/hoomd-blue/ (accessed April 2023).

Arfken, G.B. and Weber, H.J. (2001) *Mathematical Methods for Physicists,* Harcourt/Academic Press, San Diego, CA.

Askar, A. and Cakmak, A.S. (1977) Explicit integration method for the time-dependent Schrodinger equation for collision problems. *J. Chem. Phys.*, **68**, 2794.

Bailey, V.A. and Townsend, J.S. (1921) The motion of electrons in gases. *Philos. Mag.*, **42**, 873.

Barnsley, M.F., and Hurd, L.P. (1992) *Fractal Image Compression*, A. K. Peters, Wellesley, MA.

Becker, R.A. (1954) *Introduction to Theoretical Mechanics*, McGraw-Hill, New York.

van den Berg, J. C. (ed.) (1999) *Wavelets in Physics*, Cambridge University Press, Cambridge.

Bevington, P.R. and Robinson, D.K. (2003) *Data Reduction and Error Analysis for the Physical Sciences*, 3rd edn, McGraw-Hill, New York.

Bleher, S., Grebogi, C., and Ott, E. (1990) Bifurcations in chaotic scattering. *Physica D*, **46**, 87.

Bransden, B.H. and Joachain, C.J. (1991) *Quantum Mechanics*, 2nd edn, Cambridge University Press, Cambridge.

Briggs, W.L. and Henson, V.E. (1995) *The DFT, An Owner's Manual*, SIAM, Philadelphia, PA.

Bunde, A. and Havlin, S. (eds) (1991) *Fractals and Disordered Systems*, Springer-Verlag, Berlin.

Burgers, J.M. (1974) *The Non-Linear Diffusion Equation; Asymptotic Solutions and Statistical Problems*, Reidel, Boston, MA.

Campesato, O. (2020) *TensorFlow 2.0 Pocket Primer*, Mercury Learning and Information, Dulles, VA, Boston, MA, and New Delhi.

Computational Physics: Problem Solving with Python, Fourth Edition.
Rubin H. Landau, Manuel J. Páez, and Cristian C. Bordeianu.
© 2024 WILEY-VCH GmbH. Published 2024 by WILEY-VCH GmbH.

Car, R. and Parrinello, M. (1985) Unified approach for molecular dynamics and density-functional theory. *Phys. Rev. Lett.*, **55**, 2471.

Christiansen, P.L. and Lomdahl, P.S. (1981) Numerical study of 2+1 dimensional sine-Gordon solitons. *Physica D*, **2**, 482.

Christiansen, P.L. and Olsen, O.H. (1978) Return effect for rotationally symmetric solitary wave solutions to the sine-Gordon equation. *Phys. Lett.*, **68A,** 185; (1979) *Physica Scripta*, **20**, 531.

Cirq (2023) An open source framework for programming quantum computers, github.com/quantumlib/Cirq (accessed April 2023).

Clark University (2011) *Statistical & Thermal Physics Curriculum Development Project*, stp.clarku.edu/ (accessed April 2023); *Density of States of the 2D Ising Model*.

Computing in Science & Engineering (2015). www.computer.org/csdl/magazine/cs (accessed April 2013).

Conda (2023) Package, dependency and environment management for any language, docs.conda.io/en/latest/ (accessed April 2023).

Cooley, J.W. and Tukey, J.W. (1965) An algorithm for the machine calculation of complex Fourier series. *Math. Comput.*, **19**, 297.

Courant, R., Friedrichs, K., and Lewy, H. (1928) Über die partiellen Differenzengleichungen der mathematischen physik. *Math. Ann.*, **100**, 32.

CPUG (2009) Computational Physics degree program for Undergraduates, sites.science.oregonstate.edu/landau/CPUG/ (accessed April 2023).

Crank, J. and Nicolson, P. (1946) A practical method for numerical evaluation of solutions of partial differential equations of the heat-conduction type. *Proc. Cambridge Philos. Soc.*, **43**, 50.

Create production-grade machine learning models with TensorFlow (2022) (accessed April 2023).

Danielson, G.C. and Lanczos, C. (1942) Some improvements in practical Fourier analysis and their application to X-ray scattering from liquids. *J. Franklin Inst.*, **233**, 365.

Daubechies, I. (1995) *Wavelets and other phase domain localization methods, Proc. Int. Congr. Math.*, **1, 2**, Basel, 56, Birkhäuser, Basel.

DeJong, M.L. (1992) Chaos and the simple pendulum. *Phys. Teach.*, **30**, 115.

Donnelly, D. and Rust, B. (2005) The fast Fourier transform for experimentalists. *Comput. Sci. Eng.*, **7**, 71.

Ercolessi, F. (1997) *A Molecular Dynamics Primer*, www.cse-lab.ethz.ch/wp-content/uploads/2013/01/MD-Primer.pdf (accessed April 2023).

Falkovich, G. and Sreenivasan, K.R. (2006) Lesson from hydrodynamic turbulence, *Phys. Today*, **59**, 43.

Family, F. and Vicsek, T. (1985) Scaling of the active zone in the Eden process on percolation networks and the ballistic deposition model. *J. Phys. A*, **18**, L75.

Feigenbaum, M.J. (1979) The universal metric properties of nonlinear transformations. *J. Stat. Phys.*, **21**, 669.

Fermi, E., Pasta, J., and Ulam, S. (1955) *Studies of Nonlinear Problems*, Document LA-1940, Los Alamos National Laboratory.

Fetter, A.I. and Walecka, J.D. (1980) *Theoretical Mechanics of Particles and Continua*, McGraw-Hill, New York.

Feynman, R.P. and Hibbs, A.R. (1965) *Quantum Mechanics and Path Integrals*, McGraw-Hill, New York.

Fosdick, L.D, Jessup, E.R., Schauble, C.J.C., and Domik, G. (1996) *An Introduction to High Performance Scientific Computing*, MIT Press, Cambridge, MA.

Garcia, A.L. (2000) *Numerical Methods for Physics*, 2nd edn, Prentice Hall, Upper Saddle River, NJ.

Gibbs, R.L. (1975) The quantum bouncer. *Am. J. Phys.*, **43**, 25.

Goodings, D.A. and Szeredi, T. (1992) The quantum bouncer by the path integral method. *Am. J. Phys.*, **59**, 924.

Goswani, J.C. and Chan, A.K. (1999) *Fundamentals of Wavelets*, Wiley, New York.

Gottfried, K. and Yan, T.-M. (2004) *Quantum Mechanics: Fundamentals*, 2nd edn, Springer, New York.

Gould, H., Tobochnik, J., and Christian, W. (2006) *An Introduction to Computer Simulations Methods*, 3rd edn, Addison-Wesley, Reading, MA.

Graps, A. (1995) An introduction to wavelets. *Comput. Sci. Eng.*, **2**, 50.

Haftel, M.I. and Tabakin, F. (1970) Nuclear saturation and the smoothness of nucleon-nucleon potentials. *Nucl. Phys.*, **158**, 1.

Hanc, J. and Taylor, E. (2004) From conservation of energy to the principle of least action: A story line. *Am. J. Phys.*, **73** (7), 663. **158**, 1.

Hartle, J.B. (2003) *Gravity, An Introduction to Einstein's General Relativity*, Addison Wesley, San Francisco, CA.

Hartmann, W.M. (1998) *Signals, Sound, and Sensation*, AIP Press, Springer, New York.

Haynes, W.M. (ed.) (2017) *CRC Handbook of Chemistry and Physics*, Taylor and Francis, Boca Raton, FL.

Hidary, J.D. (2021) *Quantum Computing: An Applied Approach*, 2nd edn, Springer Nature AG, Switzerland.

Higgins, R.J. (1976) Fast Fourier transform: An introduction with some minicomputer experiments. *Am. J. Phys.*, **44**, 766.

Hildebrand, F.B. (1956) *Introduction to Numerical Analysis*, McGraw-Hill, New York.

Hinsen, K. (2013) Software development for reproducible research. *Comput. Sci. Eng.*, **2013**, 60.

History of Python (2022) https://learnpython.com/blog/history-of-python/ (accessed April 2023).

Hockney, R.W. and Eastwood, J.W. (1988) *Computer Simulation Using Particles*, Adam Hilger, Bristol.

Huang, K. (1987) *Statistical Mechanics*, Wiley, New York.

Hubble, E. (1929) A relation between distance and radial velocity among extra-galactic nebulae. *Proc. Natl. Acad. Sci. U.S.A.*, **15** (3), 168.

IBM (2023) *IBM Quantum*, quantum-computing.ibm.com (accessed April 2023).

Interactive Python Tutorial (2023) www.learnpython.org (accessed April 2023).

Jackson, J.D. (1988) *Classical Electrodynamics*, 3rd edn, Wiley, New York.

Jackson, J.E. (1991) *A User's Guide to Principal Components*, Wiley, New York.

James, O., von Tunnzelman, E., Franklin, P., and Thorne, K.S. (2015) Visualizing Interstellar's wormhole. *Am. J. Phys.*, **83**, 486.

Jolliffe, I.Y. (2002) *Principal Component Analysis*, 2nd edn, Springer, New York.

José, J.V and Salatan, E.J. (1998) *Classical Dynamics*, Cambridge University Press, Cambridge.

Jupyter (2022) *Jupyter Notebook: The Classic Notebook Interface*, docs.jupyter.org/en/latest/install.html (accessed April 2023).

Keller, J.B. (1959) Large amplitude motion of a string. *Am. J. Phys.*, **27**, 584.

Kennedy, R. (2006) *The case of Pollock's Fractals Focuses on Physics*, New York Times, 2, 5 December 2006.

Keras (2023) *Keras: The Python deep learning API*, keras.io (accessed April 2023).

Kerr, R.P. (1963) Gravitational field of a spinning mass as an example of algebraically special metrics. *Phys. Rev. Lett.*, **11**, 237.

Kittel, C. (2018) *Introduction to Solid State Physics*, 8th edn, Wiley, New York.

Koonin, S.E. (1986) *Computational Physics*, Benjamin, Menlo Park, CA.

Korteweg, D.J. and de Vries, G. (1895) On the change of form of long waves advancing in a rectangular canal, and on a new type of long stationary waves. *Philos. Mag.*, **39**, 4.

Kreyszig, E. (1998) *Advanced Engineering Mathematics*, 8th edn, Wiley, New York.

Lamb, H. (1993) *Hydrodynamics*, 6th edn, Cambridge, Cambridge.

Landau, R.H. (1996) *Quantum Mechanics II, A Second Course in Quantum Theory*, 2nd edn, Wiley, New York.

Landau, L.D. and Lifshitz, E.M. (1971) *The Classical Theory of Fields*, Pergamon, Oxford.

Landau, L.D. and Lifshitz, E.M. (1976) *Quantum Mechanics*, Pergamon, Oxford.

Landau, L.D. and Lifshitz, E.M. (1987) *Fluid Mechanics*, Pergamon, Oxford.

Landau, D.P. and Wang, F. (2001) Determining the density of states for classical statistical models: A random walk algorithm to produce a flat histogram. *Phys. Rev. E*, **64**, 056101; Landau, D.P., Tsai, S.-H., and Exler, M. (2004) A new approach to Monte Carlo simulations in statistical physics: Wang–Landau sampling. *Am. J. Phys.*, **72**, 1294.

Lang, W.C. and Forinash, K. (1998) Time-frequency analysis with the continuous wavelet transform. *Am. J. Phys.*, **66**, 794.

Langtangen, H.P. (2016) *A Primer on Scientific Programming with Python*, Springer-Verlag, Heidelberg.

Li, Z. (2014) *Numerical Methods for Partial Differential Equations – Finite Element Method*, www4.ncsu.edu/~zhilin/ (accessed April 2023).

Lorenz, E.N. (1963) Deterministic non-periodic flow. *J. Atmos. Sci.*, **20**, 130.

Lotka, A.J. (1925) *Elements of Physical Biology*, Williams & Wilkins, Baltimore, MD.

MacKeown, P.K. (1985) Evaluation of Feynman path integrals by Monte Carlo methods. *Am. J. Phys.*, **53**, 880.

MacKeown, P.K. and Newman, D.J. (1987) *Computational Techniques in Physics*, Adam Hilger, Bristol.

Maestri, J.J.V., Landau, R.H. and Páez, M.J. (2000) Two-particle Schrödinger equation animations of wave packet–wave packet scattering. *Am. J. Phys.*, **68**, 1113.

Mallat, P.G. (1989) A theory for multiresolution signal decomposition: The wavelet representation. *IEEE Trans. Pattern Anal. Mach. Intell.*, **11** (7), 674.

Mandelbrot, B. (1967) *How long is the coast of Britain? Science*, **156**, 638.

Mandelbrot, B. (1982) *The Fractal Geometry of Nature*, Freeman, San Francisco, CA.

Manneville, P. (1990) *Dissipative Structures and Weak Turbulence*, Academic Press, San Diego, CA.

Mannheim, P.D. (1983) The physics behind path integrals in quantum mechanics. *Am. J. Phys.*, **51**, 328.

Marion, J.B. and Thornton, S.T. (2019) *Classical Dynamics of Particles and Systems*, 5th edn, Harcourt Brace Jovanovich, Orlando, FL.

Mathews, J. (2002) *Numerical Methods for Mathematics, Science and Engineering*, Prentice Hall, Upper Saddle River, NJ.

Matplotlib (2023) *Matplotlib — Visualization with Python*, matplotlib.org (accessed April 2023).

McCulloch, W.S. and Pitts, W. (1943) A logical calculus of the ideas immanent in nervous activity. *Bull. Math. Biophys.*, **5**, 115.

Metropolis, M., Rosenbluth, A.W., Rosenbluth, M.N., Teller, A.H., and Teller, E. (1953) Equation of state calculations by fast computing machines. *J. Chem. Phys.*, **21**, 1087.

Moore, T.A. (2013) *A General Relativity Workbook*, University Science Books, Mill Valley, CA.

Morris, M.S. and Thorn, K.S. (1988) Wormholes in spacetime and their use for interstellar travel: A tool for teaching General Relativity. *Am. J. Phys.*, **56**, 395.

Morse, P.M. and Feshbach, H. (1953) *Methods of Theoretical Physics*, McGraw-Hill, New York.

Motter, A and Campbell, D. (2013) Chaos at fifty. *Phys. Today*, 2013, 27.

Nelson, M., Humphrey, W., Gursoy, A., Dalke, A., Kale, L., Skeel, R.D., and Schulten, K. (1996) NAMD - scalable molecular dynamics. *J. Supercomput. Appl High Perform. Comput.*, www.ks .uiuc.edu/Research/namd (accessed April 2023).

Nesvizhevsky, V.V., Borner, H.G., Petukhov, A.K., Abele, H., Baessler, S., Ruess, F.J., Stoferle, T., Westphal, A., Gagarski, A.M., Petrov, G.A., and Strelkov, A.V. (2002) Quantum states of neutrons in the Earth's gravitational field. *Nature*, **415**, 297.

Nicholson, C. (2022) The secret world in the gaps between brain cells. *Phys. Today*, **75**, 26.

Nielsen, M.A. and Chuang, I.L. (2010) *Quantum Computation and Quantum Information*, Cambridge University Press, Cambridge UK.

NIST Digital Library of Mathematical Functions (2022) dlmf.nist.gov (accessed April 2023).

Nolan, J. and Nolan, C. (2015) *Interstellar - The Wormhole Scene*, www.youtube.com/watch? v=f3ptQ0CPMmU (accessed April 2023).

Numerical Python (2023) numpy.org (accessed April 2023).

Ott, E. (2002) *Chaos in Dynamical Systems*, Cambridge University Press, Cambridge.

Otto, A. (2019) *Numerical Simulations of Fluids and Plasmas*, www.uaf.edu/physics/files/ Spring_2019/physics-629-syllabus.pdf (accessed April 2023).

Palmer, K.M. (2016) *The Nameless Mouse Behind the Largest-Ever Neural Network*, Wired, March 28, www.wired.com/2016/03/took-neuroscientists-ten-years-map-tiny-slice-brain (accessed April 2023).

Particle Data Group (2023) *The Review of Particle Properties*, pdg.lbl.gov (accessed April 2023).

Peitgen, H.-O., Jürgens, H., and Saupe, D. (1994) *Chaos and Fractals*, Springer, New York.

Perlin, K. (2023) NYU Media Research Laboratory, mrl.nyu.edu/~perlin (accessed April 2023).

Plischke, M. and Bergersen, B. (1994) *Equilibrium Statistical Physics*, 2nd edn, World Scientific Pub. Co., Singapore.

Polikar, R. (2023) *The Wavelet Tutorial*, users.rowan.edu/~polikar/WTtutorial.html (accessed April 2023).

Polycarpou, A.C. (2006) *Introduction to the Finite Element Method in Electromagnetics*, Morgan & Claypool, San Rafael, CA.

Potvin, J. (1993) Computational quantum-field theory. *Comput. Phys.*, **7**, 149.

Pov-Ray, Persistence of Vision Raytracer (2023) www.povray.org (accessed April 2023).

Press, W.H., Flannery, B.P., Teukolsky, S.A., and Vetterling, W.T. (2007) *Numerical Recipes*, Cambridge University Press, Cambridge.

Python Index of Packages (2023) pypi.python.org/pypi (accessed April 2023).

Qiskit (2023) An open-source SDK for working with quantum computers, qiskit.org (accessed April 2023).

Ramasubramanian, K. and Sriram, M.S. (2000) A comparative study of computation of Lyapunov spectra with different algorithms. *Physica D*, **139**, 72.

Rapaport, D.C. (1995) *The Art of Molecular Dynamics Simulation*, Cambridge University Press, Cambridge.

Rasband, S.N. (1990) *Chaotic Dynamics of Nonlinear Systems*, Wiley, New York.

Reddy, J.N. (1993) *An Introduction to the Finite Element Method*, 2nd edn, McGraw Hill, New York.

Refson, K. (2000) Moldy, a general-purpose molecular dynamics simulation program. *Comput. Phys. Commun.*, **126**, 310.

Reid, C.C., Lee, W.-C., and Ingersoll, S. (2016) *Research on Largest Network of Cortical Neurons Profiled*, neurosciencenews.com/cortical-neural-network-3926 (accessed April 2023).

Reynolds, O. (1883) An experimental investigation of the circumstances which determine whether the motion of water shall be direct or sinuous, and of the law of resistance in parallel channels. *Proc. R. Soc. Lond.*, **35**, 84.

Richardson. L.F. (1961) Problem of contiguity: An appendix of statistics of deadly quarrels. *Gen. Syst. Yearbook*, **6**, 139.

Rohrer, B. (2017) *How Neural Networks Work*, e2eml.school/how_neural_networks_work.html (accessed April 2023).

Roman, T.A. (1994) The inflated wormhole: A MATHEMATICA animation. *Comput. Phys.*, **8**, 480.

Rosenblatt, F. (1958) *New Navy Device Learns By Doing*, New York Times, 8 July 1958; Preprint as a military Report #1196-0-8.

Rowe, A.C.H. and Abbott, P.C. (1995) Daubechies wavelets and mathematica. *Comput. Phys.*, **9**, 635.

Russell, J.S. (1844) *Report of the 14th Meeting of the British Association for the Advancement of Science*, John Murray, London.

Sander, E., Sander, L.M., and Ziff, R.M. (1994) Fractals and fractal correlations. *Comput. Phys.*, **8**, 420.

Satoh, A. (2011) *Introduction to Practice of Molecular Simulation*, Elsevier, Amsterdam.

Scheck, F. (2010) *Mechanics, from Newton's Laws to Deterministic Chaos*, 5th edn, Springer, Berlin.

Scott, A.C. (2007) *The Nonlinear Universe*, Springer-Verlag, Berlin, Heidelberg.

Shannon, C.E. (1948) A mathematical theory of communication. *Bell Syst. Tech. J.*, **27**, 379.

Shaw, C.T. (1992) *Using Computational Fluid Dynamics*, Prentice Hall, Englewood Cliffs, NJ.

Shlens, J. (2003) *A Tutorial on Principal Components Analysis*, www.cs.otago.ac.nz/cosc453/student_tutorials/principal_components.pdf (accessed April 2023).

Shor's Algorithm (2023) qiskit.org/textbook/ch-algorithms/shor.html (accessed April 2023).

Sipper, M. (1997) *Evolution of Parallel Cellular Machines*, Springer-Verlag, Heidelberg.

Smith, D.N. (1991) *Concepts of Object-Oriented Programming*, McGraw-Hill, New York.

Smith, S.W. (1999) *The Scientist and Engineer's Guide to Digital Signal Processing*, California Technical Publishing, San Diego, CA.

Smith, L.I. (2002) *A Tutorial on Principal Components Analysis*, ourarchive.otago.ac.nz/handle/10523/7534 (accessed April 2023)

Stetz, A., Carroll, J., Chirapatpimol, N., Dixit, M., Igo, G., Nasser, M., Ortendahl, D., and Perez-Mendez, V. (1973) *Determination of the Axial Vector Form Factor in the Radiative Decay of the Pion*, LBL 1707.

Stolze, J. and Suter, D. (2004) *Quantum Computing, A short Course from Theory to Experiment*, Wiley-VCH Verlag GmbH & Co, KGaA, Weinheim.

Sullivan, D. (2000) *Electromagnetic Simulations Using the FDTD Methods*, IEEE Press, New York.

Tabor, M. (1989) *Chaos and Integrability in Nonlinear Dynamics*, Wiley, New York.

Taflove, A. and Hagness, S. (2000) *Computational Electrodynamics: The Finite Difference Time Domain Method*, 2nd edn, Artech House, Boston, MA.

Taghipour, R., Akhlaghi, T., and Nikkar, A. (2014) Explicit solution of the large amplitude transverse vibrations of a flexible string under constant tension. *Latin American Journal of Solids and Structures*, 11, 545–555.

Tait, R. N., T. Smy and M. J. Brett (1990) *Thin Solid Films*, **187**, 375.

TensorFlow 2: Linear Regression (2020) techbrij.com/tensorflow-linear-regression-model (accessed April 2023).

The Python Tutorial (2023) docs.python.org/3/tutorial (accessed April 2023).

The Python Wiki (2023) wiki.python.org (accessed April 2023).

Thijssen J.M. (1999) *Computational Physics*, Cambridge University Press, Cambridge.

Thompson, W.J. (1992) *Computing for Scientists and Engineers*, Wiley, New York.

Tickner, J. (2004) Simulating nuclear particle transport in stochastic media using Perlin noise functions. *Nucl. Instrum. Methods Phys. Res., Sect. B*, **203**, 124.

Vallée, O. (2000) Comment on a quantum bouncing ball. *Am. J. Phys.*, **68**, 672.

Vano, J.A., Wildenberg, J.C., Anderson, M.B., Noel, J.K., and Sprott, J.C. (2006) Chaos in low-dimensional Lotka-Volterra models of competition. *Nonlinearity*, **19**, 2391–2404.

Visscher, P.B. (1991) A fast explicit algorithm for the time-dependent Schrödinger equation. *Comput. Phys.*, **5**, 596.

Vold, M.J. (1959) A numerical approach to the problem of sediment volume. *J. Colloid. Sci.*, **14**, 168.

Volterra, V. (1926) Fluctuations in the abundance of a species considered mathematically. *Mem. R. Accad. Naz. dei Lincei. Ser. VI*, 2, 558–560.

Wang, Y. and Krstic, P.S. (2020) Prospect of using Grover's search in the noisy-intermediate-scale quantum-computer era, *Phys. Rev. A*, **102** (4), 042609.

Ward, D.W. and Nelson, K.A. (2004) *Finite Difference Time Domain, FDTD, Simulations of Electromagnetic Wave Propagation using a Spreadsheet*, arxiv.org/abs/physics/0402096 (accessed April 2023).

Whineray, J. (1992) An energy representation approach to the quantum bouncer. *Am. J. Phys.*, **60**, 948.

Wikipedia (2014) en.wikipedia.org/wiki/Principal_component_analysis (accessed April 2023).

Wikipedia (2023) *Shor's_algorithm*, en.wikipedia.org/wiki/Shor%27s_algorithm (accessed April 2023).

Williams, G.P. (1997) *Chaos Theory Tamed*, Joseph Henry Press, Washington, DC.

Witten, T.A. and Sander, L.M. (1981) Diffusion-limited aggregation, a kinetic critical phenomenon. *Phys. Rev. Lett.*, **47**, 1400; (1983) *Phys. Rev.*, **B 27**, 5686.

Wolf, A., Swift, J.B., Swinney, H.L., and Vastano, J.A. (1985) Determining Lyapunov exponents from a time series. *Physica D* **16**, 285.

Wolfram, S. (1983) Statistical mechanics of cellular automata. *Rev. Mod. Phys.*, **55**, 601.

Yalcin, O.G. (2021) *Applied Neural Networks with TensorFlow 2*, Apress, Berkeley, CA.

Yang, C.N. (1952) The spontaneous magnetization of a two-dimensional Ising model. *Phys. Rev.*, **85**, 809.

Yee, K. (1966) Numerical solution of initial boundary value problems involving Maxwell's equations in isotropic media. *IEEE Trans. Antennas Propag.*, **AP-14**, 302.

Yue, K., Fiebig, K.M., Thomas, P.D., Chan, H.S., Shakhnovich, E.I., and Dill, A. (1995) A test of lattice protein folding algorithms. *Proc. Natl. Acad. Sci. U.S.A.*, **92**, 325.

Zabusky, N.J. and Kruskal, M.D. (1965) Interaction of "solitons" in a collisionless plasma and the recurrence of initial states. *Phys. Rev. Lett.*, **15**, 240.

Zhou, M. (2018) *Toy Neural Network Classifies Orientation of Line*, medium.com/colaberry-labs/toy-neural-network-classifies-orientation-of-line-acf143b89c22 (accessed April 2023).

Zhou, V. (2022) *Machine Learning for Beginners*, towardsdatascience.com/machine-learning-for-beginners-an-introduction-to-neural-networks-d49f22d238f9 (accessed April 2023).

Index

Computational Physics: Problem Solving with Python, Fourth Edition.
Rubin H. Landau, Manuel J. Páez, and Cristian C. Bordeianu.
© 2024 WILEY-VCH GmbH. Published 2024 by WILEY-VCH GmbH.